AI and Big Data's Potential for Disruptive Innovation

Moses Strydom
Independent Researcher, South Africa

Sheryl Buckley
University of South Africa, South Africa

A volume in the Advances in Computational
Intelligence and Robotics (ACIR) Book Series

Published in the United States of America by
IGI Global
Engineering Science Reference (an imprint of IGI Global)
701 E. Chocolate Avenue
Hershey PA, USA 17033
Tel: 717-533-8845
Fax: 717-533-8661
E-mail: cust@igi-global.com
Web site: http://www.igi-global.com

Library of Congress Cataloging-in-Publication Data

Names: Strydom, Moses, 1944- editor. | Buckley, Sheryl, 1959- editor.
Title: AI and big data's potential for disruptive innovation / Moses Strydom
  and Sheryl Buckley, editors.
Description: Hershey, PA : Engineering Science Reference, an imprint of IGI
  Global, [2020] | Includes bibliographical references and index.
Identifiers: LCCN 2019006762| ISBN 9781522596875 (hardcover) | ISBN
  9781522596899 (ebook) | ISBN 9781522596882 (softcover)
Subjects: LCSH: Artificial intelligence--Industrial applications. | Big
  data--Industrial applications. | Disruptive technologies.
Classification: LCC TA347.A78 A387 2020 | DDC 338/.064--dc23 LC record available at https://lccn.loc.gov/2019006762

This book is published in the IGI Global book series Advances in Computational Intelligence and Robotics (ACIR) (ISSN: 2327-0411; eISSN: 2327-042X)

British Cataloguing in Publication Data
A Cataloguing in Publication record for this book is available from the British Library.

All work contributed to this book is new, previously-unpublished material. The views expressed in this book are those of the authors, but not necessarily of the publisher.

For electronic access to this publication, please contact: eresources@igi-global.com.

Advances in Computational Intelligence and Robotics (ACIR) Book Series

Ivan Giannoccaro
University of Salento, Italy

ISSN:2327-0411
EISSN:2327-042X

MISSION

While intelligence is traditionally a term applied to humans and human cognition, technology has progressed in such a way to allow for the development of intelligent systems able to simulate many human traits. With this new era of simulated and artificial intelligence, much research is needed in order to continue to advance the field and also to evaluate the ethical and societal concerns of the existence of artificial life and machine learning.

The **Advances in Computational Intelligence and Robotics (ACIR) Book Series** encourages scholarly discourse on all topics pertaining to evolutionary computing, artificial life, computational intelligence, machine learning, and robotics. ACIR presents the latest research being conducted on diverse topics in intelligence technologies with the goal of advancing knowledge and applications in this rapidly evolving field.

COVERAGE

- Pattern Recognition
- Robotics
- Heuristics
- Automated Reasoning
- Cognitive Informatics
- Cyborgs
- Computer Vision
- Intelligent control
- Natural Language Processing
- Artificial Life

IGI Global is currently accepting manuscripts for publication within this series. To submit a proposal for a volume in this series, please contact our Acquisition Editors at Acquisitions@igi-global.com or visit: http://www.igi-global.com/publish/.

Titles in this Series

For a list of additional titles in this series, please visit:
https://www.igi-global.com/book-series/advances-computational-intelligence-robotics/73674

Edge Computing and Computational Intelligence Paradigms for the IoT
G. Nagarajan (Sathyabama Institute of Science and Technology, India) and R.I. Minu (SRM Institute of Science and Technology, India)
Engineering Science Reference • copyright 2019 • 347pp • H/C (ISBN: 9781522585558) • US $285.00 (our price)

Semiotic Perspectives in Evolutionary Psychology, Artificial Intelligence, and the Study of Mind Emerging Research and Opportunities
Marcel Danesi (University of Toronto, Canada)
Information Science Reference • copyright 2019 • 205pp • H/C (ISBN: 9781522589242) • US $175.00 (our price)

Handbook of Research on Human-Computer Interfaces and New Modes of Interactivity
Katherine Blashki (Victorian Institute of Technology, Australia) and Pedro Isaías (The University of Queensland, Australia)
Engineering Science Reference • copyright 2019 • 488pp • H/C (ISBN: 9781522590699) • US $275.00 (our price)

Machine Learning and Cognitive Science Applications in Cyber Security
Muhammad Salman Khan (University of Manitoba, Canada)
Information Science Reference • copyright 2019 • 321pp • H/C (ISBN: 9781522581000) • US $235.00 (our price)

Multi-Criteria Decision-Making Models for Website Evaluation
Kemal Vatansever (Alanya Alaaddin Keykubat University, Turkey) and Yakup Akgül (Alanya Alaaddin Keykubat University, Turkey)
Engineering Science Reference • copyright 2019 • 254pp • H/C (ISBN: 9781522582380) • US $185.00 (our price)

Handbook of Research on Deep Learning Innovations and Trends
Aboul Ella Hassanien (Cairo University, Egypt) Ashraf Darwish (Helwan University, Egypt) and Chiranji Lal Chowdhary (VIT University, India)
Engineering Science Reference • copyright 2019 • 355pp • H/C (ISBN: 9781522578628) • US $295.00 (our price)

Computational Intelligence in the Internet of Things
Hindriyanto Dwi Purnomo (Satya Wacana Christian University, Indonesia)
Engineering Science Reference • copyright 2019 • 342pp • H/C (ISBN: 9781522579557) • US $225.00 (our price)

701 East Chocolate Avenue, Hershey, PA 17033, USA
Tel: 717-533-8845 x100 • Fax: 717-533-8661
E-Mail: cust@igi-global.com • www.igi-global.com

To our grandchildren, Emile, Alma and Anna.
You, as an end-user, are educating big data intelligence free: big data intelligence makes you pay for its services.
You virtually know nothing about big data intelligence: big data intelligence knows everything about you.
Big data intelligence can easily live without you: the relationship between end-users and big data intelligence is asymmetric.

Editorial Advisory Board

Table of Contents

Detailed Table of Contents

Chapter 1

    Moses John Strydom, Independent Researcher, South Africa
    Sheryl Beverley Buckley, University of South Africa, South Africa

The convergence of big data and artificial intelligence, namely big data intelligence, seems inevitable at an epoch just as the automation of smart decision making becomes the future digital disruptor. Every industry will be confronted with the same Darwinian pressure of excellence and adaptation, and must conjointly be supported by the major stakeholder, the ultimate client. Authenticated by the hypothesis that big data intelligence has the potential of Darwinian disruption, the objective of this chapter was to identify the most recent worldwide research trends in the field of big data intelligence and its most relevant research areas. A social network analysis tool was employed to interpret the interrelationship between generated keywords and key phrases. The resulting taxonomy of published peer-reviewed scientific papers was bibliographically analyzed. This investigation permitted all manner of social and business interests underpinned by this technology to understand what to embrace, what to ignore, and how to adapt.

Chapter 2

    Imadeddine Mountasser, Moulay Ismaïl University, Morocco
    Brahim Ouhbi, Moulay Ismaïl University, Morocco
    Ferdaous Hdioud, Sidi Mohamed Ben Abdellah University, Morocco
    Bouchra Frikh, Sidi Mohamed Ben Abdellah University, Morocco

Tourism is an information-intensive industry that requires the interconnection of the stakeholders to make strategic decisions for both tourism organizations and tourists. For instance, trip planning as a high-engagement, time and effort-consuming decision-making process, allows choosing the most suitable travel destination, mode, and activities. Tourists often need a considerable amount of information to develop a travel plan and to build expectations for their trip. Therefore, they need a technological platform on which information relating to tourism activities could be interoperated to respond to the pre-trip tourists' information sourcing behavior. In contrast, given the substantial number of actors providing open data in the field of tourism, tourism service providers aim to explore its significant potentials for sustainable tourism development. Thus, this chapter investigates the capacity to build a centralized information platform using diverse open data sources to support travelers during their trip planning by providing more prominent and better-tailored information.

Chapter 3

Gayathri Rajendran, Pondicherry University, India
Uma Vijayasundaram, Pondicherry University, India

Robotics has become a rapidly emerging branch of science, addressing the needs of humankind by way of advanced technique, like artificial intelligence (AI). This chapter gives detailed explanation about the background knowledge required in implementing the software robots. This chapter has an in-depth explanation about different types of software robots with respect to different applications. This chapter would also highlight some of the important contributions made in this field. Path planning algorithms are required for performing robot navigation efficiently. This chapter discusses several robot path planning algorithms which help in utilizing the domain knowledge, avoiding the possible obstacles, and successfully accomplishing the tasks in lesser computational time. This chapter would also provide a case study on robot navigation data and explain the significant of machine learning algorithms in decision making. This chapter would also discuss some of the potential simulators used in implementing software robots.

Chapter 4

Anupama Hoskoppa Sundaramurthy, BMS Institute of Technology and Management, India
Nitya Raviprakash, Rashtreeya Vidyalaya College of Engineering, India
Divija Devarla, Rashtreeya Vidyalaya College of Engineering, India
Asmitha Rathis, Rashtreeya Vidyalaya College of Engineering, India

This chapter proposes a cost-effective and scalable approach to obtain information on the current living standards and development in rural areas across India. The model utilizes a CNN to analyze satellite images of an area and predict its land type and level of development. A decision tree classifies a region as rural or urban based on the analysis. A summary describing the area is generated from inferences made on the recorded statistics. The CNN is able to predict the land and development distribution with an accuracy of 95.1%. The decision tree predicts rural areas with a precision of 99.6% and recall of 88.9%. The statistics obtained for a dataset of more than 1000 villages in India are cross-validated against the Census of India 2011 data. The proposed technique is in contrast to traditional door-to-door surveying methods as the information retrieved is relevant and obtained without human intervention. Hence, it can aid efforts in tracking poverty at a finer level and provide insight on improving the economic livelihood in rural areas.

Chapter 5

Omar F. El-Gayar, Dakota State University, USA
Loknath Sai Ambati, Dakota State University, USA
Nevine Nawar, Alexandria University, Egypt

Common underlying risk factors for chronic diseases include physical inactivity accompanying modern sedentary lifestyle, unhealthy eating habits, and tobacco use. Interestingly, these prominent risk factors fall under what is referred to as modifiable behavioral risk factors, emphasizing the importance of self-care to improve wellness and prevent the onset of many debilitating conditions. In that regard, advances in wearable devices capable of pervasively collecting data about oneself coupled with the analytic capability provided by artificial intelligence and machine learning can potentially upend how we care for ourselves. This chapter aims to assess the current state and future implications of using big data and artificial intelligence in wearables for health and wellbeing. The results of the systematic review capture key developments and emphasize the potential for leveraging AI and wearables for inducing a paradigm shift in improving health and wellbeing.

Chapter 6

Gopala Krishna Behara, Wipro Technologies, India
Tirumala Khandrika, Wipro Technologies, India

Blockchain is a digital, distributed, and decentralized network to store information in a tamper-proof way with an automated way to enforce trust among different participants. An open distributed ledger can record all transactions between different parties efficiently in a verifiable and permanent way. It captures and builds consensus among participants in the network. Each block is uniquely connected to the previous blocks via a digital signature which means that making a change to a record without disturbing the previous records in the chain is not possible, thus rendering the information tamper-proof. Blockchain holds the potential to disrupt any form of transaction that requires information to be trusted. This means that all intermediaries of trust, as they exist today, exposed to disruption in some form with the initiation of Blockchain technology. Blockchain works by validating transactions through a distributed network in order to create a permanent, verified, and unalterable ledger of information.

Chapter 7

Omar F. El-Gayar, Dakota State University, USA
Martinson Q. Ofori, Dakota State University, USA

The United Nations (UN) Food and Agriculture (FAO) estimates that farmers will need to produce about 70% more food by 2050. To accommodate the growing demand, the agricultural industry has grown from labor-intensive to smart agriculture, or Agriculture 4.0, which includes farm equipment that are enhanced using autonomous unmanned decision systems (robotics), big data, and artificial intelligence. In this chapter, the authors conduct a systematic review focusing on big data and artificial intelligence in agriculture. To further guide the literature review process and organize the findings, they devise a framework based on extant literature. The framework is aimed to capture key aspects of agricultural processes, supporting supply chain, key stakeholders with a particular emphasis on the potential, drivers, and challenges of big data and artificial intelligence. They discuss how this new paradigm may be shaped differently depending on context, namely developed and developing countries.

Chapter 8

Tawanda Mushiri, University of Zimbabwe, Zimbabwe
Liberty Tende, University of Zimbabwe, Zimbabwe

The rate of production of horticultural produce had been seen increasing from the past century owing to the increase of population. Manual sorting and grading of tomatoes had become a challenge in market places and fruit processing firms since the demand of the fruit had increased. Considering grading of tomatoes, color is of major importance when it comes to the maturity of the tomatoes. Hence, there is a need to accurately classify them according to color. This process is very complicated, tiresome, and laborious when it is done manually by a human being. Apart from being labor-demanding, human sorting, and grading results in inaccuracy in classifying of tomatoes which is a loss to both the farmer and customer. This chapter had been prepared focusing on the automatic and effective tomato fruit grading system using artificial intelligence particularly using artificial neural network in Matlab. The system makes use of the image processing toolbox and the ANN toolbox to process and classify the tomatoes images according to color and size.

Chapter 9

Tahir Cetin Akinci, Istanbul Technical University, Turkey

The production, transmission, and distribution of energy can only be made stable and continuous by detailed analysis of the data. The energy demand needs to be met by a number of optimization algorithms during the distribution of the generated energy. The pricing of the energy supplied to the users and the change for investments according to the demand hours led to the formation of energy exchanges. This use costs varies for active or reactive powers. All of these supply-demand and pricing plans can only be achieved by collecting and analyzing data at each stage. In the study, an electrical power line with real parameters was modeled and fault scenarios were created, and faults were determined by artificial intelligence methods. In this study, both the power flow of electrical power systems and the methods of meeting the demands were investigated with big data, machine learning, and artificial neural network approaches.

Chapter 10

Dhanalakshmi Senthilkumar, Malla Reddy Engineering College (Autonomous), India

Blockchain is the process of development in bitcoin. It's a digitized, decentralized, distributed ledger of cryptocurrency transactions. The central authorities secure that transaction with other users to validate transactions and record data, data is encrypted and immutable format with secured manner. The cryptography systems make use for securing the process of recording transactions in private and public key pair with ensuring secrecy and authenticity. Ensuring bitcoin transaction, to be processed in network, and ensuring transaction used for elliptic curve digital signature algorithm, all transactions are valid and in chronological order. The blockchain systems potential to transform financial and model of governance. In Blockchain, databases hold their information in an encrypted state, that only the private keys must be kept, so these AI algorithms are expected to increasingly be used, whether financial transactions are fraudulent, and should be blocked or investigated.

Chapter 11

Jayapandian Natarajan, Christ University, India

The main objective of this chapter is to enhance security system in network communication by using machine learning algorithm. Cyber security network attack issues and possible machine learning solutions are also elaborated. The basic network communication component and working principle are also addressed. Cyber security and data analytics are two major pillars in modern technology. Data attackers try to attack network data in the name of man-in-the-middle attack. Machine learning algorithm is providing numerous solutions for this cyber-attack. Application of machine learning algorithm is also discussed in this chapter. The proposed method is to solve man-in-the-middle attack problem by using reinforcement machine learning algorithm. The reinforcement learning is to create virtual agent that should predict cyber-attack based on previous history. This proposed solution is to avoid future cyber middle man attack in network transmission.

Chapter 12

This chapter discusses businesses, key technology implementations, case studies, limitations, and trends. It also presents recommendations to improve data analysis, data-driven innovation, and big data project implementation. Small-to-large-scale project inefficiencies present unique challenges to both public and private sector institutions and their management. Data analytics management, data-driven innovation, and related project initiatives have grown in scope, scale, and frequency. This evolution is due to continued technological advances in analytical methods and computing technologies. Most public and private sector organizations do not deliver on project benefits and results. Many organizational and managerial practices emphasize these technical limitations. Specialized human and technical resources are essential for an organization's effective project completion. Functional and practical areas affecting analytics domain and ability requirements, stakeholder expectations, solution infrastructure choices, legal and ethical concerns will also be discussed in this chapter.

Foreword

I believe artificial intelligence (AI) is one of the most significant innovations humans have ever created. From my own work with companies and governments all across the world, it is evident that AI is going to change our world in ways we can't even imagine today.

Our increasingly digital world has seen an unprecedented explosion in the amount of data, which, in turn, is fueling AI. The vast majority of all the data we now have in this world was only generated in the past few years, and the exponential growth of data is likely to continue, from around five zettabytes today to over 175 zettabytes in the coming five years. The availability of data, combined with networked and more powerful computers has boosted leading-edge evolutions of AIs, such as machine learning and deep learning, which have given machines the ability to see, hear, smell, taste and touch, which, in turn, have given rise to algorithms that can read, speak, understand our emotions and even be creative.

Advances in AI seem to accelerate, and it clear that this will transform business and society. In my work with many of the leading and most innovative companies across the globe, I see how fast the field is making progress and how AI is being turned into more intelligent products, smarter services, and transformed business operations.

The leaders of today's most successful businesses are fully embracing AI and make sure they grab the massive opportunities it offers. Amazon CEO Jeff Bezos believes we have entered the 'golden age' of AI that allows us to solve problems that once were the realm of sci-fi (Bezos, 2017). Google co-founder Sergey Brin believes "The new spring in AI is the most significant development in computing in my lifetime" (Brin, 2018), and Microsoft CEO Satya Nadella argues AI is the "defining technology of our times" (Nadella, 2018). I regularly contribute to the World Economic Forum and have listed many times to their founder and executive chairman Klaus Schwab, who argues that AI and big data (especially when combined with all other technological innovations such as robotics, the internet of things, and blockchain) have triggered a 4th Industrial Revolution that is going to transform all parts of business and society (Schwab, 2016).

The world's political leaders are also waking up to the transformative powers of big data and AI. In the US, the White House has released numerous policy documents that emphasize the strategic significance of AI. In 2016, under President Barack Obama, the White House issued the first report "Preparing for the Future of Artificial Intelligence" (White House, 2016), which laid the foundation for a US AI strategy. In 2018, under Donald Trump, following an AI summit at the White House, the administration issued "Artificial Intelligence for the American People" (White House, 2018a), in which President Trump states, "We're on the verge of new technological revolutions that could improve virtually every aspect of our lives, create vast new wealth for American workers and families, and open up bold, new frontiers in science, medicine, and communication." The goal of the US Administration is to maintain American leadership in AI by accelerating AI research and deployment, and by training the future American workforce to take full advantage of the benefits of AI (White House, 2018b). Russia's President Putin said, "Artificial intelligence is the future, not only for Russia, but for all humankind. . . . Whoever becomes the leader in this sphere will become the ruler of the world" (Putin, 2017). China has arguably developed the most ambitious AI strategy to become the world leader in AI by 2030 (Chinese State Council, 2017). The European Commission has also issued an AI strategy in which it states: "Like the steam engine or electricity in the past, AI is transforming our world, our society, and our industry. Growth in computing power, availability of data, and progress in algorithms have turned AI into one of the most strategic technologies of the 21st century. The stakes could not be higher. The way we approach AI will define the world we live in" (Europa, 2018).

While AI and big data are at the top of the political and business agendas, there are valid reservations and increasingly loud calls to control the exploitation of big data and the use of AI with proper regulation. Many people are concerned with biased and unregulated AIs that could potentially do more harm than good. Just recently 42 countries officially took a step in the right direction by adopting the brand-new OECD "Principles on Artificial Intelligence (AI)," agreeing to uphold international standards that aim to ensure AI systems are designed to be robust, safe, fair and trustworthy.

In this context, where the world's leading businesses are steaming ahead with data and AI fueled innovations, and world political powers are wrangling over AI dominance and regulation, it couldn't be timelier to publish a book that explores the potential innovative disruptions of AI and big data.

This book brings together some of the most interesting and relevant voices within their fields to highlight new directions in contemporary research in artificial intelligence and big data. With scholars from 11 different countries and an editorial board spanning much of the globe, this book provides a broad and heterogeneous view covering topics including intelligent robots, security, agriculture, wearable technology, blockchain, and much more. The editors and contributors have done an excellent job in bringing together such a fascinating collection of contributions that provides a state-of-the-art overview of research in AI and big data.

Bernard Marr
Independent Researcher, UK
Milton Keynes, May 2019

REFERENCES

Bezos, J. (2017). AI is in a 'golden age' and solving problems that were once in the realm of sci-fi, Jeff Bezos says. *CNBC*. Retrieved from https://www.cnbc.com/2017/05/08/amazon-jeff-bezos-artificial-intelligence-ai-golden-age.html

Brin, S. (2018). *Google's Sergey Brin warns of the threat from AI in today's 'technology renaissance'.* Retrieved from https://www.theverge.com/2018/4/28/17295064/google-ai-threat-sergey-brin-founders-letter-technology-renaissance

Chinese State Council. (2017). *A Next Generation Artificial Intelligence Development Plan and Three-Year Action Plan to Promote the Development of New-Generation Artificial Intelligence Industry.* Retrieved from http://www.miit.gov.cn/n1146295/n1652858/n1652930/n3757016/c5960820/content.html

Europa. (2018). *Communication from the Commission to the European Parliament*. The European Council. Retrieved from https://ec.europa.eu/digital-single-market/en/news/communication-artificial-intelligence-europe

Nadella, S. (2018). *Microsoft CEO Satya Nadella on the rise of AI: 'The future we will invent is a choice we make'*. Retrieved from https://www.cnbc.com/2018/05/24/microsoft-ceo-satya-nadella-on-the-rise-of-a-i-the-future-we-will-invent-is-a-choice-we-make.html

Putin, V. (2017). *'Whoever leads in AI will rule the world': Putin to Russian children on Knowledge Day*. Retrieved from https://www.rt.com/news/401731-ai-rule-world-putin/

Schwab, K. (2016). *The Fourth Industrial Revolution: what it means, how to respond*. World Economic Forum. Retrieved from https://www.weforum.org/agenda/2016/01/the-fourth-industrial-revolution-what-it-means-and-how-to-respond/

White House. (2016). *Preparing for the Future of Artificial Intelligence. Executive Office of the President*. National Science and Technology Council, National Science and Technology Council Committee on Technology. Retrieved from https://obamawhitehouse.archives.gov/sites/default/files/whitehouse_files/microsites/ostp/NSTC preparing_for_the_future_of_ai.pdf

White House. (2018a). *Artificial Intelligence for the American People*. The White House. Retrieved from https://www.whitehouse.gov/briefings-statements/artificial-intelligence-american-people/

White House. (2018b). *Summary of the 2018 White House Summit on Artificial Intelligence for American Industry*. The White House. Office of Science and Technology Policy. Retrieved from https://www.whitehouse.gov/wp-content/uploads/2018/05/Summary-Report-of-White-House-AI-Summit.pdf

Preface

The editors take great pleasure in prefacing *Artificial Intelligence and Big Data's Potential for Disruptive Innovation*.

Artificial intelligence and big data are two burgeoning technologies, full of promise for businesses in all industries. However, the real revolutionary potential of these two technologies is probably their convergence, which would lead to a new paradigm of information and knowledge. This convergence seems inevitable as the automation of smart decision-making becomes the future digital disruptor.

Creative destruction—the digital disruption—refers to the incessant product and process innovation mechanism by which new production units replace outdated ones. This restructuring process permeates major aspects of macroeconomic performance, not only long-run growth but also economic fluctuations, structural adjustment and the functioning of factor markets. Over the long run, the process of creative destruction accounts for over 50 per cent of productivity growth (Caballero, 2008). At business cycle frequency, restructuring typically declines during recessions, and this add a significant cost to downturns. Obstacles to the process of creative destruction can have severe short- and long-run macroeconomic consequences. According to Christensen a disruptive innovation is a product or service that is of "inferior performance" and "lower quality" than that of incumbent companies, and that is offered to a niche market segment (Christensen, 1997). This offering continues to improve with time until it reaches a level of quality and performance that is acceptable and a fit for many of the mainstream consumers, and, as such, disrupts the incumbent firms. Prior to Christensen, Schumpeter in 1942, writing in *Capitalism, Socialism and Democracy* asserted:

[I]n capitalist reality as distinguished from its textbook picture, it is not [textbook] competition which counts but the competition from the new commodity, the new technology, the new source of supply, the new type of organization (the largest-scale unit of control for instance)–competition which commands a decisive cost or quality advantage and which strikes not at the margins of the profits and the outputs of the existing firms but at their foundations and their very lives. (McCraw, 2010)

This triumvirate—Artificial intelligence, Big data and Digital disruption—becomes for many, the magic solution to all problems and is thus impacting all sectors, from healthcare to energy and transport, from finance and insurance to retail. It has assumed a disruptive, ubiquitous macrocosm, and its positive transformational potential has already been acknowledged in multitudinous key sectors.

- Is this potential plausible?
- Is this a consequence of recent trends?

- Is this a legitimate body of knowledge?
- What are the consequences in our daily lives of this irruption?

The editors thus invited relevant voices within their fields of expertise to highlight new directions in contemporary research and business in intelligence artificial and big data's potential for disruptive innovation. Each author, with the objective of demystifying the disruptive innovation phenomena and its economic and societal impacts, has brought her/his own share of valuable lessons in analyzing this theme. In this manner, each chapter is intended to afford fully the benefits of sharing expertise from different organizations and contexts.

The book, where each chapter has been double blind reviewed, subsequently reaps these multifaceted benefits.

Against this background, the twelve chapters of the manuscript have indicated that mainstream consumers and associated ecosystems have shifted from incumbent firms to the disruptors.

Written primarily by academics, the target audience of this book will be composed of educators, academics, professionals and researchers working in the field of innovation in big data governance in various disciplines, which include, but are not limited to, education, engineering, information technology, medical science, finance, government, and knowledge management in general.

It furthermore has as an objective to amalgamate data owners, data analysts, skilled data professionals, cloud service providers, companies from industry, venture capitalists, entrepreneurs, research institutions, and universities, and will be useful to participants from both the industry and academia working in all the domains of big data, artificial intelligence, robotics and engineering.

Moreover, the book will provide insights and support executives concerned with the management of expertise, knowledge, information and organizational development in different types of work communities and environments.

The book is organized into 12 chapters concentrating on different aspects of the book's themes.

A brief overview of the chapters is presented below.

CHAPTER 1: BIG DATA INTELLIGENCE AND PERSPECTIVES IN DARWINIAN DISRUPTION

The convergence of big data and artificial intelligence, namely big data intelligence, seems inevitable at an epoch just as the automation of smart decision-making becomes the future digital disruptor. Every industry will be confronted with the same Darwinian pressure of excellence and adaptation, and must conjointly be supported by the major stakeholder, the ultimate client. Authenticated by the hypothesis that big data intelligence has the potential of Darwinian disruption, the objective of Chapter 1 is to identify the most recent worldwide research trends in the field of big data intelligence, and its most relevant research areas.

CHAPTER 2: ONTOLOGY-BASED OPEN TOURISM DATA INTEGRATION FRAMEWORK – TRIP PLANNING PLATFORM

Tourism is an information-intensive industry that requires the interconnection of the stakeholders to make strategic decisions for both tourism organizations and tourists. There is thus a need for a technological platform on which information relating to tourism activities could be inter-operated, to respond to the pre-trip tourists' information sourcing behavior. The objective of chapter 2 is to investigate the capacity to build a centralized information platform using diverse Open Data sources to support travelers during their trip planning by providing more prominent and better-tailored information.

CHAPTER 3: ARTIFICIAL INTELLIGENCE FOR EXTENDED SOFTWARE ROBOTS, APPLICATIONS, ALGORITHMS, AND SIMULATORS

Robotics has become a rapidly emerging branch of science, addressing the needs of humankind by way of advanced techniques, like artificial intelligence. Chapter 3 provides a detailed explanation about the background knowledge required in implementing software robots with respect to different applications and path planning algorithms. A case study on robot navigation data is also provided where the significance of machine learning algorithms in decision-making is explained.

CHAPTER 4: MACHINE LEARNING AND ARTIFICIAL INTELLIGENCE – RURAL DEVELOPMENT ANALYSIS USING SATELLITE IMAGE PROCESSING

In Chapter 4, the authors propose a cost-effective and scalable approach to obtain information on the current living standards and development in rural areas across India. The model utilizes a convolutional neural network to analyze satellite images of an area and predict its land type and level of development. The proposed technique contrasts with traditional door-to-door surveying methods as the information retrieved is relevant and obtained without human intervention.

CHAPTER 5: WEARABLES, ARTIFICIAL INTELLIGENCE, AND THE FUTURE OF HEALTHCARE

Common underlying risk factors for chronic diseases include physical inactivity accompanying modern sedentary life style, unhealthy eating habits and tobacco use. In that regard, advances in wearable devices capable of pervasively collecting data about oneself coupled with the analytic capability provided by artificial intelligence and machine learning can potentially upend how we care for ourselves. Chapter 5 aims to assess the current state and future implications of using big data and artificial intelligence in wearables for health and well-being.

CHAPTER 6: BLOCKCHAIN AS A DISRUPTIVE TECHNOLOGY – ARCHITECTURE, BUSINESS SCENARIOS, AND FUTURE TRENDS

Blockchain is a digital, distributed and decentralized network storing information in an automated manner to enforce trust among different participants. An open distributed ledger can record all transactions between different parties efficiently in a verifiable and permanent way. It captures and builds consensus among participants in the network. Chapter 6 indicates how blockchain holds the potential to disrupt any form of transaction that requires information to be trusted.

CHAPTER 7: DISRUPTING AGRICULTURE – THE STATUS AND PROSPECTS FOR AI AND BIG DATA IN SMART AGRICULTURE

The Food and Agriculture Organization of the United Nations estimates that farmers will need to produce about 70% more food by 2050. To accommodate the growing demand, the agricultural industry has grown from labor-intensive to Smart Agriculture, or Agriculture 4.0 which includes farm equipment that are enhanced using autonomous unmanned decision systems (robotics), big data and artificial intelligence. In Chapter 7, the authors conduct a systematic literature review focusing on big data and artificial intelligence in agriculture. A framework is aimed to capture key aspects of agricultural processes, supporting supply chain, key stakeholders with an emphasis on the potential drivers, challenges of big data and artificial intelligence.

CHAPTER 8: AUTOMATED GRADING OF TOMATOES USING ARTIFICIAL INTELLIGENCE CASE OF ZIMBABWE – ARTIFICIAL INTELLIGENCE-BASED RESEARCH

The increase of the world population has been accompanied by an equally accelerated increase in the rate of production of horticultural produce especially tomatoes, which presently represent the world's second largest produced fruits by volume. Because of their perishability and restricted shelf-life, it is preferable that, immediately after being harvested, they are be washed, graded and packed. This process is complicated, tiresome and laborious when done manually. Chapter 8 focuses on the automatic and effective tomato grading system using Matlab's artificial neural network.

CHAPTER 9: APPLICATIONS OF BIG DATA AND ARTIFICIAL INTELLIGENCE IN ELECTRIC POWER SYSTEMS ENGINEERING

The production, transmission and distribution of energy can only be made stable and continuous by the detailed analysis of the data. The energy demand needs to be met by several optimization algorithms during the distribution of the generated energy. In Chapter 9, both the power flow of electrical power systems and the methods of meeting the demands were investigated utilizing big data technology, machine learning and artificial neural network approaches.

CHAPTER 10: BLOCKCHAIN AND ITS INTEGRATION AS A DISRUPTIVE TECHNOLOGY

A blockchain is essentially a distributed database of records or public ledger of all transactions or digital events that have been executed and shared among participating parties. Each transaction in the public ledger is verified by consensus of the majority of the participants in the system. And, once entered, information can never be erased. The blockchain contains a certain and verifiable record of every single transaction ever made. In chapter 10, the author considers the recent surge in blockchain interest as an alternative to traditional centralized systems and considers the emerging applications thereof. Key techniques required for blockchain implementation are assessed, offering a primer to guide research practitioners.

CHAPTER 11: CYBER SECURE MAN-IN-THE-MIDDLE ATTACK INTRUSION DETECTION USING MACHINE LEARNING ALGORITHM

With an immense number of threats pouring in from nation states and hacktivists as well as terrorists and cybercriminals, the requirement of a globally secure infrastructure becomes a major obligation. The increased complexity and inter-connectivity of Supervisory Control and Data Acquisition (SCADA) systems in the Smart Grid has exposed them - the systems - to a wide range of cybersecurity issues, and there are a multitude of potential access points for cyber attackers including man-in-the-middle assaults. The objective of Chapter 12 is to enhance security in network communication by using machine learning algorithms.

CHAPTER 12: THE INTERSECTION OF DATA ANALYTICS AND DATA-DRIVEN INNOVATION

Data analytics management, data-driven innovation and related project initiatives have grown in scope, scale and frequency. This evolution is due to continued technological advances in analytical methods and computing technologies. Chapter 13 discusses businesses, key technology implementations, case studies, limitations and trends. It also presents recommendations to improve data analysis, data-driven innovation, and big data project implementation.

REFERENCES

Caballero, R. J. (2008). The New Palgrave Dictionary of Economics (S. N. Durlauf & L. E. Blume, Eds.; 2nd ed.). Academic Press. Retrieved from https://economics.mit.edu/files/12606

Christensen, C. (1997). *The Innovator's Dilemma: When New Technologies Cause Great Firms to Fail.* Harvard Business School Press.

McCraw, T. K. (2010). *Prophet of Innovation: Joseph Schumpeter and Creative Destruction.* Cambridge, MA: Harvard University Press. doi:10.4159/9780674040779

Acknowledgment

The editors would like to acknowledge the help of all the people involved in this project and, more specifically, to the authors and reviewers that took part in the double blind review process.

Without their support, this book would not have become a reality.

Firstly, the editors would like to thank each one of the authors for their contributions. Our sincere gratitude goes to the chapter's authors who contributed their time and expertise to this book.

Secondly, the editors wish to acknowledge the valuable contributions of the reviewers regarding the improvement of quality, coherence, and content presentation of chapters.

Most of the authors also served as referees; we highly appreciate their double task.

Moses Strydom
Independent Researcher, South Africa

Sheryl Beverley Buckley
University of South Africa, South Africa

Chapter 1
Big Data Intelligence and Perspectives in Darwinian Disruption

Moses John Strydom
Independent Researcher, South Africa

Sheryl Beverley Buckley
ⓘD https://orcid.org/0000-0002-2393-4741
University of South Africa, South Africa

ABSTRACT

The convergence of big data and artificial intelligence, namely big data intelligence, seems inevitable at an epoch just as the automation of smart decision making becomes the future digital disruptor. Every industry will be confronted with the same Darwinian pressure of excellence and adaptation, and must conjointly be supported by the major stakeholder, the ultimate client. Authenticated by the hypothesis that big data intelligence has the potential of Darwinian disruption, the objective of this chapter was to identify the most recent worldwide research trends in the field of big data intelligence and its most relevant research areas. A social network analysis tool was employed to interpret the interrelationship between generated keywords and key phrases. The resulting taxonomy of published peer-reviewed scientific papers was bibliographically analyzed. This investigation permitted all manner of social and business interests underpinned by this technology to understand what to embrace, what to ignore, and how to adapt.

INTRODUCTION

In this fast-paced universe, the web is inundated with an exponential production of huge amounts of data that is rapidly transforming the manner in which business is concluded throughout all industries and societal sectors.

DOI: 10.4018/978-1-5225-9687-5.ch001

Technology is disrupting[1] everything, and everything is disruptable and everything must be disrupted (Christensen, 1997): the companies, their business models, their products and services, the expertise they offer to their customers, but also our models of organization, our public institutions, our political leaders, our ways of thinking, learning, communicating, working, our representations of the world, our values and even as far as the very fiber of our beings. Today's world, as we know it, should consequently be animated and inspired by these, sometimes prodigious developments – but paradoxically it rarely feels that way. Deplorably, the proximity of abundance and health abruptly rubs against despair.

Moreover, in this hyper-competitive market where everything is offered, the selection is ultra-fast.

Every product, service, idea or content that is adapted to its environment, that responds to an existing demand, is very quickly spotted and selected by consumers that make it emerge and likewise diffuse it. This selection can be classified as being Darwinian because it allows to impose what was not planned intentionally, often after several tests of random characteristics: no buzz is expected, no innovation is anticipated (Christensen, 1997; Wang et al., 2018). Chance and selection by the environment give clues to the innovator to develop his vision and then direct his efforts to make it happen. Predictably, these expansive set of circumstances combined with the unlimited education available on the internet, heralds a period of massive debate. In order to cope with this acceleration of globalization, which is making the old world disappear violently, it is urgent to understand the dynamics that address disruption. As a stratagem, this chapter proposes the following investigative tasks:

- *To explain* how these new technologies, especially big data intelligence, come to challenge humans about what makes them special - their intelligence.
- *To understand*, follow and let the new organizational models, the culture and the requirements of this new paradigm take hold.
- And finally, *to know* the state of mind, the aspirations and the techniques of these new professional actors of the disruption who have no limits to reach their objectives and draw their ideas in science-fiction.

Faced with disruption, there is now only one option: to disrupt oneself to avoid being disrupted.

This chapter further ambitions to provide the keys to apprehend this world being invented, not to fear it, but to prevent others from building it for us. The chapter aspires to increase the body of knowledge of a complex domain and, by so doing, encourage real conversations about bigger issues. Likewise, the authors resolve to connect the dots, to see commonalities and differences based on data from a widespread literature review as well as conversations with colleagues and friends.

The cornerstone of this investigation is, after a holistic reflection, the creation of a taxonomy of research areas which would aid in analyzing and synthesizing of normative literature on artificial intelligence and big data to support the signposting of future research directions.

BACKGROUND

We live in an era of disruption in which powerful global forces are changing how we live and work. The rise of several emerging economies, the rapid spread of digital technologies, growing challenges to globalization, and, in some countries, the splintering of long-held social contracts are all disrupting business, the economy, and society. These trends offer considerable new opportunities to companies, sectors, countries, and individuals that embrace them successfully.

This study focused on the intense competitive and societal challenges we all face in this era of technological ferment. The impact of these and other disruptive forces are being felt worldwide, touching all countries, sectors, companies, and, increasingly, workers and the environment.

History of Thinking Machines

Fundamentally, artificial intelligence [2] and big data[3] - big data intelligence - and their associated technologies are far from being novel ideas. *Automat* [4] is a term used to describe moving machines, especially those that have been made to resemble human or animal actions. For centuries (see Table 1), whether it be with calculating machines or analytical engines, the history of modern thinking machines was planted by classical philosophers who attempted to describe the process of human thinking as the mechanical manipulation of symbols, long before the invention of the first computers (Adkins, 2013; Breton, 1995; Dalakov, 2018; Offray, 1912)

From the 15th century BC, the Egyptian Amenhotep, son of Hapu, had made a statue of Memnon, King of Ethiopia, near Thebes in Egypt, which uttered a melodious sound when struck by the suns rays in the morning and during sunset. It was suggested that a divine power was partly responsible as the mechanisms were far too simple to sustain the noise (Table 1).

From the 8th century BC, in ancient Greece, Homer (Table 1) describes in Iliad's Verse XVIII how Hephaestus, the blacksmith god, built tripods with casters that could commute autonomously from the palace of the gods to his studio. This same Hephaestus, whose ideas were unequivocally consistent, also made women in gold, able to work and talk, and assist him in his daily tasks. Other texts from ancient Greece mention the existence of talking heads, including a mask of Orpheus that gave oracles to Lesbos. At the same time, in Egypt, articulated statues, animated by steam and fire, shook their heads and moved their arms (Dalakov, 2018).

Internet Usage

According to the most recent United Nations estimates, elaborated by Worldometers (2018), the current world population is 7.6 billion of which 4.2 billion (55%) are connected to the internet (Table 2). Though Africa, on a world scale represents 11% internet users, the growth of her internet users, in the interval 2000-2018, reached a gargantuan count of 10,199%. The internet penetration rate of North America, with barely 5% of the world's population, has been estimated to be 95%. Half the world's internet users hail from Asia. At the same time, with billions of people still below the first rung of the digital ladder, the climb to web ownership is becoming more challenging than ever. For example, in July 2018, Amazon sold more than 100 million products to consumers worldwide during its annual Prime Day event, a $4.2 billion bonanza that included sales of table salt in India, Coke Zero in Singapore and toothpaste in China (Soper, 2018). But data like this mask the fact that for many people in developing countries, the road to the internet and to e-commerce, particular, is riddled with potholes. Simply put, the growth of e-commerce is not automatic, and the spread of its benefits is not guaranteed.

At the core of this vast scale of the internet, a minute of time goes much further than can be ever imagined (Figure 1). That is because the internet has a degree of scale on which our linear human brains are unaccustomed to operating. The numbers for these services are so enormous that they can only be advantageously shown using the 60-second time scale. Any larger dimension, and our brains would be incapable of processing these massive quantities in any useful measure. For example, in 2018, during

Table 1. The trajectory of artificial intelligence then and now. (Adapted from Adkins, 2013; Breton, 1995; Dalakov, 2018; Offray, 1912)

Pioneers	The Technology	Period
Egyptian Amenhotep	Statue of Memnon, King of Ethiopia, which uttered a melodious sound when struck by the suns rays in the morning and during sunset.	15th century BC
Yen Shi, the Chinese artificer.	Life-size, human-shaped figure of mechanical handiwork that, like a robot mimicked all human actions.	10th century BC
Solomon, King of Israel	Movable throne with mechanical animals, which followed Solomon, wherever he wished to go.	10th century BC
Hephaestus to whom Homer refers in the Iliad	The description of Hephaestus tripods with casters that could commute alone from the palace of the gods to his studio:	8th century BC
Greek mechanician Daedalus	Stone statues that seemed to breath in or move their marble feet	6th century BC
Chinese engineer King-Shu Tse	Flying bird (magpie) constructed out of wood and bamboo as well as a wooden horse worked by springs, that could jump.	6th century BC
Greek scientist, Archytas of Tarentum	A wooden flying pigeon	5th century BC
Egypt: Greek: Roman: Chinese	Counting board or abacus	3rd century BC
Greece	Antikythera mechanism: complex gear mechanism (sometimes called the first known analog computer)	1st century BC
Leonardo da Vinci	Device for Calculation: An early version of today's complicated calculator	1493
Blaise Pascal	Pascaline; for a long time considered as the first mechanical calculator in the world	1623
William Pratt	The Arithmetical Jewell: a rudimentary mechanical reconfiguration of the conventional reckoning technique	1616
John Napier	Napier's bones or rods: Calculation of logarithmic tables	1617
Tito Livio Burattini	Calculating machine: the ciclografo),	1659
Samuel Morland	Calculating Machines; one for trigonometry (1663), one for addition and subtraction (1666) and one for multiplication and division (1662).	1669
Gottfried Leibniz	Stepped Reckoner; a logical (thinking) device	1670
Giovanni Poleni	Calculating machine based on the pin-wheels	1709
Anton Braun	Calculating machine's with six-place setting mechanism	1727
Johann Reichold	Arithmetic machine	1792
Thomas de Colmar	Arithmometer; based on the stepped drum mechanism of Leibniz	1820
Charles Babbage	Analytical Engine; first digital computer in the world, which embodied in its mechanical and logical details just about every major principle of the modern digital computer	1830
Pafnuty Lvovich Chebyshev	An Adding Machine of Continuous Motion	1876
Alexander Rechnitzer	The world's first motor-driven calculating machine and it was also the first machine to embody full automatic multiplication and division.	1900
Christel Hamann	Mechanical calculator; based on the stepped drum of Leibniz, mounted in the center of the machine,	1905
Alan Mathison Turing	Turing machine is an idealized computing device, consisting of a read/write head (or scanner) with a paper tape passing through it.	1936
John Louis von Neumann	Created the field of cellular automata without the aid of computers, constructing the first self-replicating automata with pencil and graph paper.	1948
Joseph Carl Robnett Licklider	Intergalactic Computer Network which was the forerunner of the internet.	1962

Table 2. World Population and Internet Usage, June 2018 (Adapted from Worldometers 2018)

World Region	Population, 2018 Estimate	Population, % of World-Population	Internet Users, June 20, 2018	Internet Users, %	Penetration Rate, % of Population	Growth, 2000-2018, %
Africa	1,287,914,329	17%	464,923,169	11%	35%	10,199%
Asia	4,207,588,157	55%	2,062,197,366	49%	49%	1,709%
Europe	827,650,849	11%	705,064,923	17%	85%	570%
Latin America&Caribbean	652,047,996	8%	438,248,446	10%	67%	2,35%
Middle East	254,438,981	3%	164,037,259	4%	65%	4,894%
North America	363,844,662	5%	345,660,847	8%	95%	219%
Oceania&Australia	41,273,454	1%	28,439,277	1%	69%	273%
World	7,634,758,428	100%	4,208,571,287	100%	55%	1,066%

an internet minute, Google recorded 3.7 million search queries. The comparative recording for the year 2017 was 3.5 million. On the other hand, the Whatsapp score was respectively 18 million text messages in 2018, against 16 million in 2017. These almost inexhaustible quantities of data inspired Roger Magoulas (Ross, 2010) to coin the term " *big data* ", which refers to a wide range of large data sets and which, due to their size, but also their complexity, are almost impossible to manage and process using traditional data management tools.

Are we already living in the future, in Society 4.0, where a world recently belonging to science fiction is progressively becoming reality?

Figure 1. Milestone Events in an Internet Minute. 2017 and 2018 (© Copyright 2018, Desjardins 2018)

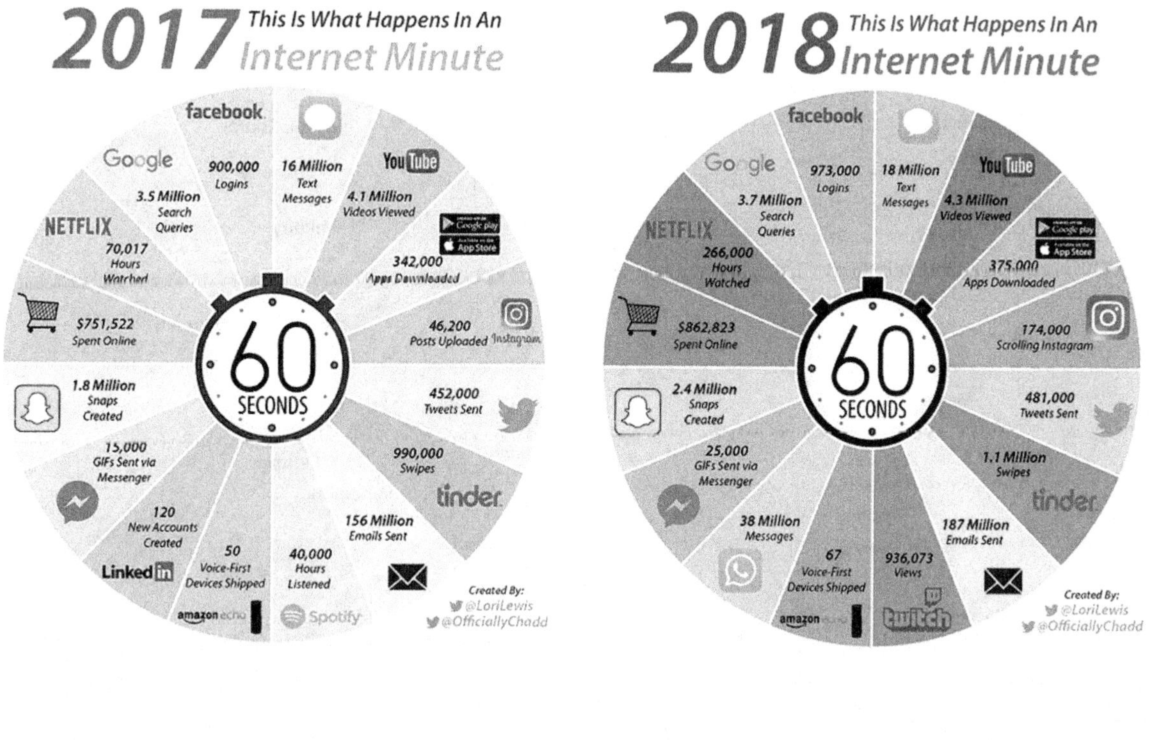

Characteristics of Big Data

Big data is not a single technology or initiative. Rather, it is a contiguous derivative of all our interactions with rapidly evolving technologies. This latter reality presages that big data is not a stable platform but is as dynamic as is the technology that defines it (Strydom & Buckley, 2018, pp. 56). Much of the literature stresses the opportunities afforded by big data technologies by referring to the so-called 3 V's: volume, velocity, and variety (Laney, 2001). Though the evolution of the V's in big data it is beyond the scope of this chapter, it will suffice to say that the 3V's were found inadequate to fully describe big data hence this surge to define other big data V-based characterization, where the V is merely used as a mnemonic device to label the different challenges. In this manner, since the conception of the 3 V's, other scholars have added veracity, specifically, how much noise is in the data, granularity, and many other features that have largely been associated with the technological functionality of big data (Figure 2. Strydom & Buckley, 2018, pp. 61). As a countermeasure, Strydom and Buckley (2018) propose a first-principles' approach founded on the ineluctable conviction that big data is not a stable platform but, on the contrary, is as dynamic as the technology that underpins it.

Figure 2. Eye-weighted spring conjugated edge betweenness diagram showing the evolution of the V's in big data from 3V's to 42V's (Strydom & Buckley, 2018, pp. 61)

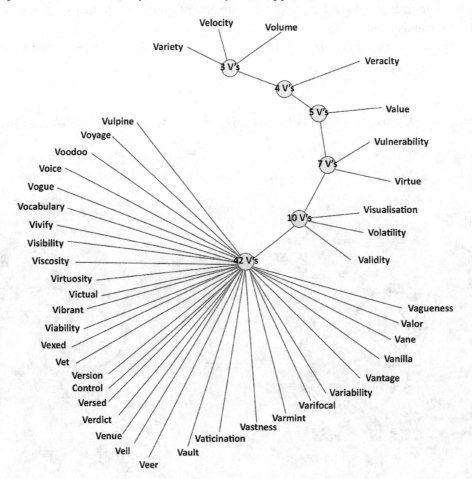

Artificial Intelligence

Artificial Intelligence [5] itself has been studied for at least 6 decennaries and is still one of the most elusive subjects in computer science. This is partly due to the topics' amplitude and ambiguity. It ranges from machines truly capable of thinking to search algorithms used to play board games. It has applications in nearly every way we use computers in society.

The term artificial intelligence was first coined by John McCarthy in 1956 when he held the first academic conference on the subject (Knapp, 2006). But the journey to understand if machines can truly think began much earlier than that date. In his seminal work, *As We May Think,* Bush (1945) proposed a system which amplifies people's own knowledge and understanding. Five years later Alan Turing (1950) wrote a paper on the notion of machines being able to simulate human beings and the ability to do intelligent things, such as play chess.

In any event, is was in 1956, that John McCarthy (Knapp, 2006) invited many of the leading researchers of the time in a wide range of advanced research topics such as complexity theory, language simulation, neuron nets, abstraction of content from sensory inputs, relationship of randomness to creative thinking, and learning machines to Dartmouth in New Hampshire to discuss a subject so new to the human imagination that he had to coin a new term for it: "artificial intelligence". Accordingly, the 1956 Dartmouth Artificial Intelligence conference gave birth to the field of artificial intelligence, and gave succeeding generations of scientists their first sense of the potential for information technology to be of benefit to human beings in a profound way.

Big Data Intelligence

Artificial intelligence and big data are two burgeoning technologies, full of promise for businesses in all industries. However, the real revolutionary potential of these two technologies is probably their convergence, which would lead to a new paradigm of information and knowledge. This convergence seems inevitable as the automation of smart decision-making becomes the future digital disruptor (Dubé, Du, McRae, Sharma, Jayaraman, & Nie, 2018). Chandiok and Chaturvedi (2018) posit that every industry will be confronted with the same Darwinian pressure of excellence and adaptation must additionally be supported by the major stakeholder, the ultimate client. Soaring agility, smarter business processes and higher productivity are the most likely benefits of this convergence. These two technologies are so inextricably linked, to the point that we can justifiably refer to this convergence as big data intelligence. This architecture includes mechanisms for ingesting, protecting, processing, and transforming data into file systems or database structures. Big data intelligence tools and analyst queries run in the environment to mine intelligence from data, which outputs to a variety of different vehicles. Furthermore, big data intelligence has been found to be the chief vector of digital transformation (Sousa & Rocha, 2019). Over the last several decades, big data intelligence technologies - computing methods for automated perception, learning, understanding, and reasoning - have become commonplace in our lives. We plan trips using global positioning systems that rely on artificial intelligence to crosscut the complexity of millions of routes to find the most appropriate. Thanks to speech-based natural user interfaces (NUI) our smartphones comprehend our speech, and internationally used NUI's, Siri, Cortana, and Google Now (Lopez et al., 2017) are getting better at understanding, and even anticipating, our intentions. Artificial intelligence algorithms detect faces as we take images with our smartphones and, in addition, recognize the faces of individuals when we post those photos to Facebook (Makhija & Sharma, 2019). Internet search engines, such as Google and Bing, rely on a fabric of artificial intelligence subsystems. In like manner, big data intelligence provides hundreds of millions of people with search results, traffic predictions, and recommendations about books and movies. This technology effectuates translations among languages in real-time and accelerates the operation of our laptops by attempting to anticipate what we will do next. Several companies, such as Google, BMW, and Tesla, are working on cars that are self-driven - either with partial human oversight or entirely autonomous (Badue et al., 2019).

Beyond influences in our daily lives, big data intelligence techniques are playing a major role in science and medicine. This technology is employed in hospitals helping physicians understand the complications of high-risk patients and where artificial intelligence algorithms help to find important proverbial needles in massive data haystacks. This substantial progress also raises a number of near-term concerns: privacy, bias, inequality, safety and security. To this end, a growing body of experts within and outside the field of artificial intelligence has raised concerns that future developments may pose long-term safety and security risks (Cave & OhEigeartaigh, 2018; Dietterich, 2015). There is worldwide evidence that big data intelligence technologies have demonstrated their potential for supporting regimes and subverting the relationship between citizen and state, thereby accelerating a global resurgence of authoritarianism (Van Wynsberghe & Robbins, 2018; Feldstein, 2019). During 2018, a proliferation of " deep-fake " videos (Shahbaz & Phukan, 2018) - worse than fake news - showed how easy it is becoming to create fake clips using artificial intelligence. We already note the technology being used for fake celebrity porn, lots of weird movie mashups, and, potentially, virulent political smear campaigns with a real risk in deep-fake being weaponized for more nefarious purposes. Generative adversarial networks (GANs), which involve two dueling neural networks, can conjure extraordinarily realistic but completely concocted images and video (Goodfellow et al., 2014). Nvidia recently showed how GANs can generate photo-realistic faces of whatever race, gender, and age desired (Wang, Liu, Zhu, Tao, Kautz, & Catanzaro, 2018). Some comments raise the possibility of dystopian futures where artificial intelligence systems become " super-intelligent " and threaten the survival of humanity (Murphy, 2018). It is organic that new technologies trigger exciting new capabilities and applications. This is, however, counterbalanced by the generation of new anxieties (Obozintsev, 2018; Remarczyk, Narayanan, Mitrovic, & Black, 2018).

The precipitation towards broad applicability results in algorithms and social platforms being under-regulated, and, thus, collaterally being easily weaponized resulting in cybersecurity becoming a veritable and accelerating threat, so much so that cybersecurity and privacy are now considered to be converging technologies (Shackelford, 2018). In fact, 2018 can be considered as the year of this duo. In May, the European Union's General Data Protection Regulation, the world's most stringent privacy law, came into effect. By the end of the year, even Apple's and Microsoft's CEOs were calling for new national privacy standards in the United States. What was once an abstract concept designed to protect expectations about our own data is now becoming more concrete, and more critical - on par with the threat of adversaries accessing our data without authorization.

The Five Eyes alliance (United States, Britain, Canada, Australia and New Zealand, to be soon joined by Japan and Germany) that share intelligence to combat espionage, terrorism and global crime, recently categorized Huawei, a Chinese state sponsored technology company, as a national security risk resulting in Huawei being banned from supplying 5G equipment to the alliance (Cilluffo & Cardash, 2018). In a similar manner, Google and Microsoft sell their services to the United State's government, many times against the wishes of their own employees and best ethical judgment. Due to protests, one of these project, Project Maven (Pellerin, 2017), which was utilized by the United States Air Force for classifying drone imagery was abandoned and, for remedial measures, an artificial intelligence code of ethics was created.

The trend of putting profits over ethics and integrity is very dangerous for humanity. Instead of thinking about the hype of big data intelligence, we need to be more skeptical about where this technology could be heading us as a civilization. We are living in an era where the existential threats to our survival will be increasingly exponentially with our technological experiments. World-renowned entrepreneurs like Bill Gates and Elon Musk in conjunction with scientists like Stephen Hawking and Andrew McAfee, from MIT, as well as 8,000 other signatories to the open letter (Tegmark, 2018) mention the existing risks if we loose control of this phenomenon. What would happen if artificial intelligence (even in its infancy) is not used properly? What if artificial intelligence became an immortal dictator (if it modeled itself on the Chinese Government, it is a believable scenario). Russia's president Vladimir Putin said: " *Artificial intelligence is the future, not only for Russia, but for all humankind. It comes with enormous opportunities, but also threats that are difficult to predict. Whoever becomes the leader in this sphere will become the ruler of the world. When one party's drones are destroyed by drones of another, it will have no other choice but to surrender.* " (Vincent, 2017). What if the dictators of the future were not men on power trips, but actual artificial intelligence with the capability to hack any system? The world needs to prepare for this possibility, that machine learning would evolve more independently and aggressively in the future. How would a comparable system be effectively deactivated? Killer drones and other are an early example, but lethal autonomous weapons systems (LAWS) are a pretty broad spectrum in a world where we are manipulated and exploited for profit on all sides (Hale 2018). The military organizations and national security agencies of this world are not likely to use artificial intelligence in a benign manner (Coeckelbergh, 2018). Facebook has been reeling from a series of privacy scandals, triggered by revelations that a political consulting firm, Cambridge Analytica, improperly used Facebook data to build tools, like malicious chatbots to influence the 2016 United States presidential election (Granville, 2018). The resulting uproar showed how the algorithms that decide what news and information to surface on social media can be gamed to amplify misinformation, undermine healthy debate, and isolate citizens with different views from one another. Indeed, internationally, we already face a major issue with the big data and deep learning technologies quasi oligopolistically and that have effectively given connectivity a new name, which is in effect several names: GAFAM (Google, Apple, Facebook, Amazon, and Microsoft), and the Chinese-centric BATX (Baidu, Alibaba, Tencent and Xiaomi). (Roberge & Chantepie, (2017),

Big data intelligence narratives, which have the potential to herald both the good and the bad, hope and despair, as well as poverty and wealth, are paralleled by the famous lines from Charles Dickens' " A Tale of Two Cities ",

It was the best of times, it was the worst of times, it was the age of wisdom, it was the age of foolishness, it was the epoch of belief, it was the epoch of incredulity, it was the season of Light, it was the season of Darkness, it was the spring of hope, it was the winter of despair, we had everything before us, we had nothing before us, we were all going direct to heaven, we were all going direct the other way - the period was so far like the present period, that some of its noisiest authorities insisted upon its being received, for good or for evil, in the superlative degree of comparison only. (Dickens, 1859).

Digital Transformation

Every product, service, idea or content that is adapted to its environment, that responds to an existing demand, is very quickly spotted and selected by consumers that make it emerge and likewise diffuse it. This selection can be identified as being synonymous to the Darwinian theory of evolution which advances that organisms that are better adapted to their environment tend to survive longer and transmit more of their genetic characteristics to succeeding generations than do those that are less well adapted. "*Darwin considered and found some evidence for the possibility that variations, no matter when caused, may appear "at all periods of life....and that every change is not necessarily effect of some change or condition in embryo – the embryo itself not necessarily affected, excepting that its future growth at different periods is to vary"* " (Ospoval, 1995). Thus it is not the strongest of the species that survives, nor the most intelligent that is successful. It is the one that is most adaptable to change and business models that were fashioned in the bygone era of stability, linearity and predictability are not expected to survive the digital disruption (Goodwin, 2018). Disruption is rampart fundamentally because it allows to impose what was not planned intentionally, often after several tests of random characteristics: no buzz is expected, no innovation is anticipated (Schumpeter (as cited in McCraw, 2010); Christensen, 1997). Piecemeal digital solutions are to be precluded, at all costs. Instead what needs to be undertaken is a complete revision of businesses and their technology infrastructures.

Unlike Darwin's theory of evolution, where it is external factors and the environment which have the largest influence on natural selection, digital Darwinism would seem to be driven by our own inexorable hunger for new technology and new innovations. It appears to be a race without a finish line, but a truly exciting one nonetheless.

For the past two decades, drivers of digital transformation have come in the form of information accessibility, initially by way of the internet and subsequently via mobile devices, principally smartphones (Iapichino, De Rosa, & Liberace, 2018). The most recent wave of transformation is enabled by information insight, providing consumers with products and services that meet their needs when and where they need them. To provide such products and services requires the ability to collect and analyze vast amounts of structured and unstructured data, and to use those insights to inform business decisions and take action in real time. Moreover, it requires the ability to fuse together multiple technologies and strategic solutions.

In a similar manner, with the amount of information in the world nearly doubling each year, it is no surprise that data complexity is the top challenge standing in the way of digital transformation (Gampfer, 2018).

One of the fastest growing uses of this technology is to " *listen* " to all customer communications, both directly with a company and about that company in the market at large, ranging from call center conversations to chat sessions and even social media activity (Popescu 2018). Big data intelligence has become ubiquitous in all business enterprises where decision-making is morphed by intelligent machines. The need for smarter decisions and big data management are the criteria that drive this trend.

The world is changing at a phenomenal pace and nothing illustrates this better than analyzing changes in rankings of the world's largest companies (Table 3). These technology giants - GAFAM - are achieving world domination by investing heavily in developing new products and services leading to an explosion in innovation and faster growth. As these companies become more dominant, they are disrupting established companies. Consider the impact of Amazon on the retail sector (Soper, 2018). These large technology companies which benefit from globalization, are additionally accessing new markets. Any promising small company that could compete is quickly gobbled up by its larger rival. For example, Facebook acquired upcoming threats Whatsapp and Instagram, while Google has made more than 120 acquisitions in the last 10 years (Galston and Hendrickson, 2018).

Table 3. Largest global companies in 2018 versus 2008 (Adapted from Bloomberg, Google)

Year 2018				Year 2008			
Rank	Company	Founded	Value, $bn	Rank	Company	Founded	Value, $bn
1	Apple	1976	890	1	PetroChina	1999	728
2	Google	1998	768	2	Exxon	1870	492
3	Microsoft	1975	680	3	General Electric	1892	358
4	Amazon	1994	592	4	China Mobile	1997	344
5	Facebook	2004	545	5	Industrial&Commercial Bank of China	1984	336

Critical Analysis of the Literature

Due to the nature of this chapter, the selected scholars and their literature concerning the inter-relatedness between disruptive technology and business, all approvingly referenced Christensen's (1997) thesis of the disruptive innovation (Alberti-Alhtaybat, Al-Htaybat, & Hutaibat, 2019; Snihur, & Wiklund, 2018; Mario, 2018; Zhang, Daim. & Zhang, 2018; Bolesnikov et al., 2018: Li, Porter, & Suominen, 2018; Sivathanu, & Pillai, 2018). That said, there is a dearth of research which refutes this synergy. Kelly (2016) discusses significant future effects of novel technology on our lives but does not mention Schumpeter (McGraw, 2010) nor Christensen (1997), at all. Furthermore, none of the above-mentioned references stress the difference between technological sustainability and disruption. Could disruption, which is illustrated by rapid intermittent changes, be considered sustainable? This unquestionably calls on future research to deal with this conundrum.

Furthermore, the opportunities created by rapid technological development sustainability influence how organizations interact with big data intelligence, develop organizational models, and deal with stakeholder interests. That said, future research should reflect this state of affairs by critically examining how they are empirically exploited by organizations in today's hyper-competitive world. By the same token, this calls for investigations on how to create co-action between professional practices and their models. This would consequentially permit the creation of more flexible organizational models with the advantage of encouraging cross-disciplinary interactions at all levels. Organizations face a supplementary challenge of functioning efficiently today, in a world of future compression, while innovating effectively for tomorrow. Forthcoming research can usefully focus on how organizations address this dilemma which is not only riskier but is also more volatile, uncertain, complex, and ambiguous.

Given theses premises, the authors identified that all this literature does not give the reader a sufficient granular comparison and understanding of the current big data intelligence ecosystem; hence the need for this chapter.

METHODOLOGY

Technology is everywhere. Whether in our smartphone, computer, television, car, fridge, watch…we can see intelligence invading our workspace and our home. Everything will be soon " smart ". This body of knowledge will highlight the unusual and unexpected connections across many fields artificial intelligence is taking us.

In order to cope with this acceleration of globalization, and to preclude being or becoming a blind determinist, it is urgent to understand the dynamics that work towards disruption, which is making the old world disappear violently. Of utmost importance is that all the actors must be fully cognizant of the trends in this state-of-the-art ecosystem.

A successful data ecosystem, which should be the most conspicuous constituent of any data-driven economy, has as objective to amalgamate data owners, data analytics, skilled data professionals, cloud service providers, companies from industry, venture capitalists, entrepreneurs, research institutions and universities. An efficient understanding and utilization of big data as an economic asset carries great financial and collective potential.

The cornerstone of this scholarship was founded on an exhaustive and unambiguous assemblage of pertinent key words or key phrases (Baker, 2016; Boell & Cecez-Kecmanovic, 2015; Kulage & Larson, 2016). The root cause analysis operation is principally utilized for identifying a wide range of approaches, tools and techniques to uncover causes of problems (Rooney & VanDen Heuvel, 2004). The authors employed this approach to generate keywords and key phrases. This latter was the rostrum of the categorization of research areas in the big data intelligence domain. To this end, TRIZ - a Russian acronym meaning " Theory of Inventive Problem Solving " (Apte, Shah, & Mann, 2001) as well as 5W1H – 5 W's and 1 H- event structures (Zhang et al., 2018) were employed. The generation of keywords or key phrases was articulated on the questions " What? ", " When? ", " Where? ", " Why? ", " Who? ", and " How? ".

What is the topic? What is hoped to be accomplished by this study? What is the end goal? What does the customer want? What is the ideal situation?

When do people realize the signification of big data intelligence? When does awareness occur?

Where is the disruption happening? Where do complications occur (the zones of conflict)?

Why are people concerned? Why is everybody concerned about ultra-intelligence? Why does the harmful action occur? (root-cause of the problem)?

Who is concerned? Who are the present advocates? Who is affected by harmful actions? By good actions?

How will this theme affect all avenues of our daily lives? How does the harmful action arise? How to improve these heterogeneous technologies? How to control and hence automate the tasks? How to provide control and consistency? How would inconveniences disappear? Under what conditions?

Collaboratively, keywords and key phrases were additionally generated from Coverage Book, (n.d), K-Meta (n.d.) and Keyword Tool (n.d). Finally, the authors also utilized, as benchmark, the World Economic Forum dynamic knowledge tool that catalogs all the issues and forces driving transformational change across economies, industries, global issues and other system initiatives.

The resulting taxonomy was arranged in 3 broad categories, or levels (see Table 4, Table 5 and Table 6), namely, the Macro level (6 items), the Meso level (29 items) and the Micro level (177 items). Because of its rigor, this was the research area classification schema that was adopted. Furthermore, this classification permitted to identify the most important and the most neglected research areas in big data intelligence.

This chapter employs these three categories of keywords and key phrases (Macro-, Meso- and Micro-level) to study empirical documentation. Following the content analysis, the findings were reported using descriptive analysis. Subsequently, a social network analysis was employed to further analyze keywords selected to to achieve a deeper comprehension and distinguish the links between them. The proposed methodology thus vigorously depends on the quality analysis of publications in the field that were selected based on Elsevier's - Science Direct and Scopus databases - " title, abstract or keywords " search engine covered in the predefined taxonomy of key words and phrases.

These keywords and phrases were utilized with the assumption that they provide a holistic reflection of the research domains in big data intelligence. The proposed methodology thus vigorously depends on the quality analysis of publications in the field that were selected based on Elsevier's - Science Direct and Scopus databases - " title, abstract or keywords " search engine covered in the predefined taxonomy of key words and phrases.

Because of the rapid evolution of big data technology, a situation that predisposes the field to research in very brief time spans, the year 2015 was chosen as the cut-off date for this study and the authors, as follow-up research, also examined articles published until 2018.

The occurrence or frequency of these indicators – keywords – were indexed or mapped in a network. Networks that represent liaisons between elements can be a valuable analytical tool. Cytoscape (n.d.) is a the worldwide open-source platform for network visualization and data integration, funded by the National Institute of General Medical Sciences and licensed directly from the Max Planck Institute for Informatics that interprets the interrelationship between keywords. Although at the onset designed for molecular components and interactions (Schwikowski, 2017), it is now a general platform for any type of network including that used in this chapter, that is, the big data attributes. Operationally, Cytoscape permitted to create a cloud network layout of an active sub-network with each individual node beginning its own cluster. This pattern enabled identifying sub-networks characterized by specific features such as the presence of dense interconnections with respect to the rest of the network.

FINDINGS AND DISCUSSION

The scope of this section was not to evaluate the benefits or otherwise of innovative technology, but, rather, to identify research trends which all possessed the potential of provoking Darwinian-type disruptions. The initial findings of the research were based on the analysis of descriptive keywords, which, it was assumed, provided a holistic reflection of research paper topics. There was a total of 212 keywords – 6 at the macro-level, 29 meso level and 177 micro-level (Table 4., Table 5. and Table 6.). The clusters of research articles pertaining to these big data intelligence keywords, after a research database search, resulted in a total of 9,402 occurrences or frequencies (Figure 3).

The spread of the 6 macro-level keywords was as follows.

Macro-Level: Smart Sensors and Autonomous Robots

The frequency count for these macro-level keywords - " Smart sensors and autonomous robots " - tallied 1,958, or 21% of the total occurrences (Figure 3).

Table 4. Taxonomy of keywords and key phrases attributed to big data intelligence with the macro-level "Smart sensors&autonomous robots" which comprise 7 meso-levels and 38 micro-level, the macro-level "Autonomous transportation", which comprises 7 at the meso-level and 47 at the micro-level, and the macro-level "Ethics&Values" which comprise 7 meso-levels and 37 micro-level.

Macro-Level	Meso-Level	Micro-Level
Smart sensors and autonomous robots	Digital economy&society	Access&adaption: Informed governance: Emerging technologies: Securing people&processes: Responsible transformation: Robust digital identities: Data sharing&permissions
	Future of health&healthcare	Global health governance: Environmental health&climate change: Healthcare delivery systems: Keeping populations healthy: Healthcare technology
	Information technology	Digital policy&governance: Cyber resilience: New disruptive technologies: Trust&privacy: Future of work
	International security	Future conflict domains: Rise of non-state actors: Technological arms race: Geo-strategic competition
	Internet of Things	Convergent design: Scalability is the goal: Privacy&security: New businesses: The right amount of standardization: Government assistance: Digital meets physical
	Neuroscience	Mental disorders: Neurological disorders: Mapping the brain: Neuro-ethics: Neutral tools: Behavior&decision-making
	Values	Environmental stewardship: Human dignity: The common good: Maintaining trust
Autonomous transportation	Automotive	Business model shifts: Self-driving cars: Data, security&privacy: Automotive standardization: Emissions&pollution: Market divergence
	Aviation, travel&tourism	Infrastructure investment: Demographic shifts: Data-enabled business: Safety&security risks: Global governance&regulation: Human capital: Sustainability: Travel&trade barriers
	Digital economy&society	Access&adaption: Informed governance: Emerging technologies: Securing people&processes: Responsible transformation: Robust digital identities: Data sharing&permissions
	Drones	AI&machine learning: Airspace management&infrastructure: Policy&social impact: Logistics&delivery: Aerial data capture: Environmental presentation: Urban aerial mobility
	Fourth industrial revolution	Ethics&identity: Agile governance: Fusing technologies: Disruption to jobs&skills: Security&conflict: Inequality: Business disruption: Innovation&productivity
	Internet of Things	Convergent design: Scalability is the goal: Privacy&security: New businesses: The right amount of standardization: Government assistance: Digital meets physical
	Values	Environmental stewardship: Human dignity: The common good: Maintaining trust
Ethics&values	Digital economy&society	Access&adaption: Informed governance: Emerging technologies: Securing people&processes: Responsible transformation: Robust digital identities: Data sharing&permissions
	Education&skills	Relevant specialized education: Quality basic education: Education innovation: 21st century curriculum: Digital fluency&STEM skills: Lifelong learning pathways
	Future of government	New governance models: The technological imperative: Mega disruptions: Ultra-urbanization: Demographic divide: Political turbulence
	Global governance	Rising multi-polarity: Transnational actors: Anti-globalism: Deepening interdependence: Institutional pluralism
	Global risks	Geopolitical risks: Economic risks: Environmental risks: Technological risks: Societal risks
	International security	Future conflict domains: Rise of non-state actors: Technological arms race: Geo-strategic competition
	Values	Environmental stewardship: Human dignity: The common good: Maintaining trust

Table 5. Taxonomy of keywords and key phrases attributed to big data intelligence with the macro-level "Machine cooperation&coordination" which comprise 6 meso-levels and 37 micro-level, and macro-level "Machine learning&predictive systems" which comprise 8 meso-levels and 52 micro-level.

Macro-Level	Meso-Level	Micro-Level
Machine cooperation&coordination	Digital economy&society	Access&adaption: Informed governance: Emerging technologies: Securing people&processes: Responsible transformation: Robust digital identities: Data sharing&permissions
	Electronics	Intellectual property: The promise of the Internet of Things: Future product development: Trade&investment
	Future of production	Global economy&trade: Technology&innovation: Human capital&skills: Consumer expectations: Regulation&governance: Accelerating sustainability: Locations for production
	Information technology	Digital policy&governance: New disruptive technologies: Trust&privacy: Future of work
	Supply chain&transport	e-Commerce&demand chains: Logistics skill shortages: Digitalization of supply chains: Logistics property&infrastructure: Restructuring global value chains: Supply chain sustainability: Supply risk&recovery
	Workforce&employment	Disruption to jobs&skills: Inclusive labor markets: Social protection: Training&certification systems: Job creation&entrepreneurship: Labor market demographics: New work models
Machine learning&predictive systems	Behavioral sciences	Human development: Sustainability&behavior change: Technology&digitalization: Economics&finance: Society&governance: Organizations&business
	Digital economy&society	Access&adaption: Informed governance: Emerging technologies: Securing people&processes: Responsible transformation: Robust digital identities: Data sharing&permissions
	Education&skills	Relevant specialized education: Quality basic education: Education innovation: 21st century curriculum: Digital fluency&STEM skills: Lifelong learning pathways
	Fourth industrial revolution	Ethics&identity: Agile governance: Fusing technologies: Disruption to jobs&skills: Security&conflict: Inequality: Business disruption: Innovation&productivity
	Healthcare delivery systems	Access to care: Health policy: Promise of precision medicine: Healthcare data: Healthcare technology: Healthcare human capital: Value-based healthcare
	Information technology	Digital policy&governance: New disruptive technologies: Trust&privacy: Future of work
	Neuroscience	Mental disorders: Neurological disorders: Mapping the brain: Neuro-ethics: Neutral tools: Behavior&decision-making
	Workforce&employment	Disruption to jobs&skills: Inclusive labor markets: Social protection: Training&certification systems: Job creation&entrepreneurship: Labor market demographics: New work models

Macro-Level: Autonomous Transportation

Research items in the category autonomous transportation represented a frequency of 1,736, or, 18% of the total cohort (Figure 3).

Macro-Level: Ethics and Values

New developments naturally raise serious ethical questions: here, there were 1,501 scholars or 16% of the total occurrences who dealt with this topic (Figure 3).

Table 6. Taxonomy of keywords and key phrases attributed to big data intelligence with the macro-level " Robots at work " which comprise 8 at the meso-level and 57 at the micro-level, and the macro-level " Human enhancement and assistance " which comprise 10 at the meso-level and 62 at the micro-level.

Macro-Level	Meso-Level	Micro-Level
Robots at work	Entrepreneurship	Human capital&mindsets: Ecosystem for innovation-driven entrepreneurship: Corporate entrepreneurship: Social impact-driver entrepreneurship: Regulatory frameworks: Emerging market entrepreneurship
	Fourth industrial revolution	Ethics&identity: Agile governance: Fusing technologies: Disruption to jobs&skills: Security&conflict: Inequality: Business disruption: Innovation&productivity
	Future of production	Global economy&trade: Technology&innovation: Human capital&skills: Consumer expectations: Regulation&governance: Accelerating sustainability: Locations for production
	Humanitarian action	Prevent&end conflict: Respect for the rules of war: Leave no one behind: Humanitarian assistance delivery: Work differently to en need: Invest in humanity: Data&evidence
	Internet of Things	Convergent design: Scalability is the goal: Privacy&security: New businesses: The right amount of standardization: Government assistance: Digital meets physical
	Mining and minerals	Technological innovation: Mining&metals economics: Geopolitical uncertainty: Environmental sustainability: Skills mismatch: Access to resources: Cross-industry disruptions: Social contract
	Space	Spurring growth: Applications for society: Risks to space sustainability: Space conflict: Policy&governance: Increased access: Business model disruption
	Workforce&employment	Disruption to jobs&skills: Inclusive labor markets: Social protection: Training&certification systems: Job creation&entrepreneurship: Labor market demographics: New work models
Human enhancement and assistance	Aging	Workplace&workforce change: Health&functional ability: Elder care-giving: Innovation readiness: Social welfare programs: Age-friendly infrastructure: The silver economy
	Electronics	Intellectual property: The promise of the Internet of Things: Future product development: Trade&investment
	Fourth industrial revolution	Ethics&identity: Agile governance: Fusing technologies: Disruption to jobs&skills: Security&conflict: Inequality: Business disruption: Innovation&productivity
	Future of production	Global economy&trade: Technology&innovation: Human capital&skills: Consumer expectations: Regulation&governance: Accelerating sustainability: Locations for production
	Healthcare delivery systems	Access to care: Health policy: Promise of precision medicine: Healthcare data: Healthcare technology: Healthcare human capital: Value-based healthcare
	Human enhancement	Humanness&autonomy: Enhanced longevity: Enhanced minds: Enhanced bodies: Equity&social justice: Enhanced genes
	Information technology	Digital policy&governance: New disruptive technologies: Trust&privacy: Future of work
	Retail, consumer goods&lifestyle	New patterns of consumption: Artisan production: Immaterial consumption: The global middle class: The global luxury market
	Virtual&augmented reality	Technological barriers: Digital teleportation: Healthcare application&side effects: Immersive journalism&environmental stewardship: New ways to make, do&buy: Immersive media&art: The gamification of life&reality distortion
	Workforce&employment	Disruption to jobs&skills: Inclusive labor markets: Social protection: Training&certification systems: Job creation&entrepreneurship: Labor market demographics: New work models

Macro-Level: Machine Cooperation and Coordination

This macro level accounted for 1,419 occurrences or 15% (Figure 3).

Macro-Level: Machine Learning and Predictive Systems

Here, the frequency of research was 1,302 or 14% of the total (Figure 3).

Macro-Level: Robots at Work

There were 784 (8%) scholars whose research was focused on diverse forms of robots (Figure 3).

Macro-Level: Human Enhancement and Assistance

Last but not least, this macro-level was researched by 702 scholars or 7% (Figure 3).

The score 1,958 occurrences for the macro-level keywords, "Smart sensors and autonomous robots", during the period 2015-2018, delineated the most widespread research area and as such, according to scholars was the domain where research was the most intense (Figure 3). Should this axiom be viewed as a surprise! As a matter of course, living creatures receive data about the real world through senses like vision and smell; researchers in the fields of robotics and computer science are now endeavoring to give machines an ability to acquire real world data in similar ways. The end result could be machines that can autonomously adapt to circumstances as their environments change, or even disrupt.

Associated to the above-mentioned 7 macro-level keywords were their respective 29 affiliated meso-level keywords (Figures 4 to 17). Figure 4 and Figure 5, which pertain to the macro-level keywords, "Smart sensors and autonomous robots" had 7 constituent meso-level keywords, namely, Digital economy&society (43%), Values (20%), Future of health&healthcare (17%), Sensors (8%), Information technology (7%), Internal security (4%) and Neuroscience (1%).

Figure 3. Spread of macro level occurrences during the period of study, that is, 2015 to 2018. " Smart sensors and autonomous robots " were the key phrases that represented the most active area (1,958 occurrences or 21% of the total research efforts-9,402 occurrences) with " Human enhancement and assistance " being the least favored domain (702 occurrences, or 7% of total research efforts).

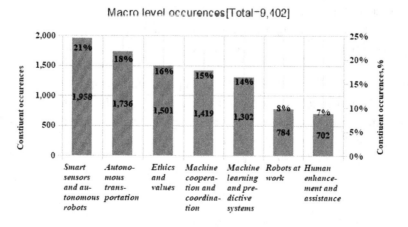

Figure 4. Mapping of the macro-level key words, " Smart sensors and autonomous robots " with 7 conjugated meso-level key words. Edge-weighted directed selected nodes – macro- and meso-level -with betweenness centrality. The hierarchical lower level macro- and meso-level key words are invoked elsewhere in the chapter.

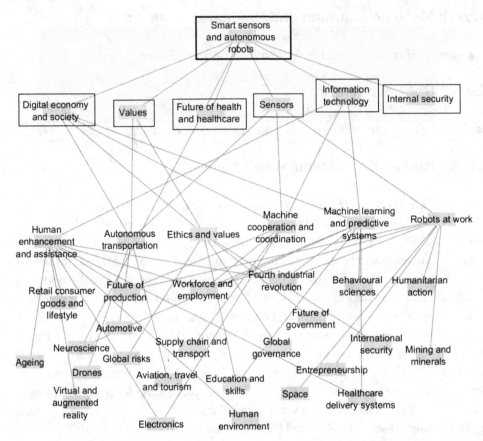

The meso-level keywords, Digital economy&society, - at 43% the most fecund research area - sequentially generated the following micro-level keywords (Table 4): Access&adaption, Informed governance, Emerging technologies, Securing people&processes, Responsible transformation, Robust digital identities and Data sharing&permissions. In this sample, this combination of micro-level keywords represented the most prevalent research topics. Intellectually, it is the enterprises who adjust to the growing importance of these technologies, and who, in turn, develop a coherent strategy for integrating these potentially disruptive enablers, would be successful. The digital technology has the potential to enhance economies, societies, and the environment. However, if related risks outweigh these benefits it could further exacerbate exclusion, the concentration of power and wealth, and result in social instability. Coordinated, multi-stakeholder action is required to create a digital future that is inclusive and sustainable. Artificial intelligence principles outlined by Google, for example, include not only a prerequisite to be socially beneficial, but also to avoid reinforcing unfair bias, to be accountable to people, and to uphold high standards of scientific excellence.

Figure 5. The distribution of macro level occurrences for the key phrases, " Smart sensors and autonomous robots ", with the 7 associated meso-level key phrases along the abscissa. Digital economy and society were the key phrases that represented the most active research area (840 occurrences or 43% of the total research efforts – 1,958 occurrences) with Neuroscience being the least favored domain (28 occurrences, or 1% of total research efforts).

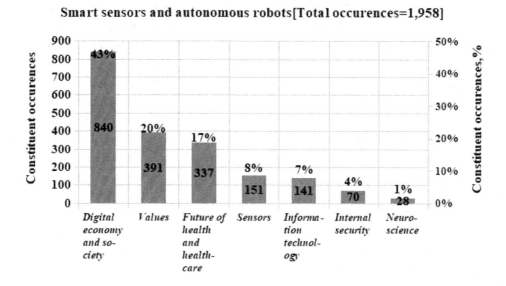

On the other hand, the meso-level keywords, Values (20%) (Figure 5) comprise the micro-level keywords, or key issues, Environmental stewardship, Human dignity, The common good and Maintaining trust (Table 4).

Values are the basis of personal and collective judgments about what is important in life - influenced by culture, religion, and laws. They are factors in our decision-making on social, environmental and political matters, and on the most effective utilization of our time, money, and valuable materials. One of the most essential values in the midst of the disconcerting changes wrought by the Fourth Industrial Revolution is trust; between the governed and governments, and between civil society and the private sector (Jelen & Milana, 2018).

The key issues determined by scholars for the category, " Future of health&healthcare ", were Global health governance, Environmental health&climate change, Healthcare delivery systems, Keeping populations healthy and Healthcare technology (Table 4).

Neuroscience, which concerned Mental disorders, Neurological disorders, Mapping the brain, Neuroethics, Neutral tools, Behavior&decision-making (Table 4.) were only pursued by 1% of scholars. This negligible frequency constituted another surprise because healthy brain function is an essential foundation for childhood development, education, social interaction, and professional life, Furthermore, the prevalence of brain disorders has imposed heavy financial and emotional burdens throughout the world, and will only degenerate as global populations age (Trautmann, Rehm, & Wittchen, 2016).

Referring to Figure 6 and Figure 7, the macro-level keywords, " Autonomous transportation " had 7 constituent meso-level keywords, namely, Digital economy&society (48%), Values (23%), Automotive/car/vehicle (16%), Sensors (9%), Aviation, travel and tourism (4%), Drones (<1%).

The latter - drones - has the following micro-level keywords; AI&machine learning, Airspace management&infrastructure, Policy&social impact, Logistics&delivery, Aerial data capture, Environmental presentation and Urban aerial mobility (Table 4). Because of the keywords associated with this category of technology, the result of less than 1% was a surprise. Were different keywords or phrases used for the study of drones? This might probably be the case.

Self-driving vehicles are poised to become mainstream, but safety issues must first be addressed (Taeihagh & Si MinLim, 2018). Large drones, originally invented for military missions, are almost ready for general transportation use. These drones are expected to be integrated into civil airspace after 2020, when the International Civil Aviation Organization (ICAO, 2011) releases rules for worldwide operation (Lykou, Iakovakis, & Gritzalis, 2019; ICAO, 2011). Meanwhile smaller drones, thanks to their low operating costs and low risk, are already gaining popularity. Agencies such as the National Aeronautics and Space Administration (NASA, 2015) in the United States are developing a new drone traffic management system, and we may soon see crowds of drones in the sky that resemble scenes from a science fiction film. These machines are expected to create new job opportunities in some cases, and to take away jobs in others. Autonomous cars, unlike drones, are not yet ready for prime time (Bansal & Kockelman, 2018). The latest self-driving cars, however, are equipped with sensors that can see in night

Figure 6. Mapping of the macro-level key words, "Autonomous transportation" with 6 conjugated meso-level key words. Edge-weighted directed selected nodes – macro- and meso-level - with betweenness centrality. The hierarchical lower level macro- and meso-level key words are invoked elsewhere in the chapter.

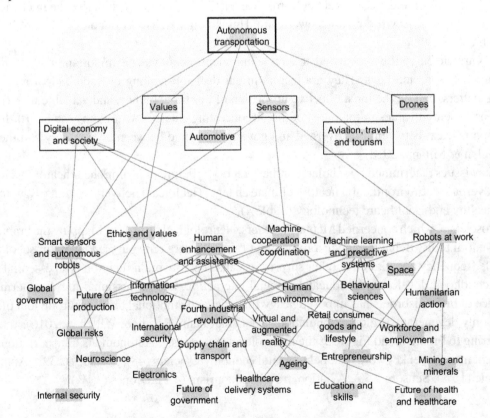

Figure 7. The distribution of macro level occurrences for the key phrases, " Autonomous transportation ", with the 6 associated meso-level key phrases along the abscissa. Digital economy and society were the key phrases that represented the most active research area (840 occurrences or 48% of the total research efforts – 1,736 occurrences) with Drones being the least favored domain (6 occurrences, or less than 1% of total research efforts).

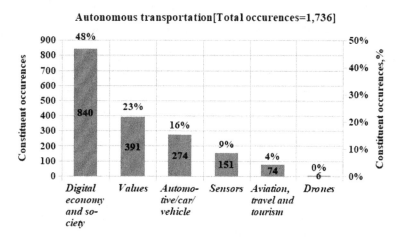

time darkness and scan their surroundings more than 10 times per second, with extreme accuracy. Their brains are state-of-the-art computers that can make trillions of calculations per second, and communicate both with other cars and with the cloud in order to receive the latest traffic updates. Self-driving cars compute an optimal path by sensing their environment and predicting the movement of obstacles. (Um, 2019). During this process, artificial intelligence views human life as a cost in its calculations. But it is very difficult to represent the value of human lives as a cost function in a mathematical equation. And no matter how sophisticated the on-board artificial intelligence, it is very difficult to assess how much collateral damage a potential collision will cause beyond impact. Only when autonomous cars' accident rate is proven to be significantly lower than that of human drivers, will they be broadly welcomed, and become safe enough to carry human passengers (Elliott, Keen, & Miao, 2019). In 2016, Airbus announced plans for an autonomous air taxi (Sun, Wandel, & Stump, 2018). Not to be outdone, Uber also announced its own plans for an autonomous air taxi. In the meantime, however, Uber has had to deal with safety issues; an Uber autonomous car struck and killed a pedestrian in Arizona. If and when autonomous drones and cars are granted equal right of way in civil airspace and on roads, it will prompt an interesting question: should robotic workers be granted similarly equal privileges at work, which protect them from discrimination, for example. (Van den Hoven van Genderen, 2019).

Pertaining to Figure 8 and Figure 9, the macro-level keywords, " Ethics and values " had 7 constituent meso-level keywords, namely, Digital economy&society (48%), Values (23%), Global risks (10%), Education and skills (5%), Future of government (2%), Global governance (1%) and International security (<1%).

The meso-level, Education and skills had the following micro-level keywords, Relevant specialized education, Quality basic education, Education innovation, 21st century curriculum, Digital fluency&STEM skills and Lifelong learning pathways (Table 4).

Figure 8. Mapping of the macro-level key words, " Ethics and values " with 7 conjugated meso-level key words. Edge-weighted directed selected nodes – macro- and meso-level - with betweenness centrality. The hierarchical lower level macro- and meso-level key words are invoked elsewhere in the chapter.

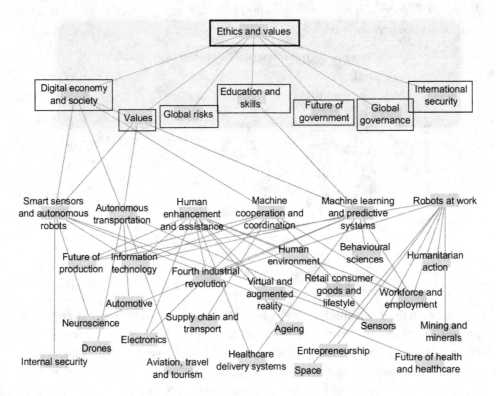

Figure 9. The distribution of macro level occurrences for the key phrases, " Ethics and values ", with the 7 associated meso-level key phrases along the abscissa. Digital economy and society were the key phrases that represented the most active research area (840 occurrences or 56% of the total research efforts – 1,501 occurrences) with Internal security being the least favored domain (4 occurrences, or less than 1% of total research efforts).

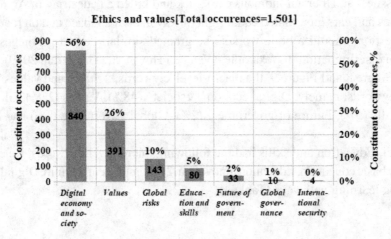

Big data intelligence is raising new questions about the manner we live, fight, and feel.

Many systems equipped with artificial intelligence do not pose an immediate risk, should something go wrong. Others, however, have been placed in command of safety-critical and military systems. Even seemingly benign automated assistance provided by so-called chatbots, or artificial intelligence that interacts with human computer users, can bring important questions about ethics and values to the fore. In early 2016, for example, a Microsoft chatbot was appropriated by users who taught it ethnic slurs (Perez, 2016). What might happen if technology - for example, an artificial intelligence system powering a deadly weapon - was equipped with the means to act on such ugly sentiments? Lethal weapons capable of zeroing in on targets faster and more accurately are on track to gain ever greater autonomy, and even if we could instruct such technology in ethics and human values, it's not immediately clear which ethics and values we might select. And, how could we ensure that a robot instructed to kill during a war will understand that the same behavior is unacceptable during peacetime? When humans are in charge of weapons, tragic mistakes can result; the question is therefore whether we should always place more trust in a human, or sometimes leave the decision making to artificial intelligence with access to better information, and lower known mistake rates (Helbing, 2018).

Another consideration is how we should treat intelligent machines. There are cases where humans have been found to empathize with robotic systems, especially when the systems resemble humans or pets. When Boston Dynamics (Cho, Kim, Kim, Park, & Kim, 2016) tested a robot that resembles a dog by kicking it, the company faced criticism. In other instances people have had very different reactions to robots; in 2018, the co-founder of a drone delivery service revealed that people had been kicking the company's machines. Other questions to ponder include whether artificial intelligence should have to reveal that it is artificial intelligence. In 2018, Google demonstrated an artificial intelligence tool that could chat so naturally with people on the phone that they believed they were speaking with another human - eliciting awe, and criticism that this may cross an ethical line (Welch, 2018). What is clear is that artificial intelligence should be created and deployed with caution. Specialists and experts in the field should be educated in ethics. And clear standards for the industry, which emphasize accountability, should be considered.

According to the 2018 Edelman Trust Barometer (ETB), a survey conducted in 28 countries, public trust in the technology sector declined slightly compared with the prior year. In a supplementary report published by Edelman, the public relations firm noted that trust in technology among the " informed public " actually fell sharply in regions including the US, Hong Kong, and France; according to the supplementary report, a fear of fake news being used as a weapon was on the minds of seven in 10 respondents, and trust in newer, emerging technologies such as blockchain and self-driving cars commanded less public confidence than more established technologies. In addition to privacy and security concerns, there are broader ethical questions about the ways that organizations have used digital technology in a way that threatens to erode trust in institutions (Milhorance & Singer, 2018). There are also concerns about the effects of new digital technologies on the environment, related to power-hungry data centers, electronic waste, and the significant amounts of energy needed to power blockchain. Establishing new norms of ethical behavior regarding digital technology, and attaining higher levels of customer trust, will be critical for a successful digital transformation.

Figure 10. Mapping of the macro-level key words, " Machine cooperation and coordination " with 7 conjugated meso-level key words. Edge-weighted directed selected nodes – macro- and meso-level - with betweenness centrality. The hierarchical lower level macro- and meso-level key words are invoked elsewhere in the chapter.

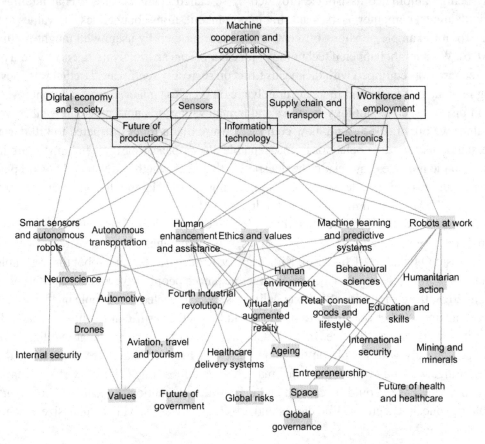

With reference to Figure 10 and Figure 11, the macro-level keywords, " Machine cooperation and coordination " had 7 constituent meso-level keywords, namely, Digital economy&society (59%), Future of production (13%), Sensors (11%), Information technology (10%), Supply chain and transport (5%), Workforce and employment (1%) and Electronics (1%). The micro-level that corresponded to each of these 7 meso-levels are illustrated in Table 5.

Robots are increasingly being programmed to work with other robots and humans.

There are entire fields of research focused on enabling robots to cooperate. If robots are needed outside of controlled environments like factory floors, for example, they have to be able to quickly adapt to their surroundings in order to work both with other robots and with humans. Visual sensors help the machines to understand complex environments, while artificial intelligence aids their understanding of human gestures, facial expressions and even intentions. One of the forerunners in this area is Baxter, an industrial robot developed in 2011 that can be easily taught new tasks, thanks to natural language processing and synthesis using so-called deep learning, or training computers to learn in a way that is similar to a human brain absorbing information (Al-Abdulqader & Mohan, 2018). This variety of robot

Figure 11. The distribution of macro level occurrences for the key phrases, "Machine cooperation and coordination", with the 7 associated meso-level key phrases along the abscissa. Digital economy and society were the key phrases that represented the most active research area (840 occurrences or 59% of the total research efforts – 1,419 occurrences) with Electronics being the least favored domain (11 occurrences, or 1% of total research efforts).

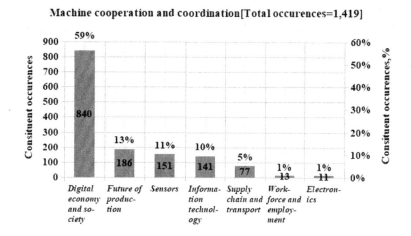

can be easily communicated with using voice commands, and can be visually taught in a way that is much faster than the traditional method of entering commands into a handheld device. In addition to staffing an assembly line, robots can be useful as companions. Their powerful processors and advanced sensors allow them to recognize faces, understand voice commands, and do some pretty complex tricks. Amazon's Alexa and Google Home are examples of potentially helpful companion robots (Zhang, Mi, Feng, Wang, Tian, & Qian, 2018), as is Jibo, developed at the Massachusetts Institute of Technology (Singh, 2018) while Pepper, a humanoid robot from SoftBank (Varrasi, Lucas, Soranzo, McNamara & Di Nuovo 2018), is designed to perceive human emotions.

In recent years, robots have begun to compete against each other in soccer games, by autonomously adapting offensive and defensive strategies (Li, Feng, Huang, & Chu, 2018). One of the greatest advantages that robots have is a capability for high-speed communications. Whereas a typical human can speak 130 to 200 words per minute, which would translate into tens of bytes per second, robots can communicate as fast as one gigabyte (1 billion bytes) per second. Groups of robots can " wirelessly " work together even when separated by great distances, and can be synchronized by GPS technology with a great degree of accuracy (Matusov et al., 2018). Extreme computation speed (more than one trillion operations per second) can enable these machines to cooperate regardless of the size of their grouping, or the distance between them, making humans far less competitive when it comes to many jobs. Just as data servers have made some aspects of traditional archives and libraries obsolete, robots, like data servers that can move around and behave autonomously, promise to impact on our daily lives beyond anyone's imagination.

Referring to Figure 12 and Figure 13, the macro-level keywords, " Machine learning and predictive systems " had 8 constituent meso-level keywords, namely, Digital economy&society (59%), Information technology (11%), Fourth industrial revolution (10%), Education and skills (6%), Health delivery systems (4%), Neuroscience (2%), Behavioral sciences (1%) and Workforce and employment (1%).

The micro-level that corresponded to each of these 8 meso-levels are illustrated in Table 5.

Latterly, it has become commonplace to effectively distinguish patterns in large, evolving data sets in order to better understand them and create appropriate predictive systems (Winskell, Singleton, & Sabben, 2018). Machine learning involves creating algorithms that can recognize patterns in large, evolving data sets, and drawing conclusions from past experience by using that data - in order to make machines smarter (Biggio & Roliab, 2018). When people refer to " artificial intelligence ", they often really mean machine learning. Examples of tasks that make use of machine learning include the following:

- Decision making and control
- Sensory information storage and data compression
- Repeatable pattern detection and classification
- Regression analysis of noisy pattern sequences to find new, hidden patterns in complex data

Figure 12. Mapping of the macro-level key words, " Machine learning and predictive systems " with 8 conjugated meso-level key words. Edge-weighted directed selected nodes – macro- and meso-level - with betweenness centrality. The hierarchical lower level macro- and meso-level key words are invoked elsewhere in the chapter.

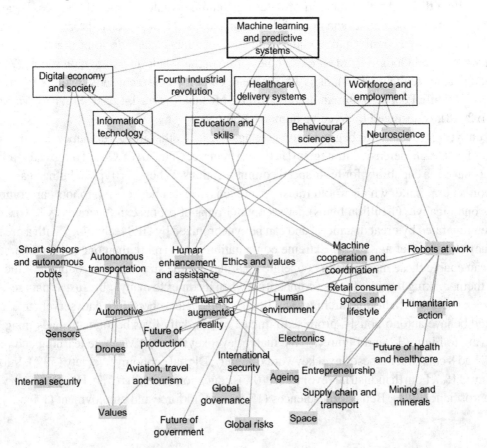

Figure 13. The distribution of macro level occurrences for the key phrases, " Machine learning and predictive systems ", with the 8 associated meso-level key phrases along the abscissa. Digital economy and society were the key phrases that represented the most active research area (840 occurrences or 65% of the total research efforts – 1,419 occurrences) with Workforce and employment being the least favored domain (13 occurrences, or 1% of total research efforts).

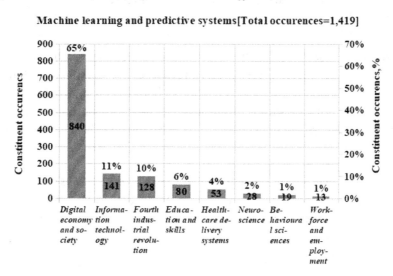

Recently, an aspect of machine learning dubbed " deep learning algorithms " has received a great deal of attention (Larson, 2018). That is because advances in computing power and masses of large-scale data, referred to as big data, have led to deep learning-based algorithms that are faster and more accurate than the human eye. In 2015, DeepMind, a UK-based firm that shares a parent company with Google, put the power of such algorithms on display when it pitted its AlphaGo computer program against a top human player of the board game Go. The computer program won (Holcomb, Porter, Ault, Mao, and Wang, 2018). More recently, in 2018, a company called Preferred Networks became the most valuable startup in Japan, as investors including Toyota bet that the company's deep learning algorithms can eventually be applied to a number of manufacturing uses (Clavera et al., 2018). Machine learning is expected to have a profound impact on the job market. Experts predict that both menial and professional jobs will be taken over by computers and robots equipped with learning algorithms. According to a study published by researchers Haenssle et al., (2018) machine learning can do a better job of detecting skin cancer than a dermatologist, for example - though it would do a significantly poorer job of explaining to a patient why a lesion is cancerous or not. The potential advantages of the technology are clear, however: once machines learn they never forget (Villaronga, Kieseberg, & Li, 2018); a learning pattern can be efficiently copied from one machine to another; and, learning can be done in a parallel manner in order to improve and share (Chen, & Liu, 2018). For example, if one unit masters the art of driving, that learning pattern can be copied into millions of other cars within a very short time. Cars connected through a network can therefore continuously share experiences to improve overall performance. Humans, on the other hand, take a long time to learn (Ishiguro & Nishio, 2018). Their experiences cannot be shared in the same way, and the value of their individual knowledge and experience can perish at death.

Referring to Figure 14 and Figure 15, the macro-level keywords, " Robots at work " had 8 constituent meso-level keywords, namely, Future of production (24%), Space (21%), Sensors (19%), Fourth industrial revolution (16%), Mining and minerals (16%), Workforce and employment (2%), Humanitarian action (1%) and Entrepreneurship (1%). The most prevalent meso-level, Future of production (24%), included the micro-level categories, Global economy&trade, Technology&innovation, Human capital&skills, Consumer expectations, Regulation&governance, Accelerating sustainability and Locations for production (Table 6).

Robot surrogates are increasingly being used for dangerous work and in extreme environments. They have long been used on factory floors for welding and painting. Now, they are making custom-ordered hamburgers, navigating through crowded hotel lobbies, hopping on elevators, and delivering room service. Amazon has been at the forefront of introducing robots to our daily lives, with its clerk-free markets that enable automatic payment via credit card, Kiva merchandise handling systems, and drone

Figure 14. Mapping of the macro-level key words, " Robots at work " with 8 conjugated meso-level key words. Edge-weighted directed selected nodes - macro- and meso-level - with betweenness centrality. The hierarchical lower level macro- and meso-level key words are invoked elsewhere in the chapter.

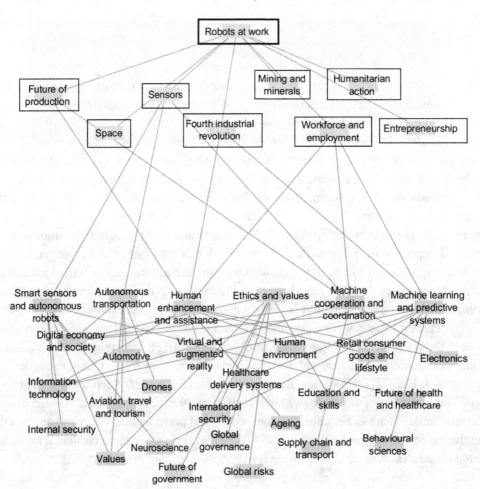

Figure 15. The distribution of macro level occurrences for the key phrases, " Robots at work ", with the 8 associated meso-level key phrases along the abscissa. Future of production were the key phrases that represented the most active research area (186 occurrences or 24% of the total research efforts – 784 occurrences) with Entrepreneurship being the least favored domain (6 occurrences, or 1% of total research efforts).

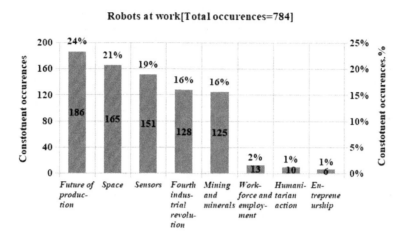

delivery (Tavakoli, Viegas, Sgrigna, & de Almeida, 2018). Robots also enable exploration, by probing the depths of the ocean floor, deep space, the moon, and asteroids. also being developed to perform crucial jobs during humanitarian crises. For example, the Defense Advanced Research Projects Agency (DARPA) Robotics Challenge (Krotkov et al., 2018) is developing robots that would have been capable of saving lives during the 2011 Fukushima nuclear disaster in Japan. Other robots in development may perform hazardous deep mining jobs, work with toxic substances, and clean sewers. Drones in particular have captured the popular imagination; they are a popular option for aerial photography, and Amazon and others are testing their use for delivery. The latest space-ready robots can work side-by-side with astronauts in the International Space Station (ISS), and US and Russian teams are working on " *robo-astronauts* " that could explore other planets (Naids et al., 2018).

In less dramatic fashion, robots can use an electronic brain located in the cloud to become trust-worthy butlers, silently and faithfully listening to queries about traffic, the weather, and turning on the air conditioning. Next-generation robots will be performing tasks that require even more complex decision-making, aided by continuously increasing computing power. In January 2016, the company Nvidia unveiled a supercomputer capable of up to 24 trillion operations per second, equal to the power of 150 MacBook Pro notebook computers (Wang, 2016). This could enable a car to learn to drive by itself, through so-called reinforcement learning, rather than the conventional approach of finding a safe path by using real-time optimization. As related technologies mature, self-driving cars will become common, and will likely fundamentally change the job market by eliminating occupations like the bus driver. Critical tasks that go beyond physical labor, such as reading x-ray films and magnetic resonance imaging (MRI) scans, are already being performed by computers. Discovering new medicines might even be a possibility; after all, artificial intelligence is already helping to dispense financial advice to many people, rapidly and with considerable accuracy.

Figure 16. Mapping of the macro-level key words, " Human enhancement and assistance " with 10 conjugated meso-level key words. Edge-weighted directed selected nodes - macro- and meso-level - with betweenness centrality. The hierarchical lower level macro- and meso-level key words are invoked elsewhere in Chapter 6.

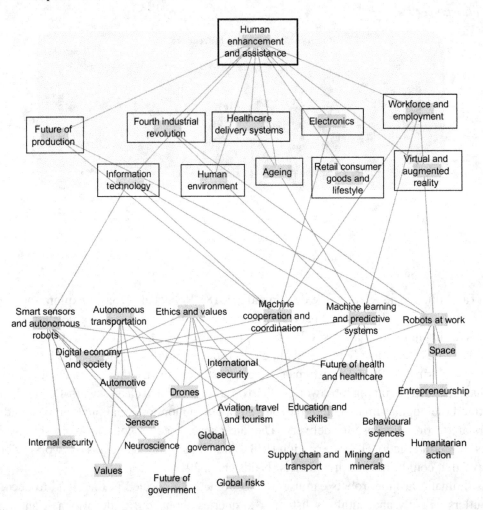

With reference to Figure 16 and Figure 17, the macro-level keywords, " Human enhancement and assistance " had 10 constituent meso-level keywords, namely, Future of production (26%), Information technology (20%), Fourth industrial revolution (18%), Human environment (11%), Health delivery systems (8%), Aging/age (7%), Virtual and augmented reality (4%), Retail consumer goods and lifestyle (2%), Workforce and employment (2%) and Electronics (2%).

By definition, the enhancement of big data intelligence is helping humans exceed their natural limitations. Recent breakthroughs in wearable robotic systems may enable humans to perform physical feats that would ordinarily be beyond their capabilities, like walking extreme distances or carrying heavy loads. In addition, research is underway that could result in better prosthetic limbs, equipped with sensors, motors, and sophisticated algorithms, and capable of receiving commands directly from the human nervous system. These technologies could one day mean the end of physical disability, or the

Figure 17. The distribution of macro level occurrences for the key phrases, " Human enhancement and assistance ", with the 10 associated meso-level key phrases along the abscissa. Future of production were the key phrases that represented the most active research area (186 occurrences or 26% of the total research efforts – 702 occurrences) with Electronics being the least favored domain (11 occurrences, or 2% of total research efforts).

Human enhancement and assistance[Total occurences = 702]

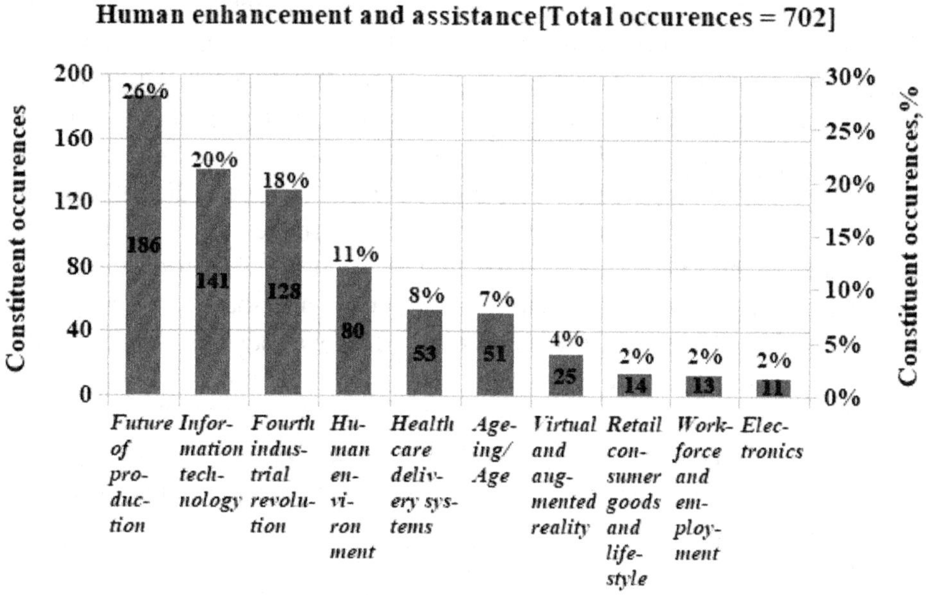

enhancement of existing physical capability and brain functions. In addition to systems that are worn on the body, smart robotics systems have been developed to assist humans in tasks that require dedicated precision and repetition, such as surgery. Meanwhile, so-called tele-presence robotics enables people to be "present" for meetings or activities via a mobile video screen.

Other forms of human enhancement include augmented reality devices that help people perceive their environment with richer relevant information. Google's Glass eye-wear drew attention as one of the first such devices that could be worn in everyday situations, though the technology was widely panned due to privacy issues and aesthetic appeal (Tag, Holz, Lukowicz, Augereau, Uema, & Kunze, 2018). Microsoft's Hololens wearable device can similarly map a user's environment and display data, though it is seen as bulky and uncomfortable (Tham, 2018). However, such devices will generally become small enough, and common enough, to be suitable for daily use. Google and other companies have also developed automatic, real-time translators.

Once mocked, they now provide accurate translation at lightning-fast speeds, thanks to recent advances in deep learning, which mimics the functions of a human brain to help computers absorb information. When combined with mobile phones or augmented reality devices, these tools could potentially eliminate the need for human translators; there may even be less of a need for people to continue trying to learn foreign languages entirely. Some have envisioned neural implants that directly connect such technology to a human brain, in order to aid not only in translation, but also computation and memory (Schwemmer et al., 2018). Some prosthetics are already controlled directly by a human nervous system; cameras can be linked to human brains in order to provide visual information to blind people, for example. While artificial intelligence systems will greatly increase our ability to interact with our environment, many jobs will be affected - including those that have generally been considered safe from the invasion of robotic workers, like translators, travel guides and news anchors.

The meso-level category, Age/Aging consisted of 7 micro-levels, Workplace&workforce change, Health&functional ability, Elder care-giving, Innovation readiness, Social welfare programs, Age-friendly infrastructure and The silver economy (Table 4). Though research for the "age-tech" represented 7%, several books have captured the magnitude of the opportunity presented in digitally-enabling the so-called, " longevity economy "; among these are the texts of Gratton & Scott (2016) and Coughlin (2017). Older people are the fastest growing demographic group (with around 2.5% annual populations growth vs. 0.7% for all population); digitization is growing across all sectors (to around 16% by 2025); and digitization in aging is converging towards the global average (Woods, 2019). Age-Tech solutions cut across all areas of the economy. They include on-site and remote senior care via blended models combining human and machine interaction; low-friction platforms for older people to transfer wealth to their younger relatives tax-efficiently, with controls and performance-based incentives; a new form of insurance that periodically adjusts to user changes in risk profile; a social network focused on enabling those displaced by automation to pursue future paths with a high sense of purpose; a digital bank adapted to the preferences of older users.

In conclusion, it was observed that the meso-level keywords, " Digital economy and society " under-pin all but 2 of the 7 macro-level keywords. This signifies that scholars unequivocally identified these meso-level keywords as been directly linked with transformation of enterprises and circumstantially as the technology that possesses the greatest disruptive potential.

While there were a number of other findings that merited to be analyzed and discussed, it is beyond the scope of this study to do so, and will be carried-over to another analogous study.

FUTURE RESEARCH DIRECTIONS

The results of this study reveal research trends and issues in big data intelligence emerging from scholarly publishing in the Elsevier research databases - Science Direct and Scopus - and provides a direction for future research. In the body of this chapter, assorted future enhancements were alluded to; they will be further discussed in this section.

Powerful innovative forces are changing our world. Propelled by high-growth in technological disruption, they include the shifting locus of economic activity and dynamism; the world's growing " connectedness " through movements in capital, people, and information; the acceleration in the scope, scale, and economic impact of technology. The impact of these and other disruptive forces is being felt worldwide, touching/embracing/affecting all countries, sectors, companies, and, increasingly, workers and the environment. These forces are also morphing in some unexpected ways and combining to create even greater impacts than we expected.

First and foremost, it is imperative that governments, businesses, academic institutions and individuals consider how to proactively shape a new, positive innovative business model - one that is desired rather than one created through inertia. Based on how different combinations of seven core variables were investigated by a variety of research scholars – The macro level, that is, Smart sensors and autonomous robots; Autonomous transportation; Ethics and values; Machine cooperation and coordination; Machine learning and predictive systems; Robots at work; Human enhancement and assistance – are likely to influence the nature of our everyday activities in the future, the chapter provides a starting point for considering a range of options around the multiple possible futures of innovative, disruptive technologies.

Each of the state-of-the-art technologies, which exemplify potential disruptive forces, and which the authors have highlighted in this study would be challenging existing on its own; taken together, they can seem daunting. Yet the opportunities for the economy, business, and society that these global forces generate are equally compelling and are already creating new prosperity for those quick to embrace them. The potential challenges cannot be ignored, chief among them being the growing social inequalities that could arise from the transformational skews. As societies, we will face challenges related to the future of work as well as inclusive growth; the two are closely linked. Embracing the trends while mitigating their negative impact on those who cannot keep abreast and on the environment is the new imperative of this era. Today, we regularly interact with machine-learning algorithms, from virtual assistants on our phones to spam filters for our e-mail. But these modern real-world applications trace their origins to a sub-field of machine learning that is concerned with the careful formalization and mathematical analysis of various machine - learning settings. Machine learning has matured as a mathematical discipline and now joins the many sub-fields of mathematics that deal with the burden of " unprovability " and the unease that comes with it. Perhaps results such as this one will bring to the field of machine learning a healthy dose of humility, even as machine-learning algorithms continue to revolutionize the world around us. Concurrently with this stipulation, scenarios have been discovered in which it is impossible to prove whether or not a machine-learning algorithm could solve a particular problem. This finding might have implications for both established and future learning algorithms.

In recent years, artificial intelligence has struggled with a major Public Relations' problem: whether or not it is intentional, developers keep programming biases into their systems, creating algorithms that reflect the same prejudiced perspectives common in society. The indispensable solution would be to develop an algorithm that can scrub the bias from artificial intelligence – an equivalent of sensitivity training for algorithms. The dynamism of present day economies has gone conjointly with the rise of highly competitive emerging-market firms, which are increasingly taking on incumbents in advanced economies – the GAFAM's and BAFTA's. Illustrative of this circumstance, in this study, the most wide-spread research concerns the meso-level " Digital economy and society ". The reality is underscored by the fact that by the end of 2018, about half of the world's population was connected to the internet - roughly 30 years after the invention of the World Wide Web; hence branding this family as a fecund area of research. From another standpoint, the meso-level, " Human enhancement and assistance ", though sparsely endorsed by scholars, is a domain that typically will better lend itself to disruptive innovation. Illustrative of this is the acquisition of expertise. Because its cost of acquisition is high (cost of learning, through education or experience) and its availability is rare, expertise still has a considerable value today. But like knowledge, expertise could very quickly become a simple commodity and become accessible to all. With the arrival of artificial intelligence, each expert in his respective field will train and teach the algorithm to behave like him, to reason like him, to discern like him: the expert will transfer his expertise to the algorithm. Each expert in the learning phase, will supervise the algorithm to give him progressively his expertise, until the algorithm equals, then exceeds the master. Eventually, expertise on any subject will be entirely entrusted to artificial intelligence. First the algorithm will complement the expert, then any expertise will be entrusted to him. Because even computer experts and programmers will be disrupted by artificial intelligence. The next step would be the creation of artificial intelligence capable of generating other modes of artificial intelligence including those like DeepCoder who are capable to write their own computer code (Balog, Gaunt, Brockschmidt, Nowozin, & Tarlow, 2016).

Tomorrow, in contrast to today, nobody will program machines and software with complex code, which requires an expert and adapted training. In does not mean that the code will disappear - the machines will always work with code - but that the automatism will be so advanced that the job of programmer will be to train the machine to manipulate high-level concepts, analogous to teaching math to a child. In the long term, everyone will be able to create artificial intelligence to accomplish tasks and to achieve objectives in natural language or being shown how to do it. As for enterprises, their role will be to invent new applications and new services by using artificial intelligence algorithms provided by the giants of the Tech, as today they provide the operating systems of our computers and our smartphones. The Tech giants will continue to compete fiercely to develop the world's most powerful artificial intelligence algorithms for businesses, and they will fight for companies to use their resources.

Yet artificial intelligence is not a silver bullet. Significant bottlenecks, especially relating to data accessibility and talent, will need to be overcome, and AI presents risks that will need to be mitigated. It could introduce or exacerbate social challenges, for example through malicious use or abuse, bias, privacy invasion, or lack of transparency.

Technology-based companies such as Uber and Airbnb have disrupted more than just their business sectors. They have raised complicated questions about how they should be regulated and by whom. Decision-makers will have to sort out the answers.

Online disinformation is only going to get more sophisticated and false information would become more convincing, and sometimes alarmingly to the detriment of the democratic process. As a counter-measure, legislation will have to made more robust and clearer.

CONCLUSION

This chapter does not pretend to be a scientific endeavor neither can it draw any firm conclusions regarding the fundamental relevance of Darwinism in big data intelligence. However, the authors make the case for the evolution of this technology and the necessity to apprehend this world being invented and to be cognizant of its influence on all business processes.

During the period 2015-2018, a multitude of scholars have investigated innovations which all possess the potential of being disruptive. The social network analysis conducted in this study contributes to the extant body of knowledge on the trends of big data intelligence technologies and how to cope with this acceleration of globalization. These scenarios provide a starting point for considering a range of options around the multiple possible futures of work. In addition, they are designed to help identify and prioritize key actions that are likely to promote the kind of future that maximizes opportunities for people to fulfill their potential across their entire lifetimes.

It is noted that an explosion in algorithmic capabilities, computing capacities, and data is potentially enabling beyond-human machine competencies and a new generation of system-level innovation.

Big data intelligence could also contribute to tackling pressing societal challenges, from healthcare to climate change to humanitarian crises. The authors' research suggests that digitization, often involving the transformation of operating and business models, promises significant productivity boosting opportunities in the future.

The authors recognize the limitations of their study, and readers, and future academics and researchers should be aware of these and indeed interpret the material presented in this chapter within the context of these limitations. Axiomatically, a meta-analysis reposes on the existing, as well as, accessible research studies (both conceptual and empirical). While, to identify all possible relevant articles, the authors conducted a thorough literature search principally through the Elsevier research databases - Science Direct and Scopus, it is possible that some research articles could have been overlooked in this review especially when compared to some other leading databases, for example, the Web of Science.

Additionally, the analysis and synthesis are based on the authors' interpretation of the selected articles. The authors attempted to avoid these issues by independently cross-checking articles and thus deal with embedded bias. Despite this, though, they humbly consider this research to be robust as every effort to mitigate errors was taken, some errors might have presumably slipped through the cracks.

REFERENCES

Adkins, M. (2013). *The Idea of the Sciences in the French Enlightenment: A Reinterpretation.* University of Delaware Press.

Al-Abdulqader, O., & Mohan, V. (2018). Learning by Demonstration with Baxter Humanoid. In K. Arai, S. Kapoor, & R. Bhatia (Eds.), *Intelligent Systems and Applications. IntelliSys 2018. Advances in Intelligent Systems and Computing* (Vol. 868). Cham: Springer.

Alberti-Alhtaybat, L., Al-Htaybat, K., & Hutaibat, K. (2019). A knowledge management and sharing business model for dealing with disruption: The case of Aramex. *Journal of Business Research, 94,* 400–407. doi:10.1016/j.jbusres.2017.11.037

Apte, P. R., Shah, H., & Mann, D. (2001). 5W's and an H of TRIZ Innovation. *The TRIZ Journal.* Retrieved from https://triz-journal.com/5ws-h-triz-innovation/

Badue, C., Guidolini, R., Carneiro, R. V., Azevedo, P., Cardoso, V. B., Forechi, A., & De Souza, A. F. (2019). *Self-Driving Cars: A Survey.* Cornell University.

Baker, J. D. (2016). The Purpose, Process, and Methods of Writing a Literature Review. *AORN Journal, 103*(3), 265–269. doi:10.1016/j.aorn.2016.01.016 PMID:26924364

Balog, M., Gaunt, A. L., Brockschmidt, M., Nowozin, S., & Tarlow, D. (2016). *DeepCoder: Learning to Write Programs. Computer Science. Machine Learning.* Cornell University.

Bansal, P., & Kockelman, K. M. (2018). Are we ready to embrace connected and self-driving vehicles? A case study of Texans. *Transportation, 45*(2), 641–675. doi:10.100711116-016-9745-z

Biggio, B., & Roliab, F. (2018). Wild patterns: Ten years after the rise of adversarial machine learning. *Pattern Recognition, 84,* 317–331. doi:10.1016/j.patcog.2018.07.023

Boell, S. K., & Cecez-Kecmanovic, D. (2015). A Hermeneutic Approach for Conducting Literature Reviews and Literature Searches. *Communications of the Association for Information Systems, 34,* 12. Retrieved from http://aisel.aisnet.org/cais/vol34/iss1/12

Bolesnikov, M., Popovic-Stijacic, M., Radisic, M., Takaci, A., Borocki, J., Bolesnikov, D., ... Dziendziora, J. (2018). Development of a Business Model by Introducing Sustainable and Tailor-Made Value Proposition for SME Clients. *Sustainability.* doi:10.1111/joms.12352

Breton, P. (1995). *L'image de l'homme: du Golem aux créatures virtuelles. Open Library*. Paris: Seuil.

Bush, V. (1945). As we may think. A top US scientist foresees a possible future world in which man-made machines will start to think. Office of scientific research and development. *Atlantic Monthly*.

Cave, S., & OhEigeartaigh, S. S. (2018). An AI Race for Strategic Advantage: Rhetoric and Risks. Association for the Advancement of Artificial Intelligence.

Chandiok, A., & Chaturvedi, D. K. (2018). CIT: Integrated cognitive computing and cognitive agent technologies based cognitive architecture for human-like functionality in artificial systems. *Biologically Inspired Cognitive Architectures.*, *26*, 55–79. doi:10.1016/j.bica.2018.07.020

Chen, Z., & Liu, B. (2018). Lifelong Machine Learning (2nd ed.). Synthesis Lectures on Artificial Intelligence and Machine Learning. Morgan & Claypool Publishers.

Cho, J., Kim, J. T., Kim, J., Park, S., & Kim, K. (2016). Simple Walking Strategies for Hydraulically Driven Quadruped Robot over Uneven Terrain. *Journal of Electrical Engineering & Technology*, *11*(5), 1433–1440. doi:10.5370/JEET.2016.11.5.1433

Christensen, C. (1997). *The Innovator's Dilemma: When New Technologies Cause Great Firms to Fail*. Harvard Business School Press.

Cilluffo, F. J., & Cardash, S. L. (2018). What's wrong with Huawei, and why are countries banning the Chinese telecommunications firm? *The Conversation*. Retrieved from https://theconversation.com/whats-wrong-with-huawei-and-why-are-countries-banning-the-chinese-telecommunications-firm-109036

Clavera, I., Rothfuss, J., Schulman, J., Fujita, Y., Asfour, T., & Abbeel, P. (2018). *Model-Based Reinforcement Learning via Meta-Policy Optimization. Computer Science. Machine Learning*. Cornell University.

Coeckelbergh, M. (2018). *Should we ban fully autonomous weapons? Uni:View Magazin*. Centre for Digital Philosophy.

Coughlin, J. (2017). *The Longevity Economy: Unlocking the World's Fastest-Growing, Most Misunderstood Market*. Hachette Book Group.

Coverage Book. (n.d.). *Answer the Public*. Retrieved from https://answerthepublic.com/

Cytoscape. (n.d.). *Cytoscape*. Retrieved from: http://www.cytoscape.org/

Dalakov, G. (2018). *The History of Computers*. Retrieved from http://history-computer.com/index.html

Dickens, C. (1859). *A Tale of Two Cities*. CreateSpace Independent Publishing Platform.

Dietterich, T. G. (2015). *Benefits and Risks of Artificial Intelligence*. The Association for the Advancement of Artificial Intelligence (AAAI). Retrieved from https://medium.com/@tdietterich/benefits-and-risks-of-artificial-intelligence-460d288cccf3

Dubé, L., Du, P., McRae, C., Sharma, N., Jayaraman, S., & Nie, J. (2018). Convergent Innovation in Food through Big Data and Artificial Intelligence for Societal-Scale Inclusive Growth. The Technology Innovation Management Review.

Elliott, D., Keen, W., & Miao, L. (2019). *Recent advances in connected and automated vehicles. Journal of Traffic and Transportation Engineering.* doi:10.1016/j.jtte.2018.09.005

Feldstein, S. (2019). The Road to Digital Unfreedom: How Artificial Intelligence is Reshaping Repression. *Journal of Democracy, 30*(1), 40–52. doi:10.1353/jod.2019.0003

Galston, W. A., & Hendrickson, C. (2018). *A policy at peace with itself: Antitrust remedies for our concentrated, uncompetitive economy.* Washington, DC: Brookings Institution.

Gampfer, F. (2018). Managing Complexity of Digital Transformation with Enterprise Architecture. 31st Bled e-Conference: Digital Transformation: Meeting the Challenges.

Goodfellow, I., Pouget-Abadie, J., Mirza, M., Xu, B., Warde-Farley, D., Ozair, S., . . . Bengio, Y. (2014). Generative Adversarial Nets. *Advances in Neural Information Processing Systems 27 (NIPS 2014).* Retrieved from http://papers.nips.cc/paper/5423-generative-adversarial-nets

Goodwin, T. (2018). *Digital Darwinism: Survival of the Fittest in the Age of Business Disruption.* Kogan Page.

Granville, K. (2018). Facebook and Cambridge Analytica: What You Need to Know as Fallout Widens. *New York Times.* Retrieved from https://www.nytimes.com/2018/03/19/technology/facebook-cambridge-analytica-explained.html

Gratton, L., & Scott, A. (2016). *The 100-Year Life: Living and Working in an Age of Longevity.* Bloomsbury Institute.

Haenssle, H. A., Fink, C., Schneiderbauer, R., Toberer, F., Buhl, T., Blum, A., ... Uhlmann, L. (2018). Man against machine: Diagnostic performance of a deep learning convolutional neural network for dermoscopic melanoma recognition in comparison to 58 dermatologists. *Annals of Oncology : Official Journal of the European Society for Medical Oncology, 29*(8), 1836–1842. doi:10.1093/annonc/mdy166

Hale, S. (2018). *Death by Algorithm; Public opinion and the lethal autonomous weapons debate. European Master's Degree in Human Rights and Democratisation. A.Y. 2017/2018.* Aristotle University of Thessaloniki.

Helbing, D. (2018). Societal, Economic, Ethical and Legal Challenges of the Digital Revolution: From Big Data to Deep Learning, Artificial Intelligence, and Manipulative Technologies. In D. Helbing (Ed.), *Towards Digital Enlightenment.* Cham: Springer.

Holcomb, S. D., Porter, W. K., Ault, S. V., Mao, G., & Wang, J. (2018). Overview on DeepMind and Its AlphaGo Zero AI. *ICBDE '18 Proceedings of the 2018 International Conference on Big Data and Education,* 67-71. 10.1145/3206157.3206174

Iapichino, A., De Rosa, A., & Liberace, P. (2018). Smart Organizations, New Skills, and Smart Working to Manage Companies' Digital Transformation. In L. Pupillo, E. Noam, & L. Waverman (Eds.), *Digitized Labor. Palgrave Macmillan.* Cham: Springer Link. doi:10.1007/978-3-319-78420-5_13

ICAO. (2011). *International Civil Aviation Organization (ICAO, 2011). Unmanned Aircraft Systems (UAS).* Retrieved from https://www.icao.int/Meetings/UAS/Documents/Circular%20328_en.pdf

Ishiguro, H., & Nishio, S. (2018). Building Artificial Humans to Understand Humans. In H. Ishiguro & F. Dalla Libera (Eds.), *Geminoid Studies*. Singapore: Springer. doi:10.1007/978-981-10-8702-8_2

Jelen, L., & Milana, V. (2018). Building and nurturing trust among members in virtual project teams. Strategic Management, 23(3), 10-16.

K-Meta. (n.d.). *Keyword Research Tool*. Retrieved from https://k-meta.com/suggestions

Kelly, K. (2016). *The Inevitable: Understanding the 12 Technological Forces That Will Shape Our Future*. Viking Press.

Keyword Tool. (n.d.). Retrieved from https://keywordtool.io/

Knapp, S. (2006). *Artificial Intelligence: Past, Present, and Future. Vox of Dartmouth*. The Newspaper for the Dartmouth Faculty and Staff.

Krotkov, E., Hackett, D., Jackel, L., Perschbacher, M., Pippine, J., Strauss, J., & Orlowski, C. (2018). The DARPA Robotics Challenge Finals: Results and Perspectives. In M. Spenko, S. Buerger, & K. Iagnemma (Eds.), *The DARPA Robotics Challenge Finals: Humanoid Robots To The Rescue. Springer Tracts in Advanced Robotics* (Vol. 121). Cham: Springer. doi:10.1007/978-3-319-74666-1_1

Kulage, K. M., & Larson, E. L. (2016). Implementation and Outcomes of a Faculty-Based, Peer Review Manuscript Writing Workshop. *Journal of Professional Nursing, 32*(4), 262–270. doi:10.1016/j.profnurs.2016.01.008

Larson, M. (2018). Reality, requirements, regulation: Points of intersection with the machine-learning pipeline. In Assessing the impact of machine intelligence on human behaviour: an interdisciplinary endeavour. In *Proceedings of 1st HUMAINT workshop*. Barcelona, Spain: European Commission.

Li, M., Porter, A. L., & Suominen, A. (2018). Insights into relationships between disruptive technology/innovation and emerging technology: A bibliometric perspective. *Technological Forecasting and Social Change, 129*, 285–296. doi:10.1016/j.techfore.2017.09.032

Li, S. A., Feng, H. M., Huang, S. P., & Chu, C. Y. (2018). Fuzzy Self-Adaptive Soccer Robot Behavior Decision System Design through ROS. *The Journal of Imaging Science and Technology, 62*(3). doi:10.2352/J.ImagingSci.Technol.2018.62.3.030401

Lopez, G., Quesada, L., & Guerrero, L. A. (2017). Alexa vs. Siri vs. Cortana vs. Google Assistant: A Comparison of Speech-Based Natural User Interfaces. *Advances in Human Factors and Systems Interaction. International Conference on Applied Human Factors and Ergonomics*. Springer Link.

Lykou, G., Iakovakis, G., & Gritzalis, D. (2019). *Aviation Cybersecurity and Cyber-Resilience: Assessing Risk in Air Traffic Management* (pp. 245–260). Critical Infrastructure Security and Resilience. Retrieved from https://link.springer.com/bookseries/5540

Makhija, Y., & Sharma, R. S. (2019). *Face Recognition: Novel Comparison of Various Feature Extraction Technique. Harmony Search and Nature Inspired Optimization Algorithms. Advances in Intelligent Systems and Computing*. Springer.

Mario, C. (2018). Disruptive Firms and Technological Change. *Quaderni IRCrES-CNR, 3*(1), 3-18. Retrieved from https://ssrn.com/abstract=3103008

Matusov, E., Wilken, P., Bahar, P., Schamper, J., Golik, P., Zeyer, A., ... Peter, J. (2018). Neural Speech Translation at AppTek. *Proceedings of the 15th International Workshop on Spoken Language Translation.*

McCraw, T. K. (2010). *Prophet of Innovation: Joseph Schumpeter and Creative Destruction.* Cambridge, MA: Harvard University Press. doi:10.4159/9780674040779

Milhorance, F., & Singer, J. (2018). Media Trust and Use among Urban News Consumers in Brazil. *Ethical Space: The International Journal of Communication Ethics, 15*(3/4). Retrieved from http://www.communicationethics.net/espace/

Murphy, J. (2018). Artificial Intelligence, Rationality, and the World Wide Web. *IEEE Intelligent Systems, 33*(1).

Naids, A., Rossetti, D., Bond, T., Huang, A., Deal, A., Fox, K., ... Mikatarian, R. (2018). *The Demonstration of a Robotic External Leak Locator on the International Space Station.* Orlando, FL: AIAA SPACE and Astronautics Forum and Exposition. doi:10.2514/6.2018-5118

NASA. (2015). *National Aeronautics and Space Administration (NASA, 2015). UTM: Air Traffic Management for Low-Altitude Drones.* Retrieved from https://www.nasa.gov/sites/default/files/atoms/files/utm-factsheet-11-05-15.pdf

Obozintsev, L. (2018). *From Skynet to Siri: an exploration of the nature and effects of media coverage of artificial intelligence.* University of Delaware, Department of Communication. Retrieved from http://udspace.udel.edu/handle/19716/24048

Offray, J. (1912). *L'Homme Machine, Man A Machine: Including Frederick the Great's "Eulogy" on La Mettrie and Extracts from La Mettrie's "The Natural History of the Soul".* The Open Court Publishing Company.

Ospoval, D. (1995). *The Development of Darwin's Theory: Natural History, Natural Theology and Natural Selection, 1838-1859.* Cambridge University Press.

Pellerin, C. (2017). *Project Maven to Deploy Computer Algorithms to War Zone by Year's End.* DoD News, Defense Media Activity. Retrieved from https://dod.defense.gov/News/Article/Article/1254719/project-maven-to-deploy-computer-algorithms-to-war-zone-by-years-end/

Perez, S. (2016). *Microsoft silences its new A.I. bot Tay, after Twitter users teach it racism.* Retrieved from https://techcrunch.com/2016/03/24/microsoft-silences-its-new-a-i-bot-tay-after-twitter-users-teach-it-racism/

Popescu, C. (2018). Improvements in business operations and customer experience through data science and Artificial Intelligence. *Proceedings of the 12th International Conference on Business Excellence.* 10.2478/picbe-2018-0072

Remarczyk, M., Narayanan, P., Mitrovic, S., & Black, M. (2018). Our New Handshake with the Robot: How Our Changing Relationship with Machines Can Make Us More Human. *Proceedings of SAI Intelligent Systems Conference IntelliSys 2018: Intelligent Systems and Applications*, 839-851.

Roberge, J., & Chantepie, P. (2017). The Promised Land of Comparative Digital Cultural Policy Studies. *The Journal of Arts Management, Law, and Society, 47*(5), 295–299.

Rooney, J. J., & VanDen Heuvel, L. N. (2004). Root cause Analysis for Beginners. *Quality Basics*. Retrieved from https://www.formsbirds.com/download-root-cause-analysis-for-beginners

Ross, J.-M. (2010). Roger Magoulas on Big Data. *O'Reilly Radar*. Retrieved from http://radar.oreilly.com/2010/01/roger-magoulas-on-big-data.html

Schwemmer, M. A., Skomrock, N. D., Sederberg, P. B., Ting, J. E., Sharma, G., Bockbrader, M. A., & Friedenberg, D. A. (2018). Meeting brain–computer interface user performance expectations using a deep neural network decoding framework. *Nature Medicine, 24*(11), 1669–1676. doi:10.103841591-018-0171-y PMID:30250141

Schwikowski, B. (2015). Cytoscape: Visualization and Analysis of omis data in interaction networks, Institut Pasteur. *Gnome Research*. Retrieved from https://research.pasteur.fr/en/software/cytoscape/

Shackelford, S. (2018). *Smart Factories, Dumb Policy? Managing Cybersecurity and Data Privacy Risks in the Industrial Internet of Things*. Kelley School of Business Research Paper No. 18-80. doi:10.2139srn.3252498

Shahbaz, N., & Phukan, A. (2018). *A Legal and Ethical Examination of Photorealistic Videos Created Using Artificial Neural Networks*. Retrieved from https://www.sccur.org/sccur/Fall_2018_Conference/Multidisc_Posters/19/

Singh, N. (2018). *Talking machines: democratizing the design of voice-based agents for the home*. Program in Media Arts and Sciences (Massachusetts Institute of Technology). Retrieved from http://hdl.handle.net/1721.1/119089

Sivathanu, B., & Pillai, R. (2018). Smart HR 4.0 – how industry 4.0 is disrupting HR. *Human Resource Management International Digest, 26*(4), 7–11. doi:10.1108/HRMID-04-2018-0059

Snihur, Y., & Wiklund, J. (2018). *Searching for innovation: Product, process, and business model innovations and search behavior in established firms, Long Range Planning*. Wiley.

Soper, S. (2018). Amazon Prime Day Shopping Topped $4 Billion, Analyst Estimates. *Bloomberg*. Retrieved from https://www.bloomberg.com/news/articles/2018-07-19/amazon-prime-day-shopping-topped-4-billion-analyst-estimates

Sousa, M. J., & Rocha, A. (2019). Skills for disruptive digital business. *Journal of Business Research, 94*, 257–263. doi:10.1016/j.jbusres.2017.12.051

Spice, B. (2017). *Machine Learning Will Change Jobs*. Carnegie Mellon University.

Strydom, M., & Buckley, S. (2018). The Big Data Research Ecosystem: an Analytical Literature Study. In M. Strydom & S. Buckley (Eds.), *Big Data Governance and Perspectives in Knowledge Management*. IGI Global.

Sun, X., Wandel, S., & Stump, E. (2018). Competitiveness of on-demand air taxis regarding door-to-door travel time: A race through Europe. *Transportation Research Part E, Logistics and Transportation Review, 119*, 1–18. doi:10.1016/j.tre.2018.09.006

Taeihagh, A., & Si MinLim, H. (2018). Governing autonomous vehicles: emerging responses for safety, liability, privacy, cybersecurity, and industry risks. *Journal Transport Reviews, 39*(1). Retrieved from https://www.tandfonline.com/doi/ref/10.1080/01441647.2018.1494640?scroll=top

Tag, B., Holz, C., Lukowicz, P., Augereau, O., Uema, Y., & Kunze, K. (2018). EyeWear 2018: Second Workshop on EyeWear Computing. *UbiComp '18 Proceedings of the 2018 ACM International Joint Conference and 2018 International Symposium on Pervasive and Ubiquitous Computing and Wearable Computers*, 964-967.

Tavakoli, M., Viegas, C., Sgrigna, L., & de Almeida, A. (2018). SCALA: Scalable Modular Rail based Multi-agent Robotic System for Fine Manipulation over Large Workspaces. *Journal of Intelligent & Robotic Systems, 89*, 421. doi:10.100710846-017-0560-3

Tegmark, M. (2018). An Open Letter: Research Priorities for Robust and Beneficial Artificial Intelligence. *Future of Life Institute*. Retrieved from https://futureoflife.org/ai-open-letter/

Tham, J. (2018). Persuasive-Pervasive Technology: Rhetorical Strategies in Wearables Advertising. *International Journal of Semiotics and Visual Rhetoric*. doi:10.4018/IJSVR.2018010104

Trautmann, S., Rehm, J., & Wittchen, H. (2016). The economic costs of mental disorders: Do our societies react appropriately to the burden of mental disorders? *Science and Society*. PMID:27491723

Turing, A. M. (1950). Computing Machinery and Intelligence. *Mind, 49*(236), 433–460. doi:10.1093/mind/LIX.236.433

Um, J. (2019). *Imaging Sensors. Drones as Cyber-Physical Systems*. Singapore: Springer. doi:10.1007/978-981-13-3741-3_6

Van den Hoven van Genderen, R. (2019). Does Future Society Need Legal Personhood for Robots and AI? In Artificial Intelligence in Medical Imaging (pp. 257-290). Springer.

Van Wynsberghe, A., & Robbins, S. (2018). *Critiquing the Reasons for Making Artificial Moral Agents*. Science and Engineering Ethics. Springer Link. doi:10.100711948-018-0030-8

Varrasi, S., Lucas, A., Soranzo, A., McNamara, J., & Di Nuovo, A. (2018). IBM Cloud Services Enhance Automatic Cognitive Assessment via Human-Robot Interaction. In G. Carbone, M. Ceccarelli, & D. Pisla (Eds.), *New Trends in Medical and Service Robotics. Mechanisms and Machine Science* (Vol. 65). Cham: Springer. doi:10.1007/978-3-030-00329-6_20

Villaronga, E. F., Kieseberg, P., & Li, T. (2018). Humans forget, machines remember: Artificial intelligence and the Right to Be Forgotten. *Computer Law & Security Review, 34*(2), 304–313. doi:10.1016/j.clsr.2017.08.007

Vincent, J. (2017). *Putin says the nation that leads in AI 'will be the ruler of the world*. Retrieved from https://www.theverge.com/2017/9/4/16251226/russia-ai-putin-rule-the-world

Wang, B. (2016). Nvidia Xavier chip 20 trillion operations per second of deep learning performance and uses 20 watts which means 50 chips would be a petaOP at a kilowatt. *NextBigFuture*. Retrieved from https://www.nextbigfuture.com/2016/11/nvidia-xavier-chip-20-trillion.html

Wang, T., Liu, M., Zhu, J.-Y., Tao, A., Kautz, J., & Catanzaro, B. (2018). High-Resolution Image Synthesis and Semantic Manipulation With Conditional GANs. *The IEEE Conference on Computer Vision and Pattern Recognition (CVPR)*, 8798-8807. 10.1109/CVPR.2018.00917

Wang, Y., Zang, H., Qiu, C. H., & Xia, S. R. (2018). A Novel Feature Selection Method Based on Extreme Learning Machine and Fractional-Order Darwinian PSO. *Computational Intelligence and Neuroscience*. doi:10.1155/2018/5078268

Welch, C. (2018). *Google just gave a stunning demo of Assistant making an actual phone call*. Retrieved from https://www.theverge.com/2018/5/8/17332070/google-assistant-makes-phone-call-demo-duplex-io-2018

Winskell, K., Singleton, R., & Sabben, G. (2018). Enabling Analysis of Big, Thick, Long, and Wide Data: Data Management for the Analysis of a Large Longitudinal and Cross-National Narrative Data Set. *Qualitative Health Research*, *28*(10), 1629–1639. doi:10.1177/1049732318759658 PMID:29557295

Woods, T. (2019). *'Age-Tech': The Next Frontier Market For Technology Disruption*. Intuition Robotics in Forbes.

Worldometers. (2018). *United Nations. DESA/ Population Division. World Population Clock*. Retrieved from http://www.worldometers.info/world-population/

Zhang, J., Li, K., & Yao, C. (2018). Event-based Summarization for Scientific Literature in Chinese. 2017 International Conference on Identification, Information and Knowledge in the Internet of Things. *Procedia Computer Science*, 88-92.

Zhang, N., Mi, X., Feng, X., Wang, X., Tian, Y., & Qian, F. (2018). *Understanding and Mitigating the Security Risks of Voice-Controlled Third-Party Skills on Amazon Alexa and Google Home. Computer Science. Cryptography and Security*. Cornell University.

Zhang, W., Daim, T., & Zhang, Q. (2018). Understanding the disruptive business model innovation of E-business microcredit: A comparative case study in China. *Technology Analysis and Strategic Management*, *30*(7), 765–777. doi:10.1080/09537325.2017.1376047

ENDNOTES

[1] Disruption: Derived from the Latin *disrumpere*; the separation, the difference.

[2] Artificial intelligence: the field of computer science dedicated to solving cognitive problems commonly associated with human intelligence, such as learning, problem solving and pattern recognition (Partnership on AI to benefit People and Society).

3   Big data: The high-volume, high-velocity and/or high-variety information assets that demand cost-effective, innovative forms of information processing that enable enhanced insight, decision making, and process automation.

4   The symbiotic term "automat" is the Latinization of the Greek word αὐτόματον, which means acting of one's own will.

5   Artificial intelligence: The field of computer science dedicated to solving cognitive problems commonly associated with human intelligence, such as learning, problem solving, and pattern recognition (Partnership on AI to Benefit People and Society).

Chapter 2
Ontology–Based Open Tourism Data Integration Framework:
Trip Planning Platform

Imadeddine Mountasser
Moulay Ismaïl University, Morocco

Brahim Ouhbi
Moulay Ismaïl University, Morocco

Ferdaous Hdioud
 https://orcid.org/0000-0002-8870-2386
Sidi Mohamed Ben Abdellah University, Morocco

Bouchra Frikh
Sidi Mohamed Ben Abdellah University, Morocco

ABSTRACT

Tourism is an information-intensive industry that requires the interconnection of the stakeholders to make strategic decisions for both tourism organizations and tourists. For instance, trip planning as a high-engagement, time and effort-consuming decision-making process, allows choosing the most suitable travel destination, mode, and activities. Tourists often need a considerable amount of information to develop a travel plan and to build expectations for their trip. Therefore, they need a technological platform on which information relating to tourism activities could be interoperated to respond to the pre-trip tourists' information sourcing behavior. In contrast, given the substantial number of actors providing open data in the field of tourism, tourism service providers aim to explore its significant potentials for sustainable tourism development. Thus, this chapter investigates the capacity to build a centralized information platform using diverse open data sources to support travelers during their trip planning by providing more prominent and better-tailored information.

DOI: 10.4018/978-1-5225-9687-5.ch002

INTRODUCTION

The tourism industry thrives on information (Benckendorff, Sheldon, & Fesenmaier, 2014). Tourism stakeholders including tourism service providers, governments, local communities and tourists, increasingly seek for new ways to enhance decision-making and opportunities for destinations competitiveness and innovations (Hjalager & Nordin, 2011; Irudeen & Samaraweera, 2013; Jafari, 2001). In fact, smart tourism destinations claim stakeholders to inter-operate through centralized technological platforms to ensure the tourism-related information collection and exchange which enriches tourism experiences (Buhalis & Amaranggana, 2015; Gretzel, Werthner, Koo, & Lamsfus, 2015). Thus, the process is heavily contingent upon the availability of information integration platforms that enable multiple visualizations in a common direction (Gretzel, Sigala, Xiang, & Koo, 2015).

Certainly, and from travelers' point of view, destination-specific information assists them in the decision-making process and enables them to acquire additional and detailed knowledge to reduce their cognitive dissonance and their perception of risk and uncertainty (Quintal, Lee, & Soutar, 2010; Ross & Bettman, 1979), and to build their expectations for the upcoming trip (Gretzel, Fesenmaier, & O'Leary, 2005). In fact, trip planning is considered as a high-engagement decision-making process, inasmuch as it includes various levels of implications, large time spent planning and several decisions that must be made before traveling (Li, Law, Vu, Rong, & Zhao, 2015a). It may depend on trip associates (For example alone, couple, family with children) and varies given trip modes (annual family trip, short trip, long trip). Also, the planning can be made according to traveler preferences and expectations (Michopoulou & Buhalis, 2013), travel fees or other's past experience (Pantano, Priporas, & Stylos, 2017). Indeed, choosing the most suitable travel destination, mode and activities remain a time and effort consuming process (Bieger & Laesser, 2004; Choi, Lehto, & Oleary, 2007), especially with the overwhelming amount of information which the travelers are overexposed to (Li, Law, Vu, Rong, & Zhao, 2015b).

As a consequence, tourism service providers are starting adapting their targeting and advertising strategies to meet with the needs, expectations and desires of travelers, in such a way to improve the competitiveness of their destinations. They benefit from the recent advances in information technologies, the Internet and the technological development of services, to build intelligent systems aiming to support travelers during their trip planning, by providing more prominent and better-tailored information.

All things considered, this chapter explores the concept of Open Data in the tourism context and investigates the capacity of different stakeholders to work together through their Open Data for better, smarter and strategic operational decisions. For that, the authors demonstrate the capabilities, to which an Open Data integration strategy may provide to develop applications able to support tourists for their travel planning scenarios (destinations and services) and to assist tourism service providers with their targeting and advertising strategies. Hence, the significance of this study affects not only the tourism industry, as a way to join government agencies in their strategic plan to promote the re-use of touristic Open Data (See Figure 1) in the development of smart tourist destinations, but also incites the academics, by providing a good foundation to explore the great opportunities and potentials of using Open Data in various fields.

Obviously, in tourism industry, different large-scale data in both structured and unstructured formats are generated, gathered and stored, opening thus a new age for innovation, productivity growth and competition (Kambatla, Kollias, Kumar, & Grama, 2014). Specifically, several valuable data sources can be considered: Social networks platforms that incorporate users' interactions, notably user-generated content, deployed smart objects (sensors) in an Internet of Things (IoT) infrastructure and Open Data

Figure 1. Cross-domain Open Data Portals for Tourism

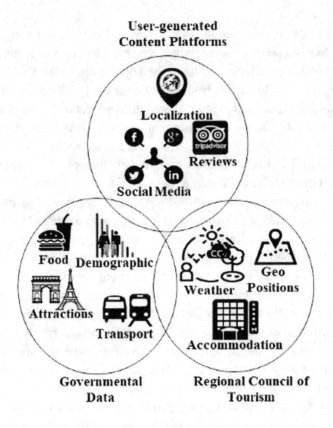

portals provided by organizations and governments (Rajapaksha, Farahbakhsh, Nathanail, & Crespi, 2017) (See Figure 1). However, for managers there is a challenge for the implementation of Open Data to support decision-making and information exchange between the stakeholders (Ivars-Baidal, Celdrán-Bernabeu, Mazón, & Perles-Ivars, 2017).

For example, during planning scenarios, tourist points of interest can be probably described given several characteristics, namely, touristic destinations, touristic events and attractions, accommodations, services and touristic route (Michopoulou & Buhalis, 2013). For that, travelers browse multiple sources and dispersed information from different systems to make prior decisions about their trip. Thus, the aim resides on combining and inter-operating these heterogeneous sources to provide a unified, integrated and consistent view of this data. This later will be exploited to develop reliable and efficient cross-language tourism-related applications able to expose the touristic interest points that are most attractive to a particular tourist.

To this end, the authors aim to build an ontology-based integration framework for Open Tourism Data over a distributed architecture and using parallel-programming techniques via Hadoop and MapReduce (See Figure 2). The framework implies first that each Open Data source should be described by its own related ontology. Then, a hybrid scalable ontology alignment approach is applied to create a shared global ontology. In this way, several kinds of semantic relations within the ontological models can be exploited to better describe touristic points of interest -in unexplored concepts- and to facilitate

Figure 2. Ontology-based Integration Framework for Open Tourism Data.

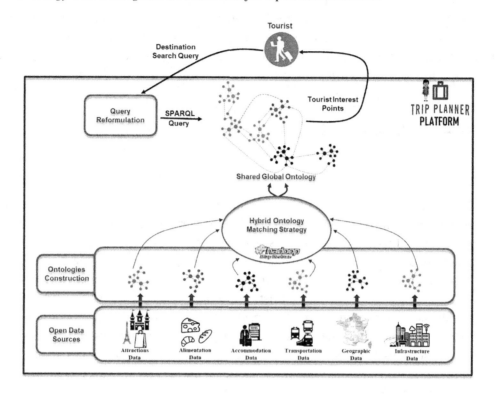

the propagation of interests between other ones. Thus, data are prepared and stored -in an integrated and appropriate RDF (Resource Description Framework) format- to be used to assist tourists in their vacation planning scenarios.

The main contributions of this chapter are as follows:

- Introducing an Open Data integration framework, based on IT and semantic mechanisms, as an enabler to successfully capture the expected potential of this data and to promote its publication and reuse. It models and integrates large tourism-related data-sets, released in several formats, located in different portals and referred to multiple overlapped domains.
- Implementing a hybrid ontology alignment strategy over a distributed architecture through leveraging parallel-programming techniques. It discovers correspondences between inter-operated local ontologies while ensuring flexibility, extensibility and scalability aspects.
- Overcoming the interoperability and heterogeneity of gathered cross-domain heterogeneous data sources, despite of their unpredicted content and their data type, by providing a unified and integrated view through a shared global ontology.
- Investigating the capabilities to which the proposed Open Data integration framework may provide, to develop applications able to support decision-making scenarios (trip planning) and to assist organizations with their targeting and advertising strategies.

The rest of this chapter is structured as follows. Section 2 describes the theoretical background and formalizes the ontology-based integration issue for Open Data. The proposed ontology-based Open Data integration framework as well as its related experimental results on ontology alignment dedicated tracks is described and demonstrated in section 3. Section 4 addresses the implementation of a trip planning platform for a specific destination using several Open Data sources. Section 7 concludes the chapter.

THEORETICAL BACKGROUND

Smart Tourism

Smart tourism, through the development of information and communication technologies, has played a critical role for the development of a sustainable tourism industry. It seems to offer promising potential for the competitiveness of tourism organizations as well as for the experience of tourists. Several studies has been elaborated to leverage smart tourism technologies during the pre-trip planning (Björk & Kauppinen-Räisänen, 2015; Fesenmaier & Jeng, 2000; Huang, Goo, Nam, & Yoo, 2017; Xiang, Magnini, & Fesenmaier, 2015). They aim to meet the tourists' expectations by collecting appropriate information suitable to provide the right offer at the right moment. Tourism stakeholders and public organizations publish data about accommodations and attractions, catering capacities and transportation options, used to build various applications for the management of their traveler destination services (Mariani, Buhalis, Longhi, & Vitouladiti, 2014). At the same time, public authorities and involved organizations in the tourism domain, especially cities, have adopted the Open Data philosophy as a principal asset of their smart city policies (Marine-Roig & Anton Clavé, 2015).

Open Data Paradigm

Concerning the gaining ground attention drawn to open innovation for smartness development, Open Data has increasingly earned more interest both in the academic community as well as in the industrial field. Actually, Open Data are mostly published by public organizations, services providers and even user-generated content platforms. Its availability has grown significantly as a concept that provides a free public access to various sources without any limitations (Maccani, Donnellan, & Helfert, 2015). the Open Knowledge Foundation[1] has defined Open Data as *data that can be freely used, shared and built on by anyone, anywhere, for any purpose*. Likewise, (Janssen, Charalabidis, & Zuiderwijk, 2012) extend this definition through three principles:

- **Availability and Access:** Data must be timely available and freely accessible as a whole, without discrimination or any copyright restrictions and barriers that would impede the re-use and redistribution of that information.
- **Re-use and Redistribution:** Data published would be platform independent and machine-readable suitable for automated redistribution and processing.
- **Universal Participation:** Everyone must be able to use, reuse and redistribute published data. It must be not controlled exclusively by a unique entity.

Notably, non-profit organizations such as institutional organisms have often recognized the importance of Open Data (European Commission, 2018; García, Linaza, Franco, & Juaristi, 2015; Longhi et al., 2014). They support the exploitation of reliable Open Data for its significant potential to open new business possibilities, to discover innovative solutions as well as for re-use in new products and services, which reduce the development cost of applications or innovative added-value services. Yet, several successful models of the efficient use of Open Data exist in various fields (Wiggins & Crowston, 2011), however its implementation for the development of smart tourism destinations has not been previously exploited or discussed. The authors are confident that tourism stakeholders will see the benefits of using Open Data, since data is indispensable for the management and the marketing of tourism destinations and businesses. Open Data can support local organizations in their planning strategies and satisfy their tourist's requirements for the purpose of earning new tourism markets and maintaining the amplitude of their destinations.

Despite these great expectations, the adoption of Open Data entails a number of practical obstacles (Janssen et al., 2012). Hence, research should be conducted to go over these limitations and investigate more strategies to explore the Open Data promises. For that, and contrary to previous studies in the Open Data trend that consisted of conceptual works (Thorsby, Stowers, Wolslegel, & Tumbuan, 2017), empirical uses (Hausenblas, 2009) and the design of procedures and approaches for promoting their potentials (Chan, 2013; Kalampokis, Tambouris, & Tarabanis, 2011), this chapter will focus on the semantic-driven exploitation of Open Data, for the purpose to discover hidden patterns and valuable insights to support traveler's in their pre-trip decision-making process.

In fact, the availability of Open Data has grown significantly as a concept that can effectively affect the decision-making process and yield to new innovative services (Eckartz, Van den Broek, & Ooms, 2016; Manyika et al., 2013; Pesonen & Lampi, 2016). To explore this potential in the tourism domain, a considerable number of involved actors (cities and governments) provides a vast amount of tourism-related data sources (ranging from demographic, geographical, economical to social data) (Longhi, Titz, & Viallis, 2014). They are freely published in a machine-readable format, through Open Data portals in order to be reused and exploited to discover hidden valuable insights and to create benefits. However, possibilities of Open Data for smart tourism development are still mostly misunderstood and unexplored.

In general, harnessing several Open Data portals (See Figure 1) for useful information on a specific field, such as tourist destinations management, may result in a large number of data-sets from multiple overlapped domains and contexts (traffic, geographical, weather, policies and touristic information) that are published in various formats and having multiple levels and characteristics. However, separately analyzing each data-set requires adequate data analysis techniques depending to each data formats (structured, semi-structured and unstructured) and could not provide a comprehensive view for the elaborated field. Hence, Open Data integration issue has to be sufficiently addressed as a significant data-driven perspective, which provides an integrated view across a complex, schema-less and heterogeneous data sources, that traditional data integration mechanisms (Doan, Halevy, & Ives, 2012; Mountasser, Ouhbi, & Frikh, 2015) fail to support.

Ontology-based Data Integration

Data integration is an effective strategy to combine and to share data that originated in various sources for providing end users with a unique point of access to those data (Lenzerini, 2002). Indeed, a considerable amount of techniques and methodologies, on several topics, have been developed (Dong & Srivastava,

2013; Ziegler & Dittrich, n.d.). However, these approaches cannot deal properly with voluminous and high variety datas-ets. Furthermore, they still fail to overcome the heterogeneity and the interoperability issues of integrated data sources and don't bring the required agility and flexibility to the integration process. The data interoperability relies on the ability of systems and services to create, exchange and consume data in a clear and a shared manner while preserving the context and the meaning of that data; whereas the data heterogeneity describes data coming from disparate data sources, with high variability of data types and formats. Therefore, semantic technologies provide a new level of integration mechanisms that have been adopted to simplify data integration and to resolve data heterogeneity and interoperability issues (Daraio et al., 2016).

Ontologies as a knowledge representational form play a significant role in the promoting and deployment of the Semantic Web. It describes in an explicit and formal manner the diverse aspects of knowledge of a specific domain. Notably, ontologies can be a prominent solution to resolve data heterogeneity and interoperability issues of overlapped domains (Virginija Uzdanaviciute, 2011; Xiaomeng, Lyu, & Terpenny, 2015). They provide a semantic ground that describes two concepts, which appear in different sources with various formats, as being the same object.

Ontology-based data integration approach relies on exploiting ontologies to combine and to aggregate data or information from various heterogeneous sources. By relying on ontologies as a semantic data model, authors are able to enhance the integration mechanisms (Mountasser, Ouhbi, & Frikh, 2016; Virginija Uzdanaviciute, 2011) by describing the specific form of the relations between concepts through an explicit and machine understandable conceptualization of the domain. Additionally, authors exploit these semantic mechanisms to overcome data sources interoperability and data heterogeneity issues (structural, syntactic, systematic or semantic conflicts) by providing a unified consistent view of the combined data sources. From another perspective, using ontologies provides certain levels of extensibility and flexibility, in such a way that for the seek to append another data source, they should describe it in its related ontology and find the set of alignments to connect this ontology to the shared global ontology created previously. This may ensure and enhance the agility and the sustainability of the Open Data integration process. Furthermore, such solution allows us to deal with a single data format (RDF triples) instead of acting with each Open Data sources formats which presents great benefits inasmuch data sources content and data types cannot be predicted. Lastly, the adoption of a semantic data model may increase the reality and the credibility of the integrated data and promote knowledge exploration and capitalization.

Among the main ontology-based strategies (Virginija Uzdanaviciute, 2011), this work implements the hybrid strategy to build an ontology-based Open Data integration framework. This strategy uses multiple local ontologies, created from each Open Data source, to be related and mapped to a global shared ontology. In fact, given the Open Data nature, this strategy may ensure a unified understanding of the same data that belongs to different domains (Santipantakis, Kotis, & Vouros, 2017). Furthermore, this strategy may allow appending new Open Data sources without modifying existing mappings. To this end, this work creates the shared global ontology by implementing an ontology alignment process over created local ontologies. This allows a flexible and extensible solution that exploits data distribution and scalable parallel processing.

Ontology Alignment

Ontology Alignment (a.k.a ontology matching) is among the major procedures used by information and knowledge-based systems for the resolution of data heterogeneity and interoperability issues (Euzenat & Shvaiko, 2007b). It consists of detecting correspondences between entities in several heterogeneous ontologies. It has been largely investigated for years, devising thus several algorithms ranging from single matching algorithms to multiple combined matching algorithms. From another side, ontology alignment approaches allow reusing existing local ontologies to develop new global ones. This will be remarkably useful for integration scenarios between different data models. It also allows reusing information between ontologies and making queries across them.

As defined in (Euzenat & Shvaiko, 2007b; Shvaiko & Euzenat, 2013), we introduce the notion of alignments between different ontologies. Given two ontologies O_1 and O_2, an alignment can be defined as a set of correspondences $\Psi(c_i)$ between their entities when c_i is a triplet $<e_1,e_2,\sigma>$ where $e_1 \in O_1$, $e_2 \in O_2$ and σ measures the degree of similarity between them, according to different criteria and metrics. Thus, among these alignments, we are only interested in equivalence alignments between concepts and properties based on the degree of similarity between them as a confidence value that reflects the real similitude between them.

Ontology alignment process can commonly manage ontologies having a huge amount of entities. Accordingly, several works have adopted several mechanisms, ranging from matching space reduction solutions that dynamically ignore some partitions of these ontologies and avoid matching them during the alignment process (Algergawy, Babalou, Kargar, & Hashem Davarpanah, 2015; Algergawy, Massmann, & Rahm, 2011; Hu, Qu, & Cheng, 2008; Shvaiko & Euzenat, 2013), to alignment solutions that implement parallelism mechanisms over a distributed infrastructure (Amin, Khan, Lee, & Kang, 2015; Gross, Hartung, Kirsten, & Rahm, 2010; Kirsten et al., 2010). However, the formers aim to perform ontology alignment far from profiting from the opportunities and the innovation of hardware architectures and parallel-computing procedures which undermines the performance, especially in real-time and large scale context. The latter implement data parallelism over parallelism-enabled platforms for effectiveness-independent performance-gain ontology alignment process, while decoupling performance and accuracy aspects.

PROPOSED FRAMEWORK

In fact, smart tourism destinations should adapt a strategy that provides information about all tourism-related activities, in order to help the travelers making decisions and building expectations for their upcoming trip. Certainly, it has a major impact on traveler's decision making criteria, especially when choosing a destination to visit. Thus, making confident decisions in the Open Data environment requires the right procedures and approaches to combine and to aggregate potentially large and complex data sets that are gathered from multiple heterogeneous information sources and having different formats. In this regard, the authors have built an Open Tourism Data integration framework, that relies on a hybrid large-scale ontology alignment strategy over distributed architecture and using parallel-programming mechanisms.

As shown in Figure 3, the proposed approach firstly models each input ontology as a directed acyclic graph by considering concepts as a finite set of nodes and including taxonomic and non-taxonomic relationships between classes as a finite set of directed edges. Then, it proceeds by decomposing these graphs into simpler resource-based subsets according to the integration strategy requirements. These personalized subsets are serialized into distributed data stores (HBase) in order to reduce the computational complexity and to improve the performance of the integration framework. After that, it applies an entity-assignment parallel partitioning strategy over a distributed infrastructure, while preserving the encoded structural knowledge within ontologies. Thus, this partitioning strategy will promote matching by parts through performing several finer matching tasks with reduced memory requirements and limited computational complexity. Furthermore, the approach identifies semantically similar clusters candidate to be matched, instead of performing the matching process between the entire clusters set. Such mechanism aims to minimize the matching overhead by ignoring entities of dissimilar clusters from the matching process. Then, a three-stage matching strategy is performed to approximate the alignments between the matched parts. Notably, the system parallelizes and distributes matching tasks between similar clusters pairs to find alignments between their entities. Finally, all the independent partial results are gathered to construct the final alignment set, used to build the shared global ontology.

Figure 3. Distributed Hybrid Ontology Alignment Strategy.

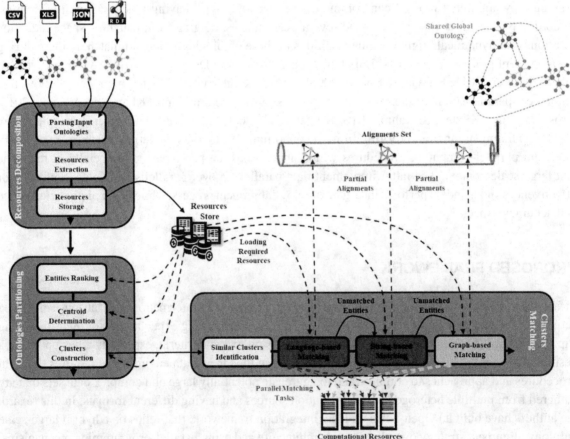

Local Ontologies Creation

Since ontologies are widely used in information sharing and knowledge management, efficient strategies for ontology development should be implemented. However, this process relies on time-consuming and labor-intensive mechanisms as most ontology development tools are not purely carried out automatically. Each ontology development solution follows the definition of the ontology as a conceptual way that defines the basic terms of a specific topic and the relationships between them (Corcho, Fernández-López, & Gómez-Pérez, 2003; Sugumaran & Storey, 2002). First, the basic terms and properties of a specific tourism-related data source are identified and transformed into a conceptual model. For example, a transportation data source may include obvious terms such as ticket, traveler, transport mode, itinerary, etc. Furthermore, ontologies are intended to provide information about a specific data source. Therefore, the relationships that must be identified are more than simple associations between terms. These relationships may include generalization (is-a), synonyms (trip and tour), the most common associations between terms (trip related to accommodations).

Therefore, considered as the automatic or semi-automatic creation and generation process of ontologies, Ontology Learning (OL) (Maedche & Staab, 2001) leverages multitude of disciplines, notably Machine Learning techniques, to extract the corresponding domain's terms and the relationships between the concepts that represent them from a corpus of natural language text. OL also alleviates human-generated biases and inconsistencies (Zhou, 2007). Furthermore, each OL strategy should rely on the nature and the format of exploited data sources. To this end, this approach uses the RDBtoOnto tool (Cerbah, 2010), as an implementation of the basic conversion rules from relational databases to OWL (Mallede, Marir, & Vassilev, 2013) for data sources with structured and semi-structured formats (For example databases, CSV and JSON files). Likewise, for data sources with an unstructured format, such as text files, the approach would use the ontology learning solutions provided by (El Idrissi Esserhrouchni, Frikh, & Ouhbi, 2015; El Idrissi Esserhrouchni, Frikh, Ouhbi, & Ibrahim, 2017).

Ontologies Decomposition & Resources Extraction

As depicted on several ontology alignment processes (Otero-Cerdeira, Rodríguez-Martínez, & Gómez-Rodríguez, 2015; Shvaiko & Euzenat, 2013), each matching algorithm loads the entire ontologies as a resource, even if the matching process will only require specific information from them, which provokes memory issues during the performed matching tasks. In this regard, the proposed ontology alignment strategy focuses on decomposing the input ontologies into simpler resource-based subsets, with performance and scalability-friendly data structures (Lists and Hashmaps) as shown in Figure 4.

In the first place, the decomposition mechanism parses the created ontologies using Jena[2] framework and represents them as labeled directed graphs while preserving their encoded knowledge. Then, it extracts smaller, simpler and finer subsets (list of nodes, hashmap of neighbors, etc.), and calculates and saves all the required metrics used during the entire stages of the ontology alignment process, notably the ontology clustering, the similar clusters identification and the matching algorithms. Indeed, this strategy mitigates computational complexity, in such a way that it conserves the effort to re-parse the same ontologies for further processing. It also favors agility aspects, by allowing the capacity to add more matching algorithms through calculating their required metrics. Furthermore, this decomposition approach supports the adaptation of the integration framework to the changing conditions of the Open Data field (updating scenarios or new data sources) by providing flexibility and extensibility aspects.

Figure 4. Resources Decomposition and Extraction Strategy

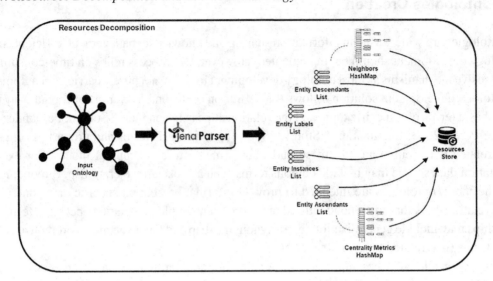

To clarify, if adding a new Open Tourism Data source, the decomposition stage parses it and creates the required resources in order to match it with previously created global ontology. From another side, this decomposition stage relies on data parallelism mechanisms provided by Hadoop, and hence serializes and stores the extracted subsets in the HBase column-oriented data store. This later provides real-time read/write access and supports parallelizing and distributing tasks over available computational resources, which deeply improves the performance of the integration framework.

All things considered, this strategy forms the kernel around which the alignment process is built, since it constructs personalized resources for each specific processing in the entire matching process. For example, for using the graph-based matching algorithm, the system loads only the required resources, namely, the list of nodes, their neighbors, and their measured cotopic distance. Furthermore, all the required metrics (centralities, similarities, etc.) are already calculated and stored for further exploitation. At the same time, if a new data source is added or new requirements are raised, this strategy will extract and encapsulate the adequate resources for the requested processing.

Ontology Partitioning

Given the high computational complexity of the ontology alignment issue, the authors apply an entity-assignment parallel partitioning approach to partition input ontologies into disjoint and exhaustive clusters - taking into account their internal cohesion - with the intention to perform matching by parts for performance-gain during the ontology alignment process. First, the clustering strategy determines the concepts (nodes) having the highest ranking as clusters centroid. Then, it assigns the remaining concepts to their adequate clusters according to their structural features (Figure 5). Certainly, using such optimization mechanism over distributed computational resources can impact the building of effective and performing ontology alignment strategy.

Figure 5. Ontology Partitioning Strategy.

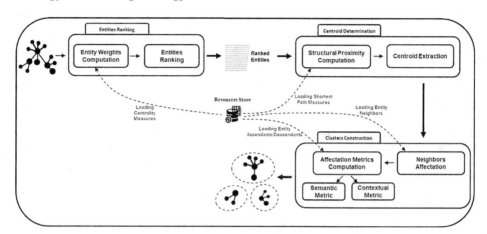

Entities Ranking

In fact, during the resources extraction step, input ontologies have been translated to their related labeled directed graphs while maintaining their encoded knowledge. Therefore, this step resides on social networks analysis theories to identify the nodes (concepts) that are significantly relevant in each ontological graph. Notably, this strategy exploits the ontological graph features by using multiple centrality metrics (Brandes, Borgatti, & Freeman, 2016; Klein, 2010; Landherr, Friedl, & Heidemann, 2010) that attribute a specific weight to each node given its position in the graph (Schuhmacher & Ponzetto, 2014). Accordingly, these metrics are essentially chosen depending to their ability to combine and provide local and global information while ensuring a performance trade-off. As a result, the relative score *RScore* for a specific node N_i is formulated by the combination of the below centrality metrics (See Table 1)

$$RScore\left(\ \right) = DC\left(\ \right) + CC\left(\ \right) + CxC\left(\ \right) + GDC\left(\ \right) \tag{1}$$

Overall, it is important to mention that during the computation of centrality metrics, this step has included the non-taxonomic relationships between nodes ignored by the state-of-the-art works, which leads to more significant and effective nodes ranking. Additionally, for performance improvement, the system parallelizes the ranking process over a distributed computational architecture and loads only the required resources from the HBase data store already created by the resources decomposition process.

Centroid Determination

After determining the most prominent nodes, the system selects the centroids based on the structural proximity between these nodes. Notably, only those having minimal proximity - given a specific threshold - are selected to create clusters. Certainly, the chosen threshold may control the outcome of this phase by determining the number and the granularity of the obtained clusters. Let N_i, N_j be two prominent nodes in a given ontological graph, the structural proximity between N_i and N_j is measured by:

Table 1. Centrality metrics

Centrality Metric	Formula	Information
Degree Centrality • Measures the neighbors a node has. • Calculates the number of input and output links of a node N_i.	$$DC\left(\ \right) = Input_{Arcs}\left(\ \right) \vee + Output_{Arcs}\left(\ \right) \vee$$	Local
Closeness Centrality • Reveals the importance of the closest nodes to the others in the graph.	$$CC\left(\ \right) = \frac{1}{\sum_{Nj} Distance\left(, Nj\right)}$$ where Distance (Ni, Nj) is the shortest path length between the nodes i and j in the graph.	Global
Context-Based Centrality • Considers the global structural context between nodes. • Determines the number of ancestors and descendants of a node N_i given a specific level L.	$$CxC\left(\ \right) = \sum_{L} ancestor\left(, L\right) \vee + descendant\left(, L\right) \vee$$	Global
Generalized Degree Centrality (Csató, 2016) • Improves the degree centrality by spreading it over the node ancestors and descendants given a specific level L.	$$GDC\left(\ \right) = DegreeC\left(\ \right) + \sum GDc\left(ancestor\left(, L\right)\right) + GDc\left(descendant\left(, L\right)\right)$$	Global

$$Sprox\left(, Nj\right) = \frac{2.\max\left(depth\left(Nij\right)\right)}{\max\left(depth\left(\ \right)\right) + \max\left(depth\left(Nj\right)\right)}$$

(2)

Where N_{ij} is the common superclass of the nodes i and j in the graph.

Clusters Construction

To achieve the clusters creation stage, the strategy injects the direct children of each centroid in their corresponding cluster depending on the encoded graph hierarchy (list of neighbors). Indeed, each centroid is responsible to attract its neighbors to its convenient cluster. After that, the remaining nodes are assigned to their adequate clusters by comparing them to the determined centroids according to the above similarity metric:

$$Global\left(, Nj\right) = Context\left(, Nj\right) + Semantic\left(, Nj\right)$$

(3)

The context-based similarity uses the nodes and their parents and children, in such a way that if two nodes have several common nodes, then their context similarity is higher. Given the nodes N_i and N_j, the contextual similarity between them can be measured as follows:

$$Context\left(, Nj\right) = \frac{\partial(\ ) \cap \partial\left(Nj\right)}{\sqrt{\partial(\ ).\partial\left(Nj\right)}} \qquad (4)$$

where $\partial(N_i)$ and $\partial(N_j)$ determine respectively the number of the neighbors of N_i and N_j, including the node itself and $\partial(N_i) \cap \partial(N_j)$ is the number of common nodes between N_i and N_j.

At the same time, the hierarchical semantic similarity depends on the node's taxonomy within the graph and can be measured using the following metric:

$$Semantic\left(, Nj\right) = \frac{2.M3}{M1 + M2 + 2.M3} \qquad (5)$$

where M_1 and M_2 are respectively the numbers of ascendants from N_i and N_j to their most specific common ancestor C, whereas M_3 is the number of descendants from the root of the taxonomy to the node C.

Parallel Clusters Matching

Actually, the aim of this stage consists of performing the parallel-based matching algorithms over formerly created clusters over a distributed infrastructure (See Figure 6). For that, the strategy aims to identify the semantically similar clusters susceptible to be individually matched. This philosophy resides on ignoring the concepts of dissimilar clusters from the matching process which can reduce matching overhead, achieves better accuracy and performance gain. After that, a sequential combination of several matching strategies, notably linguistic-based, string-based and graph-based, is performed. Each strategy is implemented in a parallel and distributed fashion over available computational resources, allowing the creation of various finer matching tasks with limited computational complexity and memory requirements. Eventually, the system gathers all the individual independent partial results to construct the final alignments set.

Identification of Similar Clusters

The adopted strategy takes advantage of the Latent Semantic Indexing (LSI) approach (Moawed, Algergawy, Sarhan, Eldosouky, & Saake, 2014) for identifying the similar clusters candidates to be individually matched. Indeed, the LSI constructs the similarity matrix between clusters instead of parsing the whole ontologies to assess the nodes similarity. The LSI needs two sets of clusters as input and aims to determine similar clusters across them over our parallel and distributed computational resources.

Clusters Matching

This phase aims to implement a combination of matching algorithms over the identified similar clusters for finding alignments between their entities. Notably, a parallel matching strategy is performed for each cluster pairs, as an individual matching task that should be independently resolved. This strategy extracts the alignments between clusters elements using a combination of several techniques of similarity computation in such a way that we extract only the alignments having similarities greater than a fixed threshold.

In fact, several strategies have been conducted to combine various matching algorithms in order to meet the challenges of the ontologies (Djeddi & Khadir, 2014; Essayeh & Abed, 2015; Song, Zacharewicz, & Chen, 2013; Steyskal & Polleres, 2013), each one of them has its strengths and weaknesses. Moreover, the quality of the matching process cannot be achieved if these strategies are not selected and combined appropriately. Therefore, this approach conducts a sequential combination of the following matching strategies (linguistic, terminological and structural) in order to improve the ontology alignment process by profiting from their power. From another perspective, these matching tasks are distributed among the available computational resources. Also, the resources management mechanisms, provided by the computational infrastructure, control the aligning and the scheduling of matching tasks which promotes reducing the matching overhead.

In detail, the matching strategy implements firstly the element-level algorithms as it determines more correspondences than structure-based ones (Euzenat & Shvaiko, 2007a). It is important to realize that each entities pair of an extracted correspondence is removed from loaded resources prior the performing the next matching algorithm. This will perfectly avoid redundant matching tasks and results, reduce the number of matching operations, since they are executed for only unmatched pairs, and improves the global performance.

Figure 6. Parallel Clusters Matching Strategy.

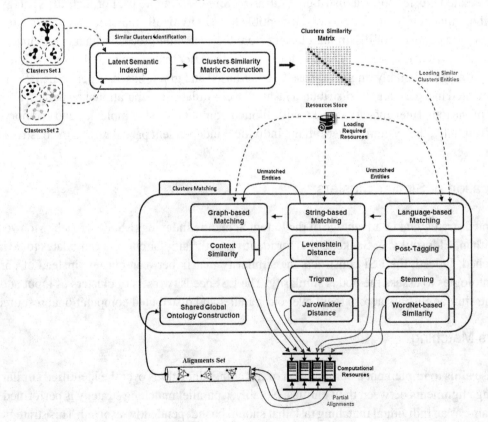

Language-Based Matching

In the first place, the system applies a language-based matching algorithm using natural language processing (NLP) procedures, namely post-tagging and stemming mechanisms, to exploit morphological characteristics of entity labels. Notably, stemming is a mechanism to strip inflectional and derivational suffixes with the aim to reduce and to merge multiple word forms into one standardized canonical form (stem). In fact, this approach is exploited before performing string-based matching algorithms, inasmuch it will obviously shorten the set of labels included in the matching space and improve their alignments. In addition, this intrinsic technique benefits from external linguistic resources (WordNet) to measure the similarity $Sim_{WordNet}$ between the stemmed labels. Immediately, the system stores the extracted alignments and proceeds to the string-based matching algorithm.

$$WordNet\left(, Nj\right) = \max \arg \sum_{Sin \in Si} \sum_{Sjm \in Sj_{sense}} \left(Sin, Sjm\right) \qquad (6)$$

In that case, each label is linked to its specific *WordNet* lexical category, called *synset*. This latter is organized into senses, providing thus synonyms, hyponym/hypernym (Is-A) and meronym/holonym (Part-Of) relationships of each label. As result, the similarity between two labels reflects the similarity between their relative senses *Sim_sense*.

String-Based Matching

The string-based algorithms aim to consider only the labels of entities or their descriptions. They extract the correspondences between entities using string equality metrics. To this end, the adopted string-based matching approach may combine the Levenshtein, Trigram and JaroWinkler metrics in order to extract the most relevant correspondences. Indeed, these metrics are applied only for the entities that have not been yet matched to any other ones during the previously conducted matching phase.

- **Levenshtein Distance:** The Levenshtein distance is a string metric for measuring difference between two strings which make it closely related to pairwise string alignments. In fact, the Levenshtein distance between two strings is the minimum number of single-character edit operations (insertions, deletions and substitutions) required for the transformation of the entity label L_1 into the entity label L_2. The greater the Levenshtein distance, the more different the strings are. Notably, using this metric aims to find correspondences for short strings situations where a small number of differences are to be expected. Certainly, the Levenshtein distance can also be applied for two longer strings. However, it remains computationally expansive as it is roughly proportional to the product of the two string lengths, hence, the role of the stemming methods that shorten the length of matched labels.
- **Trigram:** The n-gram interprets a contiguous sequence of *n* units from a string (entity label). For the sake of making a trade-off between information accuracy (reliability of the information provided by the extracted sequence) and performance (complexity of computations), this matching phase adopts trigrams as follow:

$$Tgram\left(L1, L2\right) = \frac{Trigram\left(L1\right) \sqcap Trigram\left(L2\right)}{\min\left(\left|L1\right|, \left|L2\right|\right) - 2} \tag{7}$$

- **JaroWinkler Distance:** Usually used in record linkage scenarios on different sources, this metric measures the distance between two strings (entity labels) as the minimum number of single-character transpositions needed to change one string into the other. It is scaled between 0 (not similar at all) and 1 (exact match), in such a way that, the lower the distance for two strings is, the more similar the strings are. Given the above formula, the Jaro-Winkler metric transforms the standard Jaro distance metric by putting extra weight on string differences at the start of the strings to be compared:

$$dw = dj + \delta p\left(1 - dj\right) \text{ with } dj = 1/3\left(\frac{m}{\left|S1\right|} + \frac{m}{\left|S2\right|} + \frac{m - t}{m}\right) \tag{8}$$

with δ_p: Prefix's length ; dj: Jaro's distance, m: the number of corresponding characters, $|S_i|$: Length of the string *i* and t: number of transpositions.

Graph-Based Matching

From another side, the implemented ontology alignment strategy leverages also the structural information provided by the entire taxonomy of ontologies (concepts and relations) for the extraction of correspondences between their entities. To this end, a graph-based approach is performed while considering the input ontologies as labeled graphs. Thus, based on the taxonomic structure of these graphs, we used the previous structural measure (See Equation 4) to extract the correspondences between clusters pairs, in such a way that two nodes are similar; their neighbors must also be similar. Of course, this strategy is only applied for the entities that have not been yet matched to any other ones by the previous matching algorithms.

Finally, the different independent results collected from each matching strategy are aggregated and merged to construct the final alignments set that will be established between input ontologies in order to create the shared global ontology.

TRIP PLANNING USE CASE

Obviously, tourists browse multiple sources and dispersed information, from different systems to make prior decisions about their trip. Thus, they are exposed to an overwhelming amount of information that prevent them from choosing the most suitable travel destination, mode and activities. For that, building intelligent systems aiming to assist them in the decision-making process and enables them to acquire additional and detailed knowledge, is desirable to support them during their trip planning by providing more prominent and better-tailored destination-specific information.

To this end, the authors aim to build an ontology-based integration framework for Open Tourism Data to provide an integrated view across these complex and heterogeneous data sources. This integration framework will allow developing reliable and efficient cross-language tourism-related applications able to expose the touristic interest points that are most attractive to a particular tourist and support him in its travel planning scenario.

As a typical use case, the authors aim to combine multiple Open Tourism Data sources from several domains to provide integrated, consistent and representative information on tourists' points of interest for a specific destination (Ile-de-France French Region). This trip planning scenario aims to provide a high-level representation of the possible points of interest of a tourist for this specific destination. For the purpose of gathering the major possible points of interests of a tourist during a trip planning process for the Ile-de-France region, the authors consider the following cross-domain heterogeneous data sources as depicted in Table 2.

Table 2. Multiple data sets gathered from cross-domain heterogeneous data sources

Data sets	Format	Domain
DBpedia • Large-scale multilingual encyclopedic data-set able to link other data sets to make them available on the Web. • Allows creating a corpus for different fields through sophisticated queries (SPARQL query) about any topic available in Wikipedia. • Considered as a nucleus for a web of Open Data (Auer et al., 2007) and allows the development of cross-domain applications (Musto, Lops, de Gemmis, & Semeraro, 2017).	RDF	Cross-Domains
Schema.org • Collaborative, community activity with a mission to create, maintain, promote and to share schemas for structured data on web (Guha, Brickley, & MacBeth, 2015). • Resides in providing a single integrated schema covered a variety of topics that included persons, places, events, products, offers, and so on. • The schemas are a hierarchically set of types, each associated with a set of properties (597 types, 867 properties, and 114 Enumeration values).	RDF	Cross-Domains
French communal administrative division3 • Data is provided by *OpenStreetMap* project, the mapping wiki that creates and provides free ODbL-licensed geographic data. • Contains all the French communes and boroughs given the following attributes: - Insee: INSEE code with 5 characters of the commune. - Name: name of the municipality (as shown in OpenStreetMap). - Wiki: Wikipedia entry (language code followed by the article name). - Surf_ha: area in hectares of the municipality.	CSV	Geo-Positions
French Tourist accommodations4 • Comes from a certified public service (National Institute of Statistics and Economic Studies - INSEE). • Provides statistics on tourist accommodation capacities given the following attributes: - Name, company name, address, Tel, postal code. - City, longitude, longitude, Email, category, number of rooms. - Method of payment, pricing, policies.	XLS	Accommodations
Geo-located Events5 • Open Agenda is a calendar platform for event organizers: festivals, theaters, concerts, exhibitions, exhibitions, conferences, etc. • Presents events with their precise geographic coordinates and detailed time slots given the following attributes: - Title, organization, start date, event type. - Event place, theme, style, genre. - Longitude, latitude, time slot.	JSON	Attractions

continues on following page

Table 2. Continued

Data sets	Format	Domain
<u>Geographical positions of RATP network stations</u>6 • Comes from a certified public service, namely RATP (Régie autonome des transports parisiens). • RATP provides all urban and interurban public transport modes: rail, metro, tram and bus. • Data is updated automatically approximately every 15 days and includes all transport offers in Paris.	CSV	Geo-Positions Transport
<u>PASSIM offers, services and transport data in France</u>7 • Comes from a public service certified and supported by the Ministry of Sustainable Development. • Lists and describes the offers of transport services and the information services useful for traveling in France, for all modes of transport given the following attributes: - Company name, operator, and type of transport. - Mode of transport, pricing, reservation URL.	CSV	Transport
<u>Location of Wi-Fi hotspots</u>8 • Comes from a certified public service (Mairie de Paris) and lists sites (localizations) with a Wi-Fi hotspot for a free internet connection, according to the following attributes: - Id ArcGIS, city, latitude, longitude. - Quota user max bytes and quota user max duration.	JSON	Infrastructure
<u>Population</u>9 • Gives access to the results of population censuses, time series of the INSEE Macroeconomic Data Bank. • Provides statistics on the subject of population and other data, given the following attributes: - Region code, region name, code department. - department name, municipal population.	HTML	Demographic
<u>Open Food Facts</u>10 • Lists food products around the world. • Collects Food information (photos, ingredients, nutritional composition, etc.) and made it available to everyone for any purpose.	RDF	Nutrition

Deployment Architecture

The proposed ontology-based integration framework for Open Tourism Data has been built over the Cloudera Distribution Hadoop platform which is an open source Apache Hadoop distribution. This offers a scalable, flexible and integrated platform to rapidly manage a massive amount of data. More specifically, the proposed integration framework is implemented on fully-distributed mode over a cluster of 10 slave machines and one master machine. Each one is equipped with 3.4 GHz Intel(R) Core i3(R) with 4 GB memory, Java 1.8 and Ubuntu 14.04 LTS.

Besides, the resources decomposition strategy leverages HBase, which is a No-SQL column-oriented key/value data store built to be run on top of the Hadoop Distributed File System (HDFS). It supports random, real-time read/write access and allows us to use a larger in-memory cache which effectively reduces the execution time and improves the performance of the ontology-based integration framework.

RESULTS AND DISCUSSION

The aim of this experimentation consists on investigating the capabilities to which the Open Data integration framework may provide, to assist the development of decision-making support platforms. Specifically, the authors aim to address the challenges to support travelers in their decision-making during the trip planning phase and to facilitate the development of their travel plan considering their expectations and preferences towards touristic object of interests. For instance, to plan their vacation in *Paris* (named as the top tourist destination in the world), travelers are able to explore available information (reviews, itinerary, fees, accommodations, security, etc.) on different platforms according to certain selection criteria (destination and attraction). This implies that they could exclude a large amount of information due to poorly formulated requests.

To this end, the core idea behind this experimentation consists on the combination of multiple Open Tourism Data sources through a multi-domain ontological knowledge (the created shared global ontology) to describe the touristic point of interests in a deeper and more structured manner, and to feed intelligent systems in order to predict travelers' attitudes toward a tourist point of interest. For that, the authors create an integrated semantic-based data source (the shared global ontology) as a global unified source of information, to feed a tourist research platform. This latter uses the unified ontology (See Figure 7) to expand a specific traveler interest (formulated by a query) by providing cross-domain sophisticated proposals corresponding to its query.

For instance, the traveler may first specify in a form the destination and length of stay. Then, as depicted in Figure 2, its query is translated and reformulated to a corresponding SPARQL query on the global shared ontology. This SPARQL query will select and provide the different information related to the selected destination. Specifically, it will display the accommodations and how to reach them (geographic coordinates and address), the events that will take place in the period of stay, the different types of foods that will be served and their corresponding locations (restaurants) as well as tourist services offers related to the wanted destination.

To sum up, since choosing the most suitable travel destination, mode and activities remain a time and effort consuming process; this platform will have a major impact on traveler's decision making criteria and enable multiple visualizations in a common direction, especially when choosing a destination to visit. Travelers will have more prominent and better-tailored information about all tourism-related activities for their selected destination, instead of browsing separately multiple sources with dispersed information, which help them making decisions and building expectations for their upcoming trip. From another side, this solution allows harnessing several Open Data portals with different large-scale data in both structured and unstructured formats while ensuring the scalability aspects and in-time requirements to expose the touristic interest points that are most attractive to a particular tourist.

Figure 7. Snippet of the Created Shared Global Ontology for the Ile-de-France Region.

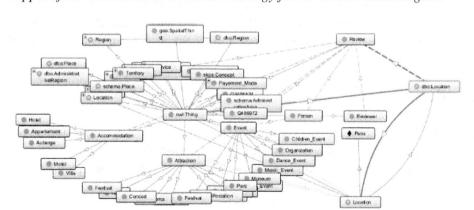

FUTURE RESEARCH DIRECTIONS

In general, to build intelligent systems on a specific field, it is important to harness several large-scale data sources, having complex features, from multiple overlapped domains and contexts. Hence, future research works will focus on the incorporation of the proposed ontology-based integration framework for Open Data to build a touristic content-based recommender system for Smart Destinations that aims to build the users' profile based on their preferences and expectations towards touristic object of interests. Thus, the recommender system will build a profile for a specific traveler according to the formulated queries on the trip planning platform which will affect their decision-making process and assist them for their travel planning scenarios by providing more personalized and targeted information. Furthermore, the authors will investigate the capabilities, to which the integration framework may provide, when different stakeholders would work together through their data in order to transform it into valuable insights for smarter and strategic operational decisions. In fact, they will support public authorities and involved organizations to adopt the potential of Open Data as a principal asset of their smart city policies, as well as promoting their advertising and targeting strategies to earn new markets and maintain the amplitude of their services. Hence, they will rely on the exploitation of the created shared ontology in various querying and analytical operations to assist stakeholders in their decision-making process and enable them to acquire additional and detailed knowledge.

CONCLUSION

Smart tourism destinations should adapt a strategy that provides information about all tourism-related activities, in order to help the travelers making decisions and building expectations for their upcoming trip. To this end, the authors aim to build a trip planning platform that exploits the available Open Tourism Data sources - commonly publish with a free public access – to support travelers in their decision-making during the trip planning phase. In fact, the authors build an ontology-based integration framework for Open Tourism Data over a distributed architecture and using parallel-programming techniques. In this way, several kinds of semantic relations within the ontological models can be exploited to better describe touristic points of interest -in unexplored concepts- and to facilitate the propagation of interests between other ones. This solution has demonstrated a significant capacity to combine and to interoperate diverse heterogeneous Open Data sources, from different overlapped domains. This will certainly affect not only the tourism industry, as a way to devise new management and marketing concepts towards maximizing both destination competitiveness and consumer satisfaction; but also incites the academics, by providing a good foundation to explore the great opportunities and potentials of using Open Data in other fields.

REFERENCES

Algergawy, A., Babalou, S., Kargar, M. J., & Hashem Davarpanah, S. (2015). Seecont: A new seeding-based clustering approach for ontology matching. In Lecture Notes in Computer Science (including subseries Lecture Notes in Artificial Intelligence and Lecture Notes in Bioinformatics) (Vol. 9282, pp. 245–258). Springer International Publishing. doi:10.1007/978-3-319-23135-8_17

Algergawy, A., Massmann, S., & Rahm, E. (2011). A clustering-based approach for large-scale ontology matching. In Lecture Notes in Computer Science (including subseries Lecture Notes in Artificial Intelligence and Lecture Notes in Bioinformatics) (Vol. 6909 LNCS, pp. 415–428). Springer. doi:10.1007/978-3-642-23737-9_30

Amin, M. B., Khan, W. A., Lee, S., & Kang, B. H. (2015). Performance-based ontology matching. *Applied Intelligence*, *43*(2), 356–385. doi:10.100710489-015-0648-z

Auer, S., Bizer, C., Kobilarov, G., Lehmann, J., Cyganiak, R., & Ives, Z. (2007). DBpedia: A nucleus for a Web of open data. In Lecture Notes in Computer Science (including subseries Lecture Notes in Artificial Intelligence and Lecture Notes in Bioinformatics) (Vol. 4825, pp. 722–735). Springer. doi:10.1007/978-3-540-76298-0_52

Benckendorff, P. J., Sheldon, P. J., & Fesenmaier, D. R. (2014). *Tourism information technology* (2nd ed.). Tourism Information Technology. doi:10.1079/9781780641850.0000

Bieger, T., & Laesser, C. (2004). Information sources for travel decisions: Toward a source process model. *Journal of Travel Research*, *42*(4), 357–371. doi:10.1177/0047287504263030

Björk, P., & Kauppinen-Räisänen, H. (2015). Contemporary insights to the dynamic pre-trip information sourcing behaviour. *Tourism and Hospitality Research*, *15*(1), 39–53. doi:10.1177/1467358414553871

Brandes, U., Borgatti, S. P., & Freeman, L. C. (2016). Maintaining the duality of closeness and betweenness centrality. *Social Networks*, *44*, 153–159. doi:10.1016/j.socnet.2015.08.003

Buhalis, D., & Amaranggana, A. (2015). Smart Tourism Destinations Enhancing Tourism Experience Through Personalisation of Services. Information and Communication Technologies in Tourism 2015. doi:10.1007/978-3-319-14343-9_28

Cerbah, F. (2010). Learning ontologies with deep class hierarchies by mining the content of relational databases. In Studies in Computational Intelligence (Vol. 292, pp. 271–286). Academic Press. doi:10.1007/978-3-642-00580-0_16

Chan, C. M. L. (2013). From Open Data to Open Innovation Strategies: Creating E-Services Using Open Government Data. *Proceedings of the 46th Hawaii International Conference on System Sciences (HICSS-46)*, 1890–1899. 10.1109/HICSS.2013.236

Choi, S., Lehto, X. Y., & Oleary, J. T. (2007). What does the consumer want from a DMO website? A study of US and Canadian tourists' perspectives. *International Journal of Tourism Research*, *9*(2), 59–72. doi:10.1002/jtr.594

Corcho, O., Fernández-López, M., & Gómez-Pérez, A. (2003). Methodologies, tools and languages for building ontologies. Where is their meeting point? *Data & Knowledge Engineering*, *46*(1), 41–64. doi:10.1016/S0169-023X(02)00195-7

Csató, L. (2016). Measuring centrality by a generalization of degree. *Central European Journal of Operations Research*, 1–20. doi:10.100710100-016-0439-6

Daraio, C., Lenzerini, M., Leporelli, C., Naggar, P., Bonaccorsi, A., & Bartolucci, A. (2016). The advantages of an Ontology-Based Data Management approach: Openness, interoperability and data quality. *Scientometrics*, *108*(1), 441–455. doi:10.100711192-016-1913-6

Djeddi, W. E., & Khadir, M. T. (2014). A novel approach using context-based measure for matching large scale ontologies. In Lecture Notes in Computer Science (including subseries Lecture Notes in Artificial Intelligence and Lecture Notes in Bioinformatics) (Vol. 8646, pp. 320–331). Springer. doi:10.1007/978-3-319-10160-6_29

Doan, A., Halevy, A., & Ives, Z. (2012). Principles of Data Integration. *Principles of Data Integration*, 95–119. doi:10.1016/B978-0-12-416044-6.00004-1

Dong, X. L., & Srivastava, D. (2013). Big data integration. *2013 IEEE 29th International Conference on Data Engineering (ICDE)*, 1245–1248. 10.1109/ICDE.2013.6544914

Eckartz, S., Van den Broek, T., & Ooms, M. (2016). Open data innovation capabilities: Towards a framework of how to innovate with open data. In Lecture Notes in Computer Science (including subseries Lecture Notes in Artificial Intelligence and Lecture Notes in Bioinformatics) (Vol. 9820, pp. 47–60). Springer. doi:10.1007/978-3-319-44421-5_4

El Idrissi Esserhrouchni, O., Frikh, B., & Ouhbi, B. (2015). Learning non-taxonomic relationships of financial ontology. In *Proceedings of the 7th International Joint Conference on Knowledge Discovery, Knowledge Engineering and Knowledge Management* (pp. 479–489). Academic Press. 10.5220/0005590704790489

El Idrissi Esserhrouchni, O., Frikh, B., Ouhbi, B., & Ibrahim, I. K. (2017). Learning domain taxonomies: The TaxoLine approach. *International Journal of Web Information Systems*, *13*(3), 281–301. doi:10.1108/IJWIS-04-2017-0024

Essayeh, A., & Abed, M. (2015). Towards ontology matching based system through terminological, structural and semantic level. Procedia Computer Science, 60, 403–412. doi:10.1016/j.procs.2015.08.154

European Commission. (2018). *Employment and social developments in Europe 2011*. Publications Office of the European Union. doi:10.2767/44905

Euzenat, J., & Shvaiko, P. (2007a). Ontology matching. Heidelberg, Germany: Springer. doi:10.1007/978-3-540-49612-0

Euzenat, J., & Shvaiko, P. (2007b). Ontology Matching. *Chemistry*, *334*. doi:10.1007/978-3-540-49612-0

Fesenmaier, D. R., & Jeng, J.-M. (2000). Assessing Structure in the Pleasure Trip Planning Process. *Tourism Analysis*.

García, A., Linaza, M. T., Franco, J., & Juaristi, M. (2015). Methodology for the Publication of Linked Open Data from Small and Medium Size DMOs. Information and Communication Technologies in Tourism 2015. doi:10.1007/978-3-319-14343-9_14

Gretzel, U., Fesenmaier, D. R., & O'Leary, J. T. (2005). The transformation of consumer behaviour. In *Tourism Business Frontiers*. Consumers, Products and Industry. doi:10.4324/9780080455914

Gretzel, U., Sigala, M., Xiang, Z., & Koo, C. (2015). Smart tourism: Foundations and developments. *Electronic Markets, 25*(3), 179–188. doi:10.100712525-015-0196-8

Gretzel, U., Werthner, H., Koo, C., & Lamsfus, C. (2015). Conceptual foundations for understanding smart tourism ecosystems. *Computers in Human Behavior, 50,* 558–563. doi:10.1016/j.chb.2015.03.043

Gross, A., Hartung, M., Kirsten, T., & Rahm, E. (2010). On matching large life science ontologies in parallel. In *International Conference on Data Integration in the Life Sciences* (pp. 35–49). Springer Berlin Heidelberg. 10.1007/978-3-642-15120-0_4

Guha, R. V., Brickley, D., & MacBeth, S. (2015). Schema.org: Evolution of Structured Data on the Web. *Queue, 13*(9), 10–37. doi:10.1145/2857274.2857276

Hausenblas, M. (2009). Exploiting linked data to build web applications. *IEEE Internet Computing, 13*(4), 68–73. doi:10.1109/MIC.2009.79

Hjalager, A. M., & Nordin, S. (2011). User-driven Innovation in Tourism-A Review of Methodologies. *Journal of Quality Assurance in Hospitality & Tourism, 12*(4), 289–315. doi:10.1080/1528008X.2011.541837

Hu, W., Qu, Y., & Cheng, G. (2008). Matching large ontologies: A divide-and-conquer approach. *Data & Knowledge Engineering, 67*(1), 140–160. doi:10.1016/j.datak.2008.06.003

Huang, C. D., Goo, J., Nam, K., & Yoo, C. W. (2017). Smart tourism technologies in travel planning: The role of exploration and exploitation. *Information & Management, 54*(6), 757–770. doi:10.1016/j.im.2016.11.010

Irudeen, R., & Samaraweera, S. (2013). Big data solution for Sri Lankan development: A case study from travel and tourism. *2013 International Conference on Advances in ICT for Emerging Regions (ICTer).* 10.1109/ICTer.2013.6761180

Ivars-Baidal, J. A., Celdrán-Bernabeu, M. A., Mazón, J. N., & Perles-Ivars, Á. F. (2017). Smart destinations and the evolution of ICTs: A new scenario for destination management? *Current Issues in Tourism.* doi:10.1080/13683500.2017.1388771

Jafari, J. (2001). The scientification of tourism. Hosts AND guests revisited: tourism issues of the 21st century.

Janssen, M., Charalabidis, Y., & Zuiderwijk, A. (2012). Benefits, Adoption Barriers and Myths of Open Data and Open Government. *Information Systems Management, 29*(4), 258–268. doi:10.1080/10580530.2012.716740

Kalampokis, E., Tambouris, E., & Tarabanis, K. (2011). A classification scheme for open government data: Towards linking decentralised data. *International Journal of Web Engineering and Technology, 6*(3), 266. doi:10.1504/IJWET.2011.040725

Kambatla, K., Kollias, G., Kumar, V., & Grama, A. (2014). Trends in big data analytics. *Journal of Parallel and Distributed Computing, 74*(7), 2561–2573. doi:10.1016/j.jpdc.2014.01.003

Kirsten, T., Kolb, L., Hartung, M., Groß, A., Köpcke, H., & Rahm, E. (2010). Data Partitioning for Parallel Entity Matching. *Strategies (La Jolla, Calif.), 3*(2), 11. Retrieved from http://arxiv.org/abs/1006.5309

Klein, D. J. (2010). Centrality measure in graphs. *Journal of Mathematical Chemistry*, 4(47), 1209–1223. doi:10.100710910-009-9635-0

Landherr, A., Friedl, B., & Heidemann, J. (2010). A Critical Review of Centrality Measures in Social Networks. *Business & Information Systems Engineering*, 2(6), 371–385. doi:10.100712599-010-0127-3

Lenzerini, M. (2002, June). Data integration: A theoretical perspective. *Pods*, 233–246. doi:10.1145/543613.543644

Li, G., Law, R., Vu, H. Q., Rong, J., & Zhao, X. (2015a). Identifying emerging hotel preferences using Emerging Pattern Mining technique. *Tourism Management*, 46, 311–321. doi:10.1016/j.tourman.2014.06.015

Li, G., Law, R., Vu, H. Q., Rong, J., & Zhao, X. (Roy). (2015b). Identifying emerging hotel preferences using Emerging Pattern Mining technique. *Tourism Management*. doi:10.1016/j.tourman.2014.06.015

Longhi, C., Titz, J. B., & Viallis, L. (2014). Open data: Challenges and opportunities for the tourism industry. In Tourism Management, Marketing, and Development: Volume I: The Importance of Networks and ICTs. Academic Press. doi:10.1057/9781137354358

Maccani, G., Donnellan, B., & Helfert, M. (2015). Open Data Diffusion for Service Innovation: An Inductive Case Study on Cultural Open Data Services. *Proceedings of the 19th Pacific-Asian Conference on Information Systems (PACIS 2015)*, 1–17.

Maedche, A., & Staab, S. (2001). Ontology Learning for the Semantic Web. *IEEE Intelligent Systems*, 16(2), 72–79. doi:10.1109/5254.920602

Mallede, W., Marir, F., & Vassilev, V. (2013). Algorithms for Mapping RDB Schema to RDF for Facilitating Access to Deep Web. *Proceedings of the First International Conference on Building and Exploring Web Based Environments*.

Manyika, J., Chui, M., Groves, P., Farrell, D., Van Kuiken, S., & Doshi, E. A. (2013). *Open Data: Unlocking Innovation and Performance with Liquid Information*. McKinsey.

Mariani, M. M., Buhalis, D., Longhi, C., & Vitouladiti, O. (2014). Managing change in tourism destinations: Key issues and current trends. *Journal of Destination Marketing & Management*, 2(4), 269–272. doi:10.1016/j.jdmm.2013.11.003

Marine-Roig, E., & Anton Clavé, S. (2015). Tourism analytics with massive user-generated content: A case study of Barcelona. *Journal of Destination Marketing & Management*, 4(3), 162–172. doi:10.1016/j.jdmm.2015.06.004

Michopoulou, E., & Buhalis, D. (2013). Information provision for challenging markets: The case of the accessibility requiring market in the context of tourism. *Information & Management*, 50(5), 229–239. doi:10.1016/j.im.2013.04.001

Moawed, S., Algergawy, A., Sarhan, A., Eldosouky, A., & Saake, G. (2014). A latent semantic indexing-based approach to determine similar clusters in large-scale schema matching. In *New Trends in Databases and Information Systems* (pp. 267–276). Springer International Publishing. doi:10.1007/978-3-319-01863-8_29

Mountasser, I., Ouhbi, B., & Frikh, B. (2015). From data to wisdom: A new multi-layer prototype for Big Data management process. In *International Conference on Intelligent Systems Design and Applications, ISDA* (pp. 104–109). IEEE. 10.1109/ISDA.2015.7489209

Mountasser, I., Ouhbi, B., & Frikh, B. (2016). Hybrid large-scale ontology matching strategy on big data environment. In *18th International Conference on Information Integration and Web-based Applications and Services* (pp. 282–287). ACM. 10.1145/3011141.3011185

Musto, C., Lops, P., de Gemmis, M., & Semeraro, G. (2017). Semantics-aware Recommender Systems exploiting Linked Open Data and graph-based features. *Knowledge-Based Systems*, *136*, 1–14. doi:10.1016/j.knosys.2017.08.015

Otero-Cerdeira, L., Rodríguez-Martínez, F. J., & Gómez-Rodríguez, A. (2015). Ontology matching: A literature review. *Expert Systems with Applications*, *42*(2), 949–971. doi:10.1016/j.eswa.2014.08.032

Pantano, E., Priporas, C. V., & Stylos, N. (2017). 'You will like it!' using open data to predict tourists' response to a tourist attraction. *Tourism Management*, *60*, 430–438. doi:10.1016/j.tourman.2016.12.020

Pesonen, J., & Lampi, M. (2016). Utilizing open data in tourism. Conference: Enter2016.

Quintal, V. A., Lee, J. A., & Soutar, G. N. (2010). Risk, uncertainty and the theory of planned behavior: A tourism example. *Tourism Management*, *31*(6), 797–805. doi:10.1016/j.tourman.2009.08.006

Rajapaksha, P., Farahbakhsh, R., Nathanail, E., & Crespi, N. (2017). ITrip, a framework to enhance urban mobility by leveraging various data sources. In Transportation Research Procedia (Vol. 24, pp. 113–122). Academic Press. doi:10.1016/j.trpro.2017.05.076

Ross, I., & Bettman, J. R. (1979). An Information Processing Theory of Consumer Choice. *Journal of Marketing*, *43*(3), 124. doi:10.2307/1250155

Santipantakis, G., Kotis, K., & Vouros, G. A. (2017). OBDAIR: Ontology-Based Distributed framework for Accessing, Integrating and Reasoning with data in disparate data sources. *Expert Systems with Applications*, *90*, 464–483. doi:10.1016/j.eswa.2017.08.031

Schuhmacher, M., & Ponzetto, S. (2014). Ranking Entities in a Large Semantic Network. In *European Semantic Web Conference* (pp. 254–258). Springer International Publishing. 10.1007/978-3-319-11955-7_30

Shvaiko, P., & Euzenat, J. (2013). Ontology Matching: State of the Art and Future Challenges. *IEEE Transactions on Knowledge and Data Engineering*, *25*(1), 158–176. doi:10.1109/TKDE.2011.253

Song, F., Zacharewicz, G., & Chen, D. (2013). An Analytic Aggregation-Based Ontology Alignment Approach with Multiple Matchers. In *Advanced Techniques for Knowledge Engineering and Innovative Applications* (pp. 143–159). Springer Berlin Heidelberg. doi:10.1007/978-3-642-42017-7_11

Steyskal, S., & Polleres, A. (2013). Mix'n'Match: An alternative approach for combining ontology matchers (short paper). In Lecture Notes in Computer Science (including subseries Lecture Notes in Artificial Intelligence and Lecture Notes in Bioinformatics) (Vol. 8185, pp. 555–563). Springer. doi:10.1007/978-3-642-41030-7_40

Sugumaran, V., & Storey, V. C. (2002). Ontologies for conceptual modeling: Their creation, use, and management. In Data and Knowledge Engineering (Vol. 42, pp. 251–271). Academic Press. doi:10.1016/S0169-023X(02)00048-4

Thorsby, J., Stowers, G. N. L., Wolslegel, K., & Tumbuan, E. (2017). Understanding the content and features of open data portals in American cities. *Government Information Quarterly, 34*(1), 53–61. doi:10.1016/j.giq.2016.07.001

Virginija Uzdanaviciute, R. B. (2011). Ontology-based Foundations for Data Integration. *The First International Conference on Business Intelligence and Technology.*

Wiggins, A., & Crowston, K. (2011). From conservation to crowdsourcing: A typology of citizen science. *Proceedings of the Annual Hawaii International Conference on System Sciences.* 10.1109/HICSS.2011.207

Xiang, Z., Magnini, V. P., & Fesenmaier, D. R. (2015). Information technology and consumer behavior in travel and tourism: Insights from travel planning using the internet. *Journal of Retailing and Consumer Services, 22,* 244–249. doi:10.1016/j.jretconser.2014.08.005

Xiaomeng, C., Lyu, Z., & Terpenny, J. (2015). Ontology development and optimization for data integration and decision-making in product design and obsolescence management. In Ontology Modeling in Physical Asset Integrity Management (pp. 87–132). Academic Press. doi:10.1007/978-3-319-15326-1_4

Zhou, L. (2007). Ontology learning: State of the art and open issues. *Information Technology Management, 8*(3), 241–252. doi:10.100710799-007-0019-5

Ziegler, P., & Dittrich, K. R. (n.d.). Data Integration — Problems, Approaches, and Perspectives. *Conceptual Modelling in Information Systems Engineering,* 39–58. doi:10.1007/978-3-540-72677-7_3

ENDNOTES

[1]   http://okfn.org/
[2]   https://jena.apache.org/
[3]   https://www.data.gouv.fr/fr/datasets/decoupage-administratif-communal-francais-issu-d-openstreet-map/
[4]   https://www.data.gouv.fr/fr/datasets/hebergements-touristiques-1/
[5]   https://www.data.gouv.fr/fr/datasets/evenements-geolocalises-et-opendata/
[6]   https://www.data.gouv.fr/fr/datasets/positions-geographiques-des-stations-du-reseau-ratp-ratp/
[7]   https://www.data.gouv.fr/fr/datasets/passim-offres-services-dinfo-et-donnees-transport-en-france-1/
[8]   https://www.data.gouv.fr/fr/datasets/liste-des-sites-municipaux-equipes-dun-point-dacces-wifi/
[9]   https://www.data.gouv.fr/fr/datasets/population/
[10]   https://www.data.gouv.fr/fr/organizations/open-food-facts/

Chapter 3
Artificial Intelligence for Extended Software Robots, Applications, Algorithms, and Simulators

Gayathri Rajendran
Pondicherry University, India

Uma Vijayasundaram
iD https://orcid.org/0000-0002-7257-7920
Pondicherry University, India

ABSTRACT

Robotics has become a rapidly emerging branch of science, addressing the needs of humankind by way of advanced technique, like artificial intelligence (AI). This chapter gives detailed explanation about the background knowledge required in implementing the software robots. This chapter has an in-depth explanation about different types of software robots with respect to different applications. This chapter would also highlight some of the important contributions made in this field. Path planning algorithms are required for performing robot navigation efficiently. This chapter discusses several robot path planning algorithms which help in utilizing the domain knowledge, avoiding the possible obstacles, and successfully accomplishing the tasks in lesser computational time. This chapter would also provide a case study on robot navigation data and explain the significant of machine learning algorithms in decision making. This chapter would also discuss some of the potential simulators used in implementing software robots.

DOI: 10.4018/978-1-5225-9687-5.ch003

INTRODUCTION

Artificial Intelligence (AI) enables a machine to perform actions like learning, reasoning, decision-making and self-correction which resemble certain aspects of human intelligence. AI additionally permits a robot to execute intelligent tasks in real-world environments over extended periods of time. Robots are widely used in assembling, packing, packaging, mining, transportation, earth and space exploration, surgery, weaponry, laboratory research, safety, security and in the mass production of consumer and industrial goods (Khan & Khan, 2017; Rath, 2018; Rath, Pati & Patayanak, 2019; Rath, 2019; Rath, Pati, Panigrahi & Sakar, 2019).

Robots based on functionality can respectively be classified as software and hardware robots. Hardware-oriented robots have a clear set of actions and enable integration of real environments with manufacturers. The first section of this chapter discusses the components of robots and the applications of hardware robots in detail. A software robot is an autonomously re-programmable and multi-functional machine which can shift and communicate like an intellectual and perform automation tasks in any environment. The purpose of software robots is to assist human beings perform intelligent tasks in a more meaningful, creative, rapid and safe manner (Khan & Khan, 2017).

Software robots can be divided into two categories, namely, industrial and service robots. Industrial robots perform the tasks in a specified order and frequently execute the tasks in well-defined environments. Service robot assists human beings in agriculture, natural disasters, medical science and military operations (Kumari, 2014; Priya, 2014; Ruiz, 2017). The different types of software robots with respect to various applications are illustrated in Figure 1 and a detailed explanation of the same is presented in the first section.

Planning, the sub-discipline of AI, will enable the robot to perform a sequence of actions that leads to the desired (target) goal location (Kunze, Hawes, Duckett, & Hanheide, 2018). Planning models, representation of planning models and planning languages are explained in the next section. Robots have to find a collision-free path in the environment and reach the desired (target) goal location. To achieve this, various path planning algorithms and machine learning algorithms are widely utilzed (Galceran & Carreras, 2013; Yang, Qi, Song, Xiao, Han & Xia, 2016; Givigi & Jardine, 2018). Path planning algorithms help in generating a suitable path that can facilitate safe and effortless movement of the robot after tackling the obstacles in virtual environments (2D or 3D). A comprehensive analysis of the different kinds of robot path planning algorithms has been performed and presented in this chapter. The most extensively utilized robotic software simulators have been listed and cross-referenced.

Figure 1. Different types and applications of software robots

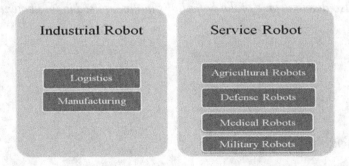

The chapter additionally commits is to provide a comprehensive understanding of the concepts with regard to software robots, their applications, various path planning algorithms and simulation software. This will enable the reader to realize the significance of path planning. Machine learning approaches when enhanced with path planning algorithms can provide better results. In order to substantiate the hypothesis, experiments have been conducted on 3 open source sensor data-sets using various machine learning algorithms. To enable a better understanding of the experimental section, a discussion of various machine learning approaches is provided. The results obtained from the experiments have been analyzed taking into account various parameters. It is found that, the support vector machine (SVM) performs better on one data-set while logistic regression (LR) techniques perform better on the other two data-sets.

ROBOTICS BACKGROUND

Planning is required for a robot to efficiently perform autonomous navigation. Planning enables a point-to-point motion of a robot in order to reach the designated goal position (Russell & Norvig, 2016). The basic principles of robotics navigation system are perception, localization, path planning and motion control. A robot can perform any type of task that can be done autonomously and monotonously. The basic principle of a robotics navigation system is shown in Figure 2.

With respect to Figure 2, it is clear that an agent interacts with the environment via perception, localization, path planning and motion control. A detailed discussion about each of these is provided below.

Environments

In order to set the path and perform actions or events, we require an environment (Russell & Norvig, 2016). The representation of robotic environments can be conducted in 2-D or 3-D work-spaces. To manage all the possible situations and to be more dynamic, robots need to be aware of the type of their environmental conditions - deterministic, non-deterministic or stochastic - in which it must function. Thus, a complete domain knowledge representation is essential to efficiently perform robot navigation. A complete environment can be specifically classified into two categories, concretely, classical and

Figure 2. Robotic navigation systems

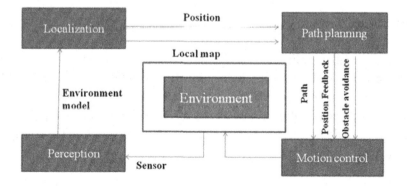

Figure 3. Classification based on environments

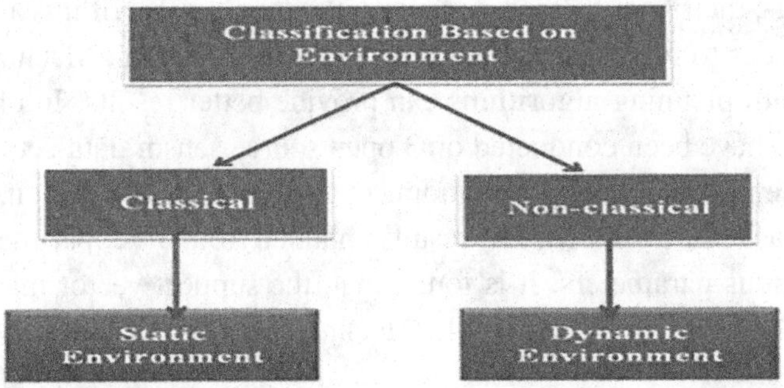

non-classical (Russell & Norvig, 2016). A classical planning environment is fully observable and can be categorized as being deterministic, finite, static or discrete. In contrast, the non-classical planning environment is partially observable and involves non-deterministic or stochastic environments.

Classification of environments is illustrated in Figure 3.

Perception

Perception involves sensing the environment and assessing its situation (Kunze, Hawes, Duckett & Hanheide, 2018). It is a technique that observes information about its environment through sensors which provide the available information to the robot. Based on that information, robots completely sense the environment and perform accordingly. There are different types of environmental sensors are categorized as being respectively active, passive or ranged. These sensors enable the robot to detect objects and obstacles that come within its path.

Localization

Localization defines where the robot is exactly placed and makes complete sense of the environment using its sensors (Galceran & Carreras, 2013). A robot can recognize extensively what is around it and detect the obstacles, and other objects in space. It performs this function by specific locations that can be identified by the coordinate values (x, y).

Path planning

Path planning can be defined as finding a path from the initial position (I) to a goal position (G) by passing through all the obstacles in the configuration space (C-space) (Russell & Norvig, 2016).

A robot can navigate in the free space (C_{free}) or avoid several obstacles in the occupied space (C_{obs}). So, path (p) of distance (d) can be calculated using the two coordinate points (x1, y1) and (x2, y2) using Euclidean distance. It is used to calculate the distance between two points in the configuration space as expressed in equation 1.

$$d(p) = \sum_{i=1}^{n-1} \sqrt{\left(x2 - x1\right)^2 + \left(y2 - y1\right)^2} \tag{1}$$

If a robot has complete knowledge about its environment, informative path planning can be accomplished. As the environment is full of ambiguous scenarios such as unstructured backgrounds, uncertain factors and unsafe motions, path planning is becoming more challenging (Russell & Norvig, 2016). Due to certain circumstances, searching a collision-free path from initial state to goal state is a tedious task faced by the robot. In these situations, finding a feasible path is a more challenging task. Such types of problems can be specified as the path planning. In order to generate a suitable path, various path planning algorithms are utilized (Yang, Qi, Song, Xiao, Han & Xia, 2016). The in-depth analysis of algorithms is provided in subsequent sections.

Motion Control

A robot, in order to achieve a goal, can encounter several type of obstacles. Robots navigate in any direction (left, right, forward or reverse) based on the feedback obtained from previous actions. A controller is a decision-maker which makes decisions based on the previous action and controls the speed, direction, and start-and-stop positions (Aparanji, Wali & Aparna, 2017). Based on the controller signals, the actuator takes the direction of movement while avoiding collision.

ROBOTIC APPLICATIONS

Currently, due to flexibility, cost efficiency and productivity, robots have played a vital role in various disciplines and fields. In this chapter, we consider specific families of robots, particularly those with industrial, agricultural, rescue, military and medical applications (Khan & Khan, 2017; Kumari, 2014; Priya, 2014).

Industrial Robot

Industrial robot play a vital role in manufacturing processes which comprise welding, assembly, processing, dispensing and handling material operations (Khan & Khan, 2017). The need for automation in the food, pharmaceutical and automobile industries is constantly increasing leading to industrial robots being widely adopted. This has lead in decreasing the time and complexity involved in manufacturing high-quality products. Recently, the continuous technical improvement in flexibility, accuracy, security and simplification of industrial robot applications has increased the number of installed robots (Perumaal & Jawahar, 2013).

Service Robot

Agricultural Robot

Agriculture is a process involving a chain rule of systematic, repetitive and time dependent tasks. In this present day and age, robots can assist in performing agricultural operations like tilling, soil analysis, seeding, transplanting, crop scouting, pest control, weed removal and harvesting (Ruiz, 2017).

Rescue Robot

Robots can help rescuers in scenarios of natural disasters and other calamities by getting to the bottom of the rubble or to the top of a mountain faster and far more efficiently. Robots could be used to encounter various obstacles in the real-world scenarios which include natural disasters, man-made disasters, earthquakes, tsunami, volcanic eruptions and transportation problems. Some of the most stable and robust of rescue robots include Unmanned Ground Vehicles (UGVs), Unmanned Aerial Vehicles (UAVs), Unmanned Underwater Vehicles (UUVs) and Unmanned Surface Vehicles (USVs) (Priya, 2014). They help rescuers in various situations where humans are incapable of doing so.

Military Robot

Military robots are used to handle complex tasks which are difficult for humans to handle. At present, military robots have been utilized by many military organizations. The military robots are equipped with an integrated system that includes video screens, sensors, gripper and cameras. An example of robots used in the military field are daksh, goalkeeper, packbot and marcbot (Kumari, 2014).

Medical Robot

Robot are employed in the medical field for taking care of a patient's pulse, scanning vital signs, taking pictures and reading case notes or even performing surgery (Khan & Khan, 2017). The most ubiquitous medical robot operates and performs surgery, treatment and diagnosis. Robots can additionally help patients rehabilitate from serious conditions. These include those utilized in tele-presence, surgical assistance, rehabilitation, medical transportation, robotic prescription dispensing systems, sanitation and disinfection (Lu & Hsu, 2015).

Figure 4. Planning model

PLANNING

Planning is arranging a sequence of actions that leads to the goal. An agent computes a plan from an initial state to goal state given a set of possible primitive actions. Planning is represented by a sequence of actions, initial state and goal state (Russell & Norvig, 2016). An example of a planning model is illustrated in Figure 4.

Planning problems can be represented in two manners, namely, a search based problem-solving agent and a logical planning agent (Russell & Norvig, 2016).

- Problem-solving agents find a possible sequence of actions that lead from the initial state to the designated goal state. The problem-solving agent is also called the goal-based agent.
- Logic based agents, also called knowledge-based agents, can select the actions based on explicit representations of previous conditions. The effect of an action specifies the resulting state.

The representation of planning problems can be performed using state space search models which are defined as states, actions and goals (Russell & Norvig, 2016). The employment of the state space search model helps in searching a state with all possible combination of actions that satisfies both preconditions and effects. The benefit of the state space search model is that it is possible to search in either direction, that is, forward from the initial state and backward from the goal state. The basic representation of classical planners is done by considering the following three categories.

1. **Representation of Initial State:** The initial state is the present state for solving a planning problem. The planner decomposes the environment into logical conditions and represents a state as conjunction of positive literals.

For example, *known - unknown* represents the state of a utility-based agent. In this agent, known and unknown utility describes how "happy" the agent is. An agent can interact with either fully observable (known) or partially observable (unknown) environments (Russell & Norvig, 2016).

2. **Representation of Goal State:** The goal state can be satisfied through the execution of appropriate actions (Russell & Norvig, 2016).
3. **Representation of Actions**: An action is specified in terms of preconditions and effects. It ensures that an agent satisfies both the conditions before the action can be executed (Russell & Norvig, 2016).

For instance,

Action: (Pick (Obj, Grasp),
Precondition: Empty (Grasp),
Effect: InGrasp (Obj), $\neg$ Empty (Grasp))

When picking up an object, an agent should not hold the object because in the resulting state, the agent holding an object cannot be empty.

The Language of Planning Problems

There are different planning languages available to address a variety of problems (Russell & Norvig, 2016). Some of these are Stanford Research Institute Problem Solver (STRIPS), Action Description Language (ADL), Planning Domain Definition Language (PDDL), First order logic (FOL), Description logic (DL), Linear Temporal Logic (LTL) and Ontology. All the languages are expressive in describing planning problems. The hierarchy of the planning languages is illustrated in Figure 5.

PDDL is a standard language for planning and provides better syntax for representing the problems when compared to STRIPS and ADL (Milicic, 2008).

Description Logic (DL) is a low-level logic-based knowledge representation formalism used to represent and reason about the conceptual knowledge in a structured and semantically well-known manner (Gayathri & Uma, 2019). DL is an extension of first order logic (FOL) which can only refer to a single subject. The extension of DL is linear temporal logic (LTL) which can be expressed as a set of possible actions based on time. Low level based knowledge representation techniques are illustrated in Figure 6.

Ontology is a high-level knowledge representation technique comprising spatial, temporal and semantic knowledge (Gayathri & Uma, 2018). These knowledge representation techniques will help in predicting the reasoning information (knowledge) and understanding the environment in a better way. High level based knowledge representation techniques are illustrated in Figure 7.

Figure 5. Planning languages

Figure 6. Low level representation

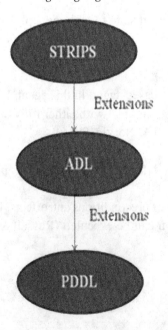

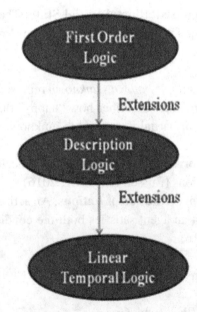

Figure 7. High level representation

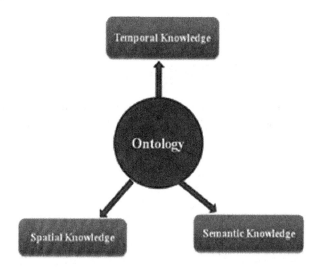

Example

A simple block world example can be expressed by using STRIPS' planners as is shown in Figure 8. The first block (Figure 8) indicates an initial state and the second block indicates the designated goal state.

1. **Initial State:** In the initial state all four blocks are present in the expression below:

Init (On(A, Table) ∧ On(B, Table) ∧ On(C, Table) ∧ On(D, Table) ∧ Block(A)

∧ Block(B) ∧ Block(C) ∧ Block(D) ∧ Clear(A) ∧ Clear(B) ∧ Clear(C) ∧ Clear(D)

2. **Goal State:** A goal state is reached when block B is on A and block D is on C.

Goal (On(A,B) ∧ On(C,D))

3. **Action:** The action of moving block B from the table can be specified in terms of precondition and effect. In the precondition state, the block B should be present on the table and it should not be placed on block A. The effect state is reached when block B is on block A.

Figure 8. Initial and Goal states of Simple block world problem

Action (Move(B,Table),

PRECOND: On(B,Table) ∧ !On(B,A) ∧ Clear(B),

EFFECT: !On(B,Table) ∧ On(B,A) ∧ Clear(Table)).

While block D is moved (lifted) from the table, it should satisfy the precondition that block D should be present on the table and it should not be placed on block C. Resulting state is reached when block D is placed on block C.

Action (Move(D,Table),

PRECOND: On(D,Table) ∧ !On(D,C) ∧ Clear(D)

EFFECT: !On(D,Table) ∧ On(D,C) ∧ Clear(Table)

PATH PLANNING ALGORITHMS

This section explains the concept of path planning algorithms that are related to robotics. The important aspect of path planning is that navigation behavior of the robot is to be controlled to generate a collision-free path from an initial position to goal position. The ojective of path planning is to reach the goal position with the minimum time and distance. Path planning algorithms use the knowledge of the robot that can facilitate safe and effortless movement of the robot after tackling the obstacles in the environment (2D or 3D) (Yang, Qi, Song, Xiao, Han & Xia, 2016). Recently, to overcome the path planning problems under different environments, there have been various path planning algorithms proposed by many researchers. Generally, path planning algorithms can be categorized as local path planning, and global path planning (Zafar & Mohanta, 2018).

Local path planning is performed in a partially known or unknown navigational environment. However, global path planning has complete knowledge about the navigational environment and generates a path based on the well-known environment. The classification of path planning algorithms based on various approaches is illustrated in Figure 9.

Classical Approach

The classical approach uses general navigation algorithms which enable a robot to move autonomously in static environments.

Sampling Based Method

The sampling based algorithm finds the shortest path in a robotic environment (Yang, Qi, Song, Xiao, Han & Xia, 2016). The fundamental idea of this algorithm is a random sampling technique which requires prior knowledge about its environment. Sampling-based motion planning is used to generate random samples in the workspace. Accordingly, the collision detector is used to check whether the samples are

Figure 9. Classification of path planning algorithms based on classical & heuristic approaches

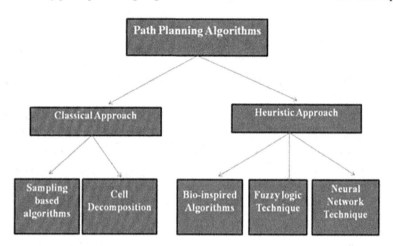

in free space C_{free} or obstacle space C_{obs}. If the samples are in obstacle space C_{obs}, they are discarded. The free space sample C_{free} is consequently considered to connect the nodes. Sampling-based algorithms are used to solve different kinds of motion planning problems in lesser time. Examples of the sampling-based methods include Probabilistic Road Map (PRM), Artificial Potential Field (APF), and the Voronoi and Rapidly Exploring Random Tree (RRT) (Yang, Qi, Song, Xiao, Han & Xia, 2016). The variations of RRT are Informed RRT*, Dynamic Domain Rapidly Exploring Random Tree (DDRRT) and Rapidly Exploring Random Graph (RRG). These planners try to find a path from the initial configuration to goal configuration in free space. The analysis of sampling based elements is presented in Table 1.

Cell Decomposition Method

The objective of the cell decomposition method is to decompose the configuration space (C-space) into multiple cells (Scheurer & Zimmermann, 2011). The space is occupied by the obstacles denoted as obstacle space $C_{obs.}$ Free space is denoted as C_{free}. These cells are connected to the corresponding

Table 1. The elements of sampling based algorithms

Method	Techniques Used	Shortcomings	Advantages	Research Work Done In Robot Path Planning
RRT	Random Sampling Approach	Single path generation, Unsuitable for dynamic environment	Low time complexity, Fast searching ability	(Nieto, Slawinski, Mut & Wagner, 2010)
Artificial Potential	Random Sampling Approach	Local optimum, oscillatory motion, Unsuitable for dynamic environment	Low time complexity, Fast Convergence	(Yang, Qi, Song, Xiao, Han & Xia,2016)
Voronoi	Random Sampling Approach	Incomplete representation, Non convergence, Unsuitable for dynamic environment	Easy to implement and removes collision checking function	(Yang, Qi, Song, Xiao, Han & Xia, 2016)
PRM	Random Sampling Approach	Generates redundant paths, Non optimal, Unsuitable for dynamic environment	Suitable for complex environments	(Rantanen, 2011)

Table 2. Analysis of each element in Cell decomposition method

Method	Techniques Used	Shortcomings	Advantages	Research Work Done in Robot Path Planning
Dijikstra's Algorithm	Breath First Search method	High time complexity, Unsuitable for dynamic environment	Easy to implement in various environments	(Verscheure, Peyrodie, Makni, Betrouni, Maouche & Vermandel, 2010)
A*	Heuristics Function	High time complexity, Nonsmoothness, Unsuitable for dynamic environment	Fast searching ability, Simplicity, modifiability	(Yang, Qi, Song, Xiao, Han & Xia, 2016)
D* (Dynamic A*)	Sensor based algorithm	Unrealistic distance	Fast searching ability, deals with dynamic environments	(Yang, Qi, Song, Xiao, Han & Xia, 2016)

adjacent cells to form the nodes of a connectivity graph G. A feasible path connecting start node and goal node exists in graph G. Such a cell path in G is known as the channel. However, the cell decomposition method is similar to sampling-based techniques which tries to find the path without building the explicit configuration space (Scheurer & Zimmermann, 2011). This method is applicable in robot path planning in high-dimensional spaces. A robot is permitted to navigate from each and every cell within configuration space (C-space) to find the shortest path without collision. The details of each algorithm, advantages and limitations are presented in Table 2.

Cell decomposition methods can be classified into two categories namely exact and approximate decomposition (Scheurer & Zimmermann, 2011).

- The exact cell decomposition techniques can find a path exactly considering the configuration space C_{free} that exists in graph G. In this approach, a path is searched along with the set of cells.
- Approximate cell decomposition techniques use rectangloids that partition the configuration space into cells and determines the adjacent cells easily when compared to exact method. The decomposition cells approximately cover the configuration free space C_{free}.

Heuristic Based Methods

Classical methods used in motion planning algorithms are not useful if the environment is dynamic (Zafar & Mohanta, 2018). Moreover, a robot collides with obstacles and it is difficult to execute a trajectory when the environment model and robot motion are uncertain. Hence, heuristic based methods are used to overcome this kind of problem. This technique deals with the problem of uncertain environment (dynamic) during the motion.

Bio-inspired Algorithms

The idea behind bio-inspired algorithm is that it is a meta-heuristic method that can efficiently deal with dynamic environment (Qu, Xing & Alexander, 2013). This method can solve different kinds of problems like NP-hard problem and optimization problems. The evolutionary algorithm, a category of Bio-inspired algorithm, stems from the behavior of certain species. Evolutionary algorithms are divided into different categories as illustrated in Table 3.

Table 3. Analysis of different elements in bio-inspired algorithms

Method	Techniques Used	Advantages	Research Work Done in Robot Path Planning
GA	Population based optimization method	Can solve NP-hard & multi objectives problems. It is flexible in dynamic environment	(Qu, Xing & Alexander, 2013)
ACO	Probabilistic technique	It deals with multi-objectives & continuous planning problems	(Chen, Kong & Fang, 2013)
PSO	Computational method	It performs faster than GA	(Yang, Qi, Song, Xiao, Han & Xia, 2016)
SFLA	Meta-heuristic optimization method	Achieves faster convergence, and performs better than PSO	(Yang, Qi, Song, Xiao, Han & Xia, 2016)
MA	Meta-heuristic search method.	Performs better than GA, and achieves Smoothness	(Yang, Qi, Song, Xiao, Han & Xia, 2016)

Fuzzy Logic Techniques

Fuzzy logic, which provides decisions as an output in the form of degrees of truth rather than true or false, is widely used in artificial intelligence. The use of fuzzy logic for autonomous robot navigation ensures efficient and safe navigation of mobile robots. Fuzzy logic provides the possible solutions to path planning problems by enabling local and global navigation of mobile robots. One of the advantages of using fuzzy logic is that it is free from local minimization issues. Fuzzy-based navigation is a reactive planning technique, wherein the immediate position and distances from obstacles are considered in computing the immediate move, without being concerned about the future (Omrane, Masmoudi & Masmoudi, 2016).

Fuzzy logic monitors a robot in three directions, concretely, reaching the target point, collision avoidance and automatic execution. The fuzzy logic controller comprises four blocks namely fuzzification, inference, rule base and defuzzification. Fuzzification transforms real value into membership function, that is, $\mu_A : X \rightarrow [0, 1]$ providing fuzzy control. Inference involves the reasoning process and fuzzification rule. The fuzzy rule based system is presented in the following form:

If <antecedents> Then <conclusion>

Advanced fuzzy sets in fuzzy logic systems are Type 1 and Type 2 (Mendel, 2017). These models are used in fuzzy systems and are represented in terms of antecedents and consequences. The type 1 fuzzy system cannot handle uncertainties. However, expanded type 2 fuzzy sets can handle uncertainties because they can be modeled to minimize their effects.

A membership function of A is defined as $\mu_A(x)$ which is called type 1 fuzzy set that is expressed as

$$A = \{(x, \mu_A (x)) \mid x \in X\} \tag{2}$$

The fuzzifier maps each input set into a membership element as [either 0 or 1] when a type 1 fuzzy set is used. Similarly, the type 2 fuzzy set is used when the circumstances are uncertain and it is difficult to determine its membership function. The type 2 fuzzy set is widely used in a number of real-world applications such as industrial robot control, mobile robot control, ambient intelligent environment, decision-making, embedded agents, pattern classification, quality control, spatial query and wireless communication.

The final objective of defuzzification block is to transform the subset of outputs using inference and rule-based systems (Mendel, 2017).

Neural Network

The idea behind neural Networks is to imitate biological nervous systems. The basic mathematics model of neural network consists of three layers, namely, the input layer, hidden layer and output layer. Neurons in each layer are interconnected by edges. Neural networks are widely used in the field of mobile robot path planning due to their properties of nonlinear mapping, learning, optimization, system identification, pattern recognition and parallel processing (Shamsfakhr & Sadeghibigham, 2017).

There are different types of neural networks that can be used in robotic path planning applications. They comprise the following, Hopfield neural network (HNN), feed-forward neural network (FFNN), recurrent neural networks (RNN), convolutional neural networks (CNN) and Radial Basis Function (RBF) neural networks. Additionally, various types of optimization techniques have been used in robotic applications such as the gradient descent method and Levenberg-Marquardt (Yu & Wilamowski, 2010). These techniques are used to avoid oscillatory motions and collisions. The techniques of different types of neural networks is presented in Table 4.

MACHINE LEARNING TECHNIQUES

In order to find the path, a robot has to understand its environment wherein it performs autonomous navigation. In such a case, learning and a decision making capability is required for a robot to efficiently navigate. To achieve this capability, a system must be able to observe its environment with the help of machine learning techniques (Aparanji, Wali & Aparna, 2017). This section discusses how machine learning techniques can utilize data to derive a decision-making strategy for autonomous robotic systems.

Table 4. Different types of neural networks

Method	Techniques Used	Shortcomings	Advantages	Research Work Done in Robot Path Planning
NN	Conjugate gradient technique	It is impossible to store the previous time step value	Achieves goal in lesser time with low number of features	(Shamsfakhr & Sadeghibigham, 2017)
RNN	Levenberg Marquart technique	Long-term time dependence.	Enables storing the previous time step value & estimates the position precisely.	(Brahmi, Ammar & Alimi, 2013)
LSTM	Softmax function.	It is not possible to calculate the distance precisely when object is far away.	LSTM generates a path shorter than MLP. It stores the long-term interval values.	(Park, Kim, Kang, Chung & Choi, 2018)
CNN	Principle component analysis (PCA), Relu function	Difficult to solve the decision –making problems and perception related issues.	CNN can adapt to various environments	(Liu, Zheng, Wang, Zhao &Li, 2017)

The concepts behind machine learning techniques that are related to path planning are also explained. Machine learning is a sub-field of artificial intelligence that encompasses the area of deep learning. Machine learning enables a machine to learn from data and effectively perform robot navigation. Two sub-fields of machine learning that are relevant to robotic applications are supervised learning and reinforcement learning. These techniques are used to extract the knowledge from robotic data and classify the robotic environments. The classification of machine learning techniques is illustrated in Figure 10.

Supervised Learning

Supervised learning can be used to train using the input data and provide output given the whole training sample (Otte, 2008). In supervised learning, class labels or target values are known for each sample. The most common supervised learning models are Naïve Bayes (NB), support vector machine (SVM), logistic regression (LR) and multi-layer perceptron (MLP). These algorithms have been used to map the unseen sample features to their correct labels or target values.

Naïve Bayes classifier (Vijayan, Singanamala, Nair, Medini, Nutakki & Diwakar, 2013) is a supervised learning algorithm which is based on the Bayes theorem and the probabilistic statistical classifier. Another significant statistical classifier is logistic regression (LR). It performs predictive analysis where the outcome is determined by one or more independent variables. The multi-layer perceptron (MLP) is a feed forward neural network model which maps sets of input data associated with a set of appropriate output data. SVM is one of the most prevalent supervised classifiers which can find a decision plane that separates the objects belonging to different classes using a margin with maximum width.

Reinforcement Learning

In a reinforcement learning algorithm, an agent interacts with the environment and an imminent feedback is obtained based on actions taken at each state at t time step(s). In order to find the optimal path, an agent must balance exploration and exploitation of the state space that maximizes the accumulated reward by interacting with the environment. An agent modifies the behavior or policy based on previous actions taken, awareness of the states and the rewards. These types of temporal sequence problems can be solved by applying reinforcement learning techniques (Altuntas, Imal, Emanet & Ozturk, 2016).

Figure 10. Classification of machine learning techniques

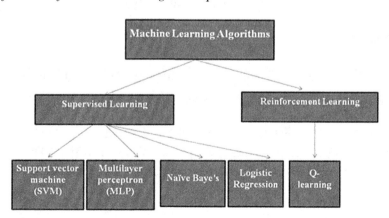

Robotic motion planning problems and predictions can be solved by a reinforcement learning algorithm namely the Q-learning approach, which is an action-value optimization function that is based on the Bellman optimality equation (Du, Zhai & Lv, 2016). Q-learning is an attempt to recognize a reward from previous actions and predict the future reward by accounting from previous trails.

In this section, we analyzed the performance of existing walls following a robot navigation data-set using supervised machine learning algorithms. The details of existing data-sets are provided in the following section.

EXPERIMENTAL RESULTS AND ANALYSIS

In this section, we conduct experiments related to robot navigation performed utilizing an open source data-set. Mining the information from the data collected by robot is termed robotic data mining. The focus of this work is to extract the knowledge from the robotic data using machine learning techniques to identify (classify) robotic directions. Supervised machine learning classifiers algorithms namely Naïve Bayes (NB), logistic regression (LR), multi-layer perceptron (MLP) and support vector machine (SVM) are used in classifying robotic data.

Data-set Description

Open source data-set named as "wall following robot navigation data-sets" is experimentally analyzed. The data-set contains three files. These three data-sets are considered for analysis. The first data-set labeled sensor_reading24 assumes readings from 24 ultrasound sensors. The second data-set labeled sensor_reading4 assumes readings from 4 ultrasound sensors and the third one labeled sensor_reading2 assumes readings from 2 ultrasound sensors. All of them are taken into account when the SCITOS G5 robot (Bley & Martin, 2010) navigates through the room following the wall in a clockwise direction. The corresponding class label in all the files are referred to as; 'move forward', 'slight right turn', 'sharp right turn' and 'slight left turn'. The SCITOS G5 robot navigates inside the room in all the directions 4 times (rounds). There are in total 5403 densely sampled instances. On an average, each scene serve as different instances for robots to navigate. The number of instances for each of the four class label is referred to as 2205, 2097, 826, 328 distribution, respectively. The corresponding percentage for each of the four class label instances is estimated based on total instances (5403) and are specifically denoted as 40.41%, 38.43%, 15.14% and 6.01% respectively (Dash, Nayak & Swain, 2015).

Results With Validation Data-Set

The accuracy of SVM is less when compared with the MLP algorithm (Gopalapillai, Gupta & Tsb, 2017). The LR model yields accuracy of 99.8% for sensor_reading4 and sensor_reading2. SVM achieves a higher accuracy of 90.08% for sensor_reading24 data-set when compared with the other algorithms. Accuracy of LR and NB is less with sensor_reading24 data-set. NB yields accuracy of 90% with sensor_reading2 and sensor_reading4 data-set which is very less compared to other classifiers. The comparison of classification algorithms when applied on three different data-sets is given in Table 5.

Table 5. Accuracy achieved using ML classifier algorithms

Data-set	Attributes	MLP Accuracy [LR=0.3, M=0.2, H=10]	SVM Accuracy [C=1, G=0]	Naïve Bayes (NB)	Logistic Regression (LR)
Sensor_reading24	24	89.69	90.28	52.45	70.49
Sensor_reading4	4	97.41	92.43	89.11	99.81
Sensor_readign2	2	97.52	93.62	90.57	99.85

ROBOTIC SIMULATION SOFTWARE

Software simulators play a vital role in autonomous robotic systems. They provide a safe, fast and efficient testing environment. Robotic simulation software systems can be classified into two categories, namely, commercial and open source software. The commercial software packages are webots and Easy-Rob (Staranowicz & Mariottini, 2011). Presently, a number of free open source robotic simulation software are available. More widely used open source software packages are Gazebo, V-rep, OpenRave, Simbad, LTLMoP, and OMPL. Various robotic simulators have been analysed and details are given in Table 6.

FUTURE RESEARCH DIRECTIONS

In future, path-planning algorithms with deep-learning approach to derive effective decision making strategy for autonomous robotic systems will be integrated. The long short term memory (LSTM) model could be utilized to supervise the deep learning approach which can predict the imminent trajectories based on the action of an agent in previous time instances. The LSTM model could then learn robot

Table 6. The individuality of autonomous robotic simulation software systems

Simulator Name	Language Used	Platform Supported	Extensibility/ External API	Open Source	Research Work Done
Simbad	Java	Mac OS, Windows XP, Linux	Plugins(Java)/ Eclipse	Yes	(Nguyen, Eguchi & Hooten, 2011)
Gazebo	C++	Linux, Mac OS, Windows	Plugins (C++)/ROS	Yes	(Kunze, Dolha, Guzman & Beetz, 2011)
Webots	C++	Linux, Mac OS, Windows	Plugins (C/C++)/ ROS	No	(Ferreira, Vargas & Oliveira, 2014)
OpenRAVE	C++, Python	Linux, Mac OS, Windows	Plugins(C++)/ Matlab/Python	Yes	(Srivastava, Riano & Abbeel, 2014)
LTLMoP	Python	Linux, Mac OS, Windows	Java/Python 2.7	Yes	(Finucane, Jing & Kress-gazit, 2010)
OMPL	Python	Linux, Mac OS, windows	ROS	Yes	(Sucan, Moll & Kavraki, 2012)
V-Rep	C/C++/Python	Linux, Mac OS, windows	ROS	Yes	(Rohmer, Singh & Freese, 2013)

navigation data, reproduce long sequences and predict their future trajectory. Integration of this approach would significantly help in predicting the next direction of movement in future instances and improve the efficiency of path planning algorithms with minimal computational complexity and memory overhead.

CONCLUSION

This chapter deals with details concerning path-planning and machine learning algorithms that are related to robotic navigation where the authors contextualized software robotic systems. Planning representation and path planning languages used in robotic system have been discussed in detail. Various types of path planning and machine learning algorithms related to the robotics have been explained. A short description of various robotic simulators mentioning their special features has been provided. The performance of various machine learning algorithms was analyzed by considering an open source robot navigation data-set. The results show that to enable the robots to navigate in a collision free path, the machine learning algorithms can be enhanced and reinforced by utilizing path planning algorithms.

REFERENCES

Altuntas, N., Imal, E., Emanet, N., & Ozturk, C. N. (2016). Reinforcement learning-based mobile robot navigation. *Turkish Journal of Electrical Engineering and Computer Sciences*, *24*(3), 1747–1767. doi:10.3906/elk-1311-129

Aparanji, V. M., Wali, U. V., & Aparna, R. (2017). Robotic Motion Control using Machine Learning Techniques. In *International Conference on Communication and Signal Processing* (pp. 6–8). Academic Press. 10.1109/ICCSP.2017.8286579

Bley, A., & Martin, C. (2010). *SCITOS G5 –A mobile platform for research and industrial applications*. Retrieved from http://download.ros.org/data/Events/CoTeSys-ROS-School/metralabs.pdf

Brahmi, H., Ammar, B., & Alimi, A. M. (2013). Intelligent path planning algorithm for autonomous robot based on recurrent neural networks. *Advanced Logistics and Transport (ICALT), 2013 International Conference on*, 199–204.

Chen, X., Kong, Y., Fang, X., & Wu, Q. (2013). A fast two-stage ACO algorithm for robotic path planning. *Neural Computing & Applications*, *22*(2), 313–319. doi:10.100700521-011-0682-7

Dash, T., Nayak, T., & Swain, R. R. (2015). Controlling Wall Following Robot Navigation Based on Gravitational Search and Feed Forward Neural Network. *Proceedings of the 2nd International Conference on Perception and Machine Intelligence*, 196–200. 10.1145/2708463.2709070

Du, X., Zhai, J., & Lv, K. (2016). Algorithm Trading using Q-Learning and Recurrent Reinforcement Learning. *Positions*, 1–7.

Ferreira, G. B. S., Vargas, P. A., & Oliveira, G. M. B. (2014). An Improved Cellular Automata-Based Model for Robot Path-Planning. In *Towards Autonomous Robotic Systems* (pp. 25–36). Cham: Springer.

Finucane, C., Jing, G., & Kress-Gazit, H. (2010). LTLMoP : Experimenting with Language, Temporal Logic and Robot Control. *Intelligent Robots and Systems (IROS), 2010 IEEE/RSJ International Conference on*, 1988–1993. 10.1109/IROS.2010.5650371

Galceran, E., & Carreras, M. (2013). A Survey on Coverage Path Planning for Robotics. *Robotics and Autonomous Systems, 61*(12), 1258–1276. doi:10.1016/j.robot.2013.09.004

Gayathri, R., & Uma, V. (2018). Ontology based knowledge representation technique, domain modeling languages and planners for robotic path planning : A survey. *ICT Express, 4*(2), 69–74. doi:10.1016/j.icte.2018.04.008

Gayathri, R., & Uma, V. (2019). A Review of Description Logic-Based Techniques for Robot Task Planning. *Integrated Intelligent Computing. Communication and Security, 771*, 101–107. doi:10.1007/978-981-10-8797-4

Gopalapillai, R., Gupta, D., & Tsb, S. (2017). Pattern Identification of Robotic Environments using Machine Learning Techniques. *Procedia Computer Science, 115*, 63–71. doi:10.1016/j.procs.2017.09.077

Khan, M. S. S., & Khan, A. S. (2017). A Brief Survey on Robotics. *International Journal of Computer Science and Mobile Computing, 6*(9), 38–45.

Kumari, S. (2014). Military Robots-A Survey. *International Journal of Advanced Research in Electrical. Electronics and Instrumentation Engineering, 3*(3), 77–80.

Kunze, L., Dolha, M. E., Guzman, E., & Beetz, M. (2011). Simulation-based Temporal Projection of Everyday Robot Object Manipulation. *International Foundation for Autonomous Agents and Multiagent Systems*, 107–114.

Kunze, L., Hawes, N., Duckett, T., Hanheide, M., & Krajnik, T. (2018). Artificial Intelligence for Long-Term Robot Autonomy : A Survey. *IEEE Robotics and Automation Letters, 3*(4), 4023–4030. doi:10.1109/LRA.2018.2860628

Liu, C., Zheng, B., Wang, C., Zhao, Y., Fu, S., & Li, H. (2017). CNN-Based Vision Model for Obstacle Avoidance of Mobile Robot. *MATEC Web of Conferences, 139*, 4–7. 10.1051/matecconf/201713900007

Lu, J. M., & Hsu, Y. L. (2015). Telepresence robots for medical and homecare applications. In M. Weijnen (Ed.), Contemporary Issues in Systems Science and Engineering (pp. 725–735). Academic Press. doi:10.1002/9781119036821.ch21

Mendel, J. M. (2017). *Uncertain Rule-Based Fuzzy Systems. Introduction and new directions*. Springer International Publishing. doi:10.1007/978-3-319-51370-6

Milicic, M. (2008). *Action*. Time and Space in Description Logics.

Nguyen, H., Eguchi, A., & Hooten, D. (2011). *Search of a Cost Effective Way to Develop Autonomous Floor Mapping Robots. In Robotic and Sensors Environments* (pp. 107–112). ROSE.

Nieto, J., Slawinski, E., Mut, V., & Wagner, B. (2010). Online Path Planning based on Rapidly-Exploring Random Trees. In *IEEE International Conference on Industrial Technology* (pp. 1451–1456). IEEE. 10.1109/ICIT.2010.5472492

Omrane, H., Masmoudi, M. S., & Masmoudi, M. (2016). Fuzzy Logic Based Control for Autonomous Mobile. *Computational Intelligence and Neuroscience.* PMID:27688748

Otte, M. W. (2008). A Survey of Machine Learning Approaches to Robotic. *The International Journal of Robotics Research, 5*(1), 90–98.

Park, S. H., Kim, B., Kang, C. M., Chung, C. C., & Choi, J. W. (2018). Sequence-to-sequence prediction of vehicle trajectory via LSTM encoder-decoder architecture. *2018 IEEE Intelligent Vehicles Symposium (IV),* 1672-1678. 10.1109/IVS.2018.8500658

Perumaal, S. S., & Jawahar, N. (2013). Automated Trajectory Planner of Industrial Robot for Pick-and-Place Task. *International Journal of Advanced Robotic Systems, 10*(2), 100. doi:10.5772/53940

Priya, S. (2014). Rescue Robot-A Study ARE USED. *International Journal of Advanced Research in Electrical. Electronics and Instrumentation Engineering, 3*(3), 158–161.

Qu, H., Xing, K., & Alexander, T. (2013). Neurocomputing An improved genetic algorithm with co-evolutionary strategy for global path planning of multiple mobile robots $. *Neurocomputing, 120,* 509–517. doi:10.1016/j.neucom.2013.04.020

Rantanen, M. T. (2011). *A Connectivity-Based Method for Enhancing Sampling in Probabilistic Roadmap Planners.* Academic Press. doi:10.100710846-010-9534-4

Rath, M. (2018). Big Data and IoT-Allied Challenges Associated With Healthcare Applications in Smart and Automated Systems. *International Journal of Strategic Information Technology and Applications, 9*(2), 18–34. doi:10.4018/IJSITA.2018040102

Rath, M., Pati, & Patanayak, B. K. (2019). An Overview on Social Networking: Design, Issues, Emerging Trends, and Security, "Social Network Analytics: Computational Research Methods and Techniques. Elsevier.

Rath, M., & Kumar, P. (2019). Security Protocol with IDS Framework Using Mobile Agent in Robotic MANET. *International Journal of Information Security and Privacy, 13*(1), 46–58. doi:10.4018/IJISP.2019010104

Rath, M., Pati, B., Panigraphi, C. R., & Sakar, J. L. (2019). QTM: A QoS Task Monitoring System for Mobile Ad hoc Networks. In P. Sa, S. Bakshi, I. Hatzilygeroudis, & M. Sahoo (Eds.), *Recent Findings in Intelligent Computing Techniques* (Vol. 707). Singapore: Springer. doi:10.1007/978-981-10-8639-7_57

Rohmer, E., Singh, S. P. N., & Freese, M. (2013). V-REP : a Versatile and Scalable Robot Simulation Framework. *Intelligent Robots and Systems (IROS), 2013 IEEE/RSJ International Conference on,* 1321-1326.

Ruiz, M. P. (2017). Task based agricultural mobile robots in arable farming : A review. *Spanish Journal of Agricultural Research, 15*(1), 1–19.

Russell, S. J., & Norvig, P. (2016). *Artificial intelligence: a modern approach.* Pearson Education Limited.

Scheurer, C., & Zimmermann, U. E. (2011). Path Planning Method for Palletizing Tasks using Workspace Cell Decomposition. In *IEEE International Conference on Robotics and Automation* (pp. 1–4). IEEE. 10.1109/ICRA.2011.5980573

Shamsfakhr, F., & Sadeghibigham, B. (2017). A neural network approach to navigation of a mobile robot and obstacle. *Turkish Journal of Electrical Engineering and Computer Sciences, 25*(3), 1629–1642. doi:10.3906/elk-1603-75

Srivastava, S., Riano, L., & Abbeel, P. (2014). Combined Task and Motion Planning Through an Extensible Planner-Independent Interface Layer. *Robotics and Automation*, 639–646.

Staranowicz, A., & Mariottini, G. L. (2011). A Survey and Comparison of Commercial and Open-Source Robotic Simulator Software. In *Proceedings of the 4th International Conference on PErvasive Technologies Related to Assistive Environments* (p. 56). ACM. 10.1145/2141622.2141689

Sucan, I. A., Moll, M., & Kavraki, L. E. (2012). The open motion planning library. *IEEE Robotics & Automation Magazine, 19*(4), 72–82. doi:10.1109/MRA.2012.2205651

Verscheure, L., Peyrodie, L., Makni, N., Betrouni, N., Maouche, S., & Vermandel, M. (2010). Dijkstra's Algorithm Applied to 3D Skeletonization of the Brain Vascular Tree: Evaluation and Application to Symbolic. *Engineering in Medicine and Biology Society (EMBC), 2010 Annual International Conference of the IEEE*, 3081–3084.

Vijayan, A., Singanamala, H., Nair, B., Medini, C., Nutakki, C., & Diwakar, S. (2013). Classification of Robotic Arm Movement using SVM and Naïve Bayes Classifiers. In *International Conference on Innovative Computing Technology* (pp. 263–268). Academic Press.

Yang, L., Qi, J., Song, D., Xiao, J., Han, J., & Xia, Y. (2016). Survey of Robot 3D Path Planning Algorithms. *Journal of Control Science and Engineering*, (5): 22.

Yu, H., & Wilamowski, B. M. (2010). Levenberg–Marquardt Training 12.1. *Industrial Electronics Handbook, 5*(12), 1–16.

Zafar, M. N., & Mohanta, J. (2018). Methodology for Path Planning and Optimization of Mobile Robots: A review. *International Conference on Robotics and Smart Manufacturing, 133*, 141–152. 10.1016/j.procs.2018.07.018

KEY TERMS AND DEFINITIONS

ADL: Action description language.
AI: Artificial intelligence.
APF: Artificial potential field (APF).
CNN: Convolutional neural networks.
DDRRT: Dynamic domain rapidly exploring random tree.
DL: Deep learning; also, description logic.
FFNN: Feed-forward neural network.
FOL: First order logic.

HNN: Hopfield neural network.

LTL: Linear temporal logic.

LTLMoP: Linear temporal logic mission planning.

LR: Logistic regression.

LSTM: Long short-term memory.

MLP: Multi-layer perceptron.

NB: Naïve bayes.

PDDL: Planning domain definition language.

PRM: Probabilistic road map.

RBF: Radial basis function.

RNN: Recurrent neural networks.

RRG: Rapidly exploring random graph.

RRT: Rapidly exploring random tree.

STRIPS: Stanford Research Institute problem solver.

SVM: Support vector machine.

UAV: Unmanned aerial vehicles.

UGV: Unmanned ground vehicles.

USV: Unmanned surface vehicles.

UUV: Unmanned underwater vehicles.

Chapter 4
Machine Learning and Artificial Intelligence:
Rural Development Analysis Using Satellite Image Processing

Anupama Hoskoppa Sundaramurthy
BMS Institute of Technology and Management, India

Nitya Raviprakash
Rashtreeya Vidyalaya College of Engineering, India

Divija Devarla
Rashtreeya Vidyalaya College of Engineering, India

Asmitha Rathis
Rashtreeya Vidyalaya College of Engineering, India

ABSTRACT

This chapter proposes a cost-effective and scalable approach to obtain information on the current living standards and development in rural areas across India. The model utilizes a CNN to analyze satellite images of an area and predict its land type and level of development. A decision tree classifies a region as rural or urban based on the analysis. A summary describing the area is generated from inferences made on the recorded statistics. The CNN is able to predict the land and development distribution with an accuracy of 95.1%. The decision tree predicts rural areas with a precision of 99.6% and recall of 88.9%. The statistics obtained for a dataset of more than 1000 villages in India are cross-validated against the Census of India 2011 data. The proposed technique is in contrast to traditional door-to-door surveying methods as the information retrieved is relevant and obtained without human intervention. Hence, it can aid efforts in tracking poverty at a finer level and provide insight on improving the economic livelihood in rural areas.

DOI: 10.4018/978-1-5225-9687-5.ch004

INTRODUCTION

Around one-fifth of the world's population is afflicted by poverty. India has one of the fastest growing economies in the world but severe poverty is prevalent across many parts of the country. According to the World Bank, as of 2012, 20% of the Indian population is poor out of which 80% is found in rural areas (Zhou, Yang, & Yu, 2013). Data released by the Socio- Economic Caste Census Survey of 2013 reveals that 75% of the rural population or 133.5 million families earn less than Rs.5000 per month (Tewari, 2015). Asset ownership is also reflective of the poor standards of living in such areas. Many houses are kutcha (made from low quality materials) and households do not have refrigerators or motorcycles. In many areas, basic facilities for education, sanitation, health, infrastructure, and transportation are lacking. The Ministry of Statistics stated that in 2013, 22.3% of males and 47.5% of females were illiterate in rural India. (DataGovIn, 2014). Due to a lack of proper education, there is insufficient skilled labor resulting in unemployment in rural areas (Cohen & Medioni, 1999). Rural India relies largely on agriculture, a low-income economy, as a means of livelihood. In addition to the many geographical factors that erratically influence this industry, there is a lack of proper road and railway transport in many localities thus destabilizing agricultural marketing (Jadhav, 2014). There is an urgent need to address sanitation issues such as improper sewage systems and shortage of toilets as well as health deficiencies like malnutrition and stunted growth in rural areas. Most of the people must travel more than 100 km just to avail a medical facility. The Ministry of Health and Family Welfare has stated that although the number of Primary Healthcare Centres (PHCs) and Sub Centres (SCs) have increased rapidly from 2005 to 2015, it still falls short by 5.21% of the overall requirement. (MHFW, 2015). It does not meet the expected standards of the World Health Organization with respect to the population norm (Shafie, Hafiz, & Ali, 2009). Many schemes are in place to propel the development of rural India and this can be advanced further by improving the quality and quantity of information regarding the problems faced by the poverty-ridden population. Measures of economic and living standards of populations influence both research and policy in a critical manner. Understanding these measures leads to informed policy making by governments such as the provision of fundamental services like water and healthcare. Although the economic data available in developing countries has improved in recent years, reliable data collection for poverty analysis is problematic. (DFID, n. d.). Manual surveying techniques are time-consuming, involve a huge expenditure of resources, and are difficult to conduct on a regular basis. For example, Census of India is only carried out once in ten years (MHA, n. d.). The complete set of statistics for every locality in a district is unavailable. For rural areas, systematic approaches for data collection is an arduous task as many regions, being remotely located, are often neglected. Often, the results are summarized at the district level making it difficult to target specific areas under poverty.

In 2015, the United Nations declared that the first Sustainable Development Goal is to eliminate poverty by 2030.(UNDESA, 2018). The main challenge behind achieving this goal is identifying all the regions that fall below the poverty line. If the exact locations are determined, it will help the government and nongovernmental organizations (NGOs) allocate the necessary resources accordingly, to improve the economic well-being of these rural areas. Using machine learning on satellite imagery, this paper proposes a technique to identify rural areas and help track poverty in India. It additionally provides a technique that reduces the manual effort and resources required in obtaining development-based statistics for rural localities.

A convolutional neural network (CNN) model is used to analyze high- resolution satellite images of a region and predict the land distribution. It is a scalable approach that returns a summary for a specified location. The information provided in the summary includes a measure of the area's development, the types of land prevalent, and the availability of nearby amenities such as schools and hospitals. As opposed to the traditional surveying methods, it does not involve any human work-force or resources. It provides statistics to assess the current development or economic scenario. This technique can be used to propose locality-specific improvements to the living standards in under-developed areas.

The rest of this paper is organized as follows; Section two briefly describes the work done in the fields of machine learning, satellite image processing, and rural development. Section three explains the proposed method. Section four illustrates the results obtained from this technique. Conclusions are drawn in Section five.

RELATED WORK

Recent progress in the field of machine learning and image processing, has made it possible to develop innovative technologies to analyze and extract useful information from satellite imagery. The following works represent some of the advancements made in this area. Horton et al. (2016) demonstrate an approach to estimate consumption expenditure and asset wealth using satellite images. Satellite data is taken from five African countries - Nigeria, Tanzania, Uganda, Malawi, and Rwanda and corresponding survey data is obtained from the publicly available living standards measurement studies (LSMS) and demographic and health surveys (DHS). A CNN is trained to detect image features that can explain up to 75% of the differences in local-level economic outcomes. A transfer learning approach, in which the model has been pre-trained on ImageNet, is utilized (Simon, Rodner & Denzler, 2016). The knowledge obtained from learning on the set of images is further tweaked to work with satellite images. The aim is to use night- time light intensity as a proxy for economic development. On providing a daytime satellite image as input, the corresponding night-time light intensity is to be predicted. Survey data can be used to indicate other features visible in daytime satellite imagery, such as roofing material and distance to urban areas that can be associated with expenditure and assist in the prediction (Aslani & Mahdavi-Nasab, 2013).

A study was conducted by Mellander et al. (2015) to determine how appropriate it is to use night-time light intensity as an economic proxy. Research shows that light intensity obtained from satellite imagery is indicative of several variables which includes urbanization, density, and economic growth. With correlation analysis and regressions that are weighted according to geographic regions, the relationship between the two is analyzed. Low- level, geo-coded micro-data is collected for various Swedish residences. This is matched against radiance and saturated light emissions. The correlation between night-time lighting and economic activity is strong enough to make it a good proxy for population and establishment density. The correlation was found to be weaker when it came to wages (Raviprakash et al., 2015).

Varshney et al. (2015) have elaborated on issues faced when selecting locations and planning out rural development activities indicated by remote sensing and satellite image analysis. The manual effort required to acquire large amounts of data for effective planning is expensive and not feasible. Hence, a faster, more reliable, and inexpensive approach would involve using machine learning techniques. A random forest classifier is pre-trained on satellite images and then used to determine whether the segmented image is part of a building or not. This helps identifying the exact location of buildings. Two cases are focused on here - Cash transfers to extremely poor villages in Africa and citing of solar-powered micro-grids across several villages in India which are remote (Vincent, 1993).

Goswami, Gakkar, and Kaur (2015) propose a technique to automatically recognize objects in satellite images using artificial neural networks. Aimed at the purpose of remote sensing applications, this method is used to detect water bodies without human intervention. The data-set created by a domain expert is used to train a multi-layer perceptron (MLP) using error back-propagation (EBP) learning algorithm (Oh & Lee, 1995). Confusion matrix and Kappa coefficient have been used to assess the accuracy (Varade, Dhotre, & Pahurkar, 2013). Katiyar and Arun (2014) propose a technique for automatic object detection from aerial and satellite images using the spectral and spatial resolution of these images. The accuracy of the system is verified against Indian resource sensing (IRS) satellite images. It is based on a neural network and the simulation of visual interpretation from remote sensing images to increase the efficiency of image analysis (Kasturirangan et al., 1996). An automatic learning approach along with intelligent interpretation and interpolation is used. The key focus is on simplicity and it bases the feature detection not only on pixel values but also the object shape. It provides the flexibility of categorizing similar objects of different shape and size. Using ground truth verification, the accuracy is evaluated (Barber, Dobkin, & Huhdanpaa, 1996).

METHODOLOGY

Using machine learning on high resolution satellite imagery to detect rural regions and provide insight on the economic development can aid efforts to track regions below the poverty line in India. By conducting an analysis of the facilities available to an area and the area's distribution among different landscape classes, an overall development summary can be generated for the selected location.

The entire process is explained briefly in the following steps. The first step involves the collection of satellite images to fully represent the selected locality (imagery taken from Google Static Maps API). In the second step, these images are sent to a convolutional neural network CNN for processing. The CNN is developed and trained for a large data-set of satellite images representing Indian landscapes. The model uses this CNN to predict what kind of land the selected locality represents and how developed it is in terms of infrastructure. Ultimately, the model determines whether the region is likely to be urban or rural using a decision tree. Further statistics obtained on rural localities include how easily accessible certain amenities including Schools, Hospitals, Buses, etc.) are. As an additional statistic, using an API for India night-lights the mean light visibility of certain villages is recorded (Subash, Kumar & Aditya, 2019). There are studies to indicate that night-light intensity may serve as a useful proxy of economic development (Shafie, Hafiz, & Ali, 2009; Aslani & Mahdavi-Nasab, 2013). Finally, a summary is generated based on the inferences made from the collected data. This is displayed to the user through a graphical user interface (GUI) for easier understanding.

For evaluation purposes, data has been collected by the proposed model for 1032 rural villages. Cumulative scores have been calculated for each district containing several villages from the data-set. These statistics are compared to Census of India, 2011 (MHA (n. d.). Several similarities and consistencies among corresponding results of the machine analysis and manual surveys indicate the accuracy of the automated approach.

The CNN Mode of Classification

The overall model comprises of two sub-models. The first sub-model is used to categorize a satellite image into different land classes based on the percentage distribution occupied by each class in the image. The CNN model developed for this purpose has been trained on a data-set containing the following classes – cultivated land, uncultivated land, forest cover, building cover, and coastal/water cover.

The second sub-model is used to categorize the same image into various development classes based on the percentage probability that the image belongs to a certain class. The CNN model developed for this purpose has been trained on a data-set containing the following classes – underdeveloped, developing, and developed. Both the CNN models are pre-trained on ImageNet which is a large image classification data-set. It can classify images into more than 1000 categories (Panda, 2018).

The entire set of satellite images representing a 10 km x 10 km area in and around a locality (specified by geographical coordinates) are passed through both the sub-models and a cumulative distribution is obtained for each land class as well as for each development class. Further, a decision tree classifier is used to predict whether the locality is rural or urban based on the development scores obtained. Using the classifier, it is found that a region having more underdevelopment it is more likely to be a part rural India.

Figure 1, Figure 2 and Figure 3 depict sample images taken from the data-sets of different land classes and development classes respectively.

Figure 1. From left to right, Building Cover, Cultivated Land, and Uncultivated Land

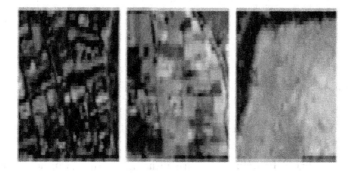

Figure 2. From left to right Forest Cover, Coastal/Water Cover

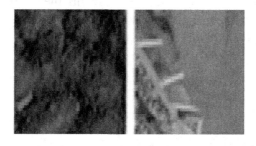

Figure 3. From left to right, Underdeveloped, Developing, Developed

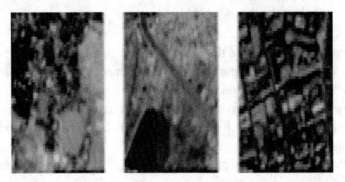

Analysis of Nearby Amenities

For a region classified as rural, the locations of certain amenities are determined to indicate the ease of access to necessary facilities in that region. This can help policy makers in evaluating the consumption poverty of a locality. Access to the following amenities has been identified in this approach.

1.   Education (Schools and Universities),
2.   Medical Provisions (Hospitals, Healthcare Facilities),
3.   Transportation (Bus and Railway Stations).

Using the Google Maps Places API, all the above-mentioned facilities falling within a 20 km radius of the selected rural locality are obtained and recorded. The Google Distance Matrix API is used to return the top three closest results with respect to each of the categories: Education, Medical Provisions, and Transportation. Based on the number of results and distances of the nearest result per category, a score is recorded (a- easy access, b- moderate access, c- difficult access). For example, if there are more than three schools within a vicinity of 5 km, the locality is given a score 'a' for Education.

Summary Generation

All the results obtained from the CNN Models and the Google Maps APIs are recorded into a spread-sheet. Certain formulas are arrived at through a detailed analysis of the village data-set to help make useful inferences and summarize the data. These formulas are applied on the newly recorded statistics of the user-specified rural locality. For example, if the percentage of a land class exceeds all others, the area is said to 'Mostly' belong to that class. It is also determined whether the building coverage in the area is 'Strong' or 'Weak'. These inferences are combined with the information on nearby amenities to provide an overall description for the region. Pie Charts depicting the land and development distribution are also generated using RStudio for easier visualization.

EXPERIMENTAL RESULTS

Data-Set for CNN Model

The data-set for training the CNN model includes a collection of satellite images downloaded from Google Static Maps. These images are categorized into eight classes. The first sub-model consists of five Land Classes namely building cover, cultivated land, uncultivated land, forest cover, and water cover. In the second sub-model, the remaining three classes to represent development i.e. underdeveloped, developing, and developed are included. Each class contains approximately 300 images. While training, bottleneck files are created for each image in each class. This bottleneck text file contains information about the image. The final test accuracy achieved by the CNN model is 95.1%.

Village Data-Set for Decision Tree Classifier

The data-set representative of the most poverty-stricken states in India included a collection of 1132 villages in the states of Assam, Bihar, Chhattisgarh, Madhya Pradesh, Jharkhand, Orissa, and Uttar Pradesh. The overall land distribution and development distribution for each village have been predicted by the CNN model.

All the districts of a state are represented by the graph. Four parameters are taken into consideration which include the total number of villages per district, the number of villages having a weak building coverage i.e. less than 15%, the number of villages having a high proportion of cultivated land i.e. greater than 75%, and the number of villages that are highly underdeveloped i.e. greater than 15%. Consider Figure 4.

This represents the state of Manipur. From the graph (Figure 4), it can be inferred that most of the land is not cultivated. The region is highly underdeveloped with very few buildings or infrastructure. Geographically, the state of Manipur consists mainly of hilly regions and the area under cultivation is limited. Studies from Census data also reveal that the state of Manipur is underdeveloped with high levels of poverty.

Decision Tree Classifier

A collection of 1132 rural localities along with their development distribution scores was sent as a part of the training set to the decision tree classifier (the remaining data-set including urban localities). The classifier correctly predicted 90% of the localities. Regarding the rural classification, 1011 localities were returned as rural out of which 1007 results were relevant. With a precision of 99.6% and recall of 88.9%, it is evident that the classifier can almost successfully distinguish between rural and urban areas.

Selection of Village: GUI

Figure 4 shows the first webpage which enables the user to select from two options; selecting a village from a dropdown list of 1007 villages or specifying geographical coordinates.

Figure 4. Readings for the State of Manipur.

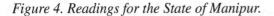

Figure 4 shows the webpage to select a village. Once the selected village is submitted, a corresponding summary is displayed to the user. Figure 4 provides the user with the flexibility to either enter coordinates of a locality or select any location on the map. Similarly, on submission, a summary is generated for the specified locality.

Summary Generation for Selected Locality

The summary contains a description of the area based on the statistics obtained by the machine learning model and nearby amenities analysis. This includes the following:

1. Information on the prominent land types such as an indication that region is mostly cultivated or that it consists of many forests.
2. Information regarding the proportion of buildings/infrastructure in the area.
3. The development level of the region such as whether it is highly underdeveloped or is developing.

(The above results are inferred by applying certain formulas to the statistics obtained.)

4. The total number of amenities and the top-three closest ones with respect to health, education, and transportation.

5. Two pie charts depicting the land distribution and the development distribution are also displayed to the user for easy visualization.

As the method is restricted to analysis of rural areas, if the locality specified by the user is detected as urban by the decision tree classifier, the summary will only contain the information. A link is provided to view the satellite images considered for processing. A link is provided to view the satellite images considered for processing. These images for a sample locality are displayed.

Cross Validation Against Census Data

To ensure the reliability of the results obtained, a scoring metric was used to compare the statistics generated by the model and the statistics present in the most recent census conducted, Census of India, 2011. The entire set of results collected for 1007 rural localities in India was summarized per district. For each of the 42 districts ranging across 8 states in India, percentage measures regarding the levels of cultivation, building coverage, and overall development were calculated. For each fundamental amenity considered, a score ranging from 1 to 4, regarding availability/accessibility was obtained (1 being the best and 4 being the worst). Scores based on similar metrics were determined for each district from the 'Village Amenities' statistics of the census data. The percentage of population below the poverty line was recorded. Comparison of the two sets of scores revealed a high level of consistency. It showed that the statistics obtained by the proposed model is indicative of whether the region is below the poverty line or not. Few discrepancies were due to some statistics being marked as unavailable in the census data and the fact that the most recent Census was conducted in 2011, thereby not precisely representing present-day conditions.

CONCLUSION

Using a CNN for satellite image processing, data regarding the land distribution and development levels can be collected. A measure of accessibility for various facilities like bus services or health centers is also made available using Google Maps APIs. The proposed method uses machine learning and automatic analysis to study the development of rural areas in India. It attempts to predict which regions are subject to poverty and thereby require attention by various schemes and policy-makers working towards bettering their living conditions and economic benefit. This paper provides a technique that reduces the manual effort and resources required in obtaining development-based statistics for rural localities. The method successfully identifies locations of interest and collects relevant data to accurately reflect the living conditions in an area. Based on the summary generated, locality-specific improvements to the identified under-developed parts of rural India can be suggested. The goal is to aid and speed up efforts to eradicate consumption poverty in India through technology.

REFERENCES

Aslani, S., & Mahdavi-Nasab, H. (2013). Optical Flow Based Moving Object Detection and Tracking for Traffic Surveillance. *International Journal of Electrical, Computer, Energetic, Electronic and Communication Eng., 7*(9), 963–967. Retrieved from https://www.semanticscholar.org/paper/Optical-Flow-Based-Moving-Object-Detection-and-for-Aslani-Mahdavi-Nasab/a3248c45cdc41417ed2ea236829a99c9783b-b3ca

Barber, C. B., Dobkin, D. P., & Huhdanpaa, H. (1996). The quickhull algorithm for convex hulls. *ACM Trans. Math. Softw., 22*(4), 469–483. Retrieved from doi:10.1145/235815.235821

Cohen, I., & Medioni, G. (1999). *Detecting and Tracking Moving Objects for Video Surveillance* (Vol. 2). IEEE Proc. Comput. Vis. Pattern Recognit. Retrieved from http://citeseerx.ist.psu.edu/viewdoc/download?doi=10.1.1.20.7779&rep=rep1&type=pdf doi:10.1109/CVPR.1999.784651

DataGovIn. (2014). *Literacy Rates in India.* Retrieved from https://data.gov.in/catalog/literacy-rate-india-nsso-and-rgi

DFID. (n.d.). *Growth: Building Jobs and Prosperity in Developing Countries.* Retrieved from https://www.oecd.org/derec/unitedkingdom/40700982.pdf

Goswami, A. K., Gakkar, S., & Kaur, H. (2014). Automatic Object Recognition from Satellite Images using Artificial Neural Network. *International Journal of Computer Applications, 95*(10). Retrieved from https://www.ijcaonline.org/archives/volume95/number10/16633-6502

Horton, M., Jean, N., & Burke, M. (2016). *Stanford scientists combine satellite data, machine learning to map poverty.* Retrieved from https://news.stanford.edu/2016/08/18/combining-satellite-data-machine-learning-to-map-poverty/

Jadhav, J. J. (2014). Moving Object Detection and Tracking for Video Survelliance. *IJERGS, 2*(4), 372–378. Retrieved from https://pdfs.semanticscholar.org/a58e/281e2d25fd8ba4e54860bf6c9b8781dd8e4a.pdf

Kasturirangan, K., Aravamudan, R., Deekshatulu, B. L., Joseph, G., & Chandrasekhar, M. H. (1996). *Indian Remote Sensing satellite IRS-1C—the beginning of a new era.* Retrieved from http://repository.ias.ac.in/88316/1/88316.pdf

Katiyar, S. K., & Arun, P. V. (2014). *An enhanced neural network based approach towards object extraction.* Retrieved from https://arxiv.org/abs/1405.6137

Mellander, C., Lobo, J., Stolarick, K., & Matheson, Z. (2015). *Night-Time Light Data: A Good Proxy Measure for Economic Activity?* Retrieved from https://www.researchgate.net/publication/283263879-Night-Time-Light-Data-A-Good-Proxy-Measure-for_Economic_Activity

MHA. (n.d.). *History of Census in India.* Retrieved from http://censusindia.gov.in/Ad_Campaign/drop_in_articles/05-History_of_Census_in_India.pdf

MHFW. (2015). *The Ministry of Health and Family Welfare. Government of India Ministry of Health and Family Welfare Statistics. Division Rural Health Statistics.* Retrieved from https://wcd.nic.in/sites/default/files/RHS_1.pdf

Oh, S.-H., & Lee, Y. (1995). *A Modified Error Function to Improve the Error Back-Propagation Algorithm for Multi-Layer Perceptrons*. Retrieved from https://onlinelibrary.wiley.com/doi/abs/10.4218/etrij.95.0195.0012

Panda, A. K. (2018). *A Case of Mutli-Label Image Classification: Build a state of the Art Multi-label Image Classifier*. Retrieved from https://towardsdatascience.com/fast-ai-season-1-episode-3-a-case-of-multi-label-classification-a4a90672a889

Raviprakash, N., Suresh, M., Rathis, A., Yadav, A., Devarla, D., & Nagaraj, G. S. (2015). Shot Segmentation for Content Based Video Retrieval. *Proceedings of the International Conference. Computational Systems for Health & Sustainability*.

Shafie, A. A., Hafiz, F., & Ali, M. H. (2009). Motion detection techniques using optical flow. World Academy of Science, Engineering and Technology. *International Journal of Electrical and Computer Engineering, 3*(8). Retrieved from http://waset.org/publications/8745

Simon, M., Rodner, E., & Denzler, J. (2016). *ImageNet pre-trained models with batch normalization*. Cornell University. Retrieved from https://arxiv.org/abs/1612.01452

Subash, S. P., Kumar, R. R., & Aditya, K. S. (2019). Satellite data and machine learning tools for predicting poverty in rural India. *Agricultural Economics Research Review, 31*(2), 231–240. doi:10.5958/0974-0279.2018.00040.X

Tewari, S. (2015). *670 Million In Rural Areas Live On Rs 33 Per Day*. Retrieved from https://archive.indiaspend.com/cover-story/670-million-in-rural-areas-live-on-rs-33-per-day-79600

UNDESA. (2018). *Transforming our world: the 2030 Agenda for Sustainable Development*. Retrieved from https://sustainabledevelopment.un.org/post2015/transformingourworld

Varade, R. R., Dhotre, M. R., & Pahurkar, A. B. (2013). A Survey on Various Median Filtering Techniques for Removal of Impulse Noise from Digital Images. *IJARCET, 2*(2), 606–609. Retrieved from https://pdfs.semanticscholar.org/030b/45e69b576f99c5c491cf8e58f911e164908c.pdf

Varshney, K. R., Chen, G. H., Abelson, B., Nowocin, K., Sakhrani, V., Xu, L., & Spatocco, B. L. (2015). *Targeting Villages for Rural Development Using Satellite Image Analysis*. Retrieved from https://www.liebertpub.com/doi/full/10.1089/big.2014.0061

Vincent, L. (1993). Morphological grayscale reconstruction in image analysis: Applications and efficient algorithms. *IEEE Transactions on Image Processing, 2*(2), 176–201. doi:10.1109/83.217222 PMID:18296207

Zhou, X., Yang, C., & Yu, W. (2013). Moving Object Detection by Detecting Contiguous Outliers in the Low-Rank Representation. *IEEE Transactions on Pattern Analysis and Machine Intelligence, 35*, 1–30. Retrieved from https://ieeexplore.ieee.org/abstract/document/6216381 PMID:22689075

Chapter 5
Wearables, Artificial intelligence, and the Future of Healthcare

Omar F. El-Gayar
Dakota State University, USA

Loknath Sai Ambati
Dakota State University, USA

Nevine Nawar
Alexandria University, Egypt

ABSTRACT

Common underlying risk factors for chronic diseases include physical inactivity accompanying modern sedentary lifestyle, unhealthy eating habits, and tobacco use. Interestingly, these prominent risk factors fall under what is referred to as modifiable behavioral risk factors, emphasizing the importance of self-care to improve wellness and prevent the onset of many debilitating conditions. In that regard, advances in wearable devices capable of pervasively collecting data about oneself coupled with the analytic capability provided by artificial intelligence and machine learning can potentially upend how we care for ourselves. This chapter aims to assess the current state and future implications of using big data and artificial intelligence in wearables for health and wellbeing. The results of the systematic review capture key developments and emphasize the potential for leveraging AI and wearables for inducing a paradigm shift in improving health and wellbeing.

INTRODUCTION

Chronic diseases such as heart disease and diabetes are conditions that last one year or more and require ongoing medical attention or limit activities of daily living or both (Centers for Disease Control [CDC], 2018). Such conditions have been a major healthcare concern as a leading cause of death and a declining quality of life. According to the Center for Disease Control (CDC), heart disease and stroke are the leading causes of deaths in the US, while more than 29 million Americans have diabetes, and another 86

DOI: 10.4018/978-1-5225-9687-5.ch005

million are at risk of type 2 diabetes (prediabetes). This results in an annual health expenditure of 86% of the nation's $2.7 trillion (almost 18% of Gross Domestic Product (GDP)) for people with chronic and mental health conditions. The situation worldwide is not much different. According to the World Health Organization, non-communicable diseases (NCDs) kill 41 million people each year with cardiovascular diseases alone accounting for almost 44% (17.9 million people), followed by cancers, respiratory, diseases, and diabetes (World Health Organization [WHO], 2018).

While poverty and socioeconomic conditions can increase the risk for such disease conditions, the common underlying risk factors are physical inactivity accompanying modern sedentary lifestyle, unhealthy eating habits, and tobacco use. Interestingly, these prominent risk factors fall under what is referred to as modifiable behavioral risk factors. According to the World Health Organization (WHO), tobacco accounts for 7.2 million deaths every year, salt intake (part of unhealthy diets) accounts for 4.1 million deaths, while 1.6 million deaths are attributed to alcohol use and a similar figure attributed to insufficient activity (WHO, 2018).

In that regard, the proliferation of wearable technologies is hyped to drive a paradigm shift on how people care for their health and well-being as a society and as individuals. Nowadays, wearable technology is used as activity tracking devices, heart rate monitors, calorie counters, glucose (blood sugar) monitoring systems, hearing aids, and pacemakers, to name a few. In the last couple of years, smart patches and smart pills are groundbreaking improvements in the wearable health domain that are approved by the US Food and Drug Administration (FDA). The ingestible sensor in the pill transmits a message to a wearable patch, the patch then transmits the information to a mobile application. Healthcare professionals such as nurses and doctors employ wearable technology to monitor patients' vitals and conditions thereby increasing efficiency and reducing the time to gather health data (Mesh, 2018).

Many companies have either emerged as startups or directed entire product lines to the wearable market. These include tech giants such as Apple, Google, Samsung, and Huawei as well as more specialized companies such as Fitbit, Garmin, Moov, and Misfit. The key drivers for this wearable trend are patient monitoring, home healthcare, and fitness. According to market research and business intelligence firm IDTechEX, the wearable technology sales are expected to increase exponentially in the coming years with the market expected to reach $150 billion by 2027 (Stefanie, 2018). The availability of more data about the current and future health of individuals, and the growth of wearables in the healthcare domain are the key features that galvanize the market right now.

The proliferation of wearables has resulted in the generation of large amounts of data (volume) at increasing velocity, and variety. Such data is commonly referred to as 'Big data' (S. Park, Chung, & Jayaraman, 2014). The data can take the form of a continuous stream of sensor data such as heart rate or blood glucose measurements as well as audio and video signals. While such data creates significant opportunities for improving health and well-being, it could also create a data rich information poor (DRIP) environment (Tien, 2013). However, recent advances in artificial intelligence and machine learning are offering new possibilities to wearable generated big data and genuinely transform healthcare.

Extant research provides a glimpse into the world of wearables with a particular emphasis on healthcare (Chan, Estève, Fourniols, Escriba, & Campo, 2012; Mukhopadhyay, 2015; S. Park et al., 2014; Swan, 2013). In these reviews, the emphasis is predominantly at the intersection of wearables and healthcare. In this chapter, we extend prior work with an emphasis on the role of artificial intelligence and machine learning in harnessing the big data generated by wearables with a focus on health and well-being. Specifically, we aim to address the following research questions:

1. What is the current state of using big data and artificial intelligence in wearables for health and well-being?
2. What are the drivers and challenges for big data and artificial intelligence in wearables for health and well-being?
3. What is the future state of using big data and artificial intelligence in wearables for health and well-being?

Addressing these research questions, we conducted a system systematic review focusing on big data and artificial intelligence in wearables for healthcare and well-being. The review spanned the last decade and covered peer-reviewed scholarly publications. To further guide the literature review process and organize our findings, we devised a framework based on extant literature. The framework aims to capture key aspects of self-management and self-care. Importantly are the potential, drivers, and challenges of big data and artificial intelligence as they relate to wearables for healthcare and well-being.

The remainder of the chapter is organized as follows: The next section provides background information regarding wearables, AI and big data, and health and well-being. This is followed by two sections where we describe the methodology for the systematic review and introduce the proposed conceptual framework for organizing the findings. The Results section provides a summary of key findings along the proposed framework. The Discussion section discusses the current state and future of Big Data and AI technologies utilization by wearables in a healthcare context, and highlights key drivers and challenges shaping the potential of Big Data and AI through integration with wearable technologies in healthcare. We conclude the chapter with a summary of key findings and recommendations for future research.

BACKGROUND

Healthcare and Well-Being

Healthcare is moving from a provider-centric model to a patient-centric model. In a provider-centric model, information predominantly emanates for the provider (practitioner, specialist clinic and hospital, are to name a few) and insurance providers. The mode is paternalistic where knowledge flows primarily from the provider to the patient, care is provided at the point of care (usually hospital or clinic), and the patient is a passive participant (despite being a primary beneficiary). The mode is also characterized as reactive which implies that most actions are in response to an illness or an ailment. The shift to a patient-centered model moves the patient or individual to the center of the 'care' ecosystem. In this model the individual is an active participant in the caring and management of one's health and well-being. Such transformation is enabled by advancements in information technology that facilitates the sharing and dissemination of medical knowledge, such as, through web portals, the accessibility of one's health information via personal health record systems, and the advancement and increased ubiquity of wearable devices capable of collecting a variety of personal and environmental data (Amft, 2018).

Along these lines one can envision a continuum of care as shown in Figure 1 (Self Care Forum, 2019). The individual assumes almost complete ownership of their own self-care and the emphasis is on disease prevention and well-being where 'wellness' according to the Oxford Dictionary refers to a 'state of being in good health'. The left-hand side represents a deliberate effort towards improved health and well-being. As we move from left to right on this continuum (Figure 1), we encounter the management of

Figure 1. The self-care continuum (Self Care Forum, 2019).

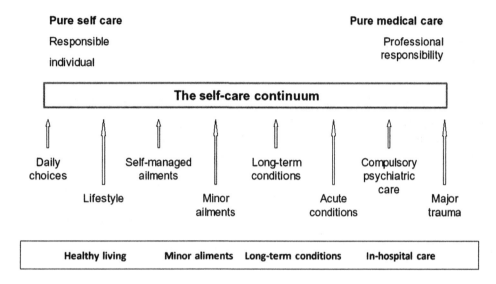

chronic (long-term) conditions and ultimately response to episodic events such as major trauma. For the purpose of this review, we explore interventions aimed broadly towards self-care and self-management. Self-care refers to processes, systems, and technologies employed by individuals to care for their health and well-being without being inflicted with a particular disease or condition, such as exercising and healthy dieting. Self-management relates to the tasks that an individual must undertake to live well with one or more chronic conditions (Adams, Corrigan, & Greiner, 2004).

Wearables

Generally speaking, 'wearable technology' or 'wearables' refer to computer systems or electronic technologies that are body-worn and utilized mostly hands-free (Gribel, 2018). Wearables can be classified using multiple parameters like features, applications, ability to sense, and many more. Park et al. (2014) develop a taxonomy and identify a number of key attributes for wearables. The taxonomy includes dimensions such as functionality (single versus multi-functional), type (passive versus active power requirements), and deployment mode (invasive versus non-invasive). Key attributes include physical attributes (lightweight, aesthetically pleasing, invisible, and shape conformable) and functional attributes (multi-functional, configurable, responsive, and bandwidth). Rizwan et al. (2018) classify wearables with respect to functionality, as predictive, preventive, personalized, and participatory based on the capabilities to perform diagnostics, monitoring, and treatment in healthcare. Alternatively, in the context of healthcare, Iqbal et al. (2018) note that wearable devices can be best classified into single-purpose or multi-purpose. Single-purpose wearable devices are built for a specific disease or purpose, such as fall detection and multi-purpose wearable devices are built to serve multiple problems, such as home health monitoring.

Applications in healthcare involve self or remote monitoring, human activity recognition, prevention, and diagnosis. The ability to monitor healthcare activity coupled with the analytic horsepower provided by AI techniques can serve as an enabler for more elaborate interventions aimed at improving health and well-being.

Artificial Intelligence (AI)

Turing (1950) demonstrated the concept of AI with the proposal of the Turing test in order to measure the machine intelligence in 1950. Later at the Dartmouth college summer 1956 AI conference, the term Artificial Intelligence was coined by John McCarthy. AI has since developed into a robust field (Beranek, Simon, & McCorduck, 1977). At the time of Turing, AI was understood as machines with the ability to think. In today's context, AI refers to the ability of a machine or a computer to perform duties that are possible by human intelligence or perform better than the human intelligence. AI techniques are essentially an efficient manner to organize and utilize the knowledge for providing the required results to fulfill the AI goals.

According to Hawas (2017), the most common AI classifications are Artificial Narrow Intelligence (ANI), Artificial General Intelligence (AGI) and Artificial Super Intelligence (ASI). Most of the literature uses ANI, which is defined as a machine to accomplish a specific task that is capable by a human being. AGI is defined as a machine with the ability to perform all the tasks that a human being is capable of. ASI is defined as a machine with the ability to outperform all the tasks that a human being is capable of. AGI and ASI have not yet been achieved. While data mining or machine learning (ML) do not encompass AI in its broadest sense, the ability to learn from data makes these techniques particularly valuable for ANI. Accordingly, in this review we pay particular attention to ML techniques that provide diagnostic, predictive, and prescriptive functionality.

Diagnostic Analytics: Diagnostic analytics is defined as the process to solve the problem of '*why did it happen?*' (Banerjee, Bandyopadhyay, & Acharya, 2013). It can be described as the deep-down analysis to answer a particular question or to find out the root cause of a phenomenon. This technique usually requires the analysis of past data to move forward.

Predictive Analytics: Predictive analytics techniques utilize the current and historic data to train a statistical model for predicting or forecasting future or unknown events. These techniques are used when the primary question is '*what is likely to happen?*' and will predict the future outcomes and the drivers associated with the observed phenomenon (Banerjee et al., 2013).

Prescriptive Analytics: Prescriptive analytics is used to get the best course of action among multiple predicted choices. It is defined as '*what courses of action may be taken*' from the predicted outcomes. It evaluates or optimizes the predicted outcomes to choose the best possible outcome. This technique works best when the problem is of the type '*what should be done about it?*' and follows descriptive and predictive analytics (Banerjee et al., 2013).

To improve the efficacy of machine learning techniques, researchers and developers can increase the complexity of the models and/or increase the size of the data sets. Wearables can serve as a pervasive and ubiquitous data acquisition layer where machine learning can provide the intelligence needed to support a particular healthcare intervention.

Table 1.

(*big data* or *artificial intelligence* or *machine learning* or *deep learning*)   AND   (*pedometers* or *heart rate monitor* or *fitness tracker* or *Wearable technology* or *activity tracker* or *wearable devices*)   AND   (*healthcare* or *well being* or *well-being* or *wellness*)

METHODOLOGY

We conducted a systematic literature review to address the above research questions. The systematic review followed guidelines in accordance with the Preferred Reporting Items for Systematic Reviews and Meta-Analyses (PRISMA) (Liberati et al., 2009). The PRISMA statement provides an evidence-based minimum set of items for reporting systematic reviews and meta-analyses with the aim of improving the quality of reporting by authors of such reviews. The review emphasizes recent innovations at the intersection of healthcare, wearables, and AI and big data during the last decade. The review targets the following databases: Web of Science, PubMed, IEEE Explorer and ABI/Inform for peer reviewed articles.

The search query captured the intersection of the three areas of concern, namely wearable technology, AI, and Healthcare. The search query is developed iteratively. Initially, a few words were selected from each of the three groups. Based on the relevance of the search results, the search query was refined by removing/adding key words so as to increase the number of relevant articles while reducing the number of irrelevant articles as shown in Table 1.

Using the finalized search query on the four databases mentioned, the resultant set of articles were analyzed for relevance based on the title, abstract and the full text using the inclusion and exclusion criteria. For articles to be included, they needed to meet the following criterion:

- Interventions leveraging AI techniques and data obtained from the wearable technology in a patient-centered healthcare context.

Articles from the review were excluded if any of the following criteria was met:

- Articles solely aimed at introducing frameworks for applying AI in healthcare rather than a specific healthcare intervention.
- Articles with an exclusive focus on wearable technology in multiple domains rather than just healthcare or well being.
- Articles describing the use of wearable technology with the combination of non-wearable technology for data acquisition like a smart belt with the combination of a video camera capturing the activity of the user.
- Articles describing the use of wearables in hospital environments like wearable electrodes or patient monitoring in the ambulatory or hospital setting.
- Systematic review, overview, and articles that lack specificity regarding the technology used or the nature of the intervention.

FRAMEWORK

Figure 2 depicts a conceptual framework that captures the interplay of artificial intelligence and wearable technology for health and well-being. The framework has three layers. The bottom layer depicts wearables primarily serving as a source for data where AI analytic techniques can be applied to support health and well-being. These wearables can potentially generate a plethora of data characterized by volume, velocity, and variety (Laney, 2001). Volume represents the amount of data and the storage concerns; variety represents the types of data which is formed by the big data, and velocity represents the speed of data processing. Over time, the three dimensions are extended to five dimensions with the introduction of veracity and value. Veracity is referred to as the quality of data or how accurate the data is and value refers to the worth of the data.

Artificial intelligence and machine learning provide the capacity to move away from a data rich information poor (DRIP) context and towards realizing the value of big data generated by these wearables and the attainment of improved health and well-being. AI diagnostic, predictive, and prescriptive capabilities can be used to make sense of raw data for human activity recognition (HAR), and to provide recommendations for improved self-care and self-management.

Encompassing the integration of wearables and AI in healthcare are drivers and challenges. On one hand, drivers extend the innovation frontier with respect to possible applications and interventions of wearable and AI technologies to the betterment of health and well-being. These drivers include advances in information technologies and telecommunication infrastructure, to name a few. On the other hand, challenges represent factors and issues that can impede the wide scale proliferation of these technologies in healthcare. Examples include data security and privacy, and user technology self-efficacy.

RESULTS

The systematic literature review involved three steps resulting in a total of 441 peer-reviewed articles (Figure 3). The first step included the removal of duplicate articles. The second step involved screening using the title and abstract for relevance using the inclusion and exclusion criteria. The third step was similar to the second step but focused on screening using the full text for relevance using the same inclusion and exclusion criteria. The resultant peer reviewed articles after step three were read thoroughly for

Figure 2. Underlying framework for wearables, AI and healthcare

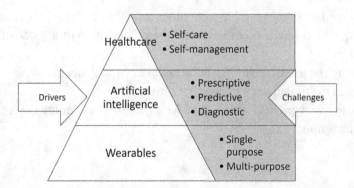

relevant information to answer the research questions stated in the introduction. The relevant information was extracted based on the proposed framework. Figure 3 depicts the article selection process and results. Figure 4 and Figure 5 show time series analysis of the literature and Table 1 illustrates the distribution of articles in the literature according to the conceptual framework. The following sections further describe the findings in multiple perspectives of the respective layers in the framework. As shown in Table 1, out of the resultant 68 articles, human activity recognition (HAR) accounted for 47.1% (32 articles) followed by self-management of specific disease conditions 26.5% (18 articles), while the remaining articles addressed self-care. Predictive analytics methods were most prevalent. Overall, 52 articles utilized predictive data analytics (76.5%), 11 articles utilized prescriptive data analytics (16.2%), and only five articles utilized diagnostic data analytics (7.3%). Likewise, two thirds of the articles utilized single purpose wearables specific to the targeted application.

Human Activity Recognition (HAR)

HAR can be defined as the recognition or classification of physical activities done by a human entity using data from a wearable device such as Fitbit that recognizes human activities like walking, running, etc. HAR can occur in the context of self-care and self-management applications alike.

Figure 3. Systematic literature review process

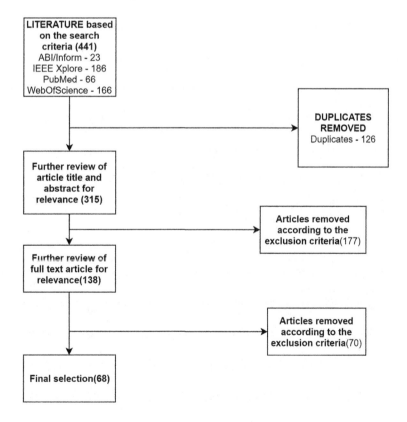

Figure 4. Time series analysis of the literature

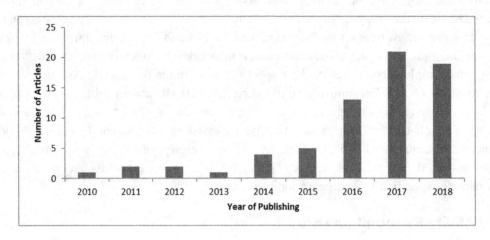

Figure 5. Time series analysis of the AI techniques

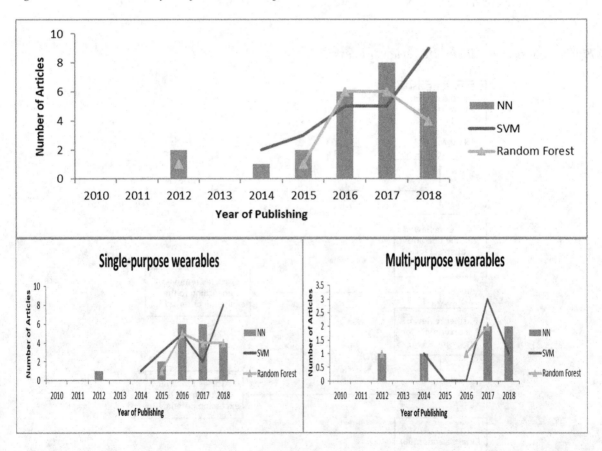

Table 2. Distribution of literature based on the proposed framework

			Artificial Intelligence		
			Diagnostic	Predictive	Prescriptive
Single-purpose wearables		HAR		23	
		Self-care	2	10	
		Self-management	2	9	
Multi-purpose wearables		HAR		6	3
		Self-care	1	3	2
		Self-management		1	6

Overall, there were 32 studies (47.1%) related to HAR. Three used prescriptive analytics to analyze data from multi-purpose wearables to identify the best possible physical activity and prescribe appropriate action (Lim, Kim, & Choi, 2017; Ngo, La, Leong, & Quek, 2017; Reiss & Stricker, 2014). Predictive analytics was used in the remaining 29 studies to predict the nature of human activity. Out of these 29 articles, six articles utilized multi-purpose wearables for health monitoring and fitness towards wellness of the user (Alelaiwi, Hassan, & Bhuiyan, 2017; Badawy et al., 2018; Y. Cheng, Jiang, & Peng, 2014; Jiang et al., 2016; Weiss et al., 2016; Yuan & Herbert, 2014), while 20 articles utilize single-purpose wearable devices (Baktir, Tunca, Ozgovde, Salur, & Ersoy, 2018; K. Liu, Hsieh, Hsu, & Chan, 2018; Lustrek et al., 2015; Razum, Seketa, Vugrin, & Lackovic, 2018; Saha et al., 2018). While the mentioned studies emphasized HAR regardless of the existence of an ailment, Tahavori et al. (2017) utilized single-purpose wearable devices to manage Parkinson disease using HAR and Nemati, Liaqat, Rahman and Kuang (2017) investigated the maintenance of health in cancer patients and elderly people based on HAR. Another study introduced new techniques in HAR for ambulant children and adolescents with cerebral palsy (Ahmadi, O'Neil, Fragala-Pinkham, Lennon, & Trost, 2018).

Self-Care

Table 1 shows that a total of 18 articles (26.5%) targeted self-care intervention. Cardiac monitoring was the most prevalent self-care application and was used in six self-care interventions (33.3%). Predictive analytics was most prevalent (13 articles) while single purpose wearable devices were the most encountered (12 articles). Three articles utilized prescriptive analytics and three articles employed diagnostic data analytics techniques. The following paragraphs detail the results with respect to these interventions.

Cardiac monitoring: Six studies focused on improving heart condition monitoring and detecting heart problems. All of the six studies employed predictive analytics, of which one article utilized multi-purpose wearables for the detection of heart rate anomalies (Zhang, Zhou, & Zeng, 2017). The remaining five studies utilized single-purpose wearables to focus on heart rate variability as a basis for heart anomaly detection (Alqaraawi, Alwosheel, & Alasaad, 2016; S. Cheng, Tamil, & Levine, 2015; Mei, Gu, Chen, & Chen, 2018; Nii, Iwamoto, Okajima, & Tsuchida, 2016; Walinjkar & Woods, 2017).

Other self-care interventions: Other than cardiac monitoring, 12 articles addressed a variety of interventions deploying prescriptive, predictive, and diagnostic methods. There were seven studies that utilized predictive methods for miscellaneous healthcare applications. Out of which five studies utilized single-purpose wearables; one study used deep learning based breath rate monitoring model and succeeded in attaining the professional monitoring accuracy using a smart phone (B. Liu et al., 2018), one study focused on detection of social isolation and loneliness based on talking in respiratory signals (Ejupi & Menon, 2018), another study focused on detection of alcohol intoxication based on gait monitoring (E. Park et al., 2017), and two other studies aimed at sleep quality prediction (Choi, Kang, & Son, 2017; Sathyanarayana et al., 2016). The other two predictive articles utilized multi-purpose wearables for real-time prediction of severe clinical events based on monitoring the vital signs (Forkan, Khalil, & Atiquzzaman, 2017) and sleep quality prediction (Hidayat, Tambunan, & Budiawan, 2018).

There were two studies that employed prescriptive methods. Both studies utilized multi-purpose wearables for detection of emotion and medical history, health monitoring for e-prescriptions (Ullah, Bhuiyan, & Das, 2017), and predicting sleep quality and efficiency to recommend changes in sleep habits (Sathyanarayana, Srivastava, & Fernandez-Luque, 2017).

Three more studies employed diagnostic methods; one study utilized single-purpose wearables for detection & diagnosing of glaucoma using intraocular pressure monitoring (Gisler, Ridi, Fauquex, Genoud, & Hennebert, 2014), and another study employed multi-purpose wearables for energy efficient real-time smart diagnosis using conditional deep learning (Parsa, Panda, Sen, & Roy, 2017). The third study also employed AI and wearables for the early identification of prediabetes (Baig, Mirza, Gholam-Hosseini, Gutierrez, & Ullah, 2018).

Self-Management

Table 1 shows 18 articles (26.4%) targeting self-management interventions. Chronic condition related interventions were the most prevalent with eight articles (44.4%) aiming to support ongoing self-management. Rehabilitation or post-hospital condition was addressed in seven articles (38.8%). Out of those 18 articles, 10 articles utilized predictive analytics technique, six articles utilized prescriptive analytics technique, and two articles utilized diagnostic analytics technique. Almost two thirds of the articles utilized single purpose wearable devices while seven utilized multi-purpose wearable devices. The following paragraphs detail the results with respect to these interventions.

Chronic conditions: Table 1 shows that, among the eight self-management chronic condition articles, four used single-purpose wearables and four used multi-purpose wearables. There were four prescriptive, two predictive, and two diagnostic studies. All of the four prescriptive analytics studies utilized multi-purpose wearable devices, of which three proposed a disease management platform for diseases like chronic obstructive pulmonary disease (COPD), chronic kidney disease (CKD) and renal insufficiency (Rosso, Munaro, Salvetti, Colantonio, & Ciancitto, 2010), and diabetes (Chen et al., 2018; Khan et al., 2015). The fourth study utilized smart phones to manage Parkinson's disease and fall detection (Mazilu et al., 2012).

The two predictive analytics studies utilized single-purpose wearable devices to manage speech disorders for patients with Parkinson disease (Borthakur, Dubey, Constant, Mahler, & Mankodiya, 2017), and manage or prognosticate COPD (Karuppanan, Vairasundaram, & Sigamani, 2012).

A diagnostic analytics study targeted the diagnosis and differentiation of Parkinson disease symptoms through multi-modal single-purpose wearable sensors (Oung et al., 2015). Another study aimed at automatically detecting anomalies in blood glucose (Zhu, 2011)

Rehabilitation or post-hospital condition: Seven articles addressed rehabilitation and post-hospital condition. Two studies employed prescriptive data analytics techniques and utilized multi-purpose wearables for rehabilitation environment to recommend physical activity via physical therapy for rehab patients and wellness for elderly people (Alcaraz, Moghaddamnia, Poschadel, & Peissig, 2018; Steffen et al., 2011). Five studies employed predictive data analytics, out of which four utilized single-purpose wearables for fall recognition and detection (Harris et al., 2016), early prediction of heart stroke for post-stroke patients (Y. Cheng, Chen, Yang, & Samani, 2016; ElSaadany, Majumder, & Ucci, 2017), and rehabilitation of motor impairments (X. Liu, Rajan, Ramasarma, Bonato, & Lee, 2018). The remaining predictive study utilized multi-purpose wearable ECG for patient monitoring after discharge from a major surgery (Randazzo, Pasero, & Navaretti, 2018).

Other Healthcare applications: Three papers fell into this category. Two predictive analytics studies utilized single-purpose wearables where one study estimated the knee joint angle for knee disorders using a fabric-based strain sensor (Gholami, Ejupi, Rezaei, Ferrone, & Menon, 2018) and another study proposed a platform to monitor the health of lower limb for people with prosthetic lower limb (Mathur et al., 2016). The Third study employed diagnostic analytics that utilized wearable single-purpose sensors for sleep apnea diagnosis using deep learning (Pathinarupothi et al., 2017).

DISCUSSION

Current State of Using Big Data and AI in Wearables for Health and Wellbeing

A significant number of studies target HAR. These articles focus on wellness and on improving the activity recognition with respect to energy efficiency, classification accuracy, and smart sensing specific to HAR (Kim, Lee, Lee, & Jeon, 2016; D. Ravi, Wong, Lo, & Yang, 2016, 2017). The studies related to energy efficiency and increasing classification accuracy have significant design and adoption implications as high accuracy and low power devices would open doors towards the rise of sleek and multi-functional wearable devices. Further, the key solution for the increase in efficiency of wearables is the efficient allocation of resources, especially when resource-intensive AI techniques like Artificial Neural Networks (ANN) are used. Along these lines, two studies aim at improving the resource allocation when deep learning is used with wearable technology (D. Ravi et al., 2016, 2017). Such studies pave the way for leveraging more complex AI techniques with wearables that can have more profound implications for improved health and well-being.

Recent contributions show that home monitoring and remote monitoring of patients are an important aspect of wearables in the field of self-care intervention. Most notable, are monitoring studies targeting frail and elderly people in order to encourage a healthy life style, e.g., engaging in physical activity (Y. Cheng et al., 2014; Jiang et al., 2016; K.-C. Liu & Chan, 2017). Another interesting trend is the early detection of severe clinical events by a study using vital signs data of the individual obtained from wearable devices (Forkan et al., 2017) and recent studies delve into using health monitoring for disease diagnosis and prognosis (Parsa et al., 2017; Ullah et al., 2017). Such studies demonstrate the potential for wearables coupled with AI technologies to transform healthcare from a reactive to a proactive (preventive) stance.

AI in wearable technology for the health and well-being demonstrate the potential to reduce healthcare cost particularly with respect to health monitoring where inexpensive wearables are used in place of expensive hospital equipment (Alcaraz et al., 2018; Alelaiwi et al., 2017; Mathur et al., 2016; Randazzo et al., 2018; Rosso et al., 2010). Another study focuses on breathing rate monitoring algorithm with smart phone using bi-directional recurrent neural network with convolutional layers. The approach maintains professional level accuracy which results in the replacement of professional medical equipment (B. Liu et al., 2018). As noted below, such studies demonstrate the potential for wearables to reduce the cost of healthcare and can be one of the most important drivers for the proliferation of AI-enabled wearable devices.

While predictive analytics dominate the AI application landscape, prescriptive analytics coupled with the increasing utilization of multi-purpose wearable devices appear as promising and can further support a paradigm shift from treatment to prevention. Most of these studies deal with providing recommendations that are based on the analysis of wearable generated data. Examples include a personalized exercise trainer for the elderly based on their physical activity (Steffen et al., 2011) and frameworks for managing of diseases such as CHRONIOUS which is an open, ubiquitous and adaptive chronic disease management platform for COPD, chronic kidney disease (CKD) and renal insufficiency (Rosso et al., 2010).

With respect to specific AI analytic techniques, Support Vector Machines (SVM), ANN and Random Forest have increased and are the most prevalent. SVM technique and ensemble modeling are very popular in HAR and fall detection interventions. Despite their computational requirements, ANN demonstrated an increasing trend possibly a reflection of the increased dimensionality and complexity of the underlying data.

Even though AI and wearables in healthcare have the potential for helping patients with chronic diseases manage their health condition, a significant share of the commercial wearable market is dedicated to fitness and basic acquisition of data, such as movement and heart rate. Regardless of the emphasis on self-care or self-management, extant research and development (while very promising) appear to be in a nascent stage with respect to the extent of intelligence afforded by the current state of artificial intelligence. AI applications can be characterized as Artificial Narrow Intelligence (ANI) focusing on the accomplishment of a specific task mostly prediction or diagnosis. While the common underlying risk factors of physical inactivity accompanying modern sedentary lifestyle, unhealthy eating habits, and tobacco use fall under what is referred to as modifiable behavioral risk factors, none of the studies leveraged AI for the purpose of leveraging big data collected from sensors to induce behavioral change. It is thus no surprise that current research shows a mixed picture. Results from small pilot studies show repeated failure when duplicated over larger samples that utilize longitudinal designs (Ellis & Piwek, 2018). For example, some studies do not find a benefit for patients that self-monitored their blood glucose while other studies point to increased levels of depression associated with these interventions (Ellis & Piwek, 2017).

DRIVERS AND CHALLENGES OF AI AND WEARABLES IN HEALTHCARE

Drivers

Healthcare Drivers

At the top of the list are rising healthcare costs, remote patient monitoring, wellness and self-care, self-management, and efficient chronic disease management. The rising healthcare costs is an important driver towards AI and wearables in healthcare. Based on the analysis of Dieleman et al. (2017), the healthcare spending on 155 health conditions and six health service categories has increased by $933.5 billion between 1996 and 2013.

Another important driver is the need for mobile healthcare monitoring and services. The need stems from the busy lives of users in addition to an increased awareness and realization of the importance of fitness and other healthcare intervention monitoring. The aforementioned awareness and realization are the result of the sensing capabilities and seamless nature of wearables, as well as the opportunity to utilize wearable technology in various complex health research problems and to help patients with their health through continuous monitoring and diagnosis. The awareness on self-care and self-management of one's own health plays a key role in today's usage of wearables. According to Mercer et al. (2016), wearable activity trackers contain several behavior change techniques that have demonstrated an increase in physical activity in older adults. AI and big data offer the possibility for personalizing these interventions in a manner that lead to behavioral changes for better maintenance of health and wellness.

AI and Big Data Drivers

There is no real impact of wearable sensors until the processing of the sensed data is converted and projected into useful insight for the user and/or the physician, hence the drive for applying AI to data from wearable devices. The need for solving complex healthcare issues like Parkinson's disease, diabetes, and other diseases can be addressed using AI techniques.

Studies such as Alelaiwi et al. (2017), Baktir et al. (2018), and Yuan and Herbert (2014) focus on proposing big data, Internet of Things (IoT), and cloud platforms for wearable data processing and decision making. The advancement and introduction of new and efficient algorithms and the increased computational power pave the way towards its usage in wearables for health and well-being.

Wearable Market Drivers

Based on the literature review, the main reasons for customers to buy a wearable device include but are not limited to, reduction in healthcare costs, home health monitoring, rise of advanced technology products, health data access to providers and user, and innovation in wearable designs. The advancements along the lines of sensing, wearability, efficiency, multi-functionality, and design innovation in the field of wearables such as Zephyr (Bio-patch), Google glass, smart watches, and smart pills have spurred the need, ease and advantages of their usage in healthcare. The developments increase the sensing capabilities and accuracy of sensors in mobile phones, which when coupled with an exponential increase in computational capabilities allow the usage of mobile smart phone as ubiquitous wearable device.

Studies like Forkan et al. (2017) demonstrate that vital signs collected by wearable devices can be successfully used to predict critical health events beforehand and diagnose the problem with high accuracy performance close to a medical professional. Similar studies demonstrate the possibility of eliminating the need for a doctor's visit effectively enacting a paradigm shift and opening doors for increasing the scope of self-awareness, self-care and self-management. AI techniques are introduced and improved in the field of wearables to combat the challenges of disease management, unavailability of healthcare professionals, and cognition of complex wearable big data.

The Motivation for Wearable Usage in a Work Environment

Studies such as Lingg, Leone, Spaulding and B'Far (2014) demonstrate the correlation between an employee's health and performance. The trend has the potential to grow into a popular practice where corporate companies and organizations provide incentives to employees who reach their physical activity goals set by their personalized activity profile. Such paradigm shift is equally advantageous to the employer because of the performance improvement and the employee because of improved health and sense of well-being. Health insurance providers may also start promoting wearable technology by reducing the premium on the health insurance for employees who participate in such practices.

Challenges

Despite the potential and factors driving the application of AI-enabled wearables in healthcare, there are a number of challenges that need to be addressed so as to realize the full potential of these developments. These challenges include:

Ownership, Privacy and Security of Wearable Data

Such a challenge relates to wearables regardless of the applications of AI techniques. Recent breaches of consumer information and abuses by corporate users have heightened awareness and concerns surrounding the ownership, privacy and security of personal wearable data. Such concerns resulted in regulations such as the European Union General Data Protection Regulation (GDRP). This is in addition to existing regulations governing healthcare data, such as the US Health Insurance Portability and Accountability Act (HIPAA) and the associated HIPAA Security Rule. Despite these regulations, there is evidence that suggests that a significant number of users do not read privacy agreements and do not change their default privacy settings (Punagin & Arya, 2015). Overall, a heightened user concern can impede the adoption by users (Gribel, 2018; Kalantari, 2017; S. Park et al., 2014; Shin et al., 2019). Further, overly restrictive regulations can limit the amount of data needed to train AI models and thereby slow the developments of these technologies.

Social Acceptance and Design Constraints

As with any disruptive technology, AI enabled wearables will have societal and design challenges that can curtail large-scale adoption. Take the example of Google Glass, despite its potential it has not been highly adopted. In fact, it is banned in places like cars, movie theatres theaters, hospitals, bars, concerts, banks, locker rooms, class rooms, and casinos (Costill, 2013) because of issues like safety, privacy and, piracy. Wearable designers need to consider the above-mentioned issues and other socio-technical and user-centered requirements to ensure large scale adoption, sustained utilization, and meaningful impact (Al-Ramahi, Liu, & El-Gayar, 2017; El-Gayar, Nasralah, & Elnoshokaty, 2019; El-Gayar, Nasralah, & Noshokaty, 2018; Wahbeh, Sarnikar, & El-Gayar, 2016).

Power Consumption and Processing Power

The main constraints associated with the use of wearables are battery life, computational power, ease of wear, the cost of the product, and data accuracy (El-Gayar et al., 2019). AI techniques are particularly resource intensive, so is the transmission of the large amounts of data generated by wearables. For example, while the use of ANN has increased over the time, the rate of utilization has been relatively slow due to the high resource requirements. Further, Park et al. (2014) explicitly identifies the reliance on lithium-ion rechargeable batteries as a limiting factor due to their rigidity in relation to the flexible nature of wearables. Recent advances in telecommunication coupled with large scale deployment of cloud environments are easing the expectations regarding processing power requirements while advances in battery technologies and alternative sources of energy, such as kinetic energy is expected to mitigate power requirements.

Device Accuracy

The value for leveraging big data and AI is impeded if the quality of the data is questionable. In a systematic review, Shin et al. (2019) report that approximately quarter of the studies (121) within the scope of the review have a technology focus that emphasizes 'data quality' provided by activity trackers. Research in this area typically focuses on accuracy, reliability, and validity. El-Gayar et al. (2019) report that device accuracy is a major concern of users of wearable devices as reported on social media. It is not uncommon for providers to dismiss the potential of wearables citing the current level of device accuracy.

The Future of AI and Wearables in Healthcare

According to Figure 4, the trend of AI in wearables for healthcare research has gradually increased over the last five years. This is expected to grow particularly as the global wearable medical device market is estimated to reach USD 58.3 billion by 2025 (Bhise, 2018). Furthermore, Clayton (2015) projects that the key for future healthcare is considered to be prescriptive analytics as it can provide solutions and suggestions to short term and long term administrative and health concerns. While this has not yet materialized where studies utilizing prescriptive analytics represent 16.2% of the studies as shown in Table 1, we contend that AI is uniquely positioned to provide prescriptive analytics not only in the context of wearables and healthcare, but to reshape the provision of healthcare in general.

While current research emphasizes the processing of data collected via wearables, there remains an untapped potential of integrating this data with other data sources to predict and prescribe personalized courses of action aimed at improving an individual's health and well-being. Examples include individual data from electronic health records (EHR), insurance providers' information systems, environmental information systems (including weather) to name a few. While there are issues with processing such big data, the challenges are mostly organizational and social rather than technical. Issues related to data ownership, privacy, and security is likely to dominate such developments. Nevertheless, the potential exists as demonstrated by Ullah et al. (2017) where they provide a glimpse of such possibilities by proposing a system that integrates emotion and medical history with biological information and family health data for improved disease prediction. The combination of utilizing all the healthcare data available for an individual and the ability to detect disease risk based on this collective data can give rise to a robust wearable recommendation system that helps in preventing and prescribing measures to prevent diseases.

The biggest hurdle for wearables to bring about behavioral change is the lack of effective interaction between the user and the device. The information presented back to the user must be easy to comprehend, rouses activity, and manages inspiration toward enhanced well-being. The fruitful use and potential medical advantages identified with wearables depend more on the outline of the engagement strategies than on the features of their innovation (Patel, Asch, & Volpp, 2015). Intelligent and personalized engagement strategies are needed. These strategies can leverage big data and artificial intelligence for the personalization of the interaction between the user and the device leading to enhancements to the interventions' efficacy with respect to enacting behavioral change and overall improvement in health and well-being.

Wearable devices are becoming more pervasive with the developments of wearable clothing, nano sensors, skin patches, and smart pills with a high computational ability and less physical footprint. Coupled with the increasing complexity and low resource requirements of AI techniques, we expect AI-enabled wearable devices to transform healthcare and well-being with a dramatic move towards preventive intervention rather than remedial cure.

CONCLUSION

This chapter provides a systematic review of the applications and recent developments of AI-enabled wearables for improved health and well-being. The focus is on self-care aimed at prevention of disease and the sustenance of health, and on the self-management of existing conditions. The results emphasize the potential of AI-enabled wearables to induce a paradigm shift from a reactive (treatment and management) focus to a proactive (preemptive and preventive) emphasis. Along these lines, there is an increasing trend towards leveraging big data from wearables and other sources with AI analytic capabilities to prescribe (or recommend) courses of action. Coupled with the incorporation of behavioral change theories, such trend has the potential of inducing sustained behavioral change which is at the crux of improved health and prevention of chronic ailment such as diabetes, obesity, and heart conditions among others.

While research and development are yet to realize the full potential for AI-enabled wearables in healthcare, the chapter identifies a number of factors that are expected to drive the growth of the market for wearables and the applications of AI in this domain. Most notable are the rising cost of healthcare, advances in technology, and ever-increasing ubiquity and pervasiveness of wearables. Along the future trajectory of AI-enabled wearables in healthcare, a number of challenges will continue to shape the agenda for future research, governing policies and regulations, and national debates. Most notable are issues surrounding the ownership, privacy, and security of data as well as the ethical considerations surrounding the use of AI techniques. Overall, the synergy between wearables (as a pervasive and ubiquitous source for real-time data) and artificial intelligence (for providing advanced analytics including diagnostic, predictive, and prescriptive capabilities) are likely to transform healthcare in the foreseeable future. Last but not least, the scope of the systematic literature review is limited to articles published in the English language and emphasizes peer-review literature indexed in scholarly databases. Accordingly, the research represents a strong academic focus capturing current research contributions and potential for future research. With the rapid advances of big data AI in healthcare, future reviews could increase reliance on 'grey literature' with an increased emphasis on commercialization endeavors of the technology.

REFERENCES

Adams, K. M., Corrigan, J. M., & Greiner, A. C. (2004). *1st Annual Crossing the Quality Chasm Summit: A Focus on Communities.* Washington, DC: National Academies Press.

Ahmadi, M., O'Neil, M., Fragala-Pinkham, M., Lennon, N., & Trost, S. (2018). Machine learning algorithms for activity recognition in ambulant children and adolescents with cerebral palsy. *Journal of Neuroengineering and Rehabilitation, 15*(1), 105. doi:10.1186/s12984-018-0456-x PubMed

Alcaraz, J. C., Moghaddamnia, S., Poschadel, N., & Peissig, J. (2018). Machine Learning as Digital Therapy Assessment for Mobile Gait Rehabilitation. 2018 IEEE 28th International Workshop on Machine Learning for Signal Processing (MLSP), 1–6. 10.1109/MLSP.2018.8517005

Alelaiwi, A., Hassan, M. M., & Bhuiyan, M. Z. A. (2017). A Secure and Dependable Connected Smart Home System for Elderly. 2017 IEEE 15th Intl Conf on Dependable, Autonomic and Secure Computing, 15th Intl Conf on Pervasive Intelligence and Computing, 3rd Intl Conf on Big Data Intelligence and Computing and Cyber Science and Technology Congress(DASC/PiCom/DataCom/CyberSciTech), 722–727. 10.1109/DASC-PICom-DataCom-CyberSciTec.2017.126

Alqaraawi, A., Alwosheel, A., & Alasaad, A. (2016). Heart rate variability estimation in photoplethysmography signals using Bayesian learning approach. Healthcare Technology Letters, 3(2), 136–142. doi:10.1049/htl.2016.0006 PubMed

Al-Ramahi, M. A., Liu, J., & El-Gayar, O. F. (2017). Discovering Design Principles for Health Behavioral Change Support Systems: A Text Mining Approach. *ACM Transactions on Management Information Systems, 8*(2–3), 1–24. doi:10.1145/3055534

Amft, O. (2018). How Wearable Computing Is Shaping Digital Health. *IEEE Pervasive Computing, 17*(1), 92–98. doi:10.1109/MPRV.2018.011591067

Badawy, R., Raykov, Y. P., Evers, L. J. W., Bloem, B. R., Faber, M. J., Zhan, A., ... Little, M. A. (2018). Automated Quality Control for Sensor Based Symptom Measurement Performed Outside the Lab. *Sensors (Basel), 18*(4), 1215. doi:10.3390/s18041215 PubMed

Baig, M., & Mirza, F. GholamHosseini, H., Gutierrez, J., & Ullah, E. (2018). Clinical Decision Support for Early Detection of Prediabetes and Type 2 Diabetes Mellitus Using Wearable Technology. 2018 40th Annual International Conference of the IEEE Engineering in Medicine and Biology Society (EMBC), 4156–4159. 10.1109/EMBC.2018.8513343

Baktir, A. C., Tunca, C., Ozgovde, A., Salur, G., & Ersoy, C. (2018). SDN-Based Multi-Tier Computing and Communication Architecture for Pervasive Healthcare. *IEEE Access : Practical Innovations, Open Solutions, 6*, 56765–56781. doi:10.1109/ACCESS.2018.2873907

Banerjee, A., Bandyopadhyay, T., & Acharya, P. (2013). Data Analytics: Hyped Up Aspirations or True Potential? *Vikalpa, 38*(4), 1–12. doi:10.1177/0256090920130401

Beranek, B., Simon, H. A., & McCorduck, P. (1977). History Of Artificial Intelligence. *IJCAI (United States), 2*, 4.

Bhise, S. (2018, October 11). Forces Driving the Growth of Wearable Medical Device Market. Retrieved January 24, 2019, from Health Works Collective website: https://www.healthworkscollective.com/forces-driving-the-growth-of-wearable-medical-device-market/

Borthakur, D., Dubey, H., Constant, N., Mahler, L., & Mankodiya, K. (2017). Smart fog: Fog computing framework for unsupervised clustering analytics in wearable Internet of Things. 2017 IEEE Global Conference on Signal and Information Processing (GlobalSIP), 472–476. doi:10.1109/GlobalSIP.2017.8308687

Center for Disease Control (CDC). (2018, September 5). About Chronic Disease. Retrieved September 25, 2018, from https://www.cdc.gov/chronicdisease/about/index.htm

Chan, M., Estève, D., Fourniols, J.-Y., Escriba, C., & Campo, E. (2012). Smart wearable systems: Current status and future challenges. *Artificial Intelligence in Medicine*, *56*(3), 137–156. doi:10.1016/j.artmed.2012.09.003 PubMed

Chen, M., Yang, J., Zhou, J., Hao, Y., Zhang, J., & Youn, C. (2018). 5G-Smart Diabetes: Toward Personalized Diabetes Diagnosis with Healthcare Big Data Clouds. *IEEE Communications Magazine*, *56*(4), 16–23. doi:10.1109/MCOM.2018.1700788

Cheng, S., Tamil, L. S., & Levine, B. (2015). A Mobile Health System to Identify the Onset of Paroxysmal Atrial Fibrillation. 2015 International Conference on Healthcare Informatics, 189–192. doi:10.1109/ICHI.2015.29

Cheng, Y., Chen, P., Yang, C., & Samani, H. (2016). IMU based activity detection for post mini-stroke healthcare. 2016 International Conference on System Science and Engineering (ICSSE), 1–4. doi:10.1109/ICSSE.2016.7551611

Cheng, Y., Jiang, P., & Peng, Y. (2014). Increasing big data front end processing efficiency via locality sensitive Bloom filter for elderly healthcare. 2014 IEEE Symposium on Computational Intelligence in Big Data (CIBD), 1–8. doi:10.1109/CIBD.2014.7011524

Choi, R., Kang, W., & Son, C. (2017). Explainable sleep quality evaluation model using machine learning approach. 2017 IEEE International Conference on Multisensor Fusion and Integration for Intelligent Systems (MFI), 542–546. doi:10.1109/MFI.2017.8170377

Clayton, C. (2015, January 27). The future of healthcare analytics is prescriptive. Retrieved January 30, 2019, from Healthcare IT News website: https://www.healthcareitnews.com/blog/future-healthcare-analytics-prescriptive

Costill, A. (2013, August 7). Top 10 Places that Have Banned Google Glass. Retrieved January 24, 2019, from Search Engine Journal website: https://www.searchenginejournal.com/top-10-places-that-have-banned-google-glass/66585/

Dieleman, J. L., Squires, E., Bui, A. L., Campbell, M., Chapin, A., Hamavid, H., ... Murray, C. J. L. (2017). Factors Associated with Increases in US Health Care Spending, 1996-2013. *Journal of the American Medical Association*, *318*(17), 1668–1678. doi:10.1001/jama.2017.15927 PubMed

Ejupi, A., & Menon, C. (2018). Detection of Talking in Respiratory Signals: A Feasibility Study Using Machine Learning and Wearable Textile-Based Sensors. *Sensors (Basel)*, *18*(8), 2474. doi:10.3390/s18082474 PubMed

El-Gayar, O., Nasralah, T., & Elnoshokaty, A. (2019). *Wearable devices for health and wellbeing: Design Insights from Twitter. In 52nd Hawaii International Conference on Systems Sciences (HICSS-52'19).* Maui, HI: IEEE Computer Society; doi:10.24251/HICSS.2019.467.

El-Gayar, O., Nasralah, T., & Noshokaty, A. E. (2018). IT for diabetes self-management - What are the patientsâ€TM expectations? AMCIS 2018 Proceedings. Retrieved from https://aisel.aisnet.org/amcis2018/DataScience/Presentations/18

Ellis, D. A., & Piwek, L. (2017). When wearable devices fail: Towards an improved understanding of what makes a successful wearable intervention. 1st GetAMoveOn Annual Symposium.

Ellis, D. A., & Piwek, L. (2018). Failing to encourage physical activity with wearable technology: What next? Journal of the Royal Society of Medicine. doi: PubMed doi:10.1177/0141076818788856

ElSaadany, Y., Majumder, A. J. A., & Ucci, D. R. (2017). A Wireless Early Prediction System of Cardiac Arrest through IoT. 2017 IEEE 41st Annual Computer Software and Applications Conference (COMP-SAC), 2, 690–695. 10.1109/COMPSAC.2017.40

Forkan, A. R. M., Khalil, I., & Atiquzzaman, M. (2017). ViSiBiD: A learning model for early discovery and real-time prediction of severe clinical events using vital signs as big data. *Computer Networks, 113,* 244–257. doi:10.1016/j.comnet.2016.12.019

Gholami, M., Ejupi, A., Rezaei, A., Ferrone, A., & Menon, C. (2018). Estimation of Knee Joint Angle Using a Fabric-Based Strain Sensor and Machine Learning: A Preliminary Investigation. 2018 7th IEEE International Conference on Biomedical Robotics and Biomechatronics (Biorob), 589–594. 10.1109/BIOROB.2018.8487199

Gisler, C., Ridi, A., Fauquex, M., Genoud, D., & Hennebert, J. (2014). Towards glaucoma detection using intraocular pressure monitoring. 2014 6th International Conference of Soft Computing and Pattern Recognition (SoCPaR), 255–260. 10.1109/SOCPAR.2014.7008015

Gribel, L. (2018). Drivers of Wearable Computing Adoption: An Empirical Study of Success Factors Including IT Security and Consumer Behaviour-Related Aspects. University of Plymouth. Retrieved from https://pearl.plymouth.ac.uk/bitstream/handle/10026.1/11662/2018gribel10508825.pdf?sequence=1&isAllowed=y

Harris, A., True, H., Hu, Z., Cho, J., Fell, N., & Sartipi, M. (2016). Fall recognition using wearable technologies and machine learning algorithms. 2016 IEEE International Conference on Big Data (Big Data), 3974–3976. doi:10.1109/BigData.2016.7841080

Hawas, M. A. (2017). Are We Intentionally Limiting Urban Planning and Intelligence? A Causal Evaluative Review and Methodical Redirection for Intelligence Systems. *IEEE Access : Practical Innovations, Open Solutions, 5,* 13253–13259. doi:10.1109/ACCESS.2017.2725138

Hidayat, W., Tambunan, T. D., & Budiawan, R. (2018). Empowering Wearable Sensor Generated Data to Predict Changes in Individual's Sleep Quality. 2018 6th International Conference on Information and Communication Technology (ICoICT), 447–452. 10.1109/ICoICT.2018.8528750

Iqbal, Z., Ilyas, R., Shahzad, W., & Inayat, I. (2018). A comparative study of machine learning techniques used in non-clinical systems for continuous healthcare of independent livings. 2018 IEEE Symposium on Computer Applications Industrial Electronics (ISCAIE), 406–411. doi:10.1109/ISCAIE.2018.8405507

Jiang, P., Winkley, J., Zhao, C., Munnoch, R., Min, G., & Yang, L. T. (2016). An Intelligent Information Forwarder for Healthcare Big Data Systems with Distributed Wearable Sensors. *IEEE Systems Journal*, *10*(3), 1147–1159. doi:10.1109/JSYST.2014.2308324

Kalantari, M. (2017). Consumers adoption of wearable technologies: Literature review, synthesis, and future research agenda. *International Journal of Technology Marketing*, *12*(1), 1. doi:10.1504/IJTMKT.2017.10008634

Karuppanan, K., Vairasundaram, A. S., & Sigamani, M. (2012, August 3). A comprehensive machine learning approach to prognose pulmonary disease from home. Academic Press. doi:10.1145/2345396.2345482

Khan, W. A., Idris, M., Ali, T., Ali, R., Hussain, S., & Hussain, M. ... Kang, B. H. (2015). Correlating health and wellness analytics for personalized decision making. 2015 17th International Conference on E-Health Networking, Application Services (HealthCom), 256–261. 10.1109/HealthCom.2015.7454508

Kim, S., Lee, K., Lee, J., & Jeon, J. Y. (2016). EPOC aware energy expenditure estimation with machine learning. *2016 IEEE International Conference on Systems, Man, and Cybernetics (SMC)*, 001585–001590. 10.1109/SMC.2016.7844465

Laney, D. (2001). 3D data management: Controlling data volume, velocity and variety. META Group Research Note, 6(70), 1.

Liberati, A., Altman, D. G., Tetzlaff, J., Mulrow, C., Gøtzsche, P. C., Ioannidis, J. P., ... Moher, D. (2009). The PRISMA statement for reporting systematic reviews and meta-analyses of studies that evaluate health care interventions: Explanation and elaboration. *PLoS Medicine*, *6*(7), e1000100. doi:10.1371/journal.pmed.1000100 PubMed

Lim, C.-G., Kim, Z. M., & Choi, H.-J. (2017). Context-based healthy lifestyle recommendation for enhancing user's wellness. *2017 IEEE International Conference on Big Data and Smart Computing (BigComp)*, 418–421. 10.1109/BIGCOMP.2017.7881748

Lingg, E., Leone, G., Spaulding, K., & B'Far, R. (2014). Cardea: Cloud based employee health and wellness integrated wellness application with a wearable device and the HCM data store. 2014 IEEE World Forum on Internet of Things (WF-IoT), 265–270. doi:10.1109/WF-IoT.2014.6803170

Liu, B., Dai, X., Gong, H., Guo, Z., Liu, N., Wang, X., & Liu, M. (2018). Deep Learning versus Professional Healthcare Equipment: A Fine-Grained Breathing Rate Monitoring Model. Mobile Information Systems, 2018, 1–9. doi:10.1155/2018/1904636

Liu, K., Hsieh, C., Hsu, S. J., & Chan, C. (2018). Impact of Sampling Rate on Wearable-Based Fall Detection Systems Based on Machine Learning Models. *IEEE Sensors Journal*, *18*(23), 9882–9890. doi:10.1109/JSEN.2018.2872835

Liu, K.-C., & Chan, C.-T. (2017). Significant Change Spotting for Periodic Human Motion Segmentation of Cleaning Tasks Using Wearable Sensors. *Sensors (Basel)*, *17*(1). doi: PubMed doi:10.339017010187

Liu, X., Rajan, S., Ramasarma, N., Bonato, P., & Lee, S. I. (2018). The Use of a Finger-Worn Accelerometer for Monitoring of Hand Use in Ambulatory Settings. IEEE Journal of Biomedical and Health Informatics. doi: PubMed doi:10.1109/JBHI.2018.2821136

Lustrek, M., Gjoreski, H., Vega, N. G., Kozina, S., Cvetkovic, B., Mirchevska, V., & Gams, M. (2015). Fall Detection Using Location Sensors and Accelerometers. *IEEE Pervasive Computing*, *14*(4), 72–79. doi:10.1109/MPRV.2015.84

Mathur, N., Paul, G., Irvine, J., Abuhelala, M., Buis, A., & Glesk, I. (2016). A Practical Design and Implementation of a Low Cost Platform for Remote Monitoring of Lower Limb Health of Amputees in the Developing World. *IEEE Access: Practical Innovations, Open Solutions*, *4*, 7440–7451. doi:10.1109/ACCESS.2016.2622163

Mazilu, S., Hardegger, M., Zhu, Z., Roggen, D., Tröster, G., Plotnik, M., & Hausdorff, J. M. (2012). Online detection of freezing of gait with smartphones and machine learning techniques. 2012 6th International Conference on Pervasive Computing Technologies for Healthcare (PervasiveHealth) and Workshops, 123–130. 10.4108/icst.pervasivehealth.2012.248680

Mei, Z., Gu, X., Chen, H., & Chen, W. (2018). Automatic Atrial Fibrillation Detection Based on Heart Rate Variability and Spectral Features. *IEEE Access: Practical Innovations, Open Solutions*, *6*, 53566–53575. doi:10.1109/ACCESS.2018.2871220

Mercer, K., Li, M., Giangregorio, L., Burns, C., & Grindrod, K. (2016). Behavior Change Techniques Present in Wearable Activity Trackers: A Critical Analysis. *JMIR mHealth and uHealth*, *4*(2), e40. doi:10.2196/mhealth.4461 PubMed

Mesh, J. (2018, August 2). How Wearables are Changing the Healthcare Industry. Retrieved September 27, 2018, from Healthcare IT Leaders website: https://www.healthcareitleaders.com/blog/how-wearables-are-changing-the-healthcare-industry/

Mukhopadhyay, S. C. (2015). Wearable Sensors for Human Activity Monitoring: A Review. *IEEE Sensors Journal*, *15*(3), 1321–1330. doi:10.1109/JSEN.2014.2370945

Nemati, E., Liaqat, D., Rahman, M. M., & Kuang, J. (2017). A novel algorithm for activity state recognition using smartwatch data. 2017 IEEE Healthcare Innovations and Point of Care Technologies (HI-POCT), 18–21. doi:10.1109/HIC.2017.8227574

Ngo, M. V., La, Q. D., Leong, D., & Quek, T. Q. S. (2017). User behavior driven MAC scheduling for body sensor networks. 2017 IEEE 19th International Conference on E-Health Networking, Applications and Services (Healthcom), 1–6. 10.1109/HealthCom.2017.8210762

Nii, M., Iwamoto, T., Okajima, S., & Tsuchida, Y. (2016). Hybridization of standard and fuzzified neural networks from MEMS-based human condition monitoring data for estimating heart rate. 2016 International Conference on Machine Learning and Cybernetics (ICMLC), 1, 1–6. doi:10.1109/ICMLC.2016.7860868

Oung, Q. W., Hariharan, M., Lee, H. L., Basah, S. N., Sarillee, M., & Lee, C. H. (2015). Wearable multi-modal sensors for evaluation of patients with Parkinson disease. 2015 IEEE International Conference on Control System, Computing and Engineering (ICCSCE), 269–274. doi:10.1109/ICCSCE.2015.7482196

Park, E., Lee, S. I., Nam, H. S., Garst, J. H., Huang, A., Campion, A., ... Sarrafzadeh, M. (2017). Unobtrusive and Continuous Monitoring of Alcohol-impaired Gait Using Smart Shoes. *Methods of Information in Medicine*, *56*(1), 74–82. doi:10.3414/ME15-02-0008 PubMed

Park, S., Chung, K., & Jayaraman, S. (2014). Wearables: Fundamentals, Advancements, and a Roadmap for the Future. In Wearable Sensors (pp. 1–23). Academic Press. doi:10.1016/B978-0-12-418662-0.00001-5

Parsa, M., Panda, P., Sen, S., & Roy, K. (2017). Staged Inference using Conditional Deep Learning for energy efficient real-time smart diagnosis. Conference Proceedings: ... Annual International Conference of the IEEE Engineering in Medicine and Biology Society. IEEE Engineering in Medicine and Biology Society. Annual Conference, 2017, 78–81. doi:10.1109/EMBC.2017.8036767

Patel, M. S., Asch, D. A., & Volpp, K. G. (2015). Wearable Devices as Facilitators, Not Drivers, of Health Behavior Change. *Journal of the American Medical Association*, *313*(5), 459–460. doi:10.1001/jama.2014.14781 PubMed

Pathinarupothi, R. K., Prathap, J. D., Rangan, E. S., Gopalakrishnan, E. A., Vinaykumar, R., & Soman, K. P. (2017). Single Sensor Techniques for Sleep Apnea Diagnosis Using Deep Learning. 2017 IEEE International Conference on Healthcare Informatics (ICHI), 524–529. doi:10.1109/ICHI.2017.37

Punagin, S., & Arya, A. (2015). Privacy in the age of Pervasive Internet and Big Data Analytics - Challenges and Opportunities - ProQuest. *International Journal of Modern Education and Computer Science*, *7*(7), 36–47. doi:10.5815/ijmecs.2015.07.05

Randazzo, V., Pasero, E., & Navaretti, S. (2018). VITAL-ECG: A portable wearable hospital. *2018 IEEE Sensors Applications Symposium (SAS)*, 1–6. 10.1109/SAS.2018.8336776

Ravi, D., Wong, C., Lo, B., & Yang, G. (2016). Deep learning for human activity recognition: A resource efficient implementation on low-power devices. 2016 IEEE 13th International Conference on Wearable and Implantable Body Sensor Networks (BSN), 71–76. 10.1109/BSN.2016.7516235

Ravi, D., Wong, C., Lo, B., & Yang, G. (2017). A Deep Learning Approach to on-Node Sensor Data Analytics for Mobile or Wearable Devices. *IEEE Journal of Biomedical and Health Informatics*, *21*(1), 56–64. doi:10.1109/JBHI.2016.2633287 PubMed

Razum, D., Seketa, G., Vugrin, J., & Lackovic, I. (2018). Optimal threshold selection for threshold-based fall detection algorithms with multiple features. 2018 41st International Convention on Information and Communication Technology, Electronics and Microelectronics (MIPRO), 1513–1516. 10.23919/MIPRO.2018.8400272

Reiss, A., & Stricker, D. (2014). Aerobic activity monitoring: Towards a long-term approach. *Universal Access in the Information Society*, *13*(1), 101–114. doi:10.1007/s10209-013-0292-5

Rizwan, A., Zoha, A., Zhang, R., Ahmad, W., Arshad, K., Ali, N. A., ... Abbasi, Q. H. (2018). A Review on the Role of Nano-Communication in Future Healthcare Systems: A Big Data Analytics Perspective. *IEEE Access : Practical Innovations, Open Solutions*, *6*, 41903–41920. doi:10.1109/ACCESS.2018.2859340

Rosso, R., Munaro, G., Salvetti, O., Colantonio, S., & Ciancitto, F. (2010). CHRONIOUS: An open, ubiquitous and adaptive chronic disease management platform for Chronic Obstructive Pulmonary Disease (COPD), Chronic Kidney Disease (CKD) and renal insufficiency. 2010 Annual International Conference of the IEEE Engineering in Medicine and Biology, 6850–6853. doi:10.1109/IEMBS.2010.5626451

Saha, S. S., Rahman, S., Rasna, M. J., Zahid, T. B., Islam, A. K. M. M., & Ahad, M. A. R. (2018). Feature Extraction, Performance Analysis and System Design Using the DU Mobility Dataset. *IEEE Access : Practical Innovations, Open Solutions*, 6, 44776–44786. doi:10.1109/ACCESS.2018.2865093

Sathyanarayana, A., Joty, S., Fernandez-Luque, L., Ofli, F., Srivastava, J., Elmagarmid, A., ... Taheri, S. (2016). Sleep Quality Prediction from Wearable Data Using Deep Learning. *JMIR mHealth and uHealth*, 4(4). doi:10.2196/mhealth.6562

Sathyanarayana, A., Srivastava, J., & Fernandez-Luque, L. (2017). The Science of Sweet Dreams: Predicting Sleep Efficiency from Wearable Device Data. *Computer*, 50(3), 30–38. doi:10.1109/MC.2017.91

Self Care Forum. (2019). What do we mean by self care and why is it good for people? Retrieved April 9, 2019, from Self Care Forum website: http://www.selfcareforum.org/about-us/what-do-we-mean-by-self-care-and-why-is-good-for-people/

Shin, G., Jarrahi, M. H., Fei, Y., Karami, A., Gafinowitz, N., Byun, A., & Lu, X. (2019). Wearable activity trackers, accuracy, adoption, acceptance and health impact: A systematic literature review. *Journal of Biomedical Informatics*, 93, 103153. doi:10.1016/j.jbi.2019.103153 PubMed

Stefanie, C. (2018, June 4). Wearable Tech is Here to Stay with a Robust Presence in the Future Healthcare Industry. Retrieved September 19, 2018, from Wearable Technologies website: https://www.wearable-technologies.com/2018/06/wearable-tech-is-here-to-stay-with-a-robust-presence-in-the-future-healthcare-industry/

Steffen, D., Bleser, G., Weber, M., Stricker, D., Fradet, L., & Marin, F. (2011). A personalized exercise trainer for elderly. 2011 5th International Conference on Pervasive Computing Technologies for Healthcare (PervasiveHealth) and Workshops, 24–31. 10.4108/icst.pervasivehealth.2011.245937

Swan, M. (2013). The Quantified Self: Fundamental Disruption in Big Data Science and Biological Discovery. *Big Data*, 1(2), 85–99. doi:10.1089/big.2012.0002 PubMed

Tahavori, F., Stack, E., Agarwal, V., Burnett, M., Ashburn, A., Hoseinitabatabaei, S. A., & Harwin, W. (2017). Physical activity recognition of elderly people and people with parkinson's (PwP) during standard mobility tests using wearable sensors. 2017 International Smart Cities Conference (ISC2), 1–4. doi:10.1109/ISC2.2017.8090858

Tien, J. M. (2013). Big Data: Unleashing information. *Journal of Systems Science and Systems Engineering*, 22(2), 127–151. doi:10.1007/s11518-013-5219-4

Turing, A. M. (1950). Computing Machinery and Intelligence. *Mind*, 49(236), 433–460. doi:10.1093/mind/LIX.236.433

Ullah, M. R., Bhuiyan, M. A. R., & Das, A. K. (2017). IHEMHA: Interactive healthcare system design with emotion computing and medical history analysis. 2017 6th International Conference on Informatics, Electronics and Vision 2017 7th International Symposium in Computational Medical and Health Technology (ICIEV-ISCMHT), 1–8. 10.1109/ICIEV.2017.8338606

Wahbeh, A., Sarnikar, S., & El-Gayar, O. (2016). Improving analysts' domain knowledge for the requirements elicitation phase: A socio-technical perspective. Presented at the AMCIS 2016: Surfing the IT Innovation Wave - 22nd Americas Conference on Information Systems.

Walinjkar, A., & Woods, J. (2017). ECG classification and prognostic approach towards personalized healthcare. 2017 International Conference On Social Media, Wearable and Web Analytics (Social Media), 1–8. doi:10.1109/SOCIALMEDIA.2017.8057360

Weiss, G. M., Lockhart, J. W., Pulickal, T. T., McHugh, P. T., Ronan, I. H., & Timko, J. L. (2016). Actitracker: A Smartphone-Based Activity Recognition System for Improving Health and Well-Being. 2016 IEEE International Conference on Data Science and Advanced Analytics (DSAA), 682–688. doi:10.1109/DSAA.2016.89

WHO. (2018). Non communicable diseases. Retrieved September 25, 2018, from World Health Organization website: http://www.who.int/news-room/fact-sheets/detail/noncommunicable-diseases

Yuan, B., & Herbert, J. (2014). A Cloud-Based Mobile Data Analytics Framework: Case Study of Activity Recognition Using Smartphone. 2014 2nd IEEE International Conference on Mobile Cloud Computing, Services, and Engineering, 220–227. 10.1109/MobileCloud.2014.29

Zhang, Q., Zhou, D., & Zeng, X. (2017). Highly wearable cuff-less blood pressure and heart rate monitoring with single-arm electrocardiogram and photoplethysmogram signals. *Biomedical Engineering Online*, *16*(1), 23. doi:10.1186/s12938-017-0317-z PubMed

Zhu, Y. (2011). Automatic detection of anomalies in blood glucose using a machine learning approach. *Journal of Communications and Networks (Seoul)*, *13*(2), 125–131. doi:10.1109/JCN.2011.6157411

ADDITIONAL READING

Carvalko, J. R. (2013). Law and policy in an era of cyborg-assisted-life1: The implications of interfacing in-the-body technologies to the outer world2. *2013 IEEE International Symposium on Technology and Society (ISTAS): Social Implications of Wearable Computing and Augmediated Reality in Everyday Life*, 204–215. 10.1109/ISTAS.2013.6613121

Chen, M., Ma, Y., Li, Y., Wu, D., Zhang, Y., & Youn, C. (2017). Wearable 2.0: Enabling Human-Cloud Integration in Next Generation Healthcare Systems. *IEEE Communications Magazine*, *55*(1), 54–61. doi:10.1109/MCOM.2017.1600410CM

Gualtieri, L., Rosenbluth, S., & Phillips, J. (2016). Can a Free Wearable Activity Tracker Change Behavior? The Impact of Trackers on Adults in a Physician-Led Wellness Group. *JMIR Research Protocols*, *5*(4), e237. doi:10.2196/resprot.6534 PMID:27903490

Lavalliere, M., Arezes, P., Burstein, A., & Coughlin, J. F. (2015). *The quantified-Self and wearable technologies in the workplace: implications and challenges for their implementations.* 161–163. Retrieved from http://repositorium.sdum.uminho.pt/handle/1822/38579

Meyer, J., Schnauber, J., Heuten, W., Wienbergen, H., Hambrecht, R., Appelrath, H., & Boll, S. (2016). Exploring Longitudinal Use of Activity Trackers. *2016 IEEE International Conference on Healthcare Informatics (ICHI)*, 198–206. 10.1109/ICHI.2016.29

Shameer, K., Badgeley, M. A., Miotto, R., Glicksberg, B. S., Morgan, J. W., & Dudley, J. T. (2017). Translational bioinformatics in the era of real-time biomedical, health care and wellness data streams. *Briefings in Bioinformatics*, *18*(1), 105–124. doi:10.1093/bib/bbv118 PMID:26876889

Vandervort, D. (2016, April 11)... *Medical Device Data Goes to Court*, *23–27*. doi:10.1145/2896338.2896341

Villalba, M. T., Buenaga, M. de, Gachet, D., & Aparicio, F. (2015). Security Analysis of an IoT Architecture for Healthcare. Internet of Things. IoT Infrastructures, 454–460. doi:10.1007/978-3-319-47063-4_48

KEY TERMS AND DEFINITIONS

Cognition: The capability of thinking and understanding or to process data through perception and learn from experience.

Descriptive Data Analytics: The process of summarizing the current and historic data to get useful information.

Electronic Health Record (EHR): Electronic health record is a collection of digital patient data in a systematic manner which can be used to share among multiple health care companies.

Gross Domestic Product (GDP): It is measure of total economic activity in a country.

Health Intervention: An action or effort trying to maintain or improve the mental or physical health condition of a person or population.

Intraocular Pressure: The measure of fluid pressure inside the eye.

Personal Health Record (PHR): Personal health record is where the patient health data is stored and managed by the patient himself.

Quantified Self: Quantified self is simply measuring certain physical metrics of an individual to track their health using on body sensors.

Sedentary Lifestyle: The lifestyle where an individual spends most of the day with very little or no physical activity.

Socioeconomic: The combination of social aspects like age, race, demographic, and economic intersection of a wide variety of individuals.

Chapter 6
Blockchain as a Disruptive Technology:
Architecture, Business Scenarios, and Future Trends

Gopala Krishna Behara
Wipro Technologies, India

Tirumala Khandrika
Wipro Technologies, India

ABSTRACT

Blockchain is a digital, distributed, and decentralized network to store information in a tamper-proof way with an automated way to enforce trust among different participants. An open distributed ledger can record all transactions between different parties efficiently in a verifiable and permanent way. It captures and builds consensus among participants in the network. Each block is uniquely connected to the previous blocks via a digital signature which means that making a change to a record without disturbing the previous records in the chain is not possible, thus rendering the information tamper-proof. Blockchain holds the potential to disrupt any form of transaction that requires information to be trusted. This means that all intermediaries of trust, as they exist today, exposed to disruption in some form with the initiation of Blockchain technology. Blockchain works by validating transactions through a distributed network in order to create a permanent, verified, and unalterable ledger of information.

DOI: 10.4018/978-1-5225-9687-5.ch006

INTRODUCTION

Satoshi Nakamoto invented the Blockchain in 2008 (Economist, 2015) to serve as the public transaction ledger of the cryptocurrency bitcoin. Nakamoto's research paper (Nakamoto, 2008), contained the blueprint that most modern cryptocurrency schemes follow (although with variations and modifications). Bitcoin was just the first of many Blockchain applications. The invention of Blockchain for bitcoin was the first digital currency to solve the double-spending problem without the need of a trusted authority or central server. This technology became fully recognized in 2009 with the launch of the Bitcoin network, the first of many modern cryptocurrencies.

The words Block and Chain were used separately in Satoshi Nakamoto's original paper (Nakamoto, 2008), and by 2016 became mainstream, as a single term, Blockchain. The label Blockchain 2.0 refers to new applications of the distributed Blockchain database (Kariappa, 2015). As of 2016, Blockchain 2.0 implementations continue to require an off-chain oracle to access any external data or events based on time or market conditions to interact with the Blockchain (Gray, 2017). In July 2016, IBM opened a Blockchain Innovation Research Center in Singapore (Williams, 2016). A working group for the World Economic Forum met in November 2016 to discuss the development of governance models related to Blockchain. According to Accenture, the Blockchains attained a 13.5% adoption rate within financial services domain in 2016 (Raconteur, 2016). In May 2018, Gartner found that only 1% of Chief Information Officer's (CIO) indicated any kind of Blockchain adoption within their organizations, and only 8% of CIOs were in the short-term *"planning or looking at active experimentation with Blockchain"* (Artificiallawyer, 2018). In November 2018, Conservative Member of the European Parliament, Emma McClarkin mooted a plan to utilize Blockchain technology to boost trade backed by the European Parliament's Trade Committee (McClarkin, 2018).

Figure 1 shows blockchain developments on a timeline.

Market Research predicts that, by 2024, global Blockchain market expected to be worth over $20 billion (Transparencymarketresearch, 2018). Recently, the Dubai government announced that they would put 100% of their records pertaining to land registry on Blockchain. Dubai Land Department (DLD), in fact, has claimed to be the first such governmental department anywhere in the world, to adopt

Figure 1. History of Blockchain

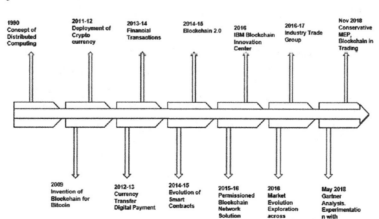

Blockchain for such high-level tasks (Das, 2017). The Republic of Georgia declared that they would use Blockchain technology to validate property-related government transactions. Countries like Sweden, Honduras and others are also developing such similar Blockchain-based systems, for enabling secured e-Governance (Shin, 2017). Gartner projects that Blockchain's business value will grow to $176 billion by 2025 (Garneto, Kandaswamy, Lovelock, & Reynolds, 2017). The European Union's commercial research group, the European Innovation Council (EIC), has launched a program to grant 2.7 billion euros to 1000 projects, that are developing systems and solutions using Blockchain technology (Shin, 2017). Andhra Pradesh has become the first state in India to pilot Blockchain technology in two departments and plans to deploy it across the administration (Firstpost, 2017).

This Chapter covers indepth the role of Blockchain technology, its importance in industry, and its architectural framework to manage data that is generated by various sources. It also addresses drivers, principles, opportunities and benefits of Blockchain implementation applicable to real world issues with detailed case studies.

BACKGROUND

There exists different Blockchain definitions, ranging from the application-specific to the excessively technical.

For example, Coinbase (2019), the world's largest cryptocurrency exchange, defines Blockchain as *"a distributed, public ledger that contains the history of every bitcoin transaction"*. According to Webopedia, Blockchain is *"a type of data structure that enables identifying and tracking transactions digitally and sharing this information across a distributed network of computers, creating in a sense a distributed trust network. The distributed ledger technology offered by Blockchain provides a transparent and secure means for tracking the ownership and transfer of assets"* (Stroud, 2015). Blockchains are distributed digital ledgers of cryptographically signed transactions that are grouped into blocks. Each block is cryptographically linked to the previous one (making it tamper evident) after validation and undergoing a consensus decision. As new blocks are added, older blocks become more difficult to modify (creating tamper resistance). New blocks are replicated across copies of the ledger within the network, and any conflicts are resolved automatically using established rules.

The simplest definition of a Blockchain is, *"A decentralized database containing sequential, cryptographically linked blocks of digitally signed asset transactions, governed by a consensus model."* (Nakamoto, 2008).

Blockchain technology is a peer-to-peer networked database governed by a set of rules. Furthermore, Blockchain represents a shift away from traditional trust agents and a move towards transparency. As a technological building block, it permits applications from a broad swath of industries to take advantage of sharing, tracking, and auditing digital assets. This technology has the potential to improve security, processes and systems in the financial services where accuracy, tamper proofing and record keeping is essential. (Sultan, Ruhi, & Lakhani, 2018).

DISRUPTIVE INNOVATION OF BLOCKCHAIN TECHNOLOGIES

Disruptive technology refers to any enhanced or completely new technology that replaces and disrupts an existing technology, rendering it obsolete. (Techopedia, 2019). Disruptive technology applies to hardware, software, networks and combination technologies. Because disruptive technology is novel, it has certain advantages, enhancements and functionalities over competitors.

Blockchain disrupts every industry. Blockchain is a disruptive technology because of its ability to digitize, decentralize, secure and incentivizes the validation of transactions. A wide range of industries is evaluating Blockchain to determine what strategic differentiators could exist for their businesses if they leverage Blockchain. Applied disrupted industries will include financial services, healthcare, aviation, global logistics and shipping, transportation, music, manufacturing, security, media, identity, automotive industries, land use and government. Blockchain is garnering a great deal off attention because as a technology it will fundamentally change many of these industries (Marr, 2018). *"The 'killer app' for the early internet was email; it's what drove adoption and strengthened the network. Bitcoin is the 'killer app' for the Blockchain."* (Ito, Narula, & Ali, 2017). As per Christensen (Varsamis, 2018), *"disruptive innovation refers to a technological or market advance that fundamentally transforms or creates entirely new markets"*. That means disruptive innovation has two capabilities, it can act as market disruptor and new market creator. Market disruptor changes the current way of operations, processes, and technologies. Uber is an example of a market disruptor. It has changed the way cab services are provided, requested and used. It never owns any cabs but acts as an aggregator (new business model). Internet is an example of new market creator. It has opened up business opportunities for segments of people who were never previously exposed to said markets. Blockchain can be market disruptor and new market creator. It holds the potential to disrupt almost every market and create more new markets. The real disruption is the establishment of trust through collaboration and code, rather than a central authority. This certitude is revolutionary and has the potential to transform nearly every industry (Afshar, 2017).

The following are the few challenges in the current market that Blockchain can solve,

- Trust
- Transparency
- Intermediaries
- Time consuming processes

Trust

Blockchain is decentralized peer-to-peer network that records transactions to blocks and writes to the distributed ledger based on consensus between all the participants in the network. This Ledger is distributed to all the nodes. These factors foster trust among the participants.

Transparency

The transparency of a blockchain stems from the fact that the holdings and transactions of each public address are open to viewing. To safeguard the data recorded on the Blockchain, all the participants in the network possess a distributed ledger copy and no party can maliciously change the data in the ledger, as the transaction will only be committed to the Blockchain once all parties come to consensus (based on applicable consensus algorithm).

Intermediaries

Blockchain technology has the potential to eliminate the need for intermediaries. Once Blockchain technology is adopted, consensus algorithms will transparently record and verify transactions, eliminating the need of manual verification by any third party. Cryptography helps in eliminating the third party intermediaries as keepers of trust.

Time Consuming Process

Every recordable transaction requires peer-to-peer verification, which can become time-consuming on account of the number of blocks involved. With transactions, executing occurs almost immediately and significantly reduces processing times (sometimes from days to hours). Blockchain helps in replacing old time-consuming process by automated verification and validation.

In cases of cross-border wire transfers concerning the banking industry, there is a huge process which is executed and which comprises compliance and technology overheads. Implementing a Blockchain-based transfer solution can significantly improve the time it takes from a couple of days to a couple of hours. Banks like the Bank of England (Eha, 2017), United Overseas Bank of Singapore and SWIFT with a consortium of banks have already implemented this solution as a pilot (Finextra, 2017). As part of the Proof of Concepts (POC) banks were able to transfer money directly from one bank to another, both connected by Blockchain. Similarly, it creates new markets to exchange any entity, goods and digital assets with anonymous or known participants in a transparent, secure and simple manner. For example, users can exchange (sell) digital assets on the internet powered by Blockchain. This was not previously possible because of the lack of trust, security and intermediaries. Blockchain has created a new market on the internet called "Internet of value" (Leonard, 2107). This refers to Blockchain enabling exchanging any digital asset over the internet in the same manner information is exchanged. Blockchain will handle the trust, security and other factors. It, in addition, enables the internet to value certain futuristic features like setting up standards for settling transactions across different ledgers and networks (like InterLedgerProtocol), connecting different Blockchain networks, requiring setting up governance.

Below are some examples of the opportunities that exist to improve processes in a variety of industries using Blockchain.

Land Use

Ownership and history of property currently requires the investigation of many different document sources such as the grantor-grantee index, land records or deed records. A grantor-grantee index is a general term for two lists of real property transfers maintained in alphabetical order of the last name of the parties transferring the property. Land records or deed records are the documents related to the ownership of the land. The goal is to find all records related to property liens, easements, Covenants, Conditions and Restrictions (CC&Rs), agreements, resolutions and ordinances. CC&Rs are limits and rules placed on a group of homes or condominium complexes by a builder, developer, neighborhood association, or homeowner's association. This is a time-consuming and laborious process in which it is easy to overlook important information. Sweden is presently leveraging Blockchain to track land registries termed the Lantmateriet (Lantmateriet, 2019). They estimate a taxpayer savings of $106 million per year based on reduction of fraud, eliminating paperwork and accelerating the process.

Identity

Across the globe, we use passports to identify people, which are paper-based identity cards similar to the driver's license. In 2013, almost 40 million travel documents were reported as lost or stolen (Serenelli, 2014). The Dubai government is working on a digital passport with a London-based company called ObjectTech. This digital passport, based on Blockchain, makes international travel quicker and safer, but also gives people back control of their personal digital data.

Global Logistics and Shipping

The second largest port in Europe, Belgium-based Antwerp, announced a Blockchain pilot to automate and streamline the port's container logistics operations (Zhao, 2017).

According to the terminal authority, moving containers from point-to-point often involves more than 30 different parties, including carriers, terminals, forwarders, haulers, drivers, shippers and more. This process results in hundreds of interactions between those parties, conducted through a mix of e-mail, phone and fax.

Automotive

German automaker Daimler AG has issued a corporate bond worth €100m as part of a Blockchain pilot project (Zhao, 2017). According to Daimler, the entire transaction cycle from origination, distribution, allocation and execution of the loan agreement, to the confirmation of repayment and of interest payments was automated digitally through the Blockchain network.

Aviation

Accenture's head of Aerospace and Defense said about Blockchain,

Through all that life cycle of the engine, the original parts, the replacement parts and configuration are all being tracked, and it is being done by a number of different companies. Blockchain is in effect a single federated ledger that everybody who uses and touches that engine could use it as a single point of truth of what has happened to the engine. (Reid, 2017).

Finance

Visa has a Blockchain effort designated "Visa B2B Connect" (Hill, 2016), which collaborates with the Chain to analyze the possibility of optimizing near real-time funds transfer system for high value bank-to-bank and corporate payments. A company called Ripple is working with banks to optimize how they send money around the world, with the goal of new revenue models, lower processing costs and better overall customer experience. IBM Global Finance is additionally working on one of the largest Blockchain implementations (Dhaliwal, 2016).

Government

The US Navy's Naval Innovation Advisory Council (NIAC) (Pimentel, 2107), will spearhead the testing of Blockchain technology in their 3D printing in order to help securely transfer data during the manufacturing process.

BLOCKCHAIN CATEGORIZATION

Blockchain networks can be categorized based on their permission model (Yaga, Mell, Roby, & Scarfone, 2018), which determines who can maintain them (for example, publish blocks). If anyone publishes a new block, it is permissionless. By contrast, if a particular user publishes blocks, it is permissioned. In simple terms, a permissioned Blockchain network is similar to a corporate intranet that regulates, while a permissionless Blockchain network is similar to the public internet, where anyone can participate. Permissioned Blockchain networks deployed for a group of organizations and individuals are typically referred to as a consortium.

Permissionless

Permissionless Blockchain networks are decentralized ledger platforms open to anyone publishing blocks (Yaga, Mell, Roby, & Scarfone, 2018) without needing permission from any authority. Permissionless Blockchain platforms are often open source software, freely available to anyone who wishes to download them. Since anyone has the right to publish blocks, this results in the property or quality that anybody can read the Blockchain as well as issue transactions on the Blockchain. Any Blockchain network user within a permissionless Blockchain network can read and write to the ledger. Since permissionless Blockchain networks are open to all to participate, malicious users could attempt to publish blocks in a way that subverts the system. To prevent this, permissionless Blockchain networks often utilize a multiparty agreement or 'consensus' system that requires users to expend or maintain resources when attempting to publish blocks.

Permissioned

Permissioned Blockchain networks are entities where users publishing blocks are authorized by some authority. To restrict read access and to restrict who can issue transactions, a permissioned Blockchain is used. Permissioned Blockchain networks may be instantiated and maintained using open source or closed source software. Organizations use permissioned Blockchain are networks that need a more rigorous control to protect their Blockchain.

PRINCIPLES OF BLOCKCHAIN

Architecture principles provide a basis for decision making when developing Blockchain solutions and design. These are enterprise specific architecture principles and requirements (McLeod, 2018; Blockchain Wikipedia, 2018). They form a structured set of ideas that collectively define and guide development of a solution architecture, from values through to design and implementation, harmonizing decision making across an organization.

Figure 2 highlights the principles to guide enterprises in their approach to Blockchain.

Integrity

Blockchain technology utilizes a distributed verification system for transactions that does not centralize power. Integrity is required, during the decision rights of Blockchain users and network's incentive structures (Figure 2). The use of complex algorithms and consensus among users ensures that transaction data once agreed cannot be tampered with. Data stored on Blockchain acts as a single version of truth for all parties involved, reducing the risk of fraud.

Figure 2. Principles of Blockchain

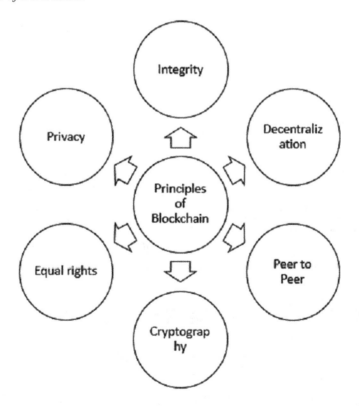

Decentralization

Blockchain distributes power among the different members using this technology (Figure 2). This means no single person or system can ever tamper with Blockchain. Blockchain is robust and is resistant to attacks and fraud due to a distributed power system.

Peer-to-Peer

All Blockchain transactions are peer-to-peer (Figure 2). It means that, at least two parties are involved. The best way to understand this is to compare a standard credit card transaction with the first practical use of the Blockchain. The benefits of the peer-to-peer aspects of the Blockchain is that two people who are involved in a transaction can both verify the transaction is correct, without the need for any other party to be involved.

Cryptographic System

Cryptography is the method of disguising and revealing, otherwise known as encrypting and decrypting, information through complex mathematics. This means that the information can only be viewed by the intended recipients and nobody else. Blockchain lends its cryptography a level of authenticity and integrity (Figure 2).

Equal Rights

Rights are reserved in a transparent manner so much so that it is self-evident to most users (Figure 2). Blockchain users know exactly what rights they are entitled to, and that the rights are equal and fair in the best way possible.

Privacy

Blockchain establishes a separate virtual identity for its users to protect their privacy, both online and offline (Figure 2). The identity consists of a series of numbers, much like code, that are difficult to profile. Hence, Blockchain fosters trust by eliminating the need for any real (or seemingly real) identity. It protects the identity of its users so that they can participate in the digital economy freely and fairly with integrity.

BLOCKCHAIN ARCHITECTURE FRAMEWORK

Blockchain is the spine of the entire cryptocurrency system. Blockchain technology not only helps the users to perform transactions using cryptocurrencies but also ensures that the security and anonymity of the users. It is a continuous growing list of records called blocks linked and secured using cryptographic techniques.

A Blockchain can serve as *"an open and distributed ledger, which can record transactions between two parties in a verifiable and permanent way."* (Bharadwaj, 2018) A ledger that is shared among anyone in the network is public for all to view. This results in the system being both transparent and trustworthy.

The simplified image, below, shows the key components of a Blockchain: users, nodes, ledgers, and the private key.

- Users - may include auditors, administrators, developers, etc.
- Nodes - make additions to the ledger by passing around transaction and block data.
- Ledger - lists who owns what.
- Private Key - is an encrypted large integer number that is used to generate a signature for each Blockchain transaction a user sends out. This private key prevents the transaction from getting altered by anyone once it has issued

It records information in a growing list of records containing a cryptographic hash of the previous block, a timestamp and the full transaction data. By its immutable design, a Blockchain is inherently resistant to unsolicited modification. Blockchain records transactions between two parties efficiently, permanently and transparently. Blockchain managed by a peer-to-peer network, which means that once the date recorded.

Blockchain is a sequence of blocks, which holds a complete list of transaction records like a conventional public ledger.

Figure 3 illustrates an example of a Blockchain. With a previous block hash contained in the block header, a block has only one parent block. It is worth noting that children block hashes would also be stored in Ethereum Blockchain. The first block of a Blockchain called genesis block that has no parent block.

Figure 3. Sequence of Blocks

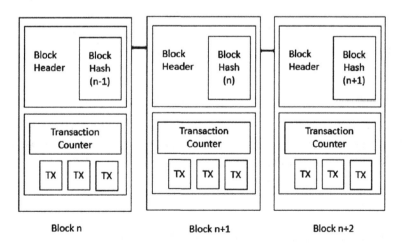

Figure 4. Block structure

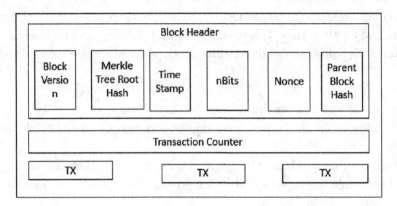

Block

A block consists of the block header and the block body (Figure 3). A hash identifies each block, a 256-bit number, created using an algorithm agreed upon by the network. A block contains header, a reference to the previous block's hash and a group of transactions. The sequence of linked hashes creates a secure, interdependent chain. Figure 4 above shows the structure of a block.

Block Header

In particular, the block header includes (Figure 4),

- **Block version:** Indicates which set of block validation rules to follow.
- **Merkle tree root hash:** The hash value of all the transactions in the block.
- **Timestamp:** Current time as seconds in universal time since January 1, 1970.
- **nBits:** Target threshold of a valid block hash.
- **Nonce:** A 4-byte field, which usually starts with 0 and increases for every hash calculation.
- **Parent block hash:** A 256-bit hash value that points to the previous block.
- **Size of Block**

Block Body

It is composed of a transaction counter and transactions. The maximum number of transactions that a block can contain depends on the block size and the size of each transaction. Blockchain uses an asymmetric cryptography mechanism to validate the authentication of transactions.

Block Time

The block time is the average time it takes for the network to generate one extra block in the Blockchain (Quora, 2017). Some Blockchains create a new block as frequently as every five seconds (Redman, 2016). By the time of block completion, the included data becomes verifiable. In cryptocurrency, this is practically when the transaction takes place, so a shorter block time means faster transactions.

Blockchain Network User

This refers to a person, organization, entity, business, government, etc. which is utilizing the Blockchain network.

Block Data

A list of transactions and ledger events included within the block

Node

An individual system within a Blockchain network

- **Full Node:** A node that stores the entire Blockchain, ensures transactions are valid.
- **Publishing Node**: A full node that also publishes new blocks.
- **Lightweight Node**: A node that does not store or maintain a copy of the Blockchain and must pass their transactions to full nodes.

Flow of Blockchain

Blockchain holds the potential to disrupt any form of transaction that requires information to be trusted (Figure 5). This means that all intermediaries of trust, as they exist today, exposed to disruption in some form with the initiation of Blockchain technology.

Transaction

It is the exchange of data between two parties (Figure 5). It could represent money, contracts, deeds, medical records, customer details or any other assets represented in digital form.

Verification

Depending on the network parameters, the transaction is either verified or recorded (Figure 5). This transaction is secure and placed in the queue.

Validation

Validate the blocks before adding to the Blockchain (Figure 5). The most accepted form of validation for open source Blockchains is proof of work. Once a block is validated, the block is distributed through the network in order to create a permanent, verified and unalterable ledger of information.

The data in any block is unaltered once recorded without the alteration of all subsequent blocks and a collusion of the network majority. Transactions once stored in the Blockchain are permanent. They cannot be hacked or manipulated.

Figure 5. The end-to-end flow of the Blockchain transactions.

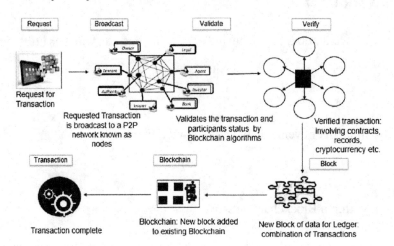

CHARACTERISTICS OF BLOCKCHAIN

Every transaction that is recorded and stored is labeled as Blockchain. The following are the main characteristics of Blockchain.

- **Digital:** Digitize all the information on Blockchain thus eliminating the need for manual documentation.
- **Distributed:** Blockchain distributes control among all users in the transaction chain, creates a shared infrastructure within an enterprise ecosystem. Indistinguishable copies of all the information shared to all the users on the Blockchain. Participants independently validate information without a centralized authority. No single point of failure because of distributed system operation. Even if one node fails, the remaining nodes continue to operate, ensuring no disruption.
- **Immutable:** All the transactions are immutable in Blockchain. Encryption is done for every transaction including time, date, participants and hash to previous block.
- **Chronology:** Each block acts like a repository that stores information pertaining to a transaction and links to the previous block in the same transaction. These connected blocks form a chronological chain providing a trail of the underlying transaction.
- **Consensus Based:** A Transaction on Blockchain executed only if all the parties on the network unanimously approve it. Also, consensus-based rules can be altered to suit various circumstances
- **Digital Signature:** Blockchain enables exchange of transactional value using unique digital signatures that rely on public keys. Public keys are decryption code known to everyone on the network. Private keys are codes known only to the owner to create proof of ownership. This is very critical in avoiding the fraud in records management
- **Consistent:** Blockchain data is complete, consistent, timely, accurate, and widely available
- **Persistence**: Transactions will not admit Invalid transactions. It is nearly impossible to delete or rollback transactions once they are included in the Blockchain. Cryptographically, the blocks created are sealed in the chain. It is impossible to delete, edit or copy already created blocks and put it on network. This leads in creation of digital assets and ensures high level of robustness and trust

- **Anonymity**: Each user can interact with the Blockchain with a generated address, which does not reveal the real identity of the user.

BLOCKCHAIN LOGICAL ARCHITECTURE

Figure 6 shows logical application architecture of the Blockchain system with key components and layers. This section explains a detailed description of these components and layers. While there exist many standard logical architectures for the Blockchain (Nomura Research Institute, 2016; Lee Kuo Chuen, 2015; Buterin, 2014; Johnson, Menezes, & Vanstone, 2001). authors introduced a detailed Blockchain logical reference architecture based on their practical experience across various domains and technologies.

Figure 6 presents an integrated view of the Blockchain infrastructure along with enterprise's architecture and a brief description of each layer.

Figure 6. Logical architecture view

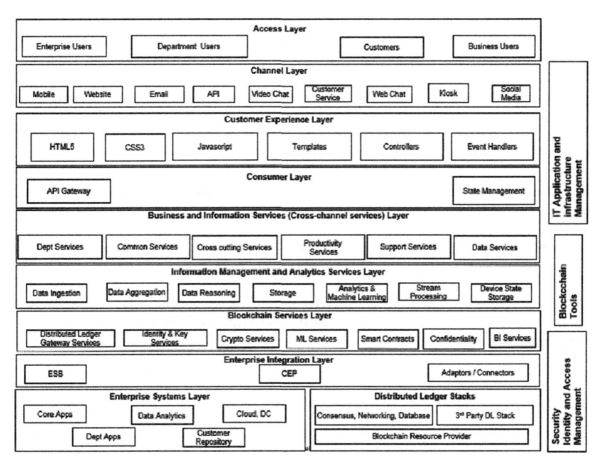

Access Layer

Various stakeholders, both internal and external will be part of this layer (Figure 6). They are the primary users of the systems.

Channel Layer

Stakeholders use channels to interact with enterprise (Figure 6). They engage with various departments or business units of the enterprise over multiple channels both physical and digital.

Customer Experience Layer

Collection of presentation components and services (Figure 6). This layer also constitutes the critical systems that are in use by enterprise and business units in engaging stakeholders.

Consumer Layer

The Consumer layer is a gateway for channels and aggregates functionalities through composition and orchestration, mediation, and routing (Figure 6).

Key capabilities are:

- **API Gateway**: a single point of entry for consumers to access back-end services. The service composition and orchestration based on customer journey and context. This capability is provided by API Management platforms
- **State Management:** manages state and transition. Control logic is decoupled from the user interface and managed at the Consumer

Business and Information Services Layer

Business and Information Services layer is designed using "Smart Architecture" principles and which will provide cross channel capabilities (Figure 6).

Information Management and Analytics Services Layer

This layer focuses on data ingestion, processing, complex event processing and real-time analytics and insights (Figure 6). A basic service system provides fundamental data services, which include data access, data processing, data fusion, data storage, identity resolution, geographic information service, user management, and inventory management.

Blockchain Services Layer

This layer acts as a gateway for reaching out to the Blockchain infrastructure (Figure 6). It consists of various services related to identity management, key management, cryptography, machine learning (ML) and business intelligence (BI). Identity, Key and cryptographic services are critical for data integrity. BI and ML are used for analysis and reporting purposes.

Enterprise Integration Layer

The Enterprise Integration Layer is a key enabler as it provides mediation, transformation, protocol and routing capabilities and acts as a gateway to integrate with core enterprise systems (Figure 6). It also provide aggregation and broker communications. The enterprise integration and presentation layer is responsible for the integration of the Blockchain infrastructure with the enterprise application stack.

Enterprise Systems Layer

The enterprise systems layer represents a collection of enterprise back office core systems and applications that house business data (Figure 6)

Distributed Ledger Stack

This layer consists of the core distributed ledger technology stack (Figure 6). Blockchain building, transaction execution and consensus will happen in this layer. The components in this stack might vary on the distributed ledger product chosen. However, the common sub-components will be consensus algorithms, data storage, transactions management and other relevant services.

Security Identity and Access Management

This meso-layer handles single sign-on, authentication, encryption and authorization capabilities (Figure 6).

IT Application and Infrastructure Management

This meso-layer illustrates the advantages of infrastructure-as-a-service and the platform-as-a-service technologies, which requires large computing capabilities (Figure 6?)

BLOCKCHAIN SECURITY ARCHITECTURE

Blockchain security ensures that the right people, internal or external, get access to the appropriate data and information at right time and place, within the right channel. Security prevents and safeguards against malicious attacks and protects enterprise data assets by securing and encrypting data while it is in-motion or at-rest. It also enables organizations to separate roles and responsibilities, protect sensitive data without compromising privileged user access.

Blockchain helps in recording the transactions of any digital assets exchanged between two unknown parties. Security aspects supported by Blockchain are critical in ensuring transparency, confidentiality and protection against fraud.

The following are the high-level security features of the Blockchain.

Ledger

Ledger records every transaction in the Blockchain. Ledger is chain of blocks and information in the block is immutable. For all the participating nodes, do the distribution of ledger.

Chain of Block

Blockchain is a chain of blocks. Each block has the hash value of the previous block and this forms a chain. Correction to data in a block (say n) will change the hash value and will not validate with the hash stored in the next block (n+1). This will be a chain reaction and affect the overall chain. Therefore, this will increase the protection of sensitive data or information.

Confidentiality

Blockchain provides confidentiality by enabling users of a ledger to be able to see authorized transactions only.

Transparency

Blockchain allows the transactions and ledger state to maintain and manage transparently by sharing the ledger to all nodes and using consensus algorithms to reach consensus among all nodes. Consensus algorithms also ensure the ordering and execution of the transactions.

Cryptology

Cryptology enables secure transactions and makes Blockchain immutable using hash-based algorithms, which produce a fixed hash based on content of the block.

Smart Contracts

Smart contract is a computer code running atop Blockchain containing a set of rules under which the parties agree to interact with each other. Nodes within the Blockchain network execute the smart contract. All nodes that execute the smart contract must derive the same results from the execution, and the results of execution are recorded on the Blockchain. If the predefined rules fulfilled, then the agreement automatically executed. No contract will execute without the network consensus.

Blockchain network users can create transactions that send data to public functions offered by a smart contract. The smart contract executes the appropriate method with the user provided data to perform a service. The code, being on the Blockchain, is also tamper evident and tamper resistant and be used as a trusted third party. A smart contract can perform calculations, store information, expose properties to reflect a publicly exposed state and, if appropriate, automatically send funds to other accounts. It does not necessarily even have to perform a financial function.

The smart contract code represents a multi-party transaction, typically in the context of a business process. In a multi-party scenario, the benefit is that this can provide attestable data and transparency that can foster trust, provide insight that can enable better business decisions, reduce costs from reconciliation that exists in traditional business-to-business applications, and reduce the time to complete a transaction.

The security at various layers for an enterprise application are classified as,

- Presentation Layer Security
- Identity and Access Management Layer
- Application/Data Layer Security
- Network Layer Security
- Infrastructure Security

Figure 7 is the layered view of the security approach with Blockchain.

Figure 7. Blockchain Security architecture view

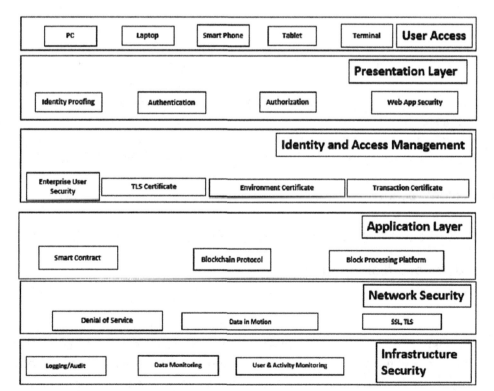

User Access

Various stakeholders, both internal and external will be part of this layer (Figure 7). They are the primary users of the systems. Stakeholders use channels to interact with enterprise. They engage with various departments or business units of the enterprise over multiple channels both physical and digital.

Presentation Layer

The front-end application security should ensure the following (Figure 7):

Authentication

Authentication is the assertion by a subscriber to prove identity (Figure 7). The authenticators and factors are:

- Something that one has to (access card)
- Something one knows (password) and
- Something one is (biometric).

Authentication mechanisms should be commensurate with the strength of the identity model, the level of access and the sensitivity of the transaction. Implement a combination of multiple authenticators (also known as Multifactor Authentication).

Authorization

Authorization is a process to establish the right to perform transactions (actions) and claim access to assets and resources by a subscriber (Figure 7). In a Blockchain application, the Authorization model should link the Identity Model and Authentication Model. A good practice is to develop a multidimensional matrix of the account associated with identity, authentication and authorization. This type of authorization model will leave room for evolution into Attribute based Access Control and Role-based Access Control at the application level and ultimately at the organizational level for enterprise Blockchain applications. The authorization model will typically address concepts such as Separation of Duties (SoD).

Web Application Security

Web application security should include protections against vulnerabilities identified in OWASP (2016).

Identity and Access Management

Digital Identity

A Digital identity is a unique representation of a person or thing engaged in a digital transaction (Figure 7). Identity proofing, or enrollment, is the physical or digital process of verifying a subject's association with their real-world identity. An identity model and its associated identity proofing should provide reasonable assurance for the identity claimed by the subject. When the trust attributes of the Blockchain application ranges from public to private, it is good practice to have multiple assurance levels for identity- proofing.

Application Layer Security

The application security should ensure the following,

- Smart Contracts and Blockchain Processing Platform are vetted to prevent calls to the unknown, valueless send, exception disorders, type casts, re-entrants, keeping secrets, immutable bugs, value (ether) lost in transfer and stack size limit
- Blockchain protocols vetted to prevent the Blockchain from not converging as expected or into an unpredictable state, safeguarding the seed (genesis) block and ensuring timestamps adequately protected.
- Consensus algorithms enable transparency by imposing transactions order to be correct for a new block added to the Blockchain. It ensures that all the nodes in the network in agreement on the new transactions added to the block.

Network Layer Security

Depending on the nature of the application and the type of network (internet-based, leased lines, virtual private network), appropriate controls extracted from the System. Controls related to boundary protection (data-in-motion) and denial of service (DoS) should be considered.

Infrastructure Security

It covers audit trail, data protection and user activity monitoring.

- **Audit Trail:** Blockchain will maintain a trail of all the transactions. Transactions ordering, and approval will be done by the consensus service on board. It ensures that the transactions are correct and transparent to other participants/nodes in the network.
- **Data Protection:** Data protection is the process of safeguarding important information or data from corruption, compromise or loss. As Blockchain is, immutable data cannot be compromise and appropriate measures like redundancy considered, to protect it from corruption and loss.
- **User & Activity Monitoring:** Track all valid transactions performed by user on Blockchain. User on-boarding and management performed by membership service shipped along with the Blockchain.

TYPES OF BLOCKCHAIN

Blockchains, depending on their application, are classified as public, private or hybrid variants. (Buterin, 2015; Mougayar, 2016):

This section focuses on the comparison between public and private Blockchain based on some recent functionalities who are expected to display many similarities as well as differences.

Public Blockchain

The Public Blockchain, is completely open to the public, namely anyone will be able to join the network as a participant. A Public Blockchain typically uses some kind of mechanism to incentivize participating parties, which encourages growing numbers of participants in network. A Public Blockchain is open for everyone to read, send transactions and participate in the consensus process. The most prominent examples of public Blockchain are the Blockchains underlying bitcoin and Ethereum.

Anyone can participate in a public Blockchain because it is open-source and public to all where no one is in charge. There is no access or rights management done for a public Blockchain and anyone can be the part of the consensus process.

Because of this, anyone at any given point of time can join or leave/read/write/audit the public Blockchain ecosystem and the network will still be mistrustful.

Private Blockchain

Private Blockchain is an absolute opposite of public Blockchain. A private Blockchain, just like the public Blockchain, mirrors the meaning of its linguistic representative. The access to a private Blockchain will be limited to the parties involved in the creation of that particular network, or those granted access to it by the creating parties. The internal mechanics of a private Blockchain can vary, from existing participants serving as a type of administrator who decide on the inclusion of future entrants to simple observers, but the public cannot access the private Blockchain.

In private Blockchains, the owner of the Blockchain is a single entity or an enterprise, which can override/delete commands on a Blockchain if needed. It is not exactly decentralized, called as distributed ledger or database with cryptography to secure it.

A hybrid Blockchain should be able to connect public Blockchain open to every single person in the world, with a private Blockchain, running in a fully permissioned environment, namely limiting the access of available information.

Comparison of Types of Blockchain

Table 1 compares features of blockchain types.

Conducting business over a decentralized hybrid Blockchain reduces transaction costs, eliminates data redundancy and speeds up transaction times. Whether a Blockchain is private or public, determine the user group that has the access to the information on that Blockchain.

BLOCKCHAIN AND ARTIFICIAL INTELLIGENCE

Intelligence is the ability to make sense of information beyond the obvious. There are two types of intelligence in nature and these are individual intelligence and group intelligence (Marwala & Hurwitz, 2017). Artificial and intelligence consists of two words, artificial and intelligence. It is Intelligence artificially

Table 1. Comparison of Blockchains.

Functionality	Public	Private
Access	Open read/write	Permissioned read and/or write
Speed	Slower	Faster
Security	Proof of Work Proof of Stake Other consensus mechanisms	Pre-approved participants
Identity	Anonymous	Know identities
Asset	Native Asset	Any asset
Run	Anyone can run BTC/LTC full node	Anyone can't run a full node
Transaction	Anyone can make transactions	Anyone can't make transactions
Audit	Anyone can review/audit the Blockchain	Anyone can't review/audit the Blockchain
Participants	Permission less (Anonymous)	Permissioned (Identified, Trusted)
Consensus	Proof of Work, Proof of Stake	Voting or Multi-party Consensus Algorithm
Transaction approval frequency	Long (10 min or more)	Short (100x msec)

made. Therefore, artificial intelligence (AI) is an umbrella term for various subsets of technological advancement that are concerned with machines to act more independently and efficiently.

From speech-pattern recognition to self-driving cars, the goal of AI is to allow machines learn and apply the knowledge gleaned from large streams of data to make them more intelligent.

Various types of artificial intelligence techniques include neural networks, support vector machines and fuzzy logic. These techniques used for missing data estimation, finite element models, modeling interstate conflict, economics and in robotics.

Life-cycle of AI application is centralized. The various steps involved are, creation of models, creation of data sets, training the model and optimization. Controlling of data and resources required by these systems are done by centralized authorities and operate under explicit trust boundaries. Knowledge acquisition is a decentralized activity, but AI systems today have a single source of authority, which leads to biased and incomplete knowledge.

Centralized AI

In centralized AI, as the knowledge is with few companies these will have greater influence and control on our daily lives. Various stages in the life cycle of centralized AI are,

- **Create Model:** An AI model is statistical or mathematical model created to extract insights from the available data used to make better business decisions. Choose a target variable based on the business requirement to gain deeper understanding. As an example, a hospital might choose "re-admitted" (which was a feature in the historical data collection) to allow hospitals to make better patient care decisions.

- **Data Creation/Collection:** This stage involves collecting and preparing all of the relevant data for use. Most relevant and accurate historical data will be gathered and transformed into suitable format for analysis. Data will be created or collected from different sources but will have single source of authority making the data biased and incomplete.
- **Training:** This stage involves training the model with the data collected in the above step to make model more accurate. Leverage this model as and when new data is collected/created to allow better forecast. AI models will change and grow over time as it processes more data and acquires more knowledge as part of training.
- **Optimize:** This stage involves optimizing the learning to make the model fit for the business requirement. Perform the identification of vulnerabilities to constantly improve or optimize the model. This will ensure that the AI model is functioning correctly.

Decentralized AI

The centralized nature of current AI systems highly contrasts with the evolution of human intelligence. Presently AI remains increasingly centralized.

With the advent of technologies like mobile computing or internet of things (IOT), they challenged the centralized notion of AI. Today, knowledge is constantly created in decentralized edges or nodes and flows towards centralized hubs. There is a shift in process where aspects such as the training, optimization, testing and knowledge creation of AI model federated across many participants/nodes.

Decentralized AI will be an essential element to guide the impact that machine intelligence will have in future generations.

In order to decentralize AI models, we need to solve a few challenges:

Privacy

Can entities train a model without having to disclose their data there by protecting the privacy of data?

Influence

Can external parties able to contribute to the behavior of knowledge of an AI model in a way that is quantitatively influential.

Economics

Can external parties be able to correctly incentivize and encourage them to contribute to the knowledge and quality of an AI model.

Transparency

Can the behavior of an AI model be transparently available to all parties without the need of trusting a centralized authority.

Blockchain can power decentralized marketplaces and coordination platforms for various components of AI, including data, algorithms, and computing power. These will foster the innovation and adoption of AI to an unprecedented level. Blockchain will also help AI's decisions be more transparent, explainable, and trustworthy. As all data on Blockchain is publicly available, AI is the key to providing users with confidentiality and privacy. Blockchain is concerned with keeping accurate records, authentication, and execution while AI helps in making decisions, assessing and understanding certain patterns and data-sets, ultimately engendering autonomous interaction.

AI and Blockchain share several characteristics that will ensure a seamless interaction in the nearest future.

The key features of Blockchain integration with AI are,

Data Sharing

AI relies greatly on big data and data sharing. It demands the analysis of open data, prediction and assessment of Machines and the algorithms generated.

Data Security

Blockchain databases hold data that are digitally signed, which means only the "respective private keys" kept secure. This allows AI algorithms to work on secure data, and thereby ensuring more trusted and credible decision outcomes. For Artificial Intelligence, the autonomous nature of the machines requires a high-level of security in order to reduce the probability of a catastrophic occurrence

Trust

To facilitate machine-to-machine communication, there is an expected level of trust. To execute certain transactions on the Blockchain network, trust is required. Any decision made by AI becomes dysfunctional when it is difficult for consumers or users to understand and trust. Blockchain records transactions in decentralized ledgers on a point-by-point basis, making it easier to accept and trust the decisions made, with confidence that the records not tampered with, during the human-involved auditing process.

High Efficiency

Multi-user business processes, which involve multiple stakeholders such as individual users, business companies, and governmental organizations, are inherently inefficient due to multiparty authorization of business transactions. The integration of AI and Blockchain technologies enables intelligent Decentralized Autonomous Agents (or DAOs) for automatic and fast validation of data/value/asset transfers among different stakeholder.

Many researchers are working to build AI systems that can apply truly intelligent decision-making processes to a restricted set of problems, some of which may positively affect our daily lives. Few emerging trends in AI applications are explainable AI (Samek, Wiegand, & Müller, 2017), digital twins (Schluse, Priggemeyer, Atorf, & Rossmann, 2018). automated machine learning (Feurer, Eggensperger, Falkner, Lindauer, & Hutter, 2018) and hybrid learning models (Lv et al., 2018), lean and augmented data learning (Peng et al., 2018) and the perceived benefits of using Blockchain technologies.

By integrating AI and Blockchain technologies, decentralized AI applications and algorithms be developed, with access to an identical view of a secure, trusted, shared platform of data, logs, knowledge, and decisions. Such platform are used to host a trusted trail of all records taken by AI algorithms before, during, and after the learning and decision making process (Panarello, Tapas, Merlino, Longo, & Puliafito, 2018).

Explainable AI

Objective is to design trustworthy and interpretable trans-parent AI algorithms to know why the algorithms reaching a specific decision. The applications that leverages this trend are Healthcare, Military and Autonomous vehicles. By adopting Blockchain, it provides the trust and reliability.

Digital Twins

Translating data and intelligence from complex physical systems into applications and simulations in digital world. The applications that leverages this trend are Wind turbines, aircraft engines. Blockchain provides trust, provenance and reliability.

Automated Machine Learning

Automating the whole process of machine learning from raw data acquisition to knowledge management in order to reduce manual work and faster application development. The applications that leverages this trend are Big Data Analytics, Industry 4.0 systems and intelligent devices. Blockchain provides permanence and immutability.

Hybrid Learning Models

Combining different machine learning models to reach better-informed decisions. The applications that leverages this trend are real time and decision agnostic. Blockchain provides trust, provenance and performance.

Lean and Augmented Data Learning

Enabling transfer learning among different AI applications to ensure high availability of relevant and accurate data. The applications that leverages this trend are low data availability applications. Blockchain provides trust, provenance and reliability.

Blockchain and AI are two technological trends, which have the potential to become more revolutionary when put together. Both serve to enhance the capabilities of the other, while also offering opportunities for better oversight and accountability.

REAL WORLD DISRUPTIVE USE-CASES OF BLOCKCHAIN

Blockchain technology is still new, lot of organizations are looking at ways to incorporate it into their businesses. The organizations are still in the process of exploring the possibility of usage of Blockchain across their business. Most of the enterprises want to take advantage of Blockchain by optimizing their network and automating it without the need for human interference. These organizations first want to understand Blockchain technology, where it fits, and then identify systems that may fit the Blockchain paradigm.

Blockchain technology solutions may be suitable if the activities or systems require features such as:

- Many participants
- Distributed participants
- Need for lack of trusted third party
- Workflow is transactional in nature (e.g., transfer of digital assets/information between parties)
- Need for a digital identifier (ex: digital land, digital property)
- Ordered registry
- Cryptographically secure system of ownership
- To reduce or eliminate manual efforts of reconciliation and dispute resolutions
- Enable real time monitoring of activity between regulators and regulated entities
- Full provenance of digital assets and a full transactional history to be shared amongst participants

Various vendors and organizations have developed frameworks, guides to determine a Blockchain is suitable for a particular system or activity, and which kind of Blockchain technology would be of most benefit.

In this section, various use-cases from different domains covering government, media, crime, transport, financial, supply chain and entertainment.

Figure 8 shows the real world disruptive Blockchain use-case view.

The following section discourses the penetration and usage of Blockchain in various industry domains.

GOVERNMENT

In government, linking the data between various departments with Blockchain ensures that the data realization in real time, when both the departments and the citizen agree on sharing data (Figure 8). Blockchain technology could improve transparency and check corruption in governments worldwide. The use-cases covering government applications such as registration, election, and finance, explained below.

Registration

By securing a unique and non-corruptible record on a Blockchain and validating changes to the status of that record across owners, a reliable property record can be created, whether for a piece of land that heretofore had no owner or as a link between stove piped systems.

Figure 8. Blockchain use cases

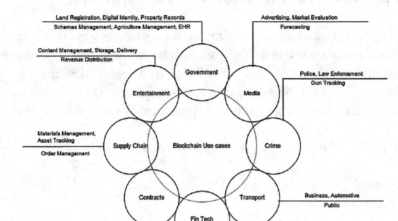

A decentralized, standardized system for land registration records could reduce the number of intermediaries required, increase trust in identity of transacting parties, increase process efficiencies, and decrease time and cost to process. Recording property rights via Blockchain would cost savings for parties through a tamper-proof ledger

Election

Citizens can cast votes the same way they initiate other secure transactions and validate that their votes were cast—or even verify the election results. Potential solutions are currently working to blend secure digital identity management, anonymous vote-casting, individualized ballot processes, and ballot casting confirmation verifiable by the voter. Potential cost savings through Blockchain-enabled voting.

Finance

Online identity has always been a time-consuming and costly process. There was always this need for registration and financial services like loans, mortgages, insurance requires always required a higher level of security, checking, authentication and authorization. This usually meant that checking official government identity documents was a required step.

One of the benefits of the Blockchain is that it has the potential to cut out the intermediaries and provide every organization/department in the network access to the same source of the right information. Blockchain technologies make tracking and managing digital identities both secure and efficient, resulting in seamless sign-on and reduced fraud. Users are able to choose how they identify themselves and with whom their identity is shared.

Town Planning

Having property records on the Blockchain makes it possible that prospective buyers shall verify the owner of a house quickly and easily.

Revenue

Seamless cross-referencing of documentations across multiple government entities. All documentations will automatically be verified by the relevant entities. The inherent features of Blockchain (digital identity) enables it not possible to make changes to the ledger.

Managing Entity

Business Unit Incorporation and governance with the use of digital signatures and programmable business rules. Approvals and Validations of the entity and its application on the Blockchain. The Distributed Ledger Technology (DLT) allows multiple parties to digitally sign for authorization and approve a transaction and documents.

Grants

An Integrated Blockchain Solution that enables seamless customer on boarding with automated Disbursements of Loans and Grants. Transactions on the Ledger are auditable real time providing greater transparency for Regulators. Highly Secured Digital Signatures which makes sure the transactions get executed only when relevant provide their signature

Agriculture

Integrated Agriculture Management System (IAMS) with a Blockchain infrastructure ensures immutability of data and a way to trace historical agricultural data as agricultural products move from production sources to the consumer. IAMS with a block chain infrastructure is a way to diversify current agricultural management practices in a way that engages the public through ownership of the agricultural production process. Other possible applications include the use of Blockchain technology to record and manage agricultural land records as well as agriculture insurance.

Healthcare

Digitization of health records is a significant task in the public health sector, which is huge, complex and associated with ethical issues. Medical records are scattered and erroneous, with inconsistent data handling processes. Sometimes, hospitals and clinics forced to work with incorrect or incomplete patient records. Blockchain technology can disrupt public health by creating a secure and flexible ecosystem for exchanging patient's electronic health records. This technology could also make the space more transparent by creating basis for critical drugs, blood, organs, etc. In addition, by putting all medical licenses on a Blockchain, fraudulent medical practitioners, druggists, chemists be prevented from practicing and selling drugs.

Education

Student records, faculty records, educational certificates, etc., are key assets in the education domain and are be shared with multiple stakeholders and it is imperative to ensure that they are trustworthy. Maintain all these records with the application of Blockchain technology. Blockchain can also simplify certificate attestation and verification.

MEDIA

Media is a multi-dollar industry that is evolving with every year (Figure 8). However, it is currently not consumer or creator friendly due to intermediates taking advantage and making tons of revenue in the process. With Blockchain, the whole landscape can change and become more accessible to the consumers as well as the content creators.

Advertising

Advertising is the core business model for many companies. To gain revenue while providing most of their services free of cost. The advertisers can place their ads and earn revenue in millions of dollars.

With Blockchain-powered grade browsers, it is now possible for consumers to turn off the advertising and protect their privacy at the same time.

Market Evaluation/Forecasting

Forecasting is the future. As things are evolving in complexity over time, it is not possible for humans to understand the underlying complexity and make decisions.

Blockchain transformation can speed up the forecasting to an entirely new level. The holistic approach can have its impact on analysis, research and other market sectors as well. As Blockchain can hold a massive amount of data that can be used in conjunction with machine learning or artificial intelligence.

CRIME MANAGEMENT

Law enforces strict guidelines, ensures that maintenance of crime at an all-time low, and affected by Blockchain transformation (Figure 8).

Police/Law

The Blockchain maintains the integrity of the data and helps in storing all the evidence. In addition, it distributed without worrying about tampering. It adds the necessary layer of security that adds meaning and trust to law enforcement.

Gun Tracking

With Blockchain, we can have a system where each gun be accounted for. The infrastructure will track each weapon that leaves the factory and helps stop illegal gun usage and export. Retrace the information once stored, whenever needed. The gun used in the crime be traced back to its original owner or its origins can be known instantly.

TRANSPORTATION

The transportation industry can transform by Blockchain adoption (Figure 8). It lowers processing and administrative costs, minimizes disputes, improving overall transportation speed and much more.

Business Transportation

Blockchain is making business transportation more secure, cost saving and helping logistic companies be top of their game. It allows easy data coordination using distributed ledger technology (DLT). They can also utilize smart contracts to automate most of their transportation pipeline or create a custom solution to enable transportation for fragile and sensitive products easily. Blockchain provides the security and authenticity. The whole network can scale to include more transportation vehicles, products, and information.

Automotive

With self-driving cars already out and shipping, it is now high time for the automotive industry to make use of the Blockchain. The revolution has already started as IOTA showed their proof-of-concept with Volkswagen and showed how their IOTA tangle system could work for the autonomous cars.

Public Transport

Public transportation always demand improvement in transport by the public and government is working towards improving it over time. With Blockchain transformation, things will change drastically.

FINTECH

Finance

With Blockchain, Fintech will also change and see a paradigm shift (Figure 8). Right now, tons of Fintech start-ups are working with Blockchain and aiming to provide a better solution.

Some of the financial systems that leverage the Blockchain technology are,

- System, which offers predictions for the investors in the volatile Blockchain environment

- Lending platform which offers a decentralized peer-to-peer lending platform
- Platform which provides a trust less, decentralized platform of liquidating crypto-assets

Blockchain is the greatest disrupter to the global financial services system of this generation (O'Donnell, 2016; Brennan, 2018). While many technologies have changed core banking system, principles and underlying banking transactions and booking keeping is complex and challenging. With Blockchain solution in place, the complex processes that once took weeks to reconcile reduce to few minutes.

Existing ledger accounting in the financial services sector is usually in the form of a centralized and private database, not too different from the paper-based versions used long time back. With Blockchain, the ledger moves to an open and distributed record shared across the participant nodes in the Blockchain. Each node maintains its own version of the ledger, and the network must collectively agree on the authenticity and correctness of transactions before carrying out an update.

CONTRACTS

Contracts (Figure 8) help to set up requirements for taking out a job, for payment and any other tasks that require two parties: examples are music, legal matters or even land or property.

Inheritances

Smart contacts empower inheritance and help Blockchain transformation. These are digital inheritance, which means to transfer digital assets to someone.

Legal Contracts

Blockchain and smart contracts helps to transform the legal industry. It will give rise to legal contracts that take advantage of the smart contracts and live on the Blockchain. This way, the owner can maintain intellectual property without being concerned about theft. The legal contracts have to adhere to the Blockchain law and cannot be externalized.

Property and Land

Legal contracts can also influence property and land transactions. Right now, it is all about having the correct documents. Blockchain will provide a suitable means to register deeds, property ownership and the like. This will remove any frauds or change of deeds without the notice of the owners themselves. In addition, the information can be backtracked with full transparency and security.

SUPPLY CHAIN MANAGEMENT

In Supply Chain, Blockchain technology can be used to store and record all the transactions and exchanges of ownership from the extraction of raw materials to the customer (Figure 8). It helps to identify, where the materials were sourced, labor type used, exact position of the goods at any given moment.

In Supply Chain management, Blockchain technology enables enterprises to,

- Verify materials
- Identify asset quantity as the assets move through the supply chain.
- Coordinate orders and shipment notifications
- Verify the constitution of materials
- Increase scalability, security, and innovation
- Link physical materials to serial numbers, barcodes, and digital tags
- Share important information about the manufacturing process, assembly, delivery, and vendors

ENTERTAINMENT

The final category for the Blockchain digital transformation is entertainment (Figure 8). Blockchain transformation is also changing entertainment

Entertainment Industry

The future of entertainment will be in the hands of the Blockchain. Blockchain will change how content is created and consumed.

Music Industry

The music industry is one of the most significant industry in the world. Blockchain can transform entertainment. It will make sharing content more accessible and use smart contracts to do. so The smart contracts will verify that the revenue is distributed to all parties according to the agreement. This simple pivotal change will make it easy for content creators to make the most out of their work. It will also remove intermediaries.

DETAILED CASE STUDY: BLOCKCHAIN IN ELECTRONIC HEALTH RECORD

The Electronic Health Records Management System (EHRMS) (Manav Sampada, n.d.; Gans, Kralewski, Hammons, & Dowd, 2005) is used for capturing data on health and health related events and parameters. This application will be a super-set of a hospital's management system and will ensure capturing of all health-related data and events of a client in the form of an Electronic Health Record (EHR). The Electronic Health Record is an electronic register of health-related information on an individual that conforms to interoperability standards. Authorized clinics and staff manage the EHR data of the individual across more than one healthcare organization.

Blockchain technology organizes the data access to be record and verified through consensus of all parties involved. The high-level functionalities of the EHRMS are,

- Patients can monitor their own health information, approving, denying or sharing changes to their data, ensuring a higher level of privacy and engagement

- Data can only be accessed by the patient's private key; even if the database is hacked, the patient's data will be unreadable
- Patients have full control over accessing their healthcare data. Patients controls the users, who access their data and what they access
- Instantaneous transfer of medical data, where every member of the distributed network of the health care Blockchain would have the same data for the patient. This leads to a reduced risk of errors, and better patient care
- For clinicians, Blockchains can support universal identifiers for any data entered into an EHRMS, which is then available to other authorized providers and care team members.
- Researchers will benefit from better data integrity and reliability, creating a 'platform of trust' for data sharing.
- Payers will have better data reconciliation, fewer errors and frauds, reduced administrative and claim processing costs, and the potential to reach under-served markets.

EHRMS system enables hospitals, physicians, laboratories, pharmacies, and other health services providers to participate and deliver faster, more accurate, safer, higher quality and less redundant medical care to patients. Automation/ IT application processes for EHR will help stakeholders in following ways as detailed in Table 2.

Blockchain technology brings the following major features to EHRMS system,

- Immutability via File Integrity
- Cybersecurity via Data Access Management
- Interoperability via Collaborative Version Control

Blockchain Technology benefits of the Healthcare are,

Data Security

Blockchain delivers access control via a shared public chain and a private chain; so for instance, only the patient would have access to their medical data using the private chain piece. If malicious parties wanted to gain access, the hackers would need to simultaneously breach every participant in the network, not just one

Data Privacy

Blockchain also supports data privacy when data updates are applied and during data transfer. For example, when a doctor prescribes a new medication for a patient, his or her records are updated, and the public ledger notes when the transaction or change was made. An encrypted link to the patient's record created for doctor and patient access only. This approach strengthens the privacy of electronic health records (EHR)

Table 2. Benefits for the stakeholders using EHRMS.

Stakeholder	Benefits
Residents	• Access to Online Electronic Health Record using his / her using unique id after proper verification process. • Online registration from anywhere through Application Portal in case of outpatient consultation or planned hospitalization • Reduction in time for registration of beneficiary • Complete record of accomplishment of patients and treatment • Improved medical care provided by hospitals by seeking guidance through telemedicine, thereby helping prevent needless travel by patients for seeking medical care through distant hospitals • Online availability of information on: o List of government and private hospitals o List of specialty services available at respective government and private hospitals o Availability of bed for admission to a hospital, whether government or private o List of investigations available at respective government and private hospitals and in Laboratories / Radio imaging Centers. o Days of availability of doctors for consultation at respective government and private hospitals o Referrals of patients to higher level hospitals
Hospitals	• Better linkage between government and private hospitals through tele-medicine and by way of referrals. • Better utilization of unutilized bed capacity of private hospitals as the patients will be able to identify availability of vacant beds in different hospitals • Improved effectiveness and efficiency of hospital services as almost all services would be available through a single software • Access to Online Electronic Health Record of an individual after proper verification process
Doctors	• Increased effectiveness and efficiency of patient care because of capturing of vital information in Electronic Health Record • Enhanced skills of doctors in remote, peripheral areas
Hospital staff	• Online registration of beneficiaries, which will increase efficiency and effectiveness of delivery of health services • Easy access to patient records by way of Electronic Health Record / Electronic Medical Record after proper verification process • Online generation and retrieval of patient history • Availability of real time MIS (daily, weekly, monthly) on number of patients registered and treated by government and private hospitals • Efficient processes removing redundant and time taking processes and saving time for hospital staff
Training Institutions and Researchers	• Improved quality of research because of availability of large volume of data on persons, wellness / risk factors / their illnesses.
Department of Medical, Health and Family Welfare	• Availability of real time data to plan and make policies for providing better health services to residents across the State through a single application • Monitoring and studying trends in medical care services provided by different levels of government and private hospitals and the trend in referral / outcome of patients • Online verification of entitlements of patients and linkage of patient care services with other schemes and with National Health Programs

Billing Management

Industry analysts from Frost & Sullivan Health Practice estimate that 5-10% of healthcare bills are fraudulent, resulting from excessive billing or billing for services that never occurred (Frost & Sullivan, 2018). Blockchain can improve the logistical tracking of reliability-centered maintenance (RCM) functions

Pharmaceuticals and Drug Tracking

Blockchain's supply chain management can track drug sourcing to reduce counterfeit drug business

Health Research

Some health studies and clinical trials are unreported and therefore unregulated, creating a safety issue for participants (Poulsen et al., 2009). Using time-stamped records and results, Blockchain advancements can address selective reporting and the manipulation of results

Healthcare IoT

Blockchain-enabled solutions can close the gap on security through preventative measures, down to the device level (Hua & Notland, 2016). In addition, if any attempted breaches should occur, the source becomes traceable; because Blockchain transactional logs cannot be changed or deleted. Hackers and malicious inside attacks can thus be more easily identified.

BENEFITS OF BLOCKCHAIN

The following are the Blockchain benefits of this technology is transforming industry.

- Greater transparency, transaction histories are becoming more transparent through the use of Blockchain technology.
- Enhanced security, there are several ways Blockchain is more secure than other record-keeping systems.
- Elimination of error handling through real-time tracking of transactions with no double spending.
- Improved traceability.
- Trusted record keeping and shared trusted process .
- Increased efficiency and speed, reduction of settlement time to mere seconds by removing intermediaries.
- Reduce cost and complexity, Material cost reduction through the elimination of expensive proprietary infrastructure.
- Full automation of transactional processes, from payment through settlement.

  Blockchain is not recommended for,

- High performance (millisecond) transactions.
- Small organization (no business network).
- Looking for a database replacement.
- Looking for a messaging solution.
- Looking for transaction processing replacement.

BLOCKCHAIN CHALLENGES TODAY

There are number of challenges preventing adoption of Blockchain in enterprises. The main challenges are;

Initial Costs

It is initially expensive to establish Blockchain at the enterprise level. The software required to run Blockchain technology in organizations must typically be developed for the specific firm and is therefore expensive to purchase

Integration with Legacy Systems

It is difficult for Blockchain solutions to handle all functions needed by enterprise/business units, initially making it difficult to eradicate legacy systems. Therefore, considerable changes done to the existing systems in order to facilitate a smooth transition. This process may take a significant amount of time, funds and human expertise to find a way to integrate their existing system with the Blockchain solution.

Inviolability

Information stored on the Blockchain is permanent and deletion is not possible. This may be an issue from a privacy perspective.

Digital Signature

Legal validity of a digital signature is increasingly diminishing with such signatures now viewed as legally binding.

Scalability

Bitcoin block size is limited to 1 MB as of now and a block is mined approximately every ten minutes. However, larger blocks mean larger storage space and slower propagation in the network. This will lead to centralization gradually as less users would like to maintain such a large Blockchain. Therefore, the trade-off between block size and security is a tough challenge.

MITM Attack

Man in The Middle Attack (MITM) is a third party interaction (Du, Li & Huang, 2011). Users are in the middle of the process and may have a forged public key. By using this key, the user can easily decrypt the sensitive data. The public key is distributed across the participating nodes in Blockchain technology. Each block is connected with a link to previous and following blocks. Because of this, the public key is immutable it should not attack by any forged keys.

Low Awareness and Understanding

This might be the principal challenge related to Blockchain technology. Many senior executives at large organizations have little or no knowledge of Blockchain. To remedy this, executives can connect with industry thought leaders, get access to existing use cases, or connect with industry associations

Lack of Standards and Best Practices

There is a lack of uniform standards on Blockchain technology even as new Blockchain-based solutions are developed (Guo & Liang, 2016). Many executives consider technical standards as critical for wider adoption. Industry players would need to collaborate better to build uniform standards and protocols rather than develop their own internal versions.

Regulatory and Legal Uncertainty

Any regulation that recognizes Blockchain applications, including smart contracts or digital identities, can provide a big boost to its adoption. Industry players can work in tandem with the regulators to devise enabling regulations in a phased manner.

FUTURE RESEARCH DIRECTIONS

Blockchain has shown its potential in industry and academia. The possible future research directions for the Blockchain technology adoption described below;

Big Data Analytics

Blockchain helps in storing important data, as this is distributed and secure. Blockchain ensures that the data is original.

Blockchain Applications

Most of the Blockchain-based solutions currently used in the financial domain. Traditional industries are also started adopting Blockchain and apply Blockchain into their fields to enhance their systems.

Blockchain as a Service (BaaS)

BaaS is a cloud-based service that allows customers to build their own Blockchain-powered products including apps, smart contracts, and use other Blockchain features without the need for setup, manage or execute Blockchain-based infrastructure.

Interoperability

As time progresses, we see new Blockchain networks increases. This leads to new chains that offer different speeds, network processing, and use-cases. Blockchain interoperability aims to improve information sharing across diverse networks or Blockchain systems. These cross-chain services improve Blockchain interoperability and makes them more practical for day-to-day usage.

Fog Computing and Blockchain

Fog computing is an emerging technology that allows customers or IoT devices to generate data for localized computing and storage close to the source of data. To augment the long delay incurred by computing and storage at the cloud environment, fog nodes are used. Fog nodes are similar to a local small-scale cloud. In the context of AI and Blockchain, future fog nodes have to be equipped with AI and machine learning capabilities as well as enable by Blockchain interface. This helps the fog nodes to perform localized management, access and control of data.

Standards and Regulations

As of today, Blockchain technology standards are evolving. Work is in progress by IEEE, NIST, ITU, and many standards' bodies to establish standards for Blockchain interoperability, governance, integration, and architecture (Anjum, Sporny, & Sill, 2017; Kakavand, Kost De Sevres, & Chilton, 2017). Also, at the local and global level, governmental and institutional guidelines, rules, laws, regulations, and policies need to be developed for Blockchain deployment, arbitration, and dispute handling. This entails research directed at devising models and proof of concepts that can play a key role in defining the right set of technical standards for Blockchain architectural models, services, deployment and interoperability.

In 2019, we will see an improvement in the technology that enables Blockchain interoperability. Few vendors are already working towards the goal of enabling or improving Blockchain interoperability using cross chain technology.

CONCLUSION

Blockchain should not be regarded as a silver bullet. It is not applicable to every situation, and enterprise leaders are researching for challenges in the fields of both technological and managerial in the form of proof of concept. Leaders need to understand the benefits of positioning the Blockchain technology in the e-governance domain.

Blockchain has shown its potential for transforming traditional industry with its key characteristics: decentralization, persistence, anonymity and verifiability.

Blockchain-based enterprise applications increase the effectiveness of companies, reduces cost of transactions and more quickly means of interaction between enterprises and its customers.

Blockchain must be viewed as a unique and a universal technology that helps streamline and automate nearly all customer services, legal contracts, while increasing the transparency and effectiveness of enterprise.

A great deal of exploration is required today in the domain area. The challenge is who will take the lead in applying Blockchain technology across various business units to minimize enterprise costs, improve security in an era of cyber uncertainty, and enhance customer delivery.

In summary, Blockchain digital transformation has just begun. We are witnessing changes in almost every sector, like government, education, health, real estate and many more.

With enterprise and startups working in parallel to introduce Blockchain strategies to new solution and improve existing ones, it is now the best time to follow the whole revolution. Currently, there is a lot of hype around the technology, and many proposed uses for it. Going forward it is likely that the hype will die down, and Blockchain technology will become just another technology leveraged across the enterprise business. It will take any time between 5-8 years for Blockchain to become ubiquitous.

ACKNOWLEDGMENT

The authors would like to thank David Kenner, Raju Alluri of Architecture Group of Wipro Technologies for giving us the required time and support in many ways in bringing this chapter as part of Global Enterprise Architecture Practice efforts.

This research received no specific grant from any funding agency in the public, commercial, or not-for-profit sectors.

REFERENCES

Afshar, V. (2017). *Blockchain Will Disrupt Every Industry*. Retrieved from https://www.huffingtonpost.com/entry/blockchain-will-disrupt-every-industry-us-5963868ce4b08f5c97d06b55

Anjum, A., Sporny, M., & Sill, A. (2017). Blockchain standards for compliance and trust. *IEEE Cloud Computing*, *4*(4), 84–90. doi:10.1109/MCC.2017.3791019

Artificialllawyer. (2018). Hype Killer – Only 1% of Companies Are Using Blockchain. *Gartner Reports*. Retrieved from: https://www.artificiallawyer.com/2018/05/04/hype-killer-only-1-of-companies-are-using-blockchain-gartner-reports

Bellare, M., & Rogaway, P. (2005). *Introduction*. Introduction to Modern Cryptography.

Bharadwaj, S. (2018). *Blockchain based solutions to everyday problems*. Retrieved from https://medium.com/swlh/blockchain-based-solutions-to-everyday-problems-7c0bb3cb83dc

Brennan, D. (2018). *The Ultimate Disruptor: How Blockchain is Transforming Financial Services*. Retrieved from https://gowlingwlg.com/getmedia/ab5aecbb-8997-4bd2-96e7-58070b33e07a/how-blockchain-is-transforming-financial-services.pdf.xml?ext=.pdf

Buterin, V. (2014). *A next-generation smart contract and decentralized application platform*. White Paper, 2014. Retrieved from https://github.com/ethereum/wiki/wiki/White-Paper

Buterin, V. (2015). *On Public and Private Blockchains*. Retrieved from https://ethereum.org/

Chuen, L. K. D. (Ed.). (2015). *Handbook of Digital Currency*. Elsevier. Available: http://EconPapers.repec.org/RePEc:eee:monogr:9780128021170

Coinbase. (2019). Available at https://en.wikipedia.org/wiki/Coinbaset

Das, S. (2017). *100%: Dubai Will Put Entire Land Registry on a Blockchain*. Retrieved from https://www.ccn.com/100-dubai-put-entire-land-registry-blockchain/

Dhaliwal, S. (2016). *IBM to Launch One of the Largest Blockchain Implementations in the World*. Retrieved from https://cointelegraph.com/news/ibm-to-launch-one-of-the-largest-blockchain-implementations-in-the-world

Du, J., Li, W., & Huang, H. (2011). *A Study of Man-in-the-Middle Attack Based on SSL Certificate Interaction*. Retrieved from https://ieeexplore.ieee.org/abstract/document/6154142/authors#authors

Economist. (2015). *The great chain of being sure about things*. Retrieved from https://www.economist.com/briefing/2015/10/31/the-great-chain-of-being-sure-about-things

Eha, B. P. (2017). *Bank of England completes cross-border payment proof of concept 2017*. Retrieved from https://www.americanbanker.com/news/bank-of-england-completes-cross-border-payment-proof-of-concept

Feurer, M., Eggensperger, K., Falkner, S., Lindauer, M., & Hutter, F. (2018). Practical automated machine learning for the automl challenge 2018. *International Workshop on Automatic Machine Learning at ICML*.

Finextra. (2017). *SWIFT Blockchain POC: enhanced cross-border payments*. Retrieved from https://www.finextra.com/blogposting/14321/swift-blockchain-poc-enhanced-cross-border-payments

Firstpost. (2017). *Andhra Pradesh To Become First State To Deploy Blockchain Technology Across The Administration*. Retrieved from https://www.firstpost.com/tech/news-analysis/andhra-pradesh-to-become-first-state-to-deploy-blockchain-technology-across-the-administration-4125897.html

Frost & Sullivan. (2018). *Frost & Sullivan's 10 Healthcare Predictions for 2018*. Retrieved from https://ww2.frost.com/frost-perspectives/frost-sullivans-10-healthcare-predictions-2018/

Gans, D., Kralewski, J., Hammons, T., & Dowd, B. (2005). *Medical Groups' Adoption Of Electronic Health Records And Information Systems*. Retrieved from https://www.healthaffairs.org/doi/full/10.1377/hlthaff.24.5.1323

Garneto, B., Kandaswamy R., Lovelock, J.D., & Reynolds, M. (2017). *Forecast: Blockchain*. Academic Press.

Gray, M. (2017). *Introducing Project "Bletchley"*. Retrieved from https://github.com/Azure/azure-blockchain-projects/blob/master/bletchley/bletchley-whitepaper.md

Guo, Y., & Liang, C. (2016). *Blockchain application and outlook in the banking industry*. Retrieved from https://jfin-swufe.springeropen.com/articles/10.1186/s40854-016-0034-9

Hill, T. (2016). *Visa Introduces Blockchain-based solution for Payment Services*. Retrieved from https://bitcoinist.com/visa-blockchain-solution-payments/

Hua, A. V., & Notland, J. S. (2016). *Blockchain enabled Trust & Transparency in supply chains*. Retrieved from https://www.academia.edu/38200535/Blockchain_enabled_Trust_and_Transparency_in_supply_chains

Ito, J., Narula, N., & Ali, R. (2017). *The Blockchain Will Do to the Financial System What the Internet Did to Media*. Retrieved from https://hbr.org/2017/03/the-blockchain-will-do-to-banks-and-law-firms-what-the-internet-did-to-media

Johnson, D., Menezes, A., & Vanstone, S. (2001). The elliptic curve digital signature algorithm (ecdsa). *International Journal of Information Security, 1*(1), 36–63. doi:10.1007102070100002

Kakavand, H., Kost De Sevres, N., & Chilton, B. (2017). *The blockchain revolution: An analysis of regulation and technology related to distributed ledger technologies*. Academic Press.

Kariappa, B. (2015). *Block Chain 2.0: The Renaissance of Money*. Retrieved from https://www.wired.com/insights/2015/01/block-chain-2-0/

Lantmateriet. (2019). Retrieved from https://www.lantmateriet.se/

Leonard, S. (2017). *The Internet of Value: What It Means and How It Benefits Everyone*. Retrieved from https://ripple.com/insights/the-internet-of-value-what-it-means-and-how-it-benefits-everyone/

Liddell, H. G., Scott, R., Jones, H. S., & McKenzie, R. (1984). *A Greek-English Lexicon*. Oxford University Press.

Lv, C., Xing, Y., Lu, C., Liu, Y., Guo, H., Gao, H., & Cao, D. (2018). Hybrid-learning-based classification and quantitative inference of driver braking intensity of an electrified vehicle. *IEEE Transactions on Vehicular Technology*.

Marr, B. (2018). *Here Are 10 Industries Blockchain Is Likely To Disrupt*. Retrieved from https://www.forbes.com/sites/bernardmarr/2018/07/16/here-are-10-industries-blockchain-is-likely-to-disrupt/#3001d98bb5a2

Marwala, T., & Hurwitz, E. (2017). *Introduction to man and machines. In Artificial intelligence and economic theory: skynet in the market* (pp. 1–14). Springer International Publishing AG.

McClarkin, E. (2018). *MEPs back plan to revolutionise international trade*. Retrieved from http://conservativeeurope.com/news/meps-back-plan-to-revolutionise-international-trade

McLeanA. (2018). Retrieved from https://www.zdnet.com/article/uob-claims-regions-first-real-time-cross-border-mobile-number-funds-transfer/

McLeod, S. (2018). *5 Principles of Blockchain CPAs Should Know*. Retrieved from: https://www.accountingweb.com/technology/trends/5-principles-of-blockchain-cpas-should-know

Menezes, A. J., van Oorschot, P. C., & Vanstone, S. A. (1997). Handbook of Applied Cryptography. Academic Press.

Mougayar, W. (2016). *The business blockchain: promise, practice, and application of the next Internet technology*. Wiley.

Nakamoto, S. (2008). *Bitcoin: A Peer-to-Peer Electronic Cash System*. Retrieved from https://www.bitcoin.com/bitcoin.pdf

Nomura Research Institute. (2016). *Survey on blockchain technologies and related services*. Tech. Rep. 2015. Retrieved from http://www.meti.go.jp/english/press/ 2016/pdf/0531 01f.pdf

O'Donnell, T. (2016). *Blockchain: Looking Beyond the Hype of Bitcoins*. Retrieved from https://www.infosys.com/blockchain/PublishingImages/pdf/hype-bitcoin.pdf

OWASP (The Open Web Application Security Project). (2016). *Top 20 OWASP Vulnerabilities And How To Fix Them.* Retrieved from https://www.upguard.com/articles/top-20-owasp-vulnerabilities-and-how-to-fix-them

Panarello, A., Tapas, N., Merlino, G., Longo, F., & Puliafito, A. (2018). Blockchain and iot integration: A systematic survey. *Sensors (Basel)*, *18*(8), 2575. doi:10.339018082575 PMID:30082633

Peng, P., Tian, Y., Xiang, T., Wang, Y., Pontil, M., & Huang, T. (2018). Joint semantic and latent attribute modelling for cross-class transfer learning. *IEEE Transactions on Pattern Analysis and Machine Intelligence*, *40*(7), 1625–1638. doi:10.1109/TPAMI.2017.2723882 PMID:28692964

Pimentel, D. (2017). *US Navy to Integrate Blockchain to Manufacturing.* Retrieved from https://block-tribune.com/us-navy-integrate-blockchain-manufacturing/

Poulsen, H. E., Andersen, J. T., Keiding, N., Schramm, T. K., Sorensen, R., Gislasson, G., ... Torp-Pedersen, C. (2009). *Why Epidemiological and Clinical Intervention Studies Often GiveDifferent or Diverging Results?* Retrieved from http://enghusen.dk/whyEpi.pdf

Quora. (2017). *What does "Block Time" mean in crypto-currency?* Retrieved from: https://www.quora.com/What-does-Block-Time-mean-in-crypto-currency

Raconteur. (2016). *The future of blockchain in 8 charts.* Retrieved from: https://www.raconteur.net/business-innovation/the-future-of-blockchain-in-8-charts

Redman, J. (2016). *Disney Reveals Dragonchain, an Interoperable Ledger.* Retrieved from: https://news.bitcoin.com/disney-dragonchain-interoperable-ledger/

Reid, D. (2017). *Blockchain set to disrupt aviation within 2 years claims Accenture.* Retrieved from https://www.cnbc.com/2017/06/27/blockchain-set-to-disrupt-aviation-within-2-years-claims-accenture.html

Rivest, R. L. (1990). Cryptography. In *Handbook of Theoretical Computer Science.* Elsevier. doi:10.1016/B978-0-444-88071-0.50018-7

Samek, W., Wiegand, T., & Müller, K.-R. (2017). *Explainable artificial intelligence: Understanding, visualizing and interpreting deep learning models.* Retrieved from https://arxiv.org/abs/1708.08296

Sampada, M. (n.d.). e-*Tool for Human Resources Management System.* Retrieved from https://ehrms.nic.in/Home/HomePageFeatures?ID=AboutUs

Schluse, M., Priggemeyer, M., Atorf, L., & Rossmann, J. (2018). Experiment table digital twins streamlining simulation-based systems engineering for industry 4.0. *IEEE Transactions on Industrial Informatics*, *14*(4), 1722–1731. doi:10.1109/TII.2018.2804917

Serenelli, L. (2014). *EU, Interpol fight epidemic of stolen, fake passports.* Retrieved from https://www.usatoday.com/story/news/world/2014/05/21/stolen-passports/9351329/

Shin, L. (2017). *The First Government To Secure Land Titles On The Bitcoin Blockchain Expands Project.* Retrieved from https://www.forbes.com/sites/laurashin/2017/02/07/the-first-government-to-secure-land-titles-on-the-bitcoin-blockchain-expands-project/#5b9dbeab4dcd

Stroud, F. (2015). *Blockchain: Webopedia Definition*. Available at: https://www.webopedia.com/TERM/B/blockchain.html

Sultan, K., Ruhi, U., & Lakhani, R. (2018). *Conceptualizing Blockchains: Characteristics & Applications*. Retrieved from https://arxiv.org/ftp/arxiv/papers/1806/1806.03693.pdf

Techopedia. (2019). *Disruptive Technology*. Retrieved from https://www.techopedia.com/definition/14341/disruptive-technology

Transparencymarketresearch. (2018). *Blockchain Technology Market to reach US$20 billion by 2024 – TMR*. Retrieved from https://globenewswire.com/news-release/2018/10/26/1627765/0/en/Blockchain-Technology-Market-to-reach-US-20-billion-by-2024-TMR.html

Varsamis, G. (2018). *Disruptive innovation: a weapon that can kneel down giants and a survival tool for difficult times*. Retrieved from https://blog.startuppulse.net/disruptive-innovation-a-weapon-that-can-kneel-down-giants-and-a-survival-tool-for-difficult-times-71af7bffb750

WikipediaB. (2018). Retrieved from https://en.wikipedia.org/wiki/Blockchain

Williams, A. (2016). *IBM to open first blockchain innovation centre in Singapore, to create applications and grow new markets in finance and trade*. Retrieved from https://www.straitstimes.com/business/economy/ibm-to-open-first-blockchain-innovation-centre-in-singapore-to-create-applications

Yaga, D., Mell, P., Roby, N., & Scarfone, K. (2018). *Blockchain Technology Overview*. Retrieved from https://nvlpubs.nist.gov/nistpubs/ir/2018/NIST.IR.8202.pdf

Zhao, W. (2017). *Europe's Second Largest Port Launches Blockchain Logistics Pilot*. Retrieved from https://www.coindesk.com/europes-second-largest-port-launches-blockchain-logistics-pilot

Zhao, W. (2017). *Daimler AG Issues €100 Million Corporate Bond in Blockchain Trial*. Retrieved from https://www.coindesk.com/daimler-ag-issues-e100-million-corporate-bond-blockchain-trial

KEY TERMS AND DEFINITIONS

Asset: Anything that is capable of being on its own or controlled to produce value is an asset. Types of assets are tangible (e.g., a house) and intangible (e.g., a mortgage). Intangible assets are subdivided into financial (e.g., bond), intellectual (e.g., patents), and digital (e.g., music).

Bitcoin: It is a digital currency, created, and held electronically. Operated as a decentralized application that directly control the transfer of digital currency. It will keep the full history of transaction.

Blockchain Network User: A person, organization, entity, business, government, etc. which is utilizing the Blockchain network.

Contract: Conditions for transaction to occur.

Cloud Computing: Cloud computing is an ICT sourcing and delivery model for enabling convenient, on-demand network access to a shared pool of configurable computing resources (e.g., networks, servers, storage, applications, and services) that can be rapidly provisioned and released with minimal management effort or service provider interaction.

Data Exhaust: Data exhaust (or digital exhaust) refers to the by-products of human usage of the internet, including structured and unstructured data, especially in relation to past interactions.

Elliptic Curve Diffie- Hellman-Merkle (ECDHM): Elliptic Curve Diffie-Hellman-Merkle (ECDHM) addresses are Bitcoin address schemes that increase privacy. ECDHM addresses be shared publicly and are used by senders and receivers to secretly derive traditional Bitcoin addresses that passive Blockchain observers cannot predict. The result is that ECDHM addresses can be "reused" without the loss of privacy that usually occurs from traditional Bitcoin address reuse.

IEEE: The Institute of Electrical and Electronics Engineers.

ITU: The International Telecommunication Union.

Ledger: It is the system of record for a business. It records asset transfer between participants. Business will have multiple ledgers for multiple business networks in which they participate.

Market: Central to the process. Can be public (fruit market, car auction) and private (supply chain financing, bonds).

MITM Attack: In cryptography and computer security, a man-in-the-middle attack is an attack where the attacker secretly relays and possibly alters the communication between two parties who believe they are directly communicating with each other.

NIST: The National Institute of Standards and Technology.

Participants: Members of a business network. These can be customer, supplier, government, regulator. They usually reside in an organization and have specific identities and roles.

Transaction: An asset transfer.

Wealth: Generated by the flow of goods and services across business networks.

Chapter 7
Disrupting Agriculture:
The Status and Prospects for AI and Big Data in Smart Agriculture

Omar F. El-Gayar
Dakota State University, USA

Martinson Q. Ofori
Dakota State University, USA

ABSTRACT

The United Nations (UN) Food and Agriculture (FAO) estimates that farmers will need to produce about 70% more food by 2050. To accommodate the growing demand, the agricultural industry has grown from labor-intensive to smart agriculture, or Agriculture 4.0, which includes farm equipment that are enhanced using autonomous unmanned decision systems (robotics), big data, and artificial intelligence. In this chapter, the authors conduct a systematic review focusing on big data and artificial intelligence in agriculture. To further guide the literature review process and organize the findings, they devise a framework based on extant literature. The framework is aimed to capture key aspects of agricultural processes, supporting supply chain, key stakeholders with a particular emphasis on the potential, drivers, and challenges of big data and artificial intelligence. They discuss how this new paradigm may be shaped differently depending on context, namely developed and developing countries.

INTRODUCTION

The Agricultural Revolution between the 17th to late 19th centuries brought about productivity through the mechanization of farm work. As the human population continues to grow, however, the demand for land, food, and resources have become more intense making it necessary to reinvent the agricultural sector. Even with the continuous advancements in agriculture, the sector is still faced with several issues such as climate change, competition for land and water resources, food waste attributed to post-harvest handling and storage, and more. The 2018 Global Report on Food Crisis stated that *"out of the 51 countries that experienced food crises in 2017, conflict and insecurity were the major drivers of food insecurity in 18*

DOI: 10.4018/978-1-5225-9687-5.ch007

countries, where almost 74 million people faced Crisis (IPC/CH Phase 3), Emergency (IPC/CH Phase 4) or Catastrophe/Famine (IPC/CH Phase 5) conditions" (Food Security Information Network, 2018). The United Nations (UN) Food and Agriculture Organization (FAO) estimates that farmers will need to produce about 70% more food by 2050. How is the sector prepared for increased production of food despite competing with humans for land? What is the optimal use for resources despite climate change? Will agriculture meet global food needs as projected by the UN?

In this regard, an emerging trend is the use of *"smart"* technologies in farming commonly referred to as Smart Agriculture. As CEMA (2017) puts it, the main difference between Precision Agriculture and its successor, Smart Agriculture, is that while the former improves the accuracy of operations and allows the management of in-field (or in-herd) variations by providing for plants (or animals) the optimal resources needed for growth, the latter uses big data analytics and artificial intelligence (AI) to act on data collected by the farm equipment. It has been suggested that Smart Agriculture solves the problem of generalization whilst providing autonomy for farm decisions enhanced by context, situation and location awareness (Wolfert et al., 2014). In this paper, we define Smart Agriculture as the use of precision agriculture technologies aided by big data and AI to make informed autonomous farm decisions that save resources in short term and increase the quality of produce in long term.

In essence, as has been done by blockchain and cryptocurrency in the payment industry, virtual reality in the entertainment industry, and trendsetters like 3D printing and augmented reality, big data and artificial intelligence (AI) in agriculture are disruptive technologies that, as defined by Christensen (1997), are changing the entire outlook of the industry through new ideas for problem solving with the hope of eventually displacing existing practices. However, as with any disruptive innovation, the impact on the target sector and society at large can have far-reaching implications. Big data and AI are already starting to reshape the manner we handle agricultural tasks such as harvesting, crop and soil management, and accounting for environmental impact on yield using predictive analytics. Examples range from robots employing advanced machine vision for harvesting pepper (Simon, 2018), to optimizing crop yields in India (Microsoft, Inc., 2018). Further, the socio-technical, and socio-economic drivers and challenges for the development and diffusion of smart agriculture are context dependent. While developed countries may be driven by a severe shortage of labor, developing countries are driven by the sheer need to support their fast-growing populations. The development and diffusion in developing countries will have to consider factors such as the nascent supporting technology infrastructure.

Being a novelty field with so much potential and rising popularity, several researchers have tried to measure the impact of Smart Agriculture and its effect on traditional agricultural practices. Big data management and analysis especially has been a major theme for most researchers in trying to understand the opportunities inherent in its incorporation into farms (Chi et al., 2016; Coble et al., 2018; Kamilaris & Prenafeta-Boldú, 2018; Nandyala & Kim, 2016; Waga & Rabah, 2014; Woodard, 2016). Wolfert et al., (2017) went further to develop a conceptual framework to analyze big data in Smart Farming applications from a socio-economic perspective. Other reviews have been more geo-specific: as has been done to understand applications of big data in developing countries (Ali et al., 2016; Misaki et al., 2018; Olaniyi et al., 2018; Protopop & Shanoyan, 2016); or more domain-specific: such as deep learning applications (Kamilaris & Prenafeta-Boldú, 2018; Zhu et al., 2018), hyper-spectral analysis (Khan et al., 2018; Thenkabail et al., 2012), and wireless technology (Mark et al., 2016). This chapter aims to contribute to existing literature by emphasizing the AI dimension of the conversation through an examination of how the relationship between these two innovations are disrupting agriculture. Specifically, this chapter addresses the following research questions:

1. What is the current status of Big Data and AI in smart agriculture?
2. What are the drivers and challenges underlying this paradigm shift?
3. What are the prospects of Big Data and AI as disruptive technology in agriculture?

Addressing these research questions, we conduct a systematic review focusing on big data and AI in agriculture. The review spans the last decade and covers peer-reviewed scholarly publications as well as, given the emerging field, *grey* literature, example industry reports. To further guide the literature review process and organize the findings, we devise a framework based on extant literature. The framework aims to capture key aspects of agricultural processes supporting supply chain and key stakeholders with a particular emphasis on the potential, drivers, and challenges of big data and artificial intelligence. We also discuss how this new paradigm may be shaped differently depending on context, namely developed and developing countries. In this chapter, Smart Agriculture, Smart Farming and Agriculture 4.0 are used to interchangeably.

The chapter is organized as follows: The following two sections describe the background and methodology for the systematic review and introduce the proposed conceptual framework, respectively. Next, we present the results followed by a discussion of the findings with respect to the current status and prospects of big data and AI in agriculture as well as the drivers and challenges moving forward.

BACKGROUND

Industrial revolutions have always gone in tandem with agriculture, and just like the Industrial Revolution 4.0, the agricultural industry has grown from the labor-intensive Agriculture 1.0, high productivity gains of the *green* Agriculture 2.0, *precision farming* through guidance systems, sensors, and telematics of Agriculture 3.0 through to Agriculture 4.0, or Smart Agriculture, which consists of farm equipment that are enhanced using autonomous unmanned decision systems (robotics), data and artificial intelligence (Cho, 2018; Wolfert et al., 2017). To understand the relevance of this slow but gradual change towards automation based on farm data, it is necessary to examine the value chain of agriculture as well as concepts related to big data and AI.

The Agriculture Value Chain

In this review we breakdown agriculture's value chain into its components as:

Pre-Production: Any activity that occurs before actual planting takes place is classified under the pre-production phase. These activities include land preparation, seed quality control, and all other planning activities, that can be very crucial to healthy crop production and potential crop yield (Krishnan & Surya Rao, 2005; Oshunsanya, 2013).

Production: The production phase covers every on-field activity such as planting, fertilizer application, irrigation, pest and weed control. Precision Agriculture delivered a plethora of mechanization improvements in this aspect of agriculture. Herein, we identify how big data and AI have further optimized production decisions – both autonomous and human-led. Cost, time and any other resource savings are brought to the forefront by examining process improvements.

Post-Production: Robotic harvesting has been around as far back as the 1980's, and this has been popularized and applied to various types of commodities: eggplants, tomatoes, cucumber, and strawberries (Arima et al., 1994; Arima et al., 2001; Hayashi et al., 2002; Kawamura et al., 1984; Kondo et al., 1996). Smart Agriculture provides an avenue for further improving accuracy of harvest, supply chain, and marketing operations, hereby reducing yield waste and increasing profit margins.

AI and Big Data

The concept and real-world application of artificial intelligence has existed since circa 1950. The concern for AI has been developing systems that act as rational agents taking the best possible action in any given situation (Russell & Norvig, 2010). Cockburn et al. (2018) groups AI applications into three interrelated separate streams: symbolic systems such as natural language processing and image recognition: have been successfully applied at various levels but also been criticized for their inability to be scaled towards commercial solutions; robotics: used heavily for industrial automation; and the learning approach: that given some inputs can predict the presence of particular physical or logical events. As demonstrated in Figure 1., these streams are not so far from the sub-specialties proposed by Michael Mills of Neota Logic (Martin, 2016). Just like the streams proposed by Cockburn et al. (2018), he divides AI into robotics, cognition tools such as Natural Language Processing and Vision systems used to capture and synthesize text, voice and images; Machine Learning consisting of deep learning and predictive analytics through supervised and unsupervised learning used to find previously unknown relationships in data; and other concepts such as Expert Systems and Planning, and Scheduling & Optimization.

Big data, on the other hand, refers to large amounts of unstructured data produced by high-performing applications (Cuzzocrea, et al., 2011). It is characterized by the 3V model proposed by Laney (2001) consisting of *volume*, *variety*, and *velocity*. Many authors have extended these 3Vs to include *value* and *veracity* (Bello-Orgaz et al., 2016; Cuzzocrea et al., 2011). Big data methodologies, such as data mining, serve up huge datasets made up of both structured and unstructured data (Fan & Bifet, 2013; Ghosh & Nath, 2016). What is inherently clear, however, is that big data is beyond the analysis capabilities of traditional relational databases (IBM, 2018).

By employing machine learning techniques, AI provides the needed analytic capabilities that support Data-Driven Decisions (DDD) (Bengio, 2009, 2013; Najafabadi et al., 2015). As the name implies, a data-driven decision is one made by using up-to-date data to analyze the facts in data and draw conclusions. As such, we consider big data and AI's value and potential as a derivative of their combined ability to provide automated DDD which result in higher productivity and market value (Brynjolfsson et al., 2011; Provost & Fawcett, 2013).

There is a plethora of underlying technologies (Hadoop, MapReduce, Spark, Mahout, Hive, etc.) and techniques (Naive Bayes, Regression Models, K-Nearest Neighbors, Decision Trees, Boosting Algorithms, Genetic Algorithms, Neural Networks etc.) used in processing, mining and analyzing big data. Our focus, however, is on value which typically is derived from deploying AI-based analytics to big data.

Overall, there are four types of analytics – descriptive, diagnostic, predictive, and prescriptive (Markkanen, 2015; Rajeswari et al., 2017). In this chapter, we focus on the latter three progressions of deriving value from big data where AI techniques are most prevalent:

Figure 1. Fields of AI. Source: (Mills, 2015)

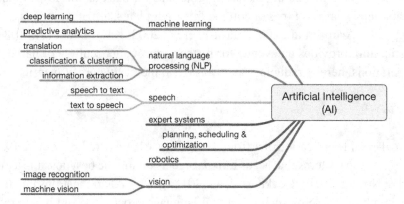

Diagnostic

This function aims at providing answers to the question: *why did it happen?* (Gartner, 2018). Diagnostic Analytics is sometimes combined with Descriptive Analytics with some researchers even calling it a step 1.5 before the next phase (Markkanen, 2015). This is because it is the most abstract of the phases: an explorative step that exists mainly to drill down on reasons behind certain trends. However, it is still key in understanding data trends, especially as machine learning models can quickly find patterns in data that human analysts might not be as quick to identify. A simple example will be where a business, before restocking due to a sudden rise in the sales of a commodity, uses this type of analytics to understand why there is an increase in sales and thereby prevent over (or under) stocking on the said commodity.

Predictive

Any analytics model that emphasizes prediction, rapid analysis, business relevance of resulting insights, and ease of use (sometimes through automated decision making) is considered Predictive Analytics. It is primarily concerned with: *what could happen* (Gartner, 2018; IBM, 2013). Some of the more common predictive models include regression analysis, decision trees, forecasting, multivariate statistics, pattern matching, and neural networks. While there are close parallels between this type of analytics and machine learning, it is important to underline that Predictive Analytics merely employs machine learning algorithms in making estimations of future outcomes however they are not the same. (Wakefield, 2018). Business uses for Predictive Analytics include risk outcomes of various business decisions, sales forecasting, deriving marketing insight for proper market segmentation, and even financial modeling aimed at forecasting business performance.

Prescriptive

Prescriptive Analytics forms part of the paradigm of advanced analytics that explores data to provide possible actions. In essence, it provides answers to the question: *what should be done?* (Gartner, 2018; IBM, 2013; Markkanen, 2015). In essence, it builds on Predictive Analytics to provide possible actions to take and the effect of such decisions. Techniques employed with this type of analytics include simulation, graph analysis, recommender systems, complex event processing, heuristics, neural networks, and machine learning. By continuously *learning* from new data, the accuracy of prescriptions improves with each iteration.

Related Work

Big data applications have been used to deal with several agricultural issues and there are many agricultural areas where big data solutions have been deployed. These include but are not limited to weather forecasting, land conservation, weed detection, biodiversity, and remote sensing (Coble et al., 2018; Kamilaris et al., 2017). Kamilaris et al., (2017) go further to state some potential areas for big data application, especially in the supply chain where there is a need for quality products and better yield and demand estimation. This view is similar to the position taken by Wolfert et al. (2017). They propose a conceptual framework to analyze big data applications in Smart Farming from a socio-economic perspective. Their findings demonstrate that big data applications in agriculture are changing the scope of the industry through push-pull mechanisms. The push factors are driven by advancements in technology, massive amounts of data generation and storage, and digital connectivity. The pull factors, on the other hand, are mainly through population and business demand: the public being interested in food sustainability, security, and nutrition; and to the farmer, smart farming being mainly about efficiency and profitability.

The recent resurgence of AI, and associated techniques like machine learning, is also pushing the limits for Smart Farming. Machine learning is often the most used technique employed by researchers for big data analysis, although other statistical models like time series analysis, and spike and regression analysis have also been used for forecasting agricultural events (Coble et al., 2018; Kamilaris et al., 2017). One branch of machine learning which has been applied, especially to computer vision and image analysis in agriculture, is deep learning. In a recent paper, Kamilaris & Prenafeta-Boldú (2018) surveyed several applications of deep learning in agriculture. Interestingly, they found that deep learning has been applied not only to imaging but sensory data and other environmental variables. Recurrent Neural Networks (RNN), a type of deep learning, were seen to offer higher performance due to their ability to capture both space and time dimension. Zhu et al. (2018) also surveyed deep learning applications in agriculture and discovered similar result. They found that RNN had been used for land cover classification, phenotype recognition, crop yield estimation, leaf area index estimation, weather prediction, soil moisture estimation, animal research, and event date estimation.

The value of big data in smart agriculture, with the right analysis, can provide a win-win situation for all stakeholders. The potential for revenue growth and increased production efficiency, however, still seem unable to drive large-scale adoption. Several challenges have been found by researchers. Wolfert et al. (2017) discussed technical and organizational challenges with the latter being the most important. One of the biggest challenges was found to be governance issues regarding privacy and security of big data leading to its underutilization in smart farming. This sentiment is echoed by several other researchers (Coble et al., 2018; Kamilaris et al., 2017; Nandyala & Kim, 2016). In developing countries, the challenge for adoption runs deeper than data security. The most glaring being the lack of adequate physical and legal infrastructure to allow collection, storage and processing of data (Kshetri, 2014; Misaki et al., 2018; Protopop & Shanoyan, 2016). Another issue raised by researchers has been the high illiteracy levels amongst farmers in these areas which has contributed to the lack of awareness and trust in smart technology. Further, Misaki et al. (2018) discuss at length issues regarding farmers' belief that integration of big data applications will result in undue advantage for their competitors and even drive up land value and seed prices.

In this chapter, we discuss the current and future prospects of smart farming, the machine learning techniques that have been most prevalent, and how the relationship between big data and AI is disrupting agriculture. This chapter aims to contribute to existing literature by emphasizing the AI dimension of the conversation through an examination of how the relationship between these two innovations are disrupting agriculture. We explore the gap inherent in current literature which have been more geo-specific (Ali et al., 2016; Misaki et al., 2018; Olaniyi et al., 2018; Protopop & Shanoyan, 2016); or domain-specific such as deep learning (Kamilaris & Prenafeta-Boldú, 2018; Zhu et al., 2018), hyper-spectral analysis (Khan et al., 2018; Thenkabail et al., 2012), and wireless technology (Mark et al., 2016).

METHODS

In this systematic review, we use guidelines in accordance with the Preferred Reporting Items for Systematic Reviews and Meta-Analyses (PRISMA) (Liberati et al., 2009). The PRISMA statement provides an evidence-based minimum set of items for reporting systematic reviews and meta-analyses with the aim of improving the quality of reporting by authors of systematic reviews.

Data Sources and Search Strategy

We systematically reviewed literature from Web of Science, IEEE Xplore Digital Library, and Science-Direct (Elsevier) using a Boolean search query consisting of three groups of keywords that is big data and AI technologies (big data, artificial intelligence, data mining, machine learning, deep learning, neural networks etc.) and associated with either the set of keywords representing smart agriculture (smart agric*, smart farm*, agriculture 4.0, digital farm*, precision farm* etc.), or agriculture in general (agric*, farm*, agronom*, cultivat* etc.). To further enrich the data retrieved, literature from ACM Digital Library, Agricola and *grey* literature from sources such as industry reports and non-peer-reviewed publications from Google Scholar are considered. Big data is a relatively new subject area; the first academic reference to big data was by Weiss & Indurkhya (1997), as such, the survey targeted literature published a decade after this first publication, that is, the period between January 2008 to September 2018. Due to language barrier, only English language publications are considered. Relevant works outside the specified timespan are considered if cited by included sources and are determined to be of importance to the chapter.

Study Selection

Papers focusing on areas other than agriculture are excluded, for example, the review excluded papers about the technologies themselves where agriculture was mentioned only as an area of application. Further, the review limits the definition of agriculture to agronomy (crop cultivation and soil management), hence livestock production and aquaculture applications are excluded as well. Other exclusion criteria included papers that are comparing technologies already covered such as reviews, surveys, and duplicates.

CONCEPTUAL FRAMEWORK

The conceptual framework (Figure 2.) serves as a background for the systematic review of big data and AI technologies in agriculture. We posit that the interaction between agriculture and smart technology, that is AI and big data, is a cross relationship between the agricultural value chain and the levels of analytics. As such, using the simple agricultural value chain and the levels of analytics described in the background, we break down agricultural production from inputs to distribution as a basis for identifying any form of disruption in the entire chain (Kaplinsky & Morris, 2001). For the purposes of this review, we emphasize the efficiencies these technologies bring to agricultural production and the magnitude of the value added, or prospects of value, as we move upstream in the agricultural value chain. Consequently, papers will be classified based on their potential value in this chain and not solely on technique alone.

Starting from the bottom, researchers employ diagnostic analytics techniques in order to understand why an event has happened or is happening. In classifying papers as having used diagnostic analytics, we refer to papers that try to provide an interpretation of environmental or societal indices and cropping using AI and big data. We look at studies that identify anomalies, do data discovery, or determine causal relationships. The middle tier, predictive analytics, identifies mainly two categories of studies: those that forecast future events or classify the existence, or non-existence, of certain agents. In each part of the value chain, there are different types of predictions to be made: in production, for example, there are advantages to both identification of weeds and yield forecasting. At the very top is prescriptive analytics where we highlight studies that adopt AI and big data to provide recommendations. Studies that provide

Figure 2. Conceptual framework for reviewing literature

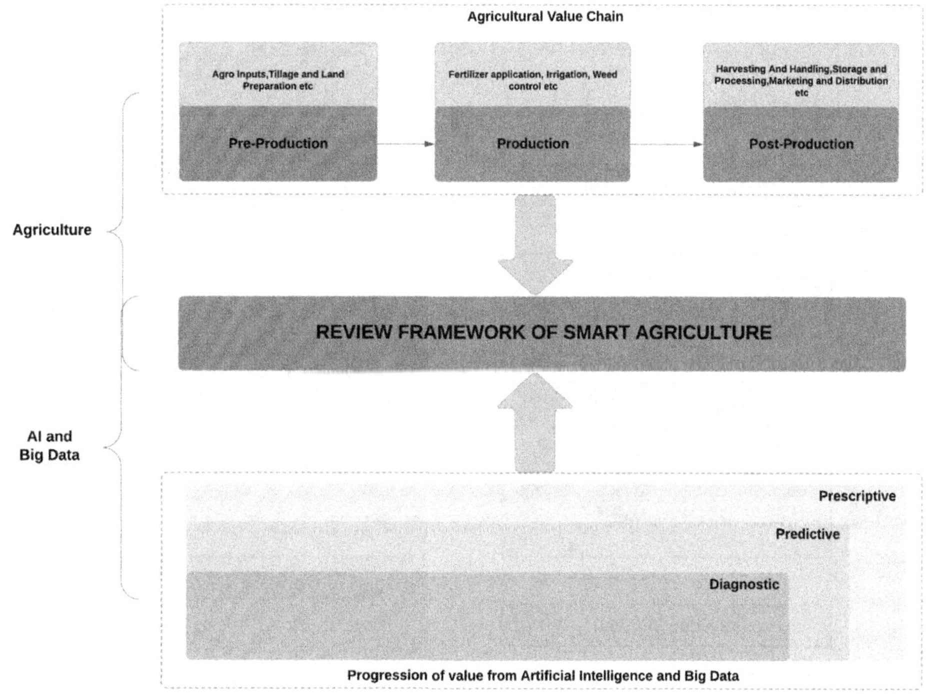

automated decisions and or apply business rules to prescribe actions. Some recommendations may not necessarily provide optimal solutions but rather simulate the possible effects of decision making. Simulations can be used, for example, to identify yield in different weather conditions.

RESULTS

As depicted in Figure 3., the search yielded 581 titles, of which 24 duplicates were excluded. The 557 remaining articles underwent careful title and abstract assessment; 59 were rejected on basis of being a survey, review, case study or comparative study; 94 excluded for being based purely on soil science; 125 articles based on animals, technology or precision agriculture were also excluded as well as 56 other off-topic articles. A total of 223 articles met our initial eligibility criteria and underwent full review out of which 153 articles covering the entire spectrum of AI and big data in agriculture were selected. Figure 4. depicts the trend line of papers over the analysis period.

Predictive (72.54%) and Prescriptive (22.88%) were the most prevalent. The majority of the work done has been during the Production phase of agriculture (61.43%) with 78 (62.17%) of that value being Predictive, and 16 (15.01%) being Prescriptive as shown in Table 1. See the Appendix (Table 2., Table 3. and Table 4.) for summary information of each included article in our study.

Figure 3. Study selection process for systematic review

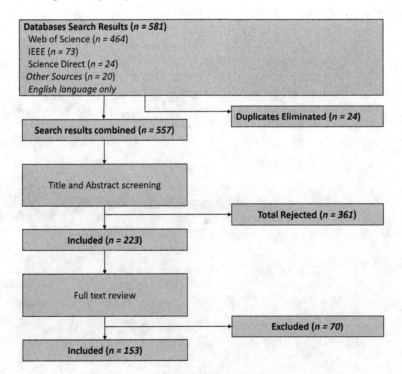

Figure 4. Distribution of papers by years

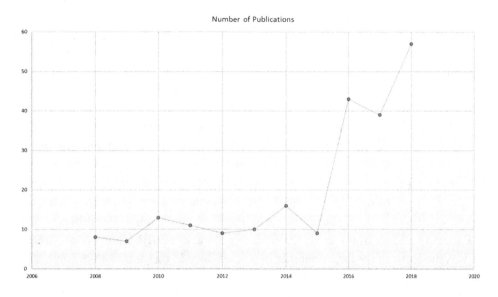

Table 1. Key findings

Agricultural Value Chain		Big Data and AI		
		Diagnostic	*Predictive*	*Prescriptive*
Pre-production	Seeds & Seedlings		4	3
	Soil, Tillage and Land Preparation	1	7	
	Other Pre-planting Activities	2	6	8
Production	Planting, Irrigation & Fertilizer Application	1	16	8
	Weed, Disease & Pest Control	2	33	7
	Yield, Weather, Energy & Other Farm Management Activities	2	29	1
Post-production	Harvest & Handling		6	7
	Storage & Processing		4	3
	Market & Consumer		10	1

Pre-Production

Seeds & Seedlings: Seven studies focused on improving the seeding process. Four studies employed the use of predictive techniques to classify infected and non-infected seedlings, sort quality of seedlings, and to predict yield (Emamgholizadeh et al., 2015; Griffel et al., 2018; Lamsal et al., 2017; Silva et al., 2013). Prescriptive methods was employed in the three remaining papers in recommending the right amount of phosphogypsum to be used when the farmer uses no-till management, optimization and seed density simulation (Caires & Guimaraes, 2018; Dornelles et al., 2018; Tesfaye et al., 2016).

Soil, Tillage and Land Preparation: Out of eight studies, one study employed diagnostic methods aimed at analyzing the effects of soil properties on traction force and traction efficiency in a bid to optimize tillage (Pentos & Pieczarka, 2017). All other papers, were predictive in nature; five focused on predicting soil properties; others focused on predicting and optimizing vitro proliferation and estimating fuel consumption during tillage and soil preparation (Ajdadi et al., 2016; Arteta et al., 2018; Borges et al., 2017; Calderano et al., 2014; Pentos & Pieczarka, 2017; Pike et al., 2009; Prasad et al., 2018; Quraishi & Mouazen, 2013).

Other Pre-planting Activities: Two papers using diagnostic methods applied supervised feature selection to define traits affecting water content in maize and Spearson's rank correlation to determine association between climatic indices and maize and sorghum yields (Byakatonda et al., 2018; Shekoofa et al., 2011). Six studies used predictive methods in various aspects; for drought assessment and water planning, for forecasting seasonal variability and temperature, and for appraising mechanization on farms to predicting the success of an agricultural enterprise based on their capital (Bakhshi et al, 2016; Osman et al., 2015; Park et al., 2016; Smith et al., 2009; Zangeneh et al., 2010; Zhong et al., 2009). Prescriptive methods was the focus of the larger number of studies done in this area with the eight studies found devoted to prescribing and simulating management routes based on seeding density, land use conversion, and crop sequence (Delgado et al., 2008; Dornelles et al., 2018; Jimenez et al., 2009; Rajeswari et al., 2017; Renaud-Gentie et al., 2014; Rizzo et al., 2014; Snow & Lovattb, 2008; Zhong et al., 2009).

Production

Planting, Irrigation & Fertilizer Application

We found twenty-five papers that addressed this area of the agricultural cycle. One paper used data mining to analyze the spatial relationship between environmental and social factors and maize cultivation (Ureta et al., 2013). Sixteen papers concentrated on prediction of evapotranspiration and irrigation, chlorophyll and nitrogen estimation, and prediction of soil properties. Eight papers were prescriptive in nature and dealt mainly with water and irrigation management, and fertilizer recommendation.

Weed, Disease & Pest Control

Almost a third (29.41%) of all studies pertained to studies directed at solving issues with weeds, diseases and pests. A total of forty-five studies were found. Two papers that employed diagnostic methods saw the introduction of electronic devices designed to quantify climatic variables complemented by a mobile application for diagnosing casual agents of wilt complex disease, and data mining techniques applied to understand the relation between crop, weather, environment and leaf spot disease. Thirty-three papers dealt in various degrees with weed detection and mapping, disease detection and diagnosis, and pest classification and identification. Appendix Table B provides a breakdown of the papers, area of application and machine learning technique applied.

Yield, Weather, Energy & Other Farm Management Activities

Of the thirty-two papers found in this area; two employed diagnostic techniques for determining appropriate environmental conditions for obtaining high yields and for modeling crop sequences (Jimenez et al., 2011; Xiao et al., 2014), twenty-nine papers were predictive in nature focusing on various agricultural improvements such as yield prediction, greenhouse management, energy input and output modeling, and temperature prediction, the remaining publication used prescriptive methods to for weather prediction and autonomous robot fruit harvesting (Sennaar, 2017).

Post-Production

Harvest & Handling

Thirteen studies examined the harvesting and handling process. Six of these studies were predictive in nature and aimed at forecasting the readiness of various agricultural produce for harvest; strawberry, apple, sugarcane, kiwi fruit, coffee, and even herbs. The seven remaining studies used prescriptive methods; six studies aimed at determining ideal time for harvesting and automated robotic harvesting, and the last study focused on sorting already harvested fruit using decision trees.

Storage & Processing

A total of seven papers, four on predictive and three on prescriptive methods were chosen. One study proposed a real-time classification method for Anthurium cultivars aimed at increasing the postharvest processing process and two studies were on predicting biomass for agricultural and industrial processes. Smith et al., (2009) introduced a system for year-round temperature prediction deployed on the website of the Georgia Automated Environmental Monitoring Network which will be suitable for managing and planning storage activity. The prescriptive papers were; Babazadeh et al., (2016) proposed a classification method for potato tubers based on solanine toxicant, Wu et al., (2014) proposed a control system for tobacco flue-curing barns, and Guine et al. (2018) evaluated the influence of production conditions and other factors on the total phenolic compounds and antioxidant activity of blueberries.

Market & Consumer

Predictive techniques were used ten papers mostly for price forecasting and measuring colorimetric properties of fruit. The one prescriptive paper found was on a decision support model aimed at improving export process.

Machine Learning Techniques

As demonstrated in Figure 5., artificial neural networks (ANN) was the most used technique by researchers with a link strength of 108 and was used in 76 of our surveyed literature. This stems from the range of application of ANN algorithms. They are used in classification of remote sensing imagery to time series forecasting. The use of Support Vector Machines (SVM) algorithm was also popular amongst researchers. It was used in 22 papers with a total link strength of 43. SVM supports both classification

Figure 5: A network of machine learning techniques and areas of application

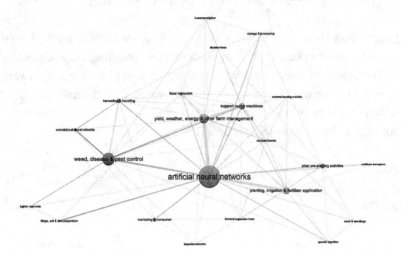

and regression problems. We also found that statistical models like Linear and Logistic Regression are still applied especially new variants of the model like Multiple Linear Regression which is used in cases where there are more than one explanatory variables. They were used in 10 and 4 papers, respectively.

DISCUSSION

Current State of Big Data and AI in Agriculture

Big data in smart agriculture originates from farm sensors, drones, satellite stations, published industry performance, price index databases, and several other sources. In line with the characteristics of big data, the volume, variety, and velocity of agricultural data allows the use of AI to solve several agricultural issues and drive efficiency along the agriculture value chain. Our findings demonstrate that the amount of relevant studies spans the entire spectrum of agriculture value chain, although evidence points to the prominence of the production phase.

The approaches used by researchers contain a range of machine learning technique that can be applied to different problems along the value chain. Figure 5. demonstrates that ANN, which is inspired by the biological neural network of the human brain, is the most popular technique used by researchers in the surveyed studies. It has shown promise in the application of deep learning models to on-farm decision making. Other techniques like Support Vector Machines, Convolutional Neural Networks (CNN), and Genetic Algorithms (GA) have been used to various degrees. The use of GA, especially, is interesting as it is also inspired by natural selection and belongs to the larger class of evolutionary algorithms. They are often used in solutions that require bio-inspired operators such as mutation, crossover, and selection. SVM, on the other hand, are used in both classification and regression analysis. GA is often used in combination with either ANN or SVM.

As pointed out earlier in this chapter, smart agriculture is a novelty field, and this may explain the large amounts of academic research available in literature. The ideal scenario is to have more commercially usable studies that employ prescriptive techniques at scale. Some potentially disruptive applications of such studies include two in Spain: a hybrid system modeled on farmer's behavior to predict irrigation depths (Perea R et al., 2018), and the automatic translation of agricultural regulations into machine-processable rules (Espejo-Garcia et al., 2018). While commercial application of big data and AI may not be ubiquitous in agriculture yet, startups such as Trace Genomics and VineView have embraced the use of machine learning for diagnosing soil defects and for crop analysis respectively. The former has already raised $19 million in funding from 6 investors, and the latter $4.9 million from 3 investors as of December 2018 (Crunchbase, 2018; Sennaar, 2017; Trace Genomics, 2018; VineView, 2018). In a recent report, McKinsey Global Institute pointed to the fact that although AI adoption in industries outside the tech sector was still at an experimental stage, external investment into the innovation had grown three times more since 2013 (Columbus, 2016). Such investments represent a win for researchers and the agricultural industry, and as more and more research efforts continue to pour in, investors will be more willing to back commercial forays into big data and AI on the farm.

As we discuss in the ensuing section, the challenges and drivers for developed and developing countries in smart farm adoption may be motivated by slightly different circumstances. While the research also appears to originate predominantly from developed countries where the technological infrastructure, availability of capital, and supporting organizational and cultural structures are more conducive to such innovations, the research agenda from the surveyed literature is very similar. Further, there has been accounts of attempts to introduce smart agriculture in lesser developed regions of the world. Examples include: drought-tolerant maize varieties used in southern Africa to evaluate their performance (Tesfaye et al., 2016), and estimating evapotranspiration in India (Kumar et al., 2016). Price index prediction has been of special interest to researchers in Brazil and South Africa (Ayankoya et al,, 2016; Correa et al., 2016; Orge Pinheiro & de Senna, 2017; Ribeiro & Oliveira, 2011). This is one such area that can drive government policy in the developing countries and bridge the gap to developed countries.

Drivers and Challenges of AI and Big Data in Agriculture

Drivers for AI and big data in agriculture can be broadly addressed along economic, business, and technology dimensions as follows:

Economic Drivers

The UN FAO (2017) lists ten important challenges in their mission to eradicating hunger and malnutrition, and ensuring food security and agricultural productivity. These challenges include but are not limited to; improving agricultural productivity and sustainability to meet increasing demand, ensuring a sustainable natural resource base, addressing climate change and intensification of natural hazards, eradicating extreme poverty and reducing inequality, and preventing transboundary and emerging agriculture and food system threats. We envision that these same challenges will be key economic drivers for the integration of technology into not only on-farm activity but the entire agricultural value chain.

Evidence from our review also suggests that the aforementioned challenges are driving academic research. Machine vision systems backed by ANN have been used in robotizing farm work in a bid to optimize previously tedious and overly manual processes. In order to conserve natural resources such as land and water, and to battle climate change, there has been focus on soil classification, evapotranspiration, and smart irrigation. Process improvements have been made to weed, diseases and pest control to assist in fighting threats to food production and ensuring food security. Again, planning sales, export and import, and determining the price point of produce through yield prediction will be important for combating economic issues such as inflation.

Business Drivers

Businesses will rely on big data and AI for informed insights into decision making. The opportunity to preemptively grow through data-driven innovation, decision making, and discovery will drive even more businesses to move from more traditional ways (relying on farmer experience) to big data-driven artificial intelligence. AI will provide agriculture with more sustainable processes that take as input environmental factors and provide recommendations that saves business resources and increase yield. From a financial standpoint, both the farmer, investors and distributors have a lot to gain from yield, weather and energy predictions in a bid to plan business models and stock supply lines. Early detection of diseases and pest have immense cost and yield benefits to everyone involved in production.

On the other hand, as available workforce continues to dwindle in the agricultural sector, automation through predictive and prescriptive methods will decrease timelines and cost and increase quality of processes and produce. Human decisions can be mimicked and optimized through algorithms that continually self-correct based on new parameters. Tasks such as sorting of seeds and weed control, which are slow when done manually and potentially hurtful when done mechanically without any insights, have the potential of being entirely overhauled through proper classification of plants.

Technology Drivers

As has been pointed out, sensors and IoT devices from the precision agriculture are generating unprecedented amounts of data that will only continue to increase with time. Cloud computing has enabled storage and compute power to become cheaper through a "pay-for-what-you-use" model; ensuring a cost-effective avenue for retrieving value from big data. With cloud computing the agricultural sector can now afford to optimize efforts without necessarily allocating up-front capital. Technological trends like Infrastructure as a Service (IaaS), Software as a Service (SaaS), and Platform as a Service (PaaS) will drive farmers to focus on what they do best, farm, while leveraging such services to grow and advance yield. Researchers such as Alipio et al. (2017) and Rajeswari et al., (2017) have integrated cloud computing into their smart agricultural models due to their scalability over traditional systems and economical cost.

Like any innovation however, we envision several challenges, most notably are:

Data Ownership and Security

Privacy is a key concern for cybersecurity associated with data generation. Innovations like the ifarma, a cloud based farm management information system (FMIS), where farmers are provided with end-user license agreement give the farmer total ownership of data (Paraforos et al., 2016). The industry, however, needs regulations around ensuring privacy and securing data as has been done for the financial and health fields through the Gramm-Leach-Bliley Act of 1999 (GLBA), Health Insurance Portability and Accountability Act of 1996 (HIPAA) and most recently the European General Data Protection Regulation (GDRP).

Data Stream Consolidation and Interoperability

While this may be a challenging task due to the volume and variability of data, it makes for a better analysis on a regional and global level when data is consolidated and curated on global scales. Especially in the wake of proper regulations and privacy controls, it will allow proper global forecasts into food sustainability.

Data Transport and Hosting

Cloud infrastructure seems to provide evidence of readiness to take on large amounts of data, however location and retrieval of data is important for building real-time decisions.

Data Processing Power: It is undoubted that computing power has increased considerably over the years, but so has big data. It is clear that faster and more reliable algorithms need to be tested to allow for commercial applications of big data in agriculture. There is a special need for more research in the areas of automated feature engineering and dimensionality reduction.

Transparent, Explainable, and Accountable AI

Similar to other domains such as healthcare, the transparency, explainability, and accountability of AI algorithms may pose a particular challenge with respect to the credibility, trustworthiness, and ultimate adoption of these techniques in the field. The extent of impact - including possible regulations (Wachter, Mittelstadt, & Floridi, 2017) is likely to vary depending on the application on hand and represents and open area for research.

Inexpensive and Commercial AI

One of the barriers to adoption of AI is the funds needed to employ the requisite talent to deploy data solutions. Data scientist are some of the most expensive talents to hire today. While most companies believe AI is plug-and-play, it requires massive amounts of domain knowledge, proper key performance indicators, and long timelines to deploy robust solutions. Largescale adoption of AI stems from the development of agriculture-friendly commercial solutions that do not require an extensive computer or statistical knowledge.

Bridging Divide Between Developed and Developing Countries

A number of papers have presented initiatives in developing countries such as Iran, South Africa, India, Brazil and Argentina (Caires & Guimaraes, 2018; Ferraro et al., 2012; Khashei-Siuki et al., 2011; Kumar et al., 2016; Tesfaye et al., 2016), it stands to reason that the problem for the adoption and diffusion of big data and AI in these developing countries is not due to lack of research or technical capability. Misaki et al., (2018) in a recent survey pointed to lack of trust and transparency between farmers, technology providers and government organizations; inappropriate cultural context; illiteracy amongst farmers; inadequate infrastructure; low awareness; bureaucracy; and a few others as challenges facing small-scale farmers in sub-Saharan Africa. Of particular significance for most developing, and some developed countries, additional challenges include:

IT Infrastructure & Power Supply

Saidu, et al. (2017) pointed to challenges such as lack of infrastructure and erratic power supply as major barriers to adoption of on-farm information technology in developing areas.

Organizational and Cultural Practices

Despite strong evidence of government in developing countries readiness to support technology adoption, there are concerns on possible disruption of the way of doing things which may act as a barrier to Smart Agriculture (Aleke et al., 2011).

As much as these and other challenges such as the ones we discussed above exist for developing countries, we propose a backward integration of big data and AI into their processes. Starting from the 'Market and Consumer' section of the agricultural value chain, AI can be used to analyze prevailing trends. There is strong evidence of the proliferation of mobile phones, internet and social media in developing countries, and they have shown increasing effects on agricultural development (Olaniyi et al., 2018). Supply chain management and development can be enhanced through analysis of big data, especially from social media, to examine trends and user perception. Availability of price indices online can also be used to shape export and import benchmarks.

Prospects of Big Data and AI Technologies in Agriculture

Years of research have gone into developing many types of sensors for recording agronomically relevant parameters, collecting enough data via Internet of Things (IoT) and sensor networks, developing farm management systems, devising AI-driven farm machinery, and the like. By 2020, it is estimated that about 75 million agricultural IoT devices will be in use with an average farm generating up to 4.1 million data points daily by 2050 as compared to the 190,000 data points generated in 2014 (Clercq et al., 2018).

Due to the fluidity and hierarchically distributed open form of big data, capturing value requires real-time information to be dredged out of said data. It is required that analysis of the data takes as little time as possible and give a better profit margin as compared to good human-made management decisions. As a result, AI algorithms are being trained to provide this value (Özdemir & Hekim, 2018; Weltzien, 2016). Table 1. and Figure 4. points to an increase in the amount of relevant research taking on this challenge. Further, it can be said that many of the problems in agriculture are represented by over-parameterized nonlinear systems that may only be solved with the help AI. AI can recognize patterns and make predictions based on historically generated data and through automation, apply to real-time systems, as such the focus on predictive and prescriptive methods by most researchers (Özdemir & Hekim, 2018; Weltzien, 2016).

The major shortfall of ANN and most AI algorithms is the need for longer training times and the risk of either unintentional (example overfitting) or intentional (example malicious intent) bias. In such cases, a solution that might work for one problem, might not generalize well to other problems. *Cognitive Neural Networks* and *Deep Reasoning* algorithms will be most beneficial in such situations. Algorithms that go beyond recognizing and identifying attributes of objects (example perception and classification) to recognizing the causal relationships inherent and basing explainable decisions on these relationships (cognitive reasoning) (Battaglia et al., 2016; IBM, 2015; Y. Wang, 2016). We envision that they will be critical going forward. AI that is explainable will be more important especially for accountability in the wake of increasing data protection regulations that penalize companies for reasoning, or lack thereof, behind decision making. While deep reasoning is still in a very experimental phase, other methods (such as *Reinforcement Learning*; *Transfer Learning*; and *Active Learning*) can cater to current needs in terms of training times and optimal use of computing power.

A Futuristic Smart Farm Scenario

In principle, AI and machines can learn to perform any agricultural task given relevant data and parameters to learn from. The future for AI and farming lies in its very definition, that given any situation will act as a rational agent to take the best action possible. In future farming scenarios, we expect that big data and AI will disrupt the entire agriculture value chain. In this section we will paint two contrasting scenarios representing the current state of agriculture and the ideal smart farm for the future.

Typically, a farmer decides how and when to till the fields. When tilling is complete, planting of crops is done, typically with a planter attached to a tractor. In the rest of the season, the farmer manually monitors for the onset of diseases or harmful insects and should there be the need, may use an aerial applicator to fly over the farm and apply fungicides, weedicides, herbicides, and other chemicals to control any such problems. Irrigation is sometimes used but most farmers rely on rainfall for crop water supply. At harvest time, a harvester is used, and crops are sent to storing barns or silos till they can be sold off to food distributors. Although some farmers have the opportunity to sell to different markets, they are usually not in control of prices offered.

In a future smart farm, an AI system will identify optimal seed varieties and optimal planting times based on soil properties and weather data. The same system will control agro-robots that till and plant the seeds. After planting and germination, a drone is programmed to fly over the farm and collect image data which is immediately processed by the system. This allows for instant detection of crop diseases and pest infestation. In other parts of the field, seedlings suffering from late shoot or some form of damage will be marked for immediate replanting. At the end of the flight, the software then delivers population counts and yield estimates. More data will be collected from satellite imagery and a prediction is made if, for instance, a swarm of insects have infested nearby farms and are headed for that general area. This prediction can be corroborated by mining social media data. Consequently, the AI will estimate potential damage and recalibrate yield for each possible action to take should there be an infestation.

Further, the AI system shall control the nutrient and water needs of each plant and perform irrigation and spraying automatically. These calibrations will be readjusted daily based on how the crops are faring. At harvest time, the system then notifies the farmer and provides the most optimal pricing and distribution channels using pricing index databases. When the farm produce is shipped off to the food distributor, the system will upload production data to their database. This means in the end; the consumer can also immediately identify which chemicals have gone into the production and under which conditions their food was produced.

These scenarios show that not only will prescriptive analytics be key to the future farm but if done right, will disrupt the entire value chain and deliver in a manner that will benefit every stakeholder in the food production chain.

CONCLUSION

Smart Agriculture is making strides in the fight towards food security and sustainability. In this chapter, we have systematically reviewed the literature on big data and AI applications in a bid to identify the status and prospects of these technologies in disrupting agriculture. The combination of big data and AI drives value through their analytics capabilities. This chapter presents a conceptual framework that examines the agricultural value chain along the lines of the different types of business analytics. Our findings demonstrate that works dedicated to weed, disease and pest control, and yield, weather and energy predictions is of most interest to researchers. As much as these technologies are a relatively new phenomenon, there are successful commercial implementations in countries such as Canada, Iran and Spain. The trajectory of the number of works in the field has picked up pace in recent years and investors have shown commitment to backing commercially viable solutions.

We expect that economic factors such as import and export planning, and price point predictions, business factors like process optimization and dwindling agricultural workforce, and technological factors such as the increase in computing power will drive the agenda for Smart Agriculture. Challenges include data ownership, stream consolidation, transport, hosting and processing power. Developing countries especially, face an uphill battle in putting in place the right infrastructure and doing away with years of skills honed through social and cultural practices.

It should be noted that the scope of the systematic literature review is limited to articles published in the English language and is dominated by peer-review literature indexed in scholarly databases. As such, the research represents a strong academic focus capturing current research contributions and potential for future research. As the proliferation of big data and AI into agriculture continues, future reviews could increase reliance on 'grey literature' with an increased emphasis on commercialization endeavors of the technology.

REFERENCES

Aghighi, H., Azadbakht, M., Ashourloo, D., Shahrabi, H. S., & Radiom, S. (2018). Machine Learning Regression Techniques for the Silage Maize Yield Prediction Using Time-Series Images of Landsat 8 OLI. *IEEE Journal of Selected Topics in Applied Earth Observations and Remote Sensing*, 1–15. doi:10.1109/JSTARS.2018.2823361

Ahmed, F., Al-Mamun, H. A., Bari, A. S. M. H., Hossain, E., & Kwan, P. (2012). Classification of crops and weeds from digital images: A support vector machine approach. *Crop Protection*, *40*, 98–104. doi:10.1016/j.cropro.2012.04.024

Ajdadi, F. R., Gilandeh, Y. A., Mollazade, K., & Hasanzadeh, R. P. R. (2016). Application of machine vision for classification of soil aggregate size. *Soil & Tillage Research*, *162*, 8–17. doi:10.1016/j.still.2016.04.012

Akbarzadeh, S., Paap, A., Ahderom, S., Apopei, B., & Alameh, K. (2018). Plant discrimination by Support Vector Machine classifier based on spectral reflectance. *Computers and Electronics in Agriculture*, *148*, 250–258. doi:10.1016/j.compag.2018.03.026

Aleke, B., Ojiako, U., & Wainwright, D. W. (2011). ICT adoption in developing countries: Perspectives from small-scale agribusinesses. *Journal of Enterprise Information Management, 24*(1), 68–84. doi:10.1108/17410391111097438

Ali, A., Qadir, J., Rasool, R., Sathiaseelan, A., Zwitter, A., & Crowcroft, J. (2016). Big data for development: Applications and techniques. *Big Data Analytics, 1*(1), 2. doi:10.118641044-016-0002-4

Ali, M., Deo, R. C., Downs, N. J., & Maraseni, T. (2018). Multi-stage committee based extreme learning machine model incorporating the influence of climate parameters and seasonality on drought forecasting. *Computers and Electronics in Agriculture, 152*, 149–165. doi:10.1016/j.compag.2018.07.013

Alipio, M. I., Dela Cruz, A. E. M., Doria, J. D. A., & Fruto, R. M. S. (2017). A smart hydroponics farming system using exact inference in Bayesian network. *2017 IEEE 6th Global Conference on Consumer Electronics (GCCE)*, 1–5. 10.1109/GCCE.2017.8229470

Ancin-Murguzur, F. J., Barbero-Lopez, A., Kontunen-Soppela, S., & Haapala, A. (2018). Automated image analysis tool to measure microbial growth on solid cultures. *Computers and Electronics in Agriculture, 151*, 426–430. doi:10.1016/j.compag.2018.06.031

Aquino, C. F., Chamhum Salomao, L. C., & Azevedo, A. M. (2016). High-efficiency phenotyping for vitamin A in banana using artificial neural networks and colorimetric data. *Bragantia, 75*(3), 268–274. doi:10.1590/1678-4499.467

Arima, S., Kondo, N., Shibano, Y., Fujiura, T., Yamashita, J., & Nakamura, H. (1994). Studies on Cucumber Harvesting Robot (Part 2). *Journal of the Japanese Society of Agricultural Machinery, 56*(6), 69–76. doi:10.11357/jsam1937.56.6_69

Arima, S., Kondo, N., Yagi, Y., Monta, M., & Yoshida, Y. (2001). Harvesting robot for strawberry grown on table top culture, 1: Harvesting robot using 5 DOF manipulator. *Journal of Society of High Technology in Agriculture*. Retrieved from http://agris.fao.org/agris-search/search.do?recordID=JP2001006293

Arteta, T. A., Hameg, R., Landin, M., Gallego, P. P., & Barreal, M. E. (2018). Neural networks models as decision-making tool for in vitro proliferation of hardy kiwi. *European Journal of Horticultural Science, 83*(4), 259–265. doi:10.17660/eJHS.2018/83.4.6

Ayankoya, K., Calitz, A. P., & Greyling, J. H. (2016). Real-Time Grain Commodities Price Predictions In South Africa: A Big Data And Neural Networks Approach. *Agrekon, 55*(4), 483–508. doi:10.1080/03031853.2016.1243060

Babazadeh, S., Moghaddam, P. A., Sabatyan, A., & Sharifian, F. (2016). Classification of potato tubers based on solanine toxicant using laser induced light backscattering imaging. *Computers and Electronics in Agriculture, 129*, 1–8. doi:10.1016/j.compag.2016.09.009

Bakhshi, M., Pourtaheri, M., & Eftekhari, A. R. (2016). Developing a Model to Predict Success of Agricultural Production Enterprises Based on Their Capitals. *Journal of Agricultural Science and Technology, 18*(6), 1443–1454.

Bakhshipour, A., & Jafari, A. (2018). Evaluation of support vector machine and artificial neural networks in weed detection using shape features. *Computers and Electronics in Agriculture*, *145*, 153–160. doi:10.1016/j.compag.2017.12.032

Bakhshipour, A., Jafari, A., Nassiri, S. M., & Zare, D. (2017). Weed segmentation using texture features extracted from wavelet sub-images. *Biosystems Engineering*, *157*, 1–12. doi:10.1016/j.biosystemseng.2017.02.002

Battaglia, P. W., Pascanu, R., Lai, M., Rezende, D., & Kavukcuoglu, K. (2016). *Interaction Networks for Learning about Objects, Relations and Physics*. Retrieved from http://arxiv.org/abs/1612.00222

Bello-Orgaz, G., Jung, J. J., & Camacho, D. (2016). Social big data: Recent achievements and new challenges. *Information Fusion*, *28*, 45–59. doi:10.1016/j.inffus.2015.08.005

Bengio, Y. (2009). Learning Deep Architectures for AI. *Foundations and Trends® in Machine Learning*, *2*(1), 1–127. doi:10.1561/2200000006

Bengio, Y. (2013). Deep Learning of Representations: Looking Forward. In A.-H. Dediu, C. Martín-Vide, R. Mitkov, & B. Truthe (Eds.), Statistical Language and Speech Processing (Vol. 7978, pp. 1–37). Academic Press. doi:10.1007/978-3-642-39593-2_1

Borges, P. H. M., Mendoza, Z. M. S. H., Maia, J. C. S., Bianchini, A., & Fernandes, H. C. (2017). Estimation Of Fuel Consumption In Agricultural Mechanized Operations Using Artificial Neural Networks. *Engenharia Agrícola*, *37*(1), 136–147. doi:10.1590/1809-4430-eng.agric.v37n1p136-147/2017

Brynjolfsson, E., Hitt, L. M., & Kim, H. H. (2011). *Strength in Numbers: How Does Data-Driven Decision-Making Affect Firm Performance?* SSRN Electronic Journal. doi:10.2139srn.1819486

Byakatonda, J., Parida, B. P., Kenabatho, P. K., & Moalafhi, D. B. (2018). Influence of climate variability and length of rainy season on crop yields in semiarid Botswana. *Agricultural and Forest Meteorology*, *248*, 130–144. doi:10.1016/j.agrformet.2017.09.016

Caires, E. F., & Guimaraes, A. M. (2018). A Novel Phosphogypsum Application Recommendation Method under Continuous No-Till Management in Brazil. *Agronomy Journal*, *110*(5), 1987–1995. doi:10.2134/agronj2017.11.0642

Calderano, B. F., Polivanov, H., da Silva Chagas, C., de Carvalho, W. J., Barroso, E. V., Teixeira Guerra, A. J., & Calderano, S. B. (2014). Artificial Neural Networks Applied for Soil Class Prediction in Mountainous Landscape of The Serra Do Mar. *Revista Brasileira de Ciência do Solo*, *38*(6), 1681–1693. doi:10.1590/S0100-06832014000600003

Camargo, A., Molina, J. P., Cadena-Torres, J., Jimenez, N., & Kim, J. T. (2012). Intelligent systems for the assessment of crop disorders. *Computers and Electronics in Agriculture*, *85*, 1–7. doi:10.1016/j.compag.2012.02.017

Cameron, M., Viviers, W., & Steenkamp, E. (2017). Breaking the "big data" barrier when selecting agricultural export markets: An innovative approach. *Agrekon*, *56*(2), 139–157. doi:10.1080/03031853.2017.1298456

Castaneda-Miranda, A., & Castano, V. M. (2017). Smart frost control in greenhouses by neural networks models. *Computers and Electronics in Agriculture, 137*, 102–114. doi:10.1016/j.compag.2017.03.024

Castro, C. A. de O., Resende, R. T., Kuki, K. N., Carneiro, V. Q., Marcatti, G. E., Cruz, C. D., & Motoike, S. Y. (2017). High-performance prediction of macauba fruit biomass for agricultural and industrial purposes using Artificial Neural Networks. *Industrial Crops and Products, 108*, 806–813. doi:10.1016/j.indcrop.2017.07.031

CEMA - European Agricultural Machinery. (2017, February 13). *Digital Farming: what does it really mean?* Retrieved September 24, 2018, from http://www.cema-agri.org/page/digital-farming-what-does-it-really-mean

Chantre, G. R., Vigna, M. R., Renzi, J. P., & Blanco, A. M. (2018). A flexible and practical approach for real-time weed emergence prediction based on Artificial Neural Networks. *Biosystems Engineering, 170*, 51–60. doi:10.1016/j.biosystemseng.2018.03.014

Chapman, R., Cook, S., Donough, C., Lim, Y. L., Vun Vui Ho, P., Lo, K. W., & Oberthür, T. (2018). Using Bayesian networks to predict future yield functions with data from commercial oil palm plantations: A proof of concept analysis. *Computers and Electronics in Agriculture, 151*, 338–348. doi:10.1016/j.compag.2018.06.006

Chaudhary, A., Kolhe, S., & Kamal, R. (2016). A hybrid ensemble for classification in multiclass datasets: An application to oilseed disease dataset. *Computers and Electronics in Agriculture, 124*, 65–72. doi:10.1016/j.compag.2016.03.026

Cheng, X., Zhang, Y., Chen, Y., Wu, Y., & Yue, Y. (2017). Pest identification via deep residual learning in complex background. *Computers and Electronics in Agriculture, 141*, 351–356. doi:10.1016/j.compag.2017.08.005

Chi, M., Plaza, A., Benediktsson, J. A., Sun, Z., Shen, J., & Zhu, Y. (2016). Big Data for Remote Sensing: Challenges and Opportunities. *Proceedings of the IEEE, 104*(11), 2207–2219. doi:10.1109/JPROC.2016.2598228

Cho, G. (2018). The Australian digital farmer: challenges and opportunities. *IOP Conference Series: Earth and Environmental Science, 185*, 012036. 10.1088/1755-1315/185/1/012036

Christensen, C. (1997). *The Revolutionary Book that Will Change the Way You Do Business (Collins Business Essentials).* New York: Harper Paperbacks.

Chung, C.-L., Huang, K.-J., Chen, S.-Y., Lai, M.-H., Chen, Y.-C., & Kuo, Y.-F. (2016). Detecting Bakanae disease in rice seedlings by machine vision. *Computers and Electronics in Agriculture, 121*, 404–411. doi:10.1016/j.compag.2016.01.008

Clercq, M. D., Vats, A., & Biel, A. (2018). Agriculture 4.0: The Future of Farming Technology. *World Government Summit*, 30.

Coble, K. H., Mishra, A. K., Ferrell, S., & Griffin, T. (2018). Big Data in Agriculture: A Challenge for the Future. *Applied Economic Perspectives and Policy, 40*(1), 79–96. doi:10.1093/aepp/ppx056

Cockburn, I., Henderson, R., & Stern, S. (2018). *The Impact of Artificial Intelligence on Innovation* (No. w24449). doi:10.3386/w24449

Columbus, L. (2016). *McKinsey's 2016 Analytics Study Defines The Future Of Machine Learning*. Retrieved November 14, 2018, from https://www.forbes.com/sites/louiscolumbus/2016/12/18/mckinseys-2016-analytics-study-defines-the-future-machine-learning/#3da708d214eb

Correa, F. E., Oliveira, M. D. B., Gama, J., Correa, P. L. P., & Rady, J. (2016). Analyzing the behavior dynamics of grain price indexes using Tucker tensor decomposition and spatio-temporal trajectories. *Computers and Electronics in Agriculture*, *120*, 72–78. doi:10.1016/j.compag.2015.11.011

Costa, A. G., Pinto, F., Motoike, S. Y., Braga Júnior, R. A., & Gracia, L. M. N. (2018). Classification of Macaw Palm Fruits from Colorimetric Properties for Determining the Harvest Moment. *Engenharia Agrícola*, *38*(4), 634–641. doi:10.1590/1809-4430-eng.agric.v38n4p634-641/2018

Crunchbase. (2018). *Crunchbase: Discover innovative companies and the people behind them*. Retrieved December 17, 2018, from Crunchbase website: https://www.crunchbase.com

Cuzzocrea, A., Song, I.-Y., & Davis, K. C. (2011). *Analytics over large-scale multidimensional data: the big data revolution!* Academic Press.

da Silva, C. A. Junior, Nanni, M. R., Teodoro, P. E., & Capristo Silva, G. F. (2017). Vegetation Indices for Discrimination of Soybean Areas: A New Approach. *Agronomy Journal*, *109*(4), 1331–1343. doi:10.2134/agronj2017.01.0003

de Barros, M. M., da Silva, F. M., Costa, A. G., Ferraz, G. A. e S., & da Silva, F. C. (2018). Use of classifier to determine coffee harvest time by detachment force. *Revista Brasileira de Engenharia Agrícola e Ambiental*, *22*(5), 366–370. doi:10.1590/1807-1929/agriambi.v22n5p366-370

Delgado, G., Aranda, V., Calero, J., Sanchez-Maranon, M., Serrano, J. M., Sanchez, D., & Vila, M. A. (2008). Building a fuzzy logic information network and a decision-support system for olive cultivation in Andalusia. *Spanish Journal of Agricultural Research*, *6*(2), 252–263. doi:10.5424jar/2008062-316

Demir, B. (2018). Application of data mining and adaptive neuro-fuzzy structure to predict color parameters of walnuts (Juglans regia L.). *Turkish Journal of Agriculture and Forestry*, *42*(3), 216–225. doi:10.3906/tar-1801-78

Demir, B., Gurbuz, F., Eski, I., Kus, Z. A., Yilmaz, K. U., & Ercisli, S. (2018). Possible Use of Data Mining for Analysis and Prediction of Apple Physical Properties. *Erwerbs-Obstbau*, *60*(1), 1–7. doi:10.100710341-017-0330-1

Diaz, I., Mazza, S. M., Combarro, E. F., Gimenez, L. I., & Gaiad, J. E. (2017). Machine learning applied to the prediction of citrus production. *Spanish Journal of Agricultural Research*, *15*(2), e0205. doi:10.5424jar/2017152-9090

Dimililer, K., & Zarrouk, S. (2017). ICSPI: Intelligent Classification System of Pest Insects Based on Image Processing and Neural Arbitration. *Applied Engineering in Agriculture*, *33*(4), 453–460. doi:10.13031/aea.12161

Ding, W., & Taylor, G. (2016). Automatic moth detection from trap images for pest management. *Computers and Electronics in Agriculture, 123*, 17–28. doi:10.1016/j.compag.2016.02.003

Dornelles, E. F., Kraisig, A. R., da Silva, J. A. G., Sawicki, S., Roos-Frantz, F., & Carbonera, R. (2018). Artificial intelligence in seeding density optimization and yield simulation for oat. *Revista Brasileira de Engenharia Agrícola e Ambiental, 22*(3), 183–188. doi:10.1590/1807-1929/agriambi.v22n3p183-188

Dos Santos Ferreira, A., Matte Freitas, D., Gonçalves da Silva, G., Pistori, H., & Folhes, M. T. (2017). Weed detection in soybean crops using ConvNets. *Computers and Electronics in Agriculture, 143*, 314–324. doi:10.1016/j.compag.2017.10.027

Du, K., Sun, Z., Li, Y., Zheng, F., Chu, J., & Su, Y. (2016). Diagnostic Model For Wheat Leaf Conditions Using Image Features And A Support Vector Machine. *Transactions of the ASABE, 59*(5), 1041–1052. doi:10.13031/trans.59.11434

Ebrahimi, E., & Mollazade, K. (2010). Integrating fuzzy data mining and impulse acoustic techniques for almond nuts sorting. *Australian Journal of Crop Science, 4*(5), 353–358.

Emamgholizadeh, S., Parsaeian, M., & Baradaran, M. (2015). Seed yield prediction of sesame using artificial neural network. *European Journal of Agronomy, 68*, 89–96. doi:10.1016/j.eja.2015.04.010

Espejo-Garcia, B., Martinez-Guanter, J., Perez-Ruiz, M., Lopez-Pellicer, F. J., & Javier Zarazaga-Soria, F. (2018). Machine learning for automatic rule classification of agricultural regulations: A case study in Spain. *Computers and Electronics in Agriculture, 150*, 343–352. doi:10.1016/j.compag.2018.05.007

Espinoza, K., Valera, D. L., Torres, J. A., Lopez, A., & Molina-Aiz, F. D. (2016). Combination of image processing and artificial neural networks as a novel approach for the identification of Bemisia tabaci and Frankliniella occidentalis on sticky traps in greenhouse agriculture. *Computers and Electronics in Agriculture, 127*, 495–505. doi:10.1016/j.compag.2016.07.008

Everingham, Y., Sexton, J., Skocaj, D., & Inman-Bamber, G. (2016). Accurate prediction of sugarcane yield using a random forest algorithm. *Agronomy for Sustainable Development, 36*(2), 27. doi:10.100713593-016-0364-z

Everingham, Y. L., Smyth, C. W., & Inman-Bamber, N. G. (2009). Ensemble data mining approaches to forecast regional sugarcane crop production. *Agricultural and Forest Meteorology, 149*(3–4), 689–696. doi:10.1016/j.agrformet.2008.10.018

Fan, W., & Bifet, A. (2013). Mining big data: Current status and forecast to the future. *ACM SIGKDD Explorations Newsletter, 14*(2), 1. doi:10.1145/2481244.2481246

Farjam, A., Omid, M., Akram, A., & Niari, Z. F. (2014). A Neural Network Based Modeling and Sensitivity Analysis of Energy Inputs for Predicting Seed and Grain Corn Yields. *Journal of Agricultural Science and Technology, 16*(4), 767–778.

Feng, Y., Peng, Y., Cui, N., Gong, D., & Zhang, K. (2017). Modeling reference evapotranspiration using extreme learning machine and generalized regression neural network only with temperature data. *Computers and Electronics in Agriculture, 136*, 71–78. doi:10.1016/j.compag.2017.01.027

Ferentinos, K. P. (2018). Deep learning models for plant disease detection and diagnosis. *Computers and Electronics in Agriculture, 145*, 311–318. doi:10.1016/j.compag.2018.01.009

Fernandes, J. L., Rocha, J. V., & Camargo Lamparelli, R. A. (2011). Sugarcane yield estimates using time series analysis of spot vegetation images. *Scientia Agrícola, 68*(2), 139–146. doi:10.1590/S0103-90162011000200002

Fernandez, R., Montes, H., Surdilovic, J., Surdilovic, D., Gonzalez-De-Santos, P., & Armada, M. (2018). Automatic Detection of Field-Grown Cucumbers for Robotic Harvesting. *IEEE Access: Practical Innovations, Open Solutions, 6*, 35512–35527. doi:10.1109/ACCESS.2018.2851376

Ferraro, D. O., Ghersa, C. M., & Rivero, D. E. (2012). Weed Vegetation of Sugarcane Cropping Systems of Northern Argentina: Data-Mining Methods for Assessing the Environmental and Management Effects on Species Composition. *Weed Science, 60*(1), 27–33. doi:10.1614/WS-D-11-00023.1

Food and Agriculture Organization of the United Nations. (Ed.). (2017). *The future of food and agriculture: trends and challenges*. Rome: Food and Agriculture Organization of the United Nations.

Food Security Information Network. (2018). *Global Report on Food Crises 2018*. Retrieved from http://www.fsincop.net/fileadmin/user_upload/fsin/docs/global_report/2018/GRFC_2018_Full_report_EN_Low_resolution.pdf

Fortin, J. G., Anctil, F., Parent, L.-E., & Bolinder, M. A. (2011). Site-specific early season potato yield forecast by neural network in Eastern Canada. *Precision Agriculture, 12*(6), 905–923. doi:10.100711119-011-9233-6

Fukuda, S., Spreer, W., Yasunaga, E., Yuge, K., Sardsud, V., & Mueller, J. (2013). Random Forests modelling for the estimation of mango (Mangifera indica L. cv. Chok Anan) fruit yields under different irrigation regimes. *Agricultural Water Management, 116*, 142–150. doi:10.1016/j.agwat.2012.07.003

Garcia-Santillan, I. D., & Pajares, G. (2018). On-line crop/weed discrimination through the Mahalanobis distance from images in maize fields. *Biosystems Engineering, 166*, 28–43. doi:10.1016/j.biosystemseng.2017.11.003

Gartner. (2018). *Gartner IT Glossary*. Retrieved November 14, 2018, from https://www.gartner.com/it-glossary/

Ghosh, K., & Nath, A. (2016). Big Data: Security Issues, Challenges and Future Scope. *International Journal of Research Studies in Computer Science and Engineering, 3*(3). doi:10.20431/2349-4859.0303001

Goldstein, A., Fink, L., Meitin, A., Bohadana, S., Lutenberg, O., & Ravid, G. (2018). Applying machine learning on sensor data for irrigation recommendations: Revealing the agronomist's tacit knowledge. *Precision Agriculture, 19*(3), 421–444. doi:10.100711119-017-9527-4

Gómez-Casero, M. T., Castillejo-Gonzalez, I. L., Garcia-Ferrer, A., Peña-Barragán, J. M., Jurado-Expósito, M., Garcia-Torres, L., & López-Granados, F. (2010). Spectral discrimination of wild oat and canary grass in wheat fields for less herbicide application. *Agronomy for Sustainable Development, 30*(3), 689–699. doi:10.1051/agro/2009052

Gonzalez-Sanchez, A., Frausto-Solis, J., & Ojeda-Bustamante, W. (2014). Predictive ability of machine learning methods for massive crop yield prediction. *Spanish Journal of Agricultural Research*, *12*(2), 313–328. doi:10.5424jar/2014122-4439

Goumopoulos, C., O'Flynn, B., & Kameas, A. (2014). Automated zone-specific irrigation with wireless sensor/actuator network and adaptable decision support. *Computers and Electronics in Agriculture*, *105*, 20–33. doi:10.1016/j.compag.2014.03.012

Griffel, L. M., Delparte, D., & Edwards, J. (2018). Using Support Vector Machines classification to differentiate spectral signatures of potato plants infected with Potato Virus Y. *Computers and Electronics in Agriculture*, *153*, 318–324. doi:10.1016/j.compag.2018.08.027

Guine, R. P. F., Matos, S., Goncalves, F. J., Costa, D., & Mendes, M. (2018). Evaluation of phenolic compounds and antioxidant activity of blueberries and modelization by artificial neural networks. *International Journal of Fruit Science*, *18*(2), 199–214. doi:10.1080/15538362.2018.1425653

Gutierrez, M., Alegret, S., Caceres, R., Casadesus, J., Marfa, O., & del Valle, M. (2008). Nutrient solution monitoring in greenhouse cultivation employing a potentiometric electronic tongue. *Journal of Agricultural and Food Chemistry*, *56*(6), 1810–1817. doi:10.1021/jf073438s PMID:18303814

Gutierrez, P. A., López-Granados, F., Peña-Barragán, J. M., Jurado-Expósito, M., Gómez-Casero, M. T., & Hervas-Martinez, C. (2008). Mapping sunflower yield as affected by Ridolfia segetum patches and elevation by applying evolutionary product unit neural networks to remote sensed data. *Computers and Electronics in Agriculture*, *60*(2), 122–132. doi:10.1016/j.compag.2007.07.011

Gutierrez, P. A., López-Granados, F., Peña-Barragán, J. M., Jurado-Expósito, M., & Hervas-Martinez, C. (2008). Logistic regression product-unit neural networks for mapping Ridolfia segetum infestations in sunflower crop using multitemporal remote sensed data. *Computers and Electronics in Agriculture*, *64*(2), 293–306. doi:10.1016/j.compag.2008.06.001

Hamedani, S. R., Liaqat, M., Shamshirband, S., Al-Razgan, O. S., Al-Shammari, E. T., & Petkovic, D. (2015). Comparative Study of Soft Computing Methodologies for Energy Input-Output Analysis to Predict Potato Production. *American Journal of Potato Research*, *92*(3), 426–434. doi:10.100712230-015-9453-9

Harada, M., Tominaga, T., Hiramatsu, K., & Marui, A. (2013). Real-Time Prediction Of Chlorophyll-A Time Series In A Eutrophic Agricultural Reservoir In A Coastal Zone Using Recurrent Neural Networks With Periodic Chaos Neurons. *Irrigation and Drainage*, *62*(1), 36–43. doi:10.1002/ird.1757

Hassanien, A. E., Gaber, T., Mokhtar, U., & Hefny, H. (2017). An improved moth flame optimization algorithm based on rough sets for tomato diseases detection. *Computers and Electronics in Agriculture*, *136*, 86–96. doi:10.1016/j.compag.2017.02.026

Hayashi, S., Ganno, K., Ishii, Y., & Tanaka, I. (2002). Robotic Harvesting System for Eggplants. *Japan Agricultural Research Quarterly: JARQ*, *36*(3), 163–168. doi:10.6090/jarq.36.163

Heim, R. H. J., Wright, I. J., Chang, H.-C., Carnegie, A. J., Pegg, G. S., Lancaster, E. K., ... Oldeland, J. (2018). Detecting myrtle rust (Austropuccinia psidii) on lemon myrtle trees using spectral signatures and machine learning. *Plant Pathology*, *67*(5), 1114–1121. doi:10.1111/ppa.12830

Hill, B. D., Kalischuk, M., Waterer, D. R., Bizimungu, B., Howard, R., & Kawchuk, L. M. (2011). An Environmental Model Predicting Bacterial Ring Rot Symptom Expression. *American Journal of Potato Research*, *88*(3), 294–301. doi:10.100712230-011-9193-4

Husin, Z., Shakaff, A. Y. M., Aziz, A. H. A., Farook, R. S. M., Jaafar, M. N., Hashim, U., & Harun, A. (2012). Embedded portable device for herb leaves recognition using image processing techniques and neural network algorithm. *Computers and Electronics in Agriculture*, *89*, 18–29. doi:10.1016/j.compag.2012.07.009

IBM. (2013). *Descriptive, predictive, prescriptive: Transforming asset and facilities management with analytics*. IBM.

IBM. (2015, September 11). *The new AI innovation equation*. Retrieved April 16, 2019, from IBM Cognitive - What's next for AI website: http://www.ibm.com/watson/advantage-reports/future-of-artificial-intelligence/ai-innovation-equation.html

IBM. (2018, October 30). *Big Data Analytics*. Retrieved January 9, 2019, from https://www.ibm.com/analytics/hadoop/big-data-analytics

Ilic, M., Ilic, S., Jovic, S., & Panic, S. (2018). Early cherry fruit pathogen disease detection based on data mining prediction. *Computers and Electronics in Agriculture*, *150*, 418–425. doi:10.1016/j.compag.2018.05.008

Jimenez, D., Cock, J., Jarvis, A., Garcia, J., Satizabal, H. F., Van Damme, P., ... Barreto-Sanz, M. A. (2011). Interpretation of commercial production information: A case study of lulo (Solanum quitoense), an under-researched Andean fruit. *Agricultural Systems*, *104*(3), 258–270. doi:10.1016/j.agsy.2010.10.004

Jimenez, D., Cock, J., Satizabal, H. F., Barreto, M. A., Perez-Uribe, A., Jarvis, A., & Van Damme, P. (2009). Analysis of Andean blackberry (Rubus glaucus) production models obtained by means of artificial neural networks exploiting information collected by small-scale growers in Colombia and publicly available meteorological data. *Computers and Electronics in Agriculture*, *69*(2), 198–208. doi:10.1016/j.compag.2009.08.008

Jokic, A., Zavargo, Z., Gyura, J., Radivojevic, S., & Seres, Z. (2010). An Artificial Neural Network Approach to Prediction of Sugar Beet Yield and Quality in Serbia. In *Sugar Beet Crops: Growth, Fertilization & Yield* (pp. 153–166). Academic Press. Retrieved from https://www.researchgate.net/publication/281874792_An_artificial_neural_network_approach_to_prediction_of_sugar_beet_yield_and_quality_in_Serbia

Kamilaris, A., Kartakoullis, A., & Prenafeta-Boldú, F. X. (2017). A review on the practice of big data analysis in agriculture. *Computers and Electronics in Agriculture*, *143*, 23–37. doi:10.1016/j.compag.2017.09.037

Kamilaris, A., & Prenafeta-Boldú, F. X. (2018). Deep learning in agriculture: A survey. *Computers and Electronics in Agriculture*, *147*, 70–90. doi:10.1016/j.compag.2018.02.016

Kaplinsky, R., & Morris, M. (2001). *A Handbook for Value Chain Research*. 113.

Kawamura, N., Namikawa, K., Fujiura, T., & Ura, M. (1984). Study on Agricultural Robot (Part 1). *Journal of the Japanese Society of Agricultural Machinery, 46*(3), 353–358. doi:10.11357/jsam1937.46.3_353

Khan, M. J., Khan, H. S., Yousaf, A., Khurshid, K., & Abbas, A. (2018). Modern Trends in Hyperspectral Image Analysis: A Review. *IEEE Access: Practical Innovations, Open Solutions, 6*, 14118–14129. doi:10.1109/ACCESS.2018.2812999

Khanal, S., Fulton, J., Klopfenstein, A., Douridas, N., & Shearer, S. (2018). Integration of high resolution remotely sensed data and machine learning techniques for spatial prediction of soil properties and corn yield. *Computers and Electronics in Agriculture, 153*, 213–225. doi:10.1016/j.compag.2018.07.016

Khashei-Siuki, A., Kouchakzadeh, M., & Ghahraman, B. (2011). Predicting Dryland Wheat Yield from Meteorological Data Using Expert System, Khorasan Province, Iran. *Journal of Agricultural Science and Technology, 13*(4), 627–640.

Khazaei, J., Naghavi, M. R., Jahansouz, M. R., & Salimi-Khorshidi, G. (2008). Yield estimation and clustering of chickpea genotypes using soft computing techniques. *Agronomy Journal, 100*(4), 1077–1087. doi:10.2134/agronj2006.0244

Khoshnevisan, B., Rafiee, S., & Mousazadeh, H. (2013). Environmental impact assessment of open field and greenhouse strawberry production. *European Journal of Agronomy, 50*, 29–37. doi:10.1016/j.eja.2013.05.003

Khoshnevisan, B., Rafiee, S., Omid, M., Mousazadeh, H., & Rajaeifar, M. A. (2014). Application of artificial neural networks for prediction of output energy and GHG emissions in potato production in Iran. *Agricultural Systems, 123*, 120–127. doi:10.1016/j.agsy.2013.10.003

Kondo, N., Nishitsuji, Y., Ling, P. P., & Ting, K. C. (1996). Visual Feedback Guided Robotic Cherry Tomato Harvesting. *Transactions of the ASAE. American Society of Agricultural Engineers, 39*(6), 2331–2338. doi:10.13031/2013.27744

Kouadio, L., Deo, R. C., Byrareddy, V., Adamowski, J. F., Mushtaq, S., & Nguyen, V. P. (2018). Artificial intelligence approach for the prediction of Robusta coffee yield using soil fertility properties. *Computers and Electronics in Agriculture, 155*, 324–338. doi:10.1016/j.compag.2018.10.014

Krishnan, P., & Surya Rao, A. V. (2005). Effects of genotype and environment on seed yield and quality of rice. *The Journal of Agricultural Science, 143*(04), 283–292. doi:10.1017/S0021859605005496

Kshetri, N. (2014). The emerging role of Big Data in key development issues: Opportunities, challenges, and concerns. *Big Data & Society, 1*(2). doi:10.1177/2053951714564227

Kumar, D., Adamowski, J., Suresh, R., & Ozga-Zielinski, B. (2016). Estimating Evapotranspiration Using an Extreme Learning Machine Model: Case Study in North Bihar, India. *Journal of Irrigation and Drainage Engineering, 142*(9), 04016032. doi:10.1061/(ASCE)IR.1943-4774.0001044

Kus, Z. A., Demir, B., Eski, I., Gurbuz, F., & Ercisli, S. (2017). Estimation of the Colour Properties of Apples Varieties Using Neural Network. *Erwerbs-Obstbau, 59*(4), 291–299. doi:10.100710341-017-0324-z

Lamsal, A., Welch, S. M., Jones, J. W., Boote, K. J., Asebedo, A., Crain, J., ... Arachchige, P. G. (2017). Efficient crop model parameter estimation and site characterization using large breeding trial data sets. *Agricultural Systems, 157*, 170–184. doi:10.1016/j.agsy.2017.07.016

Laney, D. (2001). *3D Data Management: Controlling Data Volume, Velocity, and Variety* [Technical Report]. Retrieved from https://blogs.gartner.com/doug-laney/files/2012/01/ad949-3D-Data-Management-Controlling-Data-Volume-Velocity-and-Variety.pdf

Liao, M.-S., Chuang, C.-L., Lin, T.-S., Chen, C.-P., Zheng, X.-Y., Chen, P.-T., ... Jiang, J.-A. (2012). Development of an autonomous early warning system for Bactrocera dorsalis (Hendel) outbreaks in remote fruit orchards. *Computers and Electronics in Agriculture, 88*, 1–12. doi:10.1016/j.compag.2012.06.008

Liberati, A., Altman, D. G., Tetzlaff, J., Mulrow, C., Gøtzsche, P. C., Ioannidis, J. P., ... Moher, D. (2009). The PRISMA statement for reporting systematic reviews and meta-analyses of studies that evaluate health care interventions: Explanation and elaboration. *PLoS Medicine, 6*(7). doi:10.1371/journal.pmed.1000100 PMID:19621070

Liu, Z.-Y., Wu, H.-F., & Huang, J.-F. (2010). Application of neural networks to discriminate fungal infection levels in rice panicles using hyperspectral reflectance and principal components analysis. *Computers and Electronics in Agriculture, 72*(2), 99–106. doi:10.1016/j.compag.2010.03.003

Logan, T. M., McLeod, S., & Guikema, S. (2016). Predictive models in horticulture: A case study with Royal Gala apples. *Scientia Horticulturae, 209*, 201–213. doi:10.1016/j.scienta.2016.06.033

López-Granados, F., Gómez-Casero, M. T., Peña-Barragán, J. M., Jurado-Expósito, M., & García-Torres, L. (2010). Classifying Irrigated Crops as Affected by Phenological Stage Using Discriminant Analysis and Neural Networks. *Journal of the American Society for Horticultural Science, 135*(5), 465–473. doi:10.21273/JASHS.135.5.465

Lu, J., Hu, J., Zhao, G., Mei, F., & Zhang, C. (2017). An in-field automatic wheat disease diagnosis system. *Computers and Electronics in Agriculture, 142*(A), 369–379. doi:10.1016/j.compag.2017.09.012

Mahmoud, T., Dong, Z. Y., & Ma, J. (2018). Advanced method for short-term wind power prediction with multiple observation points using extreme learning machines. *The Journal of Engineering, 2018*(1), 29–38. doi:10.1049/joe.2017.0338

Mark, T. B., Griffin, T. W., & Whitacre, B. E. (2016). The Role of Wireless Broadband Connectivity on `Big Data' and the Agricultural Industry in the United States and Australia. *International Food and Agribusiness Management Review, 19*(A), 43–56.

Markkanen, A. (2015). IoT Analytics Today and in 2020. *ABI Research*, 10.

Martin, K. (2016, April 27). How will artificial intelligence affect legal practice? *Thomson Reuters*. Retrieved April 1, 2019, from Answers On website: https://blogs.thomsonreuters.com/answerson/artificial-intelligence-legal-practice/

Mattar, M. A., El-Marazky, M. S., & Ahmed, K. A. (2017). Modeling sprinkler irrigation infiltration based on a fuzzy-logic approach. *Spanish Journal of Agricultural Research, 15*(1), e1201. doi:10.5424jar/2017151-9179

Microsoft, Inc. (2018). *Digital Agriculture: Farmers in India are using AI to increase crop yields*. Retrieved September 25, 2018, from Microsoft News Center India website: https://news.microsoft.com/en-in/features/ai-agriculture-icrisat-upl-india/

Mills, M. (2015). *Artificial Intelligence in Law – The State of Play in 2015?* Retrieved April 29, 2019, from Legal IT Insider website: https://www.legaltechnology.com/latest-news/artificial-intelligence-in-law-the-state-of-play-in-2015/

Misaki, E., Apiola, M., Gaiani, S., & Tedre, M. (2018). Challenges facing sub-Saharan small-scale farmers in accessing farming information through mobile phones: A systematic literature review. *The Electronic Journal on Information Systems in Developing Countries*, *84*(4), e12034. doi:10.1002/isd2.12034

Moller, A. B., Beucher, A., Iversen, B. V., & Greve, M. H. (2018). Predicting artificially drained areas by means of a selective model ensemble. *Geoderma*, *320*, 30–42. doi:10.1016/j.geoderma.2018.01.018

Najafabadi, M. M., Villanustre, F., Khoshgoftaar, T. M., Seliya, N., Wald, R., & Muharemagic, E. (2015). Deep learning applications and challenges in big data analytics. *Journal of Big Data*, *2*(1), 1. doi:10.118640537-014-0007-7

Nandyala, C. S., & Kim, H.-K. (2016). Big and Meta Data Management for U-Agriculture Mobile Services. *International Journal of Software Engineering and Its Applications*, *10*(2), 257–270. doi:10.14257/ijseia.2016.10.2.21

Navarro-Hellin, H., Martinez-del-Rincon, J., Domingo-Miguel, R., Soto-Valles, F., & Torres-Sanchez, R. (2016). A decision support system for managing irrigation in agriculture. *Computers and Electronics in Agriculture*, *124*, 121–131. doi:10.1016/j.compag.2016.04.003

Olaniyi, E., Оланії, Е., & Олании, Э. (2018). Digital Agriculture: Mobile Phones, Internet & Agricultural Development in Africa. *Actual Problems of Economics*, *16*.

Oo, L. M., & Aung, N. Z. (2018). A simple and efficient method for automatic strawberry shape and size estimation and classification. *Biosystems Engineering*, *170*, 96–107. doi:10.1016/j.biosystemseng.2018.04.004

Orge Pinheiro, C. A., & de Senna, V. (2017). Multivariate analysis and neural networks application to price forecasting in the Brazilian agricultural market. *Ciência Rural*, *47*(1). doi:10.1590/0103-8478cr20160077

Oshunsanya, S. O. (2013). Crop Yields as Influenced by Land Preparation Methods Established Within Vetiver Grass Alleys for Sustainable Agriculture in Southwest Nigeria. *Agroecology and Sustainable Food Systems*, *37*(5), 578–591. doi:10.1080/21683565.2012.762439

Osman, J., Inglada, J., & Dejoux, J.-F. (2015). Assessment of a Markov logic model of crop rotations for early crop mapping. *Computers and Electronics in Agriculture*, *113*, 234–243. doi:10.1016/j.compag.2015.02.015

Özdemir, V., & Hekim, N. (2018). Birth of Industry 5.0: Making Sense of Big Data with Artificial Intelligence, "The Internet of Things" and Next-Generation Technology Policy. *OMICS: A Journal of Integrative Biology*, *22*(1), 65–76. doi:10.1089/omi.2017.0194 PMID:29293405

Pallottino, F., Menesatti, P., Figorilli, S., Antonucci, F., Tomasone, R., Colantoni, A., & Costa, C. (2018). Machine Vision Retrofit System for Mechanical Weed Control in Precision Agriculture Applications. *Sustainability*, *10*(7), 2209. doi:10.3390u10072209

Pandorfi, H., Bezerra, A. C., Atarassi, R. T., Vieira, F. M. C., Barbosa Filho, J. A. D., & Guiselini, C. (2016). Artificial neural networks employment in the prediction of evapotranspiration of greenhouse-grown sweet pepper. *Revista Brasileira de Engenharia Agrícola e Ambiental*, *20*(6), 507–512. doi:10.1590/1807-1929/agriambi.v20n6p507-512

Pantazi, X. E., Tamouridou, A. A., Alexandridis, T. K., Lagopodi, A. L., Kashefi, J., & Moshou, D. (2017). Evaluation of hierarchical self-organising maps for weed mapping using UAS multispectral imagery. *Computers and Electronics in Agriculture*, *139*, 224–230. doi:10.1016/j.compag.2017.05.026

Paraforos, D. S., Vassiliadis, V., Kortenbruck, D., Stamkopoulos, K., Ziogas, V., Sapounas, A. A., & Griepentrog, H. W. (2016). A Farm Management Information System Using Future Internet Technologies. *IFAC-PapersOnLine*, *49*(16), 324–329. doi:10.1016/j.ifacol.2016.10.060

Park, S., Im, J., Jang, E., & Rhee, J. (2016). Drought assessment and monitoring through blending of multi-sensor indices using machine learning approaches for different climate regions. *Agricultural and Forest Meteorology*, *216*, 157–169. doi:10.1016/j.agrformet.2015.10.011

Peloia, P. R., & Rodrigues, L. H. A. (2016). Identification Of Commercial Blocks Of Outstanding Performance Of Sugarcane Using Data Mining. *Engenharia Agrícola*, *36*(5), 895–901. doi:10.1590/1809-4430-Eng.Agric.v36n5p895-901/2016

Pentos, K., & Pieczarka, K. (2017). Applying an artificial neural network approach to the analysis of tractive properties in changing soil conditions. *Soil & Tillage Research*, *165*, 113–120. doi:10.1016/j.still.2016.08.005

Perea, R. (2018). Prediction of applied irrigation depths at farm level using artificial intelligence techniques. *Agricultural Water Management*, *206*, 229–240. doi:10.1016/j.agwat.2018.05.019

Pike, A. C., Mueller, T. G., Schoergendorfer, A., Shearer, S. A., & Karathanasis, A. D. (2009). Erosion Index Derived from Terrain Attributes using Logistic Regression and Neural Networks. *Agronomy Journal*, *101*(5), 1068–1079. doi:10.2134/agronj2008.0207x

Pineda, M., Pérez-Bueno, M. L., & Barón, M. (2018). Detection of Bacterial Infection in Melon Plants by Classification Methods Based on Imaging Data. *Frontiers in Plant Science*, *9*, 164. doi:10.3389/fpls.2018.00164 PMID:29491881

Pour, A. S., Chegini, G., Zarafshan, P., & Massah, J. (2018). Curvature-based pattern recognition for cultivar classification of Anthurium flowers. *Postharvest Biology and Technology*, *139*, 67–74. doi:10.1016/j.postharvbio.2018.01.013

Prasad, R., Deo, R. C., Li, Y., & Maraseni, T. (2018). Soil moisture forecasting by a hybrid machine learning technique: ELM integrated with ensemble empirical mode decomposition. *Geoderma*, *330*, 136–161. doi:10.1016/j.geoderma.2018.05.035

Protopop, I., & Shanoyan, A. (2016). Big Data and Smallholder Farmers: Big Data Applications in the Agri-Food Supply Chain in Developing Countries. *International Food and Agribusiness Management Review, 19*(A, SI), 173–190.

Provost, F., & Fawcett, T. (2013). Data Science and its Relationship to Big Data and Data-Driven Decision Making. *Big Data, 1*(1), 51–59. doi:10.1089/big.2013.1508 PMID:27447038

Quraishi, M. Z., & Mouazen, A. M. (2013). Development of a methodology for in situ assessment of topsoil dry bulk density. *Soil & Tillage Research, 126*, 229–237. doi:10.1016/j.still.2012.08.009

Rajeswari, S., Suthendran, K., & Rajakumar, K. (2017). A smart agricultural model by integrating IoT, mobile and cloud-based big data analytics. *2017 International Conference on Intelligent Computing and Control (I2C2)*, 1–5. 10.1109/I2C2.2017.8321902

Ramirez-Gil, J. G., Martinez, G. O. G., & Osorio, J. G. M. (2018). Design of electronic devices for monitoring climatic variables and development of an early warning system for the avocado wilt complex disease. *Computers and Electronics in Agriculture, 153*, 134–143. doi:10.1016/j.compag.2018.08.002

Rathod, S., Singh, K. N., Patil, S. G., Naik, R. H., Ray, M., & Meena, V. S. (2018). Modeling and forecasting of oilseed production of India through artificial intelligence techniques. *Indian Journal of Agricultural Sciences, 88*(1), 22–27.

Renaud-Gentie, C., Burgos, S., & Benoit, M. (2014). Choosing the most representative technical management routes within diverse management practices: Application to vineyards in the Loire Valley for environmental and quality assessment. *European Journal of Agronomy, 56*, 19–36. doi:10.1016/j.eja.2014.03.002

Ribeiro, C. O., & Oliveira, S. M. (2011). A hybrid commodity price-forecasting model applied to the sugar-alcohol sector. *The Australian Journal of Agricultural and Resource Economics, 55*(2), 180–198. doi:10.1111/j.1467-8489.2011.00534.x

Rizzo, D., Martin, L., & Wohlfahrt, J. (2014). Miscanthus spatial location as seen by farmers: A machine learning approach to model real criteria. *Biomass and Bioenergy, 66*, 348–363. doi:10.1016/j.biombioe.2014.02.035

Romero, J. R., Roncallo, P. F., Akkiraju, P. C., Ponzoni, I., Echenique, V. C., & Carballido, J. A. (2013). Using classification algorithms for predicting durum wheat yield in the province of Buenos Aires. *Computers and Electronics in Agriculture, 96*, 173–179. doi:10.1016/j.compag.2013.05.006

Russell, S. J., & Norvig, P. (2010). *Artificial intelligence: A Modern Approach* (3rd ed.). Upper Saddle River, NJ: Prentice Hall.

Sa, I., Ge, Z., Dayoub, F., Upcroft, B., Perez, T., & McCool, C. (2016). DeepFruits: A Fruit Detection System Using Deep Neural Networks. *Sensors (Basel), 16*(8), 1222. doi:10.3390160816081222 PMID:27527168

Sabanci, K., & Aydin, C. (2017). Smart Robotic Weed Control System for Sugar Beet. *Journal of Agricultural Science and Technology, 19*(1), 73–83.

Safa, M., Samarasinghe, S., & Nejat, M. (2015). Prediction of Wheat Production Using Artificial Neural Networks and Investigating Indirect Factors Affecting It: Case Study in Canterbury Province, New Zealand. *Journal of Agricultural Science and Technology, 17*(4), 791–803.

Safavi, H. R., Mehrparvar, M., & Szidarovszky, F. (2016). Conjunctive Management of Surface and Ground Water Resources Using Conflict Resolution Approach. *Journal of Irrigation and Drainage Engineering, 142*(4), 05016001. doi:10.1061/(ASCE)IR.1943-4774.0000991

Saidu, A., Clarkson, A. M., Adamu, S. H., Mohammed, M., & Jibo, I. (2017). Application of ICT in Agriculture. *Opportunities and Challenges in Developing Countries., 3*, 11.

Sennaar, K. (2017, October 16). *AI in Agriculture - Present Applications and Impact.* Retrieved September 24, 2018, from TechEmergence website: https://www.techemergence.com/ai-agriculture-present-applications-impact/

Shamshiri, R. R., Weltzien, C., Hameed, I. A., Yule, I. J., Grift, T. E., Balasundram, S. K., ... Chowdhary, G. (2018). Research and development in agricultural robotics: A perspective of digital farming. *International Journal of Agricultural and Biological Engineering, 11*(4), 1–14. doi:10.25165/j.ijabe.20181104.4278

Sharma, N., Sharma, P., Irwin, D., & Shenoy, P. (2011). Predicting solar generation from weather forecasts using machine learning. *2011 IEEE International Conference on Smart Grid Communications (SmartGridComm)*, 528–533. 10.1109/SmartGridComm.2011.6102379

Shekoofa, A., Emam, Y., Ebrahimi, M., & Ebrahimie, E. (2011). Application of supervised feature selection methods to define the most important traits affecting maximum kernel water content in maize. *Australian Journal of Crop Science, 5*(2), 162–168.

Sideratos, G., & Hatziargyriou, N. D. (2012). Probabilistic Wind Power Forecasting Using Radial Basis Function Neural Networks. *IEEE Transactions on Power Systems, 27*(4), 1788–1796. doi:10.1109/TPWRS.2012.2187803

Silva, L. O. L. A., Koga, M. L., Cugnasca, C. E., & Costa, A. H. R. (2013). Comparative assessment of feature selection and classification techniques for visual inspection of pot plant seedlings. *Computers and Electronics in Agriculture, 97*, 47–55. doi:10.1016/j.compag.2013.07.001

Simon, M. (2018, September 25). The Creepy-Cute Robot that Picks Peppers With its Face. *Wired.* Retrieved from https://www.wired.com/story/the-creepy-cute-robot-that-picks-peppers/

Sirsat, M. S., Cernadas, E., Fernandez-Delgado, M., & Khan, R. (2017). Classification of agricultural soil parameters in India. *Computers and Electronics in Agriculture, 135*, 269–279. doi:10.1016/j.compag.2017.01.019

Smith, B. A., Hoogenboom, G., & McClendon, R. W. (2009). Artificial neural networks for automated year-round temperature prediction. *Computers and Electronics in Agriculture, 68*(1), 52–61. doi:10.1016/j.compag.2009.04.003

Snow, V. O., & Lovattb, S. J. (2008). A general planner for agro-ecosystem models. *Computers and Electronics in Agriculture, 60*(2), 201–211. doi:10.1016/j.compag.2007.08.001

Soh, Y. W., Koo, C. H., Huang, Y. F., & Fung, K. F. (2018). Application of artificial intelligence models for the prediction of standardized precipitation evapotranspiration index (SPEI) at Langat River Basin, Malaysia. *Computers and Electronics in Agriculture, 144*, 164–173. doi:10.1016/j.compag.2017.12.002

Sonobe, R., Tani, H., Wang, X., Kojima, Y., & Kobayashi, N. (2015). Extreme Learning Machine-based Crop Classification using ALOS/PALSAR Images. *Japan Agricultural Research Quarterly, 49*(4), 377–381. doi:10.6090/jarq.49.377

Stastny, J., Konecny, V., & Trenz, O. (2011). Agricultural data prediction by means of neural network. *Agricultural Economics-Zemedelska Ekonomika, 57*(7), 356–361. doi:10.17221/108/2011-AGRICECON

Suh, H. K., Ijsselmuiden, J., Hofstee, J. W., & van Henten, E. J. (2018). Transfer learning for the classification of sugar beet and volunteer potato under field conditions. *Biosystems Engineering, 174*, 50–65. doi:10.1016/j.biosystemseng.2018.06.017

Sujaritha, M., Annadurai, S., Satheeshkumar, J., Sharan, S. K., & Mahesh, L. (2017). Weed detecting robot in sugarcane fields using fuzzy real time classifier. *Computers and Electronics in Agriculture, 134*, 160–171. doi:10.1016/j.compag.2017.01.008

Sulistyo, S. B., Woo, W. L., & Dlay, S. S. (2017). Regularized Neural Networks Fusion and Genetic Algorithm Based On-Field Nitrogen Status Estimation of Wheat Plants. *IEEE Transactions on Industrial Informatics, 13*(1), 103–114. doi:10.1109/TII.2016.2628439

Tamaddoni-Nezhad, A., Milani, G. A., Raybould, A., Muggleton, S., & Bohan, D. A. (2013). Construction and Validation of Food Webs Using Logic-Based Machine Learning and Text Mining. In G. Woodward & D. A. Bohan (Eds.), Advances in Ecological Research, Vol 49: Ecological Networks in an Agricultural World (pp. 225–289). Academic Press. doi:10.1016/B978-0-12-420002-9.00004-4

Tan, D. S., Leong, R. N., Laguna, A. F., Ngo, C. A., Lao, A., Amalin, D. M., & Alvindia, D. G. (2018). AuToDiDAC: Automated Tool for Disease Detection and Assessment for Cacao Black Pod Rot. *Crop Protection, 103*, 98–102. doi:10.1016/j.cropro.2017.09.017

Tesfaye, K., Sonder, K., Cairns, J., Magorokosho, C., Tarekegn, A., Kassie, G. T., … Erenstein, O. (2016). Targeting Drought-Tolerant Maize Varieties in Southern Africa: A Geospatial Crop Modeling Approach Using Big Data. *International Food and Agribusiness Management Review, 19*(A, SI), 75–92.

Thenkabail, P. S., Lyon, J. G., & Huete, A. (2012). Advances in Hyperspectral Remote Sensing of Vegetation and Agricultural Croplands. In P. S. Thenkabail, J. G. Lyon, & A. Huete (Eds.), Hyperspectral Remote Sensing of Vegetation (pp. 3–35). Academic Press. Retrieved from https://pubs.er.usgs.gov/publication/70098951

Torkashvand, A. M., Ahmadi, A., & Nikravesh, N. L. (2017). Prediction of kiwifruit firmness using fruit mineral nutrient concentration by artificial neural network (ANN) and multiple linear regressions (MLR). *Journal of Integrative Agriculture, 16*(7), 1634–1644. doi:10.1016/S2095-3119(16)61546-0

Torres-Sospedra, J., & Nebot, P. (2014). Two-stage procedure based on smoothed ensembles of neural networks applied to weed detection in orange groves. *Biosystems Engineering, 123*, 40–55. doi:10.1016/j.biosystemseng.2014.05.005

Trace Genomics. (2018). *Trace Genomics*. Retrieved December 17, 2018, from https://www.tracege-nomics.com/#/

Tripathy, A. K., Adinarayana, J., Vijayalakshmi, K., Merchant, S. N., Desai, U. B., Ninomiya, S., ... Kiura, T. (2014). Knowledge discovery and Leaf Spot dynamics of groundnut crop through wireless sensor network and data mining techniques. *Computers and Electronics in Agriculture, 107*, 104–114. doi:10.1016/j.compag.2014.05.009

Ureta, C., Gonzalez-Salazar, C., Gonzalez, E. J., Alvarez-Buylla, E. R., & Martinez-Meyer, E. (2013). Environmental and social factors account for Mexican maize richness and distribution: A data mining approach. *Agriculture, Ecosystems & Environment, 179*, 25–34. doi:10.1016/j.agee.2013.06.017

VineView. (2018). *Aerial Vineyard Mapping - Vigor & Grapevine Disease*. Retrieved December 17, 2018, from VineView website: https://www.vineview.ca/

Wachter, S., Mittelstadt, B., & Floridi, L. (2017). Transparent, explainable, and accountable AI for robotics. *Science Robotics, 2*(6). doi:10.1126cirobotics.aan6080

Waga, D., & Rabah, K. (2014). Environmental Conditions' Big Data Management and Cloud Computing Analytics for Sustainable Agriculture. *World Journal of Computer Application and Technology*, 9.

Wakefield, K. (2018). *Predictive analytics and machine learning*. Retrieved November 14, 2018, from https://www.sas.com/en_gb/insights/articles/analytics/a-guide-to-predictive-analytics-and-machine-learning.html

Wang, L., Niu, Z., Kisi, O., Li, C., & Yu, D. (2017). Pan evaporation modeling using four different heuristic approaches. *Computers and Electronics in Agriculture, 140*, 203–213. doi:10.1016/j.compag.2017.05.036

Wang, Y. (2016). Deep reasoning and thinking beyond deep learning by cognitive robots and brain-inspired systems. *2016 IEEE 15th International Conference on Cognitive Informatics Cognitive Computing (ICCI*CC)*, 3–3. 10.1109/ICCI-CC.2016.7862095

Wang, Z., Hu, M., & Zhai, G. (2018). Application of Deep Learning Architectures for Accurate and Rapid Detection of Internal Mechanical Damage of Blueberry Using Hyperspectral Transmittance Data. *Sensors (Basel), 18*(4), 1126. doi:10.339018041126 PMID:29642454

Wang, L., Zhou, X., Zhu, X., & Guo, W. (2017). Estimation of leaf nitrogen concentration in wheat using the MK-SVR algorithm and satellite remote sensing data. *Computers and Electronics in Agriculture, 140*, 327–337. doi:10.1016/j.compag.2017.05.023

Wang, Li'ai, Zhou, X., Zhu, X., Dong, Z., & Guo, W. (2016). Estimation of biomass in wheat using random forest regression algorithm and remote sensing data. *Crop Journal, 4*(3), 212–219. doi:10.1016/j.cj.2016.01.008

Weiss, S. M., & Indurkhya, N. (1997). *Predictive data mining: a practical guide*. Retrieved from https://www.elsevier.com/books/predictive-data-mining/weiss/978-0-08-051465-9

Weltzien, C. (2016). *Digital agriculture – or why agriculture 4.0 still offers only modest returns*. Academic Press.

Wolfert, S., Ge, L., Verdouw, C., & Bogaardt, M.-J. (2017). Big Data in Smart Farming – A review. *Agricultural Systems*, *153*, 69–80. doi:10.1016/j.agsy.2017.01.023

Wolfert, S., Goense, D., & Sorensen, C. A. G. (2014). A Future Internet Collaboration Platform for Safe and Healthy Food from Farm to Fork. *2014 Annual SRII Global Conference*, 266–273. 10.1109/SRII.2014.47

Woodard, J. (2016). Big data and Ag-Analytics An open source, open data platform for agricultural & environmental finance, insurance, and risk. *Agricultural Finance Review*, *76*(1), 15–26. doi:10.1108/AFR-03-2016-0018

Wu, J., Yang, S. X., & Tian, F. (2014). A novel intelligent control system for flue-curing barns based on real-time image features. *Biosystems Engineering*, *123*, 77–90. doi:10.1016/j.biosystemseng.2014.05.008

Xiao, Y., Mignolet, C., Mari, J.-F., & Benoit, M. (2014). Modeling the spatial distribution of crop sequences at a large regional scale using land-cover survey data: A case from France. *Computers and Electronics in Agriculture*, *102*, 51–63. doi:10.1016/j.compag.2014.01.010

Yongting, T., & Jun, Z. (2017). Automatic apple recognition based on the fusion of color and 3D feature for robotic fruit picking. *Computers and Electronics in Agriculture, 142*(A), 388–396. doi:10.1016/j.compag.2017.09.019

Zangeneh, M., Omid, M., & Akram, A. (2010). Assessment of agricultural mechanization status of potato production by means of artificial Neural Network model. *Australian Journal of Crop Science*, *4*(5), 372–377.

Zhang, X., Qiao, Y., Meng, F., Fan, C., & Zhang, M. (2018). Identification of Maize Leaf Diseases Using Improved Deep Convolutional Neural Networks. *IEEE Access: Practical Innovations, Open Solutions*, *6*, 30370–30377. doi:10.1109/ACCESS.2018.2844405

Zhang, Z., Gong, Y., & Wang, Z. (2018). Accessible remote sensing data based reference evapotranspiration estimation modelling. *Agricultural Water Management*, *210*, 59–69. doi:10.1016/j.agwat.2018.07.039

Zhong, L., Hawkins, T., Holland, K., Gong, P., & Biging, G. (2009). Satellite imagery can support water planning in the Central Valley. *California Agriculture*, *63*(4), 220–224. doi:10.3733/ca.v063n04p220

Zhong, T. Y., Zhang, X. Y., & Huang, X. J. (2009). Simulation of farmer decision on land use conversions using decision tree method in Jiangsu Province, China. *Spanish Journal of Agricultural Research*, *7*(3), 687–698. doi:10.5424jar/2009073-454

Zhu, N., Liu, X., Liu, Z., Hu, K., Wang, Y., Tan, J., ... Guo, Y. (2018). Deep learning for smart agriculture: Concepts, tools, applications, and opportunities. *International Journal of Agricultural and Biological Engineering*, *11*(4), 21–28. doi:10.25165/j.ijabe.20181104.4475

KEY TERMS AND DEFINITIONS

AI: Artificial intelligence.

ANN: Artificial neural networks.
DDD: Data-driven decisions.
ELM: Extreme learning machine.
IoT: Internet of things.
RNN: Recurrent neural networks.
SVM: Support vector machine.

APPENDIX

Table 2. Breakdown of papers on Pre-Production phase

Pre-Production Category	Type	Main Study Objectives/Impact	Technique	References
Seeds & Seedlings	Predictive	Pot plant seedling classification	Feature Selection	(Silva et al., 2013)
		Seed yield prediction	Artificial Neural Networks	(Emamgholizadeh et al., 2015)
		Estimate cultivar and site-specific parameters from breeding trial data	Holographic Genetic Algorithm	(Lamsal et al., 2017)
		Classification tool for Potato Virus Y detection	Support Vector Machines	(Griffel et al., 2018)
	Prescriptive	Target drought-tolerant maize varieties	Spatial Modeling	(Tesfaye et al., 2016)
		Seeding density optimization	Artificial Neural Networks	(Dornelles et al., 2018)
		Phosphogypsum Application Recommendation	M5-Rules Algorithm, Regression Models	(Caires & Guimaraes, 2018)
Soil, Tillage and Land Preparation	Diagnostic	Analyze the effects of soil texture, moisture, compaction, horizontal deformation and vertical load on traction force and traction efficiency	Artificial Neural Networks	(Pentos & Pieczarka, 2017)
	Predictive	Erosion Index prediction to aid conservation planning	Logistic Regression, Artificial Neural Networks	(Pike et al., 2009)
		Soil classification	Artificial Neural Networks	(Ajdadi et al., 2016; Calderano et al., 2014)
		Estimation of fuel consumption in agricultural mechanized operations	Artificial Neural Networks	(Borges et al. 2017)
		Prediction of key factors for a successful kiwi micropropagation	Artificial Neural Networks: Neuro Fuzzy logic models	(Arteta et al., 2018)
		Generating soil moisture forecasts	Artificial Neural Networks	(Prasad et al., 2018)
Other Pre-planting Activities	Diagnostic	Selection of important traits contributing to maximum kernel water content of maize	Supervised Feature Selection	(Shekoofa et al., 2011)
		Determining association between climatic indices and maize and sorghum yields	Spearman's Rank Correlation, Artificial Neural Networks	(Byakatonda et al., 2018)
	Predictive	Using satellite imagery to estimate water usage	Supervised Maximum Likelihood Classification	(L. Zhong et al., 2009)
		Predicting machinery energy ratio for target farming systems	Artificial Neural Networks	(Zangeneh et al., 2010)
		Predicting crops present in a given field using crop sequence of previous years	Markov Logic Networks	(Osman et al., 2015)
		Drought assessment and monitoring	Random Forest, Boosted Regression Trees, Cubist	(Park et al., 2016)
		Year-round temperature prediction	Artificial Neural Networks	(Smith et al., 2009)
	Prescriptive	Creating rule-based information system for decision making	Fuzzy Data-Mining, Decision Trees, Simulation Models, MapReduce Algorithms	(Delgado et al., 2008; Rajeswari et al., 2017; Snow & Lovattb, 2008)
		Land use conversion simulation	Classification and Regression Tree	(Zhong et al., 2009)
		Production Models based on soil and meteorological data	Artificial Neural Networks: Multilayer Perceptron and Self-Organizing Maps	(Jimenez et al., 2009)
		Classifying technical management routes of farmers	Data Mining, Ascendant Hierarchical Clustering, K-Means Clustering	(Renaud-Gentie et al., 2014)
		Miscanthus location probabilities through farmers' criteria	Boosted Regression Trees	(Rizzo et al., 2014)
		Yield simulation	Artificial Neural Networks, Genetic Algorithm	(Dornelles et al., 2018)

Table 3. Breakdown of papers on Production phase

Production Category	Type	Main Study Objectives/Impact	Technique	References
Planting, Irrigation & Fertilizer Application	*Diagnostic*	Evaluate spatial relationships of environmental and social factors with the spatial distribution of races and that can harbor highest number of races	Data Mining	(Ureta et al., 2013)
	Predictive	Monitoring nutrient solution compositions	Artificial Neural Networks	(Gutierrez et al., 2008)
		Estimating of nutrient content of plants	Artificial Neural Networks: Recurrent Neural Networks, Multiple Linear Regression, Multiple-Kernel Support Vector Regression, Genetic Algorithm	(Harada et al., 2013; Sulistyo et al., 2017; Liai Wang et al., 2017)
		Drought forecasting, assessment and monitoring	Extreme Learning Machine, Artificial Neural Networks, Multiple Linear Regression, Random Forests, Boosted Regression Trees, Cubist	(Ali et al., 2018; Park et al., 2016)
		Prediction of evapotranspiration	Artificial Neural Networks: Generalized Regression Neural Networks, Support Vector Machines, Extreme Learning Machine, Genetic Programming, Multiple Linear Regression, Multivariate Adaptive Regression Spline, Least Square Support Vector Regression	(Feng et al., 2017; Kumar et al., 2016; Pandorfi et al., 2016; Soh et al., 2018; Wang et al, 2017; Zhang et al., 2018)
		Modeling irrigation systems	Artificial Neural Networks, Fuzzy Logic, Genetic Algorithm	(Mattar, El-Marazky, & Ahmed, 2017; Perea R et al., 2018)
		Mapping extent of artificially drained areas	Selective Ensemble	(Moller et al., 2018)
		Prediction of soil properties	Linear Regression, Random Forests, Artificial Neural Networks, Support Vector Machine, Gradient Boosting, Cubist	(Khanal et al., 2018)
	Prescriptive	Classification of irrigated crops	Artificial Neural Networks: Multilayer Perceptron and Radial Basis Function	(López-Granados et al., 2010)
		Decision support system for Automated irrigation	Partial Least Square Regression, Artificial Neural Networks: Adaptive Neuro Fuzzy Inference Systems, Gradient Boosting	(Goldstein et al., 2018; Goumopoulos et al., 2014; Navarro-Hellin et al., 2016)
		Surface and ground Water management using conflict resolution	Artificial Neural Networks, Genetic Algorithm	(Safavi et al., 2016)
		Identification of well performing crops	Regression Trees, K-Means Clustering	(Peloia & Rodrigues, 2016)
		Soil classification	Support Vector Machines, Random Forests, Extreme Learning Machine	(Sirsat et al., 2017)
		Automating the growing process of crops via smart hydroponics	Bayesian Networks	(Alipio et al., 2017)

continues on following page

Table 3. Continued

Production Category	Type	Main Study Objectives/Impact	Technique	References
Weed, Disease & Pest Control	*Diagnostic*	Understanding the relationship between crop, weather, environment and disease	Multivariate Regression	(Tripathy et al., 2014)
		Early warning system, diagnosis, and management practices of avocado wilt complex disease.	Correlation Models	(Ramirez-Gil et al., 2018)
	Predictive	Weed discrimination	Logistic Regression, Fuzzy Logic, Artificial Neural Networks, Convolutional Neural Networks, Support Vector Machines, Principal Component Analysis, Naïve Bayes, Ensemble Algorithm, Discriminant Analysis	(Akbarzadeh et al., 2018; Bakhshipour & Jafari, 2018; Bakhshipour et al., 2017; Chantre et al., 2018; Dos Santos et al., 2017; Gómez-Casero et al., 2010; Gutierrez et al., 2008; Pantazi et al., 2017; Sujaritha et al., 2017; Torres-Sospedra & Nebot, 2014)
		Fungal infection discrimination	Principal Component Analysis, Artificial Neural Networks: Learning Vector Quantization	(Chung et al., 2016; Heim et al., 2018; Liu et al., 2010)
		Prediction of bacterial infection	Artificial Neural Networks, Support Vector Machines	(Hill et al., 2011; Liao et al., 2012; Pineda et al., 2018)
		Information system for assessment of crop disorders	Support Vector Machines	(Camargo et al., 2012)
		Remote sensing crop classification	Extreme Machine Learning	(Sonobe et al., 2015)
		Leaf disease diagnoses	Support Vector Machine, Deep Convolutional Neural Networks	(Du et al., 2016; Ramirez-Gil et al., 2018; Zhang et al., 2018)
		Automatic pest detection	Convolutional Neural Networks, Artificial Neural Network,	(Cheng et al., 2017; Dimililer & Zarrouk, 2017; Ding & Taylor, 2016; Espinoza et al., 2016)
		Disease detection via agricultural datasets	Ensemble Learning – Ensemble-Vote, Support Vector Machine, Deep Learning, Linear Regression	(Chaudhary et al., 2016; Ferentinos, 2018; Hassanien et al., 2017; Ilic et al., 2018)
		Automated/Real time disease detection	Support Vector Machine, Bayesian Networks, Convolutional Neural Networks	(Ancin-Murguzur et al., 2018; Garcia-Santillan & Pajares, 2018; Lu et al., 2017; Stanley Tan et al., 2018)
	Prescriptive	Classification of crops and weeds for automated weed control	Support Vector Machine, Convolutional Neural Networks	(Ahmed et al., 2012; Suh et al., 2018)
		Robotic weed control	K-Nearest Neighbor, Machine Vision	(Pallottino et al., 2018; Sabanci & Aydin, 2017; Sennaar, 2017)
		Vision-based fruit detection system in aid of autonomous harvesting	Convolutional, Neural Networks, Support Vector Machines, Machine Vision	(Sennaar, 2017; Shamshiri et al., 2018)
		Classification of pesticide regulations	Logistic Regression	(Espejo-Garcia et al., 2018)

continues on following page

Table 3. Continued

Production Category	Type	Main Study Objectives/Impact	Technique	References
Yield, Weather, Energy & Other Farm Management Activities	*Diagnostic*	Determination of appropriate environmental conditions for obtaining high crop yields	Artificial Neural Networks	(Jimenez et al., 2011)
		Extraction of cropping patterns from time series data	Hidden Markov Models	(Xiao et al., 2014)
	Predictive	Yield mapping	Decision Trees, Region Merging Algorithm, Evolutionary Product Unit Neural Networks	(da Silva Junior et al., 2017; Gutierrez et al., 2008)
		Yield prediction	Artificial Neural Networks: Adaptive Neuro-Fuzzy Inference Systems, Fuzzy Logic, Time Series, Autoregressive Integrated Moving Average, Bayesian Networks, Model-Based Recursive Partitioning, Support Vector Machines, Boosted Regression Trees, Gaussian Process Regression, Random Forests, Multiple Linear Regression, Extreme Learning Machine, M5 Prime, Decision Trees, K-Nearest Neighbor	(Aghighi et al., 2018; Chapman et al., 2018; Diaz et al., 2017; Everingham et al., 2016; Farjam et al., 2014; Fernandes et al., 2011; Fortin et al., 2011; Fukuda et al., 2013; Gonzalez-Sanchez et al., 2014; Hamedani et al., 2015; Jokic et al., 2010; Khashei-Siuki et al., 2011; Khazaei et al., 2008; Kouadio et al., 2018; Rathod et al., 2018; Romero et al., 2013; Safa et al., 2015; Stastny et al., 2011)
		Temperature prediction	Artificial Neural Network	(Castaneda-Miranda & Castano, 2017; Smith et al., 2009)
		Solar generation prediction	Support Vector Machines	(Sharma et al., 2011)
		Wind power forecasting	Artificial Neural Networks: Adaptive Neuro-Fuzzy Inference System, Support Vector Machines	(Mahmoud et al., 2018; Sideratos & Hatziargyriou, 2012)
		Prediction of environmental indices affecting crop production	Artificial Neural Networks and Adaptive Neuro-Fuzzy Inference Systems	(Khoshnevisan et al., 2013)
		Prediction of output energy in crop production	Artificial Neural Networks	(Khoshnevisan et al., 2014)
		Prediction of soil properties	Linear Regression, Random Forest, Artificial Neural Networks, Support Vector Machines, Gradient Boosting Model, Cubist	(Khanal et al., 2018)
	Prescriptive	Weather prediction	Machine Vision and Various Machine Learning Techniques	(Sennaar, 2017)

Table 4. Breakdown of papers on Post-Production phase

Post-Production Category	Type	Main Study Objectives/Impact	Technique	References
Harvest & Handling	*Predictive*	Crop yield forecast in aid of pre-harvest sales	Lasso Approximation	(Everingham et al., 2009)
		Portable leaves recognition system	Artificial Neural Networks	(Husin et al., 2012)
		Prediction of fruit physical properties – firmness, stalk characteristics, size etc.	Data Mining: Find Laws, Artificial Neural Network, Multiple Linear Regressions	(Demir et al., 2018; Oo & Aung, 2018; Torkashvand et al., 2017)
		Determining harvest time	Artificial Neural Networks	(de Barros et al., 2018)
	Prescriptive	Post-harvest product sorting	Decision Trees, Impulse Acoustic Data Mining, Fuzzy Data Mining	(Ebrahimi & Mollazade, 2010)
		Vision-based fruit detection system in aid of autonomous harvesting	Convolutional, Neural Networks, Support Vector Machines, Machine Vision	(Fernandez et al., 2018; Sa et al., 2016; Sennaar, 2017; Shamshiri et al., 2018; Yongting & Jun, 2017)
		Harvest time estimation	Artificial Neural Networks	(Costa et al., 2018)
Storage & Processing	*Predictive*	Year-round temperature prediction	Artificial Neural Network	(Smith et al., 2009)
		Biomass Estimation	Random Forests, Support Vector Machines, Artificial Neural Networks	(Castro et al., 2017; Wang et al., 2016)
		Real time post-harvest classification	Support Vector Machines, K-Nearest Neighbors, Discriminant Analysis, Decision Trees, Naive Bayes	(Pour et al., 2018)
	Prescriptive	Autonomous real-time flue-curing barns	Artificial Neural Networks	(Wu et al., 2014)
		Classification of potato tubers based on solanine toxicant	Artificial Neural Networks	(Babazadeh et al., 2016)
		Evaluating of phenolic compounds and antioxidant activity of blueberries	Artificial Neural Networks	(Guine et al., 2018)
Market & Consumer	*Predictive*	Crop yield forecast in aid of pre-harvest sales	Lasso Approximation	(Everingham et al., 2009)
		Price forecasting	Artificial Neural Networks	(Ayankoya et al., 2016; Orge Pinheiro & de Senna, 2017; Ribeiro & Oliveira, 2011)
		Food web construction	Logic-Based Machine Learning and Text Mining	(Tamaddoni-Nezhad et al., 2013)
		Fruit vitamin phenotyping	Artificial Neural Networks	(Aquino et al., 2016)
		Predictive models in horticulture	Generalized Linear Model, Bayesian Additive Regression Tree, Boosted Classification and Regression Tree	(Logan et al., 2016)
		Estimation of color properties in aid of quality checks	Artificial Neural Networks	(Demir, 2018; Kus et al., 2017)
		Detection of internal mechanical damage of fruits	Deep Convolutional Neural Network	(Wang et al., 2018)
	Prescriptive	Trade policy making and business decision making through Big Data	TRADE-Decision Support Model	(Cameron et al., 2017)

Chapter 8
Automated Grading of Tomatoes Using Artificial Intelligence:
The Case of Zimbabwe

Tawanda Mushiri
https://orcid.org/0000-0003-2562-2028
University of Zimbabwe, Zimbabwe

Liberty Tende
University of Zimbabwe, Zimbabwe

ABSTRACT

The rate of production of horticultural produce had been seen increasing from the past century owing to the increase of population. Manual sorting and grading of tomatoes had become a challenge in market places and fruit processing firms since the demand of the fruit had increased. Considering grading of tomatoes, color is of major importance when it comes to the maturity of the tomatoes. Hence, there is a need to accurately classify them according to color. This process is very complicated, tiresome, and laborious when it is done manually by a human being. Apart from being labor-demanding, human sorting, and grading results in inaccuracy in classifying of tomatoes which is a loss to both the farmer and customer. This chapter had been prepared focusing on the automatic and effective tomato fruit grading system using artificial intelligence particularly using artificial neural network in Matlab. The system makes use of the image processing toolbox and the ANN toolbox to process and classify the tomatoes images according to color and size.

DOI: 10.4018/978-1-5225-9687-5.ch008

INTRODUCTION

The rate of production of horticultural produce has been increasing over the past century owing to the increase in the world population. Present world statistics show that tomatoes are the second largest produced fruits by volume and represent up to 100 million tonnes produced on 3.8 million hectares (Montanati et al., 2017; BRESOV, 2018; FAO, 2018; BTGA, 2018). Tomatoes are currently cultivated in 144 countries with China being the leading country in their production. The vegetable has the highest global consumption rate due the fact that most meals include the fruit. Figure 1 shows a tomato and its constituent nutrients (Haifa, 2018)

Due to the high rate of production of tomatoes and their perishability it is preferable that immediately after being harvested they are be washed, graded and packed. It is a cumbersome exercise for large-scale stakeholders such as farms and wholesalers to manually sort and grade them. Hence, there is a vital need for machines such as automatic tomato graders to perform this tiresome and repetitive task. Automated tomato graders would group the fruits by size and color; one of the critical stages in the collection and marketing of the fruit (Hernandez-Hernandez et al., 2016). Depending upon the organic attributes of the assortment, the natural product is partitioned into four, five, or more evaluations. Reviewing is manually complete - contrasting with standard examples - or on evaluating machines - as indicated by weight or size. Hand grading is usually combined with sorting or packing; machine grading is done after sorting (Londhe et al., 2013). The machine performs the task at a faster rate with minimum human supervision compared to the manual methods. This helps large farm owners to quickly harvest and dispatch the product to their consumers without delay.

The rate of manually grading tomatoes by small to medium enterprises (SMEs) is very low and inefficient compared to the demand of the fruits both nationally and globally hence there is a need to develop a system of grading tomatoes using artificial intelligence.

Figure 1. Nutritional summary for a tomato

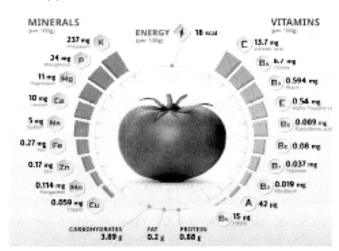

Based on this need the principal objectives of this chapter were as follows;

1. To grade tomatoes using artificial intelligence systematically.
2. Use machine vision technology to grade tomatoes.
3. Achieve process efficiency greater than 80%.

According to the authors the benefits of this methodology are manifold;

1. The machine is environmentally friendly hence it can be greatly appreciated globally,
2. Reduces the damage of the fruits during the sorting process,
3. An economy in electricity,
4. Rapid and efficient,
5. Affordable to both big and small organizations,
6. Tomatoes are going to be sorted and packed rapidly before they rot.

BACKGROUND

Tomatoes are believed to have been originally discovered in South America where they grew wild (Jones, 2017). They were domestically planted by the early explorers with some ornamental curiosities but not eaten. In some parts of Europe, they were grown as decorative plants in the gardens and the plant was regarded to being poisonous. According to Isaac (2015) tomatoes were accepted as a crop in 1840 even though in some countries they were still debating whether the plant was fit for human consumption. Global tomato production drastically increased in the past century after it was proven to be an edible fruit, and could be marketed to generate income.

Presently in Zimbabwe, when tomatoes are ripe, they are manually harvested and transported to the market usually by trucks or by vans. The actual harvesting of the tomatoes is elementary as compared to the other post-harvest processes which are sorting, grading and packing. Tomatoes being highly-perishable necessitate rapid transfer to the market place in order that the farmer sell them in optimal conditions. However, the most challenging task is to sort and grade tomatoes into the respective grades (Rao, 2015). Tomatoes are currently graded by taking three factors into account which are:

1. Color
2. Size
3. Firmness/ strength

Grading according to color is done by taking three ripening stages of the tomatoes which are mature, half ripe and ripe.

When tomatoes are graded according to the size only the following grades are considered which are:

1. Large (diameter > 50mm)
2. Medium ($40 \leq$ diameter $\leq$ 50mm)
3. Small (diameter < 40mm)

Tomatoes which are correctly graded attract customers because they consider quality as a major criterion in buying. Traditionally, tomatoes are graded manually by the farmers throughout most parts of the world. This does not only cumbersome but also results in low reaping rates both for the crop grown in fields and greenhouses. The method is very time-consuming for bulky quantities which results in low packaging rates. However, these methods had been seen to have many drawbacks which are:

1. High labor cost as many workers are required to perform the task,
2. Fatiguing exercise,
3. Inconsistency,
4. Low precision,
5. Difference in personal perception of quality,
6. Scarcity of the labor force.

In Zimbabwe tomatoes are grown in every province throughout the country. Figure 2. shows the tomato grading process at Mbare Musika in Harare.

The cultivation of tomatoes had become more appreciated in Zimbabwe because income can be generated from this crop. In every city there is a major market place where horticultural and agricultural produce are sold, for example, Mutare–Sakubva Musika and Harare-Mbare Musika (Ngonidzashe, 2015). Moreover, the production and selling of tomatoes had stimulated the need for more modernized methods to enhance this income generator. Grading of tomatoes is one of the post-harvest processes that is appreciated by the customers. In Zimbabwe grading is done manually and this resulted in some farmers incurring losses because the time taken to grade a full truck of tomatoes, which are perishables and need to be delivered to the customers as quickly as possible. In parallel, ss ripeness is the main indicator for tomato quality from customers perspective, the determination of tomato ripeness stages is a basic industrial concern regarding tomato production in order to get high quality products. Automatic ripe-

Figure 2. Manual grading of tomatoes (Ngonidzashe, 2015)

ness evaluation of tomato is an essential research topic as it may prove beneficial in ensuring optimum yield of high quality product. There is, thus, a need to examine new technologies, especially, artificial intelligence and machine learning and to benefit of their known rapid qualities (El-Bendary et al., 2015).

Tomatoes are the second most grown cultivated vegetables and China, USA, Turkey, Italy and India being the top five countries in its production and distribution (Worldatlas, 2018; Heirloom, 2017; FAO, 2016; FAO, 2018). In these countries, the implementation and appreciation of machine learning and artificial intelligence has taken the center stage in the automation of agricultural processes. Currently with over 100 million tonnes of tomatoes are being produced on 3.7 million hectares of land, China is the leading nation in the production of tomatoes.

Generally, from Figure 3., it can be concluded that tomatoes have a high annual global production compared to other vegetables hence there is a need to have more automated machines to wash, grade and pack the product as rapidly as possible due to its perishable state. The manual grading rate utilized by small to medium enterprises (SMEs) is very low and inefficient compared to the demand of the fruits both nationally and globally hence there is need to develop a more efficient system.

The major objectives of this chapter are to examine grading of tomatoes using artificial intelligence and to specifically use machine vision technology to grade tomatoes and achieve process efficiency greater than 80%.

Figure 3. Global production of vegetables in 2017, by type (in million metric tons) (Statistica, 2017)

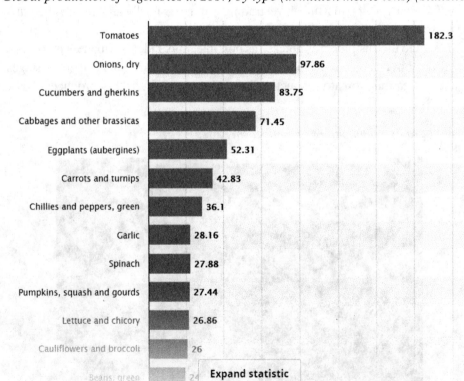

LITERATURE SURVEY ON TOMATO GRADING AND CHALLENGES FACED

Sorting and grading are processes which are generally performed manually. However, these two are labor-intensive operations in the fruit processing industry. High labor demand and lack of overall consistency in the process have led to the development of automated sorting and grading machines. Visual quality grading is one of the most difficult tasks to automatize fruit and vegetable sorting and grading (Rokunuzzaman & Jayasuriya, 2013; Singh et al., 1993). Computer vision is a crucial aspect in sensing mechanisms used for the determination of external features such as color, size, shape, and other features (Radhakrishnan 2012). However, due to limited utilization of the current available, appropriate technologies, from post-harvest, pre-packaging, storage to transportation, only an average of 40% of the total tonnage produced. Among the horticultural produce, tomato had become the second largest world total tonnage by production and accounting up to 23% from the total yielded volume commercialized by the main Brazilian vegetable wholesale market (De Castro et al., 2006).

Types of Tomatoes

There are various types of tomatoes grown in the world with their sizes ranging from marble-sized grape or cherry tomatoes, to delicious serving of mixed greens tomatoes, substantial blue tomatoes, and colossal, sweet, beefsteak tomatoes (BTGA, 2018). Their hues extend from profound ruby to orange, yellow, green, purple and chocolate (Masley, 2019; Heron & Zachariah, 1974).

- **Determinate Tomatoes** are shrubbery composes that grow 2-3 feet (60-90cm) tall, at that point the buds at the finishes of all the branches shape blossoms rather than takes off. They bloom at the same time, set and age the natural product, at that point kick the bucket (Masley, 2019).
- **Indeterminate Tomatoes** are vining composes that need confining or staking for help yet will proceed to develop and set natural product until the point where ice kills them. They are for the most part later than determinate tomatoes and deliver bigger products over a more drawn-out period (Masley, 2019). Among the grown tomatoes are shown in Figure 4.

Current Grading Methods/Systems in Zimbabwe

Local farmers and distributors sort tomatoes manually without the use of machines. The traditionally quality inspection and hand-picking focus on one or two factors, that is, size and color. The rate of sorting depends on several factors which are:

1. Workers experience and training,
2. The task's duration,
3. The working environment which depends on the temperature, light intensity.

This implies that a conducive environment is correlative to the output of the sorting and grading. Manual sorting had many drawbacks that had to be considered, the inspector cannot inspect more than two factors at the same time but rather focuses on one attribute only. Moreover, the aspect of speed cannot be ignored because the demand of tomatoes is very high over a given space of time. This implies with the utilization of manual methods there is a high correlation between the speed of grading

Figure 4. Types of tomatoes that are grown in the world (Masley, 2019).

and accuracy which means the lower the speed the higher the accuracy of process. The utilization of machine vision-based technology is expected to substitute manual based strategies for evaluating and arranging of natural product. To accelerate the procedure and keep up the consistency and exactness, a model computer vision-based programmed tomato reviewing and arranging framework was created (Chandra et al., 2012).

The manual procedure (Figure 5.) in the sorting and grading of tomatoes had been superseded by the application of computer vision technology which is accurate, has high processing speeds and, moreover, utilizes non-destructive procedures.

Challenges Faced in Grading Tomatoes

The primary problems encountered in the manual grading and sorting of tomatoes is the inability of maintaining consistency in uniformity due to fatigue and human error. Therefore, there is a need to maintain consistency, uniformity and accuracy as well as increasing the speed of the process. Enormous post-harvest misadventures in dealing with and handling tomatoes are brought about by an expanded

Figure 5. Current grading methods used in Zimbabwe.

demand for high caliber tomatoes (Mahendran et al., 2012). The technique utilized by the farmers and merchants to sort and grade agricultural and sustenance items are through conventional quality investigation and handpicking which is tedious, difficult and less effective. Manual sorting and reviewing depends on conventional visual quality examination performed by human administrators, which is tedious, cumbersome and non-predictable. Gathering generally is finished by manual tactile perceptions. Manual determination and investigation of the shade of cherry tomatoes is hazardous due to exhaustion and irregularities. Without doubt, there is an awesome contrast between computer-vision evaluating technology and human vision (Sun, 2016). In the latter case, the intrinsic issues is that the process incorporates high labor costs, laborer exhaustion, irregularity, fluctuation, and shortage of prepared work. The scarcity of accessible work and expanding business costs during the pinnacle harvesting seasons have also been distinguished as the critical components driving the interest in automation of the business.

The Automation Aspect of Tomato Grading

Automation of grading and sorting of agricultural produce is becoming more familiar due to the increased demand in different types of food required by different customers to satisfy their health needs and their living standards. This means that fruit produced from farms must be arranged according to their quality and developmental level before being marketed. An automated tomato sorting or grading machine is a machine that simultaneously performs several tasks. The machine grades and sorts tomatoes according to size, color, and weight at a faster rate compared to the manual sorting systems (Chandra et al., 2012).

In the designing of this machine (Figure 6.), there are some components, modules and constraints that must be considered (Adamu & Shehu, 2018; Omidi-Arjenaki et al., 2012)

The overall process on how the machine operates is well illustrated in Figure 7.

Figure 6. Cherry Tomato Sorter based on Shape, Maturity, Size, and Surface Defects using Machine Vision

Figure 7. The general function of sorting algorithm

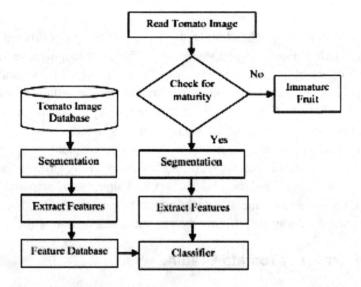

Artificial Neural Network Technique

An artificial neural network (ANN) is a representation of the human central nervous system that incorporates the interconnected bunch of neurons. The system mimics human behavior and attributes in its operations like being able to perceive the environment, adapt, learn and dispense output to/from the environment. Artificial neural networks are widely used in solving many complex computerized and non-computerized problems. Operations such as image segmentation and classification can be performed using ANN (Jackman & Sun, 2013; Mendoza & Aguilera, 2004; Soltani et al., 2015). They function from a principal of computational models that make use of internal weights to adjust and learn from its environment around. An algorithm with the required and specified commands is fed into the artificial neural network. The goal of the preparation is to look for the ideal arrangement of association weights that will make the yield of the counterfeit neural system coordinate the genuine target estimates (Ben et al., 2016; Bishop, 1995). Finding the ideal arrangement of association weights requires a preparation calculation that will have the capacity to alter the association weights of ANN with the goal that the coveted yield from the system given an arrangement of information sources can be obtained.

MATERIALS AND METHODS

How Tomatoes Are Graded

The performance of the complex grading systems depends on the input quality factors that were feed by the designer in the computer's memory chip. The tomatoes quality grading factors are classified into two distinct groups which are:

1. External quality factors
2. Internal quality factors

External Quality Factors

The external quality factors are obtained from the peripheral appearance of the tomatoes which are nearly correlated to the internal quality of the tomato. The external quality factors to be considered are: color, size, shape, gloss, surface defects and decay and texture (surface patterns) (Haralick et al., 1973).

Internal Quality Factors

Internal quality factors are the properties of the tomato within its tissues. Some of the internal properties to be considered for tomatoes are: firmness, crispness and toughness.

However, the internal quality factors are tested by the non-destructive methods such as the spectroscopic and hyperactive spectral imaging since these methods are non-destructive.

Grades of Tomatoes for the British Agricultural Market

Table 1 shows data that was collected from tomato farmers, tomato market place experts and from processing companies. This was done by means of questionnaires, observations, and interviews. Standard information was obtained from the internet and libraries.

Hardware Tools

The following were used as hardware tools; sixth or seventh generation high speed PC, Sony CCD camera DSC-RX1 / DSC-RX1R (2012, 24-megapixel Exmoor CMOS sensor), conveyor belt with power drive and illumination source.

Table 1. Grades for tomatoes: British Agricultural Market

Class	Quality Factors
Class 1	• Consists of tomatoes that are mature but not overripe or soft. • Attractive and well-shaped for the relevant cultivar. • Fairly uniform in color. • No decay or foreign matter. • Diameter is at least 50mm.
Class 2	• Comprise of tomatoes of similar from class 1 but with a fairly formed shaped but mature and not overripe. • They are free from decay and partially with foreign matter. • Reasonably uniform in color. • At least 40 mm diameter.
Class 3	• Tomatoes in this class are very firm. • Attractive and tomatoes with points that are permissible. • The size and color in this class might differs from time to time. • The diameter is at least 30mm.

Software Tools

Mat lab and Creative Webcam Live Pro were used.

RESULTS AND DISCUSSION

The computer vision system is the most efficient and reliable system that is now implemented in most commercial fruit sorting and packing companies to automatically implement sorting and packing tasks that are laborious when manually performed.

However, the researcher is going to focus on three aspects that are: color, size and shape.

Factors Considered when Choosing A Classification Method

1. **Data Size:** This is the amount of the data that the classifier can handle and able to process with very minimum errors.
2. **Accuracy:** This the degree to which the classifier can perform its operation with a stipulated of allowance of wandering from true values.
3. **Simplicity:** Refers to the ease of interpretation of the data within the classifier.
4. **Speed:** Speed is how fast the classifier can execute the given algorithm and produce the desired results with the smallest amount of error.
5. **Robustness:** It's a guarantee to the over-fitting of the data even though not theoretical.

From this point the researcher is going to focus on the sorting of tomatoes considering the three aspects which are: color, size and shape.

Research Scope

The researchers visited the major market of Harare, Mbare Musika where the data pertaining to the research was collected. Due to financial restraints the researcher did not manage to visit the fruit processing companies such as Cains Food Ltd or FAVCO in other parts of Zimbabwe to collect data regarding the grading and sorting of tomatoes. This research was limited to the marketing sector. At the market place the researchers managed to capture 250 images of tomatoes using the 12 MP Canon Camera. This is shown in Figure 8., Figure 9. and Figure 10. and with data distribution shown in Table 2.

Classification Using Artificial Neural Network

The digital images were classified according to two variables which are: classification according to color and classification according to size.

Figure 8. Grade 'A' tomatoes

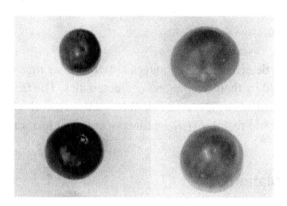

Figure 9. Grade 'B' tomatoes

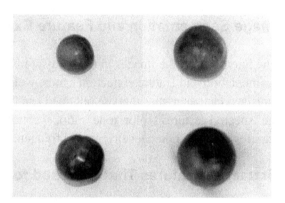

Figure 10. Grade 'C' tomatoes

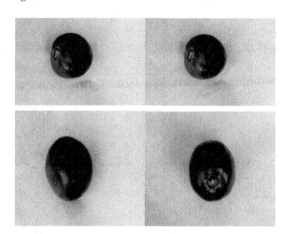

Table 2. distribution of collected data

Type of Images	Grade A	Grade B	Grade C
Training Data	50	50	50
Testing Data	40	30	30
Total	90	80	80
Grand Total		250	

Analysis of Captured Data

Pre-processing

After capturing the images manually by a camera, a data-set is created to store the images according to their grades. The images created by the camera would vary, that is, about 1225 * 3000 pixels and they need to be downsized to 300 * 300 pixels using Matlab function to make sure that the region of relative interest in the image are more pronounced.

Image resizing is done as in Figure 11.

Why the Need for Image Segmentation?

- In order to create a scale variance - a process of making the system more robust in determining the image of tomatoes placed at different angles which means 0^0, 90^0 and 180^0.
- It also permits partitioning of images into homogeneous regions, which would have common attributes such as the gray levels, mean values, and textures.

Image Segmentation and Feature Extraction

The intensities of the images in the data-set needs to be determined. The images obtained after image segmentation and feature extraction can be well described by their feature and characteristics. The features and characteristics that are obtained can be classified as in Figure 12 into three categories which are: spectral features (color, tone, ratio, spectral), geometric features (edges, lineaments), and textural features (pattern, homogeneity, spatial frequency).

Extracted Features That Are Used for Calculations

Figure 13. shows extracted features that are used for calculations. From the extracted features we can get the: Green channel pixels, Red channel pixels and Ratio of the Red to Green.

Figure 11. Image resizing. *NB* Take Note In The Down-Sampling Or Sizing You Can Lose Some Significant Features In The Images.

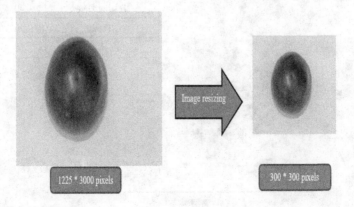

Figure 12. Segmentation and Feature extraction

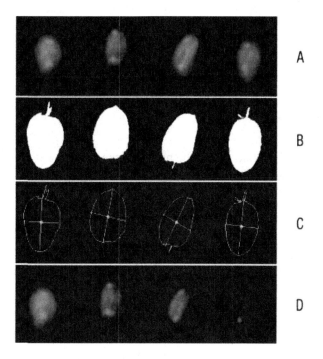

Figure 13. Extracted features that are used for calculations

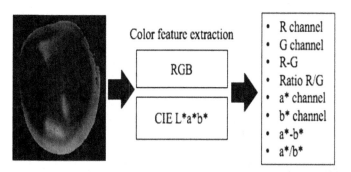

COLOR CALCULATIONS

RGB to HSV Conversion

The formula for color conversion from RGB (Red, Green, Blue) to HSV (Hue, Saturation, Value/Brightness) is shown below:

$$H = \cos^{-1}\left\{\frac{\frac{1}{2}\left[(R-G)+(R-B)\right]}{\sqrt{(R-G)^2+(R-B)(G-B)}}\right\},$$

$$S = 1 - \frac{3}{R+G+B}\Big[\min(R,G,B)\Big],$$

$$V = \frac{1}{3}(R+G+B)$$

HSV to RGB Conversion

This conversion is a bit complex. Being given the nature of the hue information, a different formula for each sector of the color triangle is given below:

Red to Green sector:

for $0° < H \leq 120°$

$$b = \frac{1}{3}(1-S), r = \frac{1}{3}\left[1 + \frac{S\cos H}{\cos(60° - H)}\right], g = 1 - (r+b)$$

Green to Blue sector:

for $120° < H \leq 240°$

$$r = \frac{1}{3}(1-S), g = \frac{1}{3}\left[1 + \frac{S\cos H}{\cos(60° - H)}\right], b = 1 - (r+b)$$

Blue to Red sector:

for $240° < H \leq 360°$

$$g = \frac{1}{3}(1-S), b = \frac{1}{3}\left[1 + \frac{S\cos H}{\cos(60° - H)}\right], r = 1 - (r+b)$$

Color Histogram Definition

An image histogram is the probability function of the image intensities. This had an extension for all color images to capture the joint probabilistic of the locally available three color channels. Table 3 and Table 4 shows the variable description for criteria and classification respectively.

$$h_{A,B,C}(a,b,c) = N \cdot \mathrm{Pr}\,ob(A = a, B = b, C = c)$$

Diameter Calculations

$$D = \sqrt{\frac{2 \cdot \left(m_{20} + m_{02} + \sqrt{\left(m_{20} - m_{02}\right)^2 + 4 \cdot m_{11}^2} \right)}{m_{00}}}$$

$$d = \sqrt{\frac{2 \cdot \left(m_{20} + m_{02} - \sqrt{\left(m_{20} - m_{02}\right)^2 + 4 \cdot m_{11}^2} \right)}{m_{00}}}$$

Where D = the major-axis distance

d = the minor-axis distance

m_{ij} = the colour moments of the image.

Detailed Design and Simulation

This gives full detail design of the grading system using the Matlab Image Processing Toolbox and Artificial Neural Network Toolbox. Tomato images are fed into the respective folders for processing. Figure 14. is the general process flowchart of the image processing.

Table 3. Criteria variable description table

Designation	Color	Size	Shape	Total	Percentage,%
Color	N/A	7	8	15	42.92
Size	6	N/A	5	11	31.2
Shape	4	5	N/A	9	25.71

Table 4. Classification and selection table

		Classifiers							
		SVM		ANN		RBS(Fuzzy)		HSB	
Selection	%	Rating	Weight	Rating	Weight	Rating	Weight	Rating	Weight
Color	42.9	4	171.60	4	171.60	4	171.60	5	214.50
Size	31.4	1	31.40	5	157.00	5	157.00	2	62.80
Shape	25.7	2	51.42	5	128.50	4	102.80	2	51.42
Totals		254.42		457.71		431.40		328.72	
Decision		No		Yes		No		No	

Figure 14. Image processing flowchart

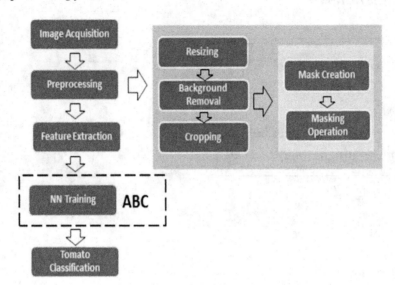

Pre-processing, Processing and Post-Processing of Captured Images

The pre-processing of captured data started in Mbare when the researchers captured tomato images using the Canon 12MP camera. The captured data was then retrieved from the camera and then grouped according to the color and texture. Creation of a folder called tomato grades.m in Matlab was done and within that same folder there are sub-folders which contains Grade 'A', Grade 'B' and Grade 'C' tomatoes. A Live Script was created to read the image in the Matlab Command window to read all the tomato images in the three folders. The images captured by the camera were very large, that is, 2056 x 3088 pixels and they could not be processed in that size therefore the image were resized as in Figure 15. The resizing process was then succeeded by the conversion to gray-scale level, which was of paramount importance since it enables us to extract the required key features of our images. Contrasting was also done after the greyscale level conversion to enhance the image contrast and to remove the filtering effects of the camera. Feature extraction Live Script was created and all the key features that were necessary for image classification were extracted from the images in Figure 16. After feature extraction the image's properties were: Surface area, Perimeter, Major-axis, Minor-axis and the mean of all.

These features are then fed into the Matlab's Artificial Neural Network to create the network for processing the images according to color and size as in Figure 17.

Artificial Neural Network (ANN)

The Neural Network Toolbox is used to create the Network for training the system. On the left side of the ANN is the input panel where the extracted data is being loaded and in the middle there is the hidden neuron panel. The hidden neural panel is responsible for processing the data into the various classification and there the more the number of hidden neuron the higher the speed of classification shown in Figure 18.

Figure 19. is the input panel showing the option of the class inputs and the target outputs. This means that the input targets are all the extracted data features of the image processed in the previous image

Figure 15. The Read.m Live Script from Matlab

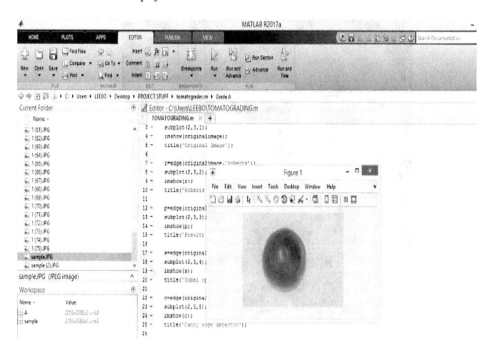

Figure 16. Feature Extraction Live Script screenshot

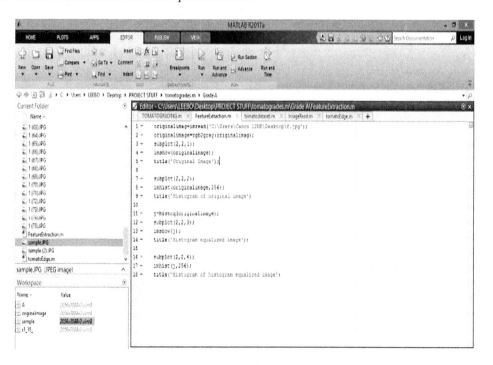

Figure 17. The Extracted Features from the image

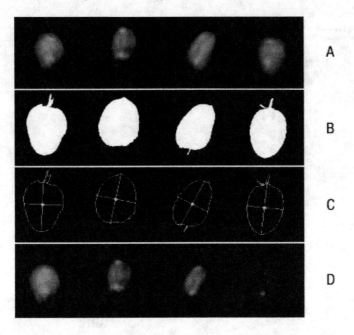

Figure 18. The Input, Hidden and Output panels

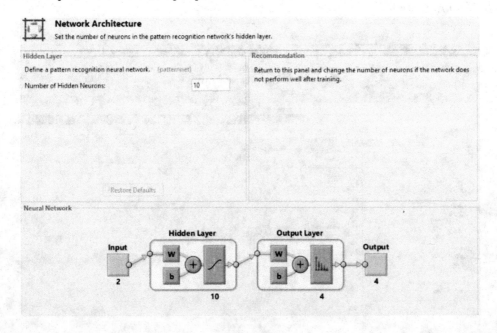

Figure 19. The input panel for selecting the targeted parameters

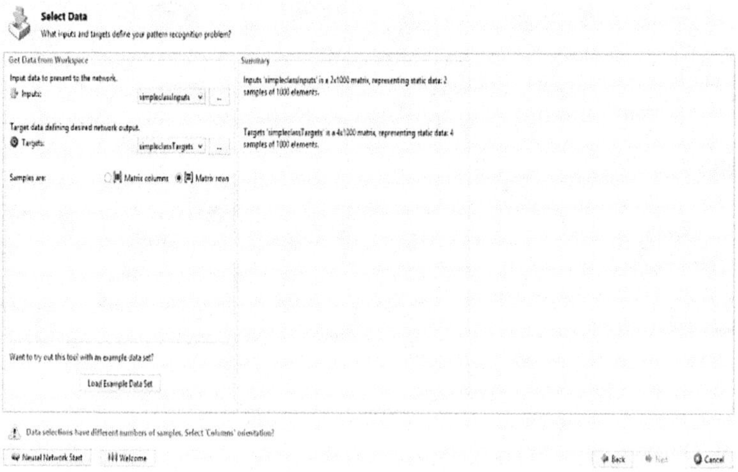

process. The select target is now our desired classification parameters and that is the range of sizes that we want and the color pixel ranges. The input classified data is then fed into the ANN input panel. The number of images used for training is then stated so that that the validation of the accuracy is going to be based on the number of trained data. The number of iterations is also classified in the training process of the ANN. The bigger the number of iterations the higher the learning rate of the network. However, the number of training data is also considerable in the learning process which means that if the training data is very large it also enhances the learning ability of the ANN system. The machine can be retrained if the percentage error is very high. The percentage error decreases as the number of training iterations is being increased. The Figure 20. shows the panel displaying the training process.

The Figure 21. is the validation of the training network. It shows the time taken to complete the classification, the total number of iterations performed, the percentage error and the gradient error.

Figure 20. Training progress of the ANN

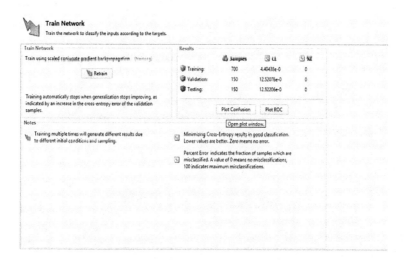

Figure 21. The validation of the training data

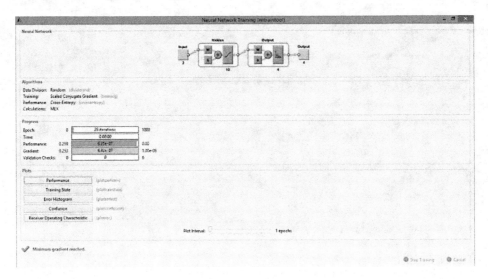

Conclusion of the Results

The ANN network process processed the data within a short space of time and the results are plotted below. The training operating characteristic of both the training, validation, and testing had a false rate. This means that the automation of image classification using ANN is very feasible as shown in Figure 22.

Figure 22. Plotted graphs of the output

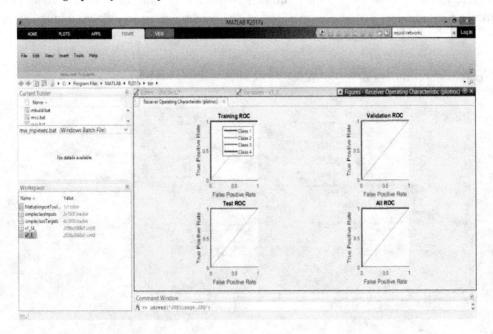

Table 5. Recommended factors and effects.

Factor	Effect
Increase light intensity in the illumination chamber.	This will reduce shading on the captured images by the camera and also the image will be of high contrast.
Using High Definition CCD camera in capturing the images of tomatoes in the illumination chamber.	The quality of image captured will be very high and this will reduce the time taken to process the image by the computer.
Taking a very large number of images and also use those to train the ANN network.	It improves the learning capability of the machine and the ability to classify images is increased.
Use the feedback propagation when training the ANN.	The feedback propagation enables that to reduce the percentage of error when classifying because the processed signal is fed back again into the processing layer before it is channeled on to the output layer.
Increase the number of iterations or the epoch number.	It gives the machine enough to learn the target classification pattern of the ANN and also reduces the percentage error.

RECOMMENDATIONS

After all been done, the major concern is the functionality of the design system. Artificial Intelligence and Machine Learning had been seen to make automation of industrial processes through their mimic of human intelligence through machine vision, reasoning and executing of commands to achieve the desired function. Industrial automation of various processes can be so economic efficient using AI. Table 5. shows the recommendations factors and effects.

CONCLUSION

The developed system had been seen to be feasible and effective. The accuracy of classification of tomatoes by color and size had been seen to about 81% that the percentage error is very little just about 19%. However, through some key adjustments to the system this percentage error can be further reduced. The system had been observed to have met all the objectives that were initially developed.

REFERENCES

Adamu, A. A., & Shehu, A. (2018). Development of an Automatic Tomato Sorting Machine Based on Color Sensor. *International Journal of Recent Engineering Research and Development, 3*(11).

Ben, X., Meng, W., Wang, K., & Yan, R. (2016). An adaptive neural networks formulation for the two-dimensional principal component analysis.(Report). *Neural Computing & Applications, 27.*

Bishop, C. (1995). *Neural Networks for Pattern Recognition.* Oxford, UK: University Press.

BRESOV. (n. d.). *Breeding for Resilient, Efficient and Sustainable Organic Vegetable Production.* Retrieved from https://www.bresov.eu/about/tomatoes

BTGA. (2018). *British Tomatoes Growers Association. Tomato Types*. Retrieved from http://www.brit-ishtomatoes.co.uk/tomato-facts/tomato-types

Chandra, N., Bipan, T., & Chiranjib, K. (2012). An Automated Machine Vision Based System for Fruit Grading and Sorting. *Six International Conference On Sensing Technology*. Retrieved from http://seat.massey.ac.nz/conferences/icst2012/files/icst2012program.pdf

De Castro, L. R., Cortez, L. A. B., & Vigneault, C. (2005). Effect of sorting, refrigeration and packaging on tomato shelf life. *World Food Journal*. Retrieved from https://www.researchgate.net/publication/266370688_Effect_of_sorting_refrigeration_and_packaging_on_tomato_shelf_life

El-Bendary, N., El-Hariri, E., Hassanien, A. E., & Badr, A. (2015). *Using Machine Learning Techniques for Evaluating Tomato Ripeness*. Retrieved from https://www.academia.edu/23217224/Using_machine_learning_techniques_for_evaluating_tomato_ripeness

FAO. (2016). *Food Agric. Organ*. Retrieved from http://faostat.fao.org

FAO. (2018). *Food and Agricultural Organisation of the United Nations, Statistics Division*. Available at http://www.fao.org/faostat/en/#data/QC/visualize

Haifa. (2018). *Crop Guide: Tomato*. Retrieved from https://www.haifa-group.com/tomato-fertilizer/crop-guide-tomato

Haralick, R. M., Shanmugam, K., & Dinstein, I. (1973). Textural Features for Image Classification. *IEEE Transactions on Systems, Man, and Cybernetics, SMC-3*(6), 610–621. doi:10.1109/TSMC.1973.4309314

Harshavardhan, N. G., Sanjeev, S. S., & Vijay, R. S., & R, A. (2012). Fruits Sorting and Grading using. *International Journal of Advanced Research in Computer Engineering & Technology, I*(6), 117–122.

Heirloom Organics Non HighBrid Seeds. (2017). Retrieved from https://www.non-hybrid-seeds.com/sp/tomato-lovers-s-pack-59.html

Hernandez-Hernandez, J. L., García-Mateos, G., González-Esquiva, J. M., Escarabajal-Henarejos, D., Ruiz-Canales, A., & Molina-Martínez, J. M. (2016). Optimal color space selection method for plant/soil segmentation in agriculture. *Computers and Electronics in Agriculture, 122*, 124–132. doi:10.1016/j.compag.2016.01.020

Heron, J. R., & Zachariah, G. L. (1974). *Automatic Sorting of Processing Tomatoes*. The American Society of Agricultural and Biological Engineers. Retrieved from doi: . doi:10.13031/2013.37013

Isaac, K. A. (2015). An overview of post-harvest challenges facing tomato production in Africa. Dunedin: African Studies Association of Australasia and the Pacific (AFSAAP).

Jackman, P., & Sun, D.-W. (2013). Recent advances in image processing using image texture features for food quality assessment. *Trends in Food Science & Technology, 29*(1), 35–43. doi:10.1016/j.tifs.2012.08.008

Jones, J. B. (2017). *Growing Tomatoes*. Retrieved from http://www.growtomatoes.com/tomato-world-production-statistics/

Londhe, D., Nalawade, S., Pawar, S., Atkari, V., & Wandkar, S. (2013). Grader: A review of different methods of grading for fruits and vegetables. *Agricultural Engineering International: CIGR Journal, 15*(3). Retrieved from https://pdfs.semanticscholar.org/7c76/5a16c27361db3a8a041742406bf9e90eb7e9.pdf

Mahendran, R., Jayashree, G. C., & Alugusandaram, K. (2012). *Application of Computer Vision Technique on Sorting and Grading of of Fruits and Vegetables. Journal of Food Process Technol S1-001.* doi:10.4172/2157-7110.S1-001

Masley, S. (2019). *Tomato Varieties and Types of Tomatoes.* Retrieved from https://www.grow-it-organically.com/tomato-varieties.html

Mendoza, F., & Aguilera, J. M. (2004). Application of image analysis for classification of ripening bananas. *Journal of Food Science, 69*(9), 471–477. doi:10.1111/j.1365-2621.2004.tb09932.x

Montanati, A., Cigognini, I. M., & Cifarelli, A. (2017). *Agri and food waste valorisation co-ops based on flexible multi-feedstocks biorefinery processing technologies for newhigh added value applications.* Retrieved from http://agrimax-project.eu/files/2017/11/AGRIMAX-D.1.2_Mapping-of-AFPW-and-their-characteristics.pdf

Ngonidzashe, C. (2015). *Retailers, supermarkets blasted over horticultural products import.* Harare: The Sunday News.

Omidi-Arjenaki, O., Maghaddam, P., & Motlagh, A. M. (2012). Online tomato sorting based on shape, maturity, size, and surface defects using machine vision. *Turkish Journal of Agriculture and Forestry, 37*(1), 62-68. Retrieved from DOI: . doi:10.3906/tar-1201-10

Radhakrishnan, M. (2012). *Application of Computer Vision Technique on Sorting and Grading of Fruits and Vegetables.* Retrieved from https://www.researchgate.net/publication/308953918_Application_of_Computer_Vision_Technique_on_Sorting_and_Grading_of_Fruits_and_Vegetables

Rao, R. S. (2015). *Cleaning, Sorting and Grading of Tomatoes.* Retrieved from https://www.slideshare.net/SUDHAKARARAOPARVATAN/tomato-grading

Rokunuzzaman, M., & Jayasuriya, H. (2013). Development of a low cost machine vision system for sorting of. *Aricultural Engineering International Journal, 15*(1), 173–180.

Singh, N., Delwiche, M. J., & Johnson, R. S. (1993). Image analysis methods for real-time color grading of stonefruit. *Computers and Electronics in Agriculture, 9*(1), 71–84. doi:10.1016/0168-1699(93)90030-5

Soltani, M., Omid, M., & Alimardani, R. (2015). Egg volume prediction using machine vision technique based on pappus theorem and artificial neural network. *Journal of Food Science and Technology, 52*(5), 3065–3071. doi:10.100713197-014-1350-6 PMID:25892810

Statistica. (2017). *Global production of vegetables in 2017, by type (in million metric tons).* Retrieved from https://www.statista.com/statistics/264065/global-production-of-vegetables-by-type/

Sun, D.-W. (2016). *Computer Vision Technology for Food Quality Evaluation. Academic Press.* Elsevier.

Worldatlas. (2018). *The World's Leading Producers of Tomatoes. Economics.* Retrieved from https://www.worldatlas.com/articles/which-are-the-world-s-leading-tomato-producing-countries.html

Chapter 9
Applications of Big Data and AI in Electric Power Systems Engineering

Tahir Cetin Akinci
Istanbul Technical University, Turkey

ABSTRACT

The production, transmission, and distribution of energy can only be made stable and continuous by detailed analysis of the data. The energy demand needs to be met by a number of optimization algorithms during the distribution of the generated energy. The pricing of the energy supplied to the users and the change for investments according to the demand hours led to the formation of energy exchanges. This use costs varies for active or reactive powers. All of these supply-demand and pricing plans can only be achieved by collecting and analyzing data at each stage. In the study, an electrical power line with real parameters was modeled and fault scenarios were created, and faults were determined by artificial intelligence methods. In this study, both the power flow of electrical power systems and the methods of meeting the demands were investigated with big data, machine learning, and artificial neural network approaches.

INTRODUCTION

In today's modern societies, electric energy is an inevitable concept. Electrical energy is a social and economic requirement for the development of society. For the last thirty years, a great deal of research has been undertaken to analyze and solve the problems of electrical power systems. Most of the research is on control theory, power electronics drivers and economic analysis. In recent years, the development of artificial intelligence and its methods has made this technology applicable in many areas. In this study, some methods have been proposed in order to provide supply balance by investigating the methods of using electric power systems with artificial intelligence techniques. Research on electrical power system can be examined in two groups as modeling and analysis. In this study, fault scenarios were created by using the energy transmission line model: these defects were then determined by artificial intelligence methods.

DOI: 10.4018/978-1-5225-9687-5.ch009

BACKGROUND

Literature Review

Recent developments in energy system energy systems seek solutions for the ongoing liberalization of energy markets, optimization of power system efficiency and power quality, emergency energy demands and challenges in dispersed energy transmission lines (Hidayatullah et al., 2011; Kadar, 2013). As a solution, the connection of renewable energy systems to the power system necessitates the control of electrical power systems by artificial intelligence techniques (Fikri et al., 2018). Artificial Intelligence techniques, along with traditional analytical techniques, can significantly contribute to the solution of related problems. Recent scientific studies emphasize that an intelligent energy transmission and distribution system should be used with evolutionary programming and other artificial intelligence methods (Akinci, 2011; Jiang et al., 2016; Bogdan et al., 2009; Paracha, 2009; Russel & Norvig, 2016).

Nowadays identified as the information age, all activities performed during the day are getting recorded with several technologies. The most valid reason for this logging is to establish-confidence for the benefit of people and society, such as security and public service. However, the smart phones and watches that people carry unconsciously are continuously recording their activities in their daily life and converting them into data (Bryant, 2014; Zimmer & Kurlanda, 2017). During a regular home-based work trip, individuals who have been able to adapt to the information age allow the system to collect data as they open the security alarm of their house when they leave for work, pass by security cameras on their route to work and use their ID card to enter the workplace. During the day, all the performance at the workplace, photos taken and shared on social network by colleagues with their smart phone and data of heart rate monitoring by smart watches are constantly being recorded. Here, copious information and copious data, from the rotational speed of the industrial machine to its temperature, the profit-loss statements of the wage paid to the workers by the company and even the company's stock exchange, are recorded. In this human-machine interactive interface world, all information has the potential to be turned into data. This data eventually becomes so large that is defined as big data. Storing and analyzing processed and unprocessed data now requires special methods and tools. The recording of all these data in daily life has led to the creation of storages in massive sizes. These storages have become centers that provide specific information called cloud. The processing of this gigantic information as well as its storage enabled an important software branch to emerge that has revealed a new profession called data analytics (Begoli & Horey, 2012; Papageorgiou, 2019; Grover et al., 2018).

Data analysts undertake critical tasks such as making companies profitable by producing meaningful results from the data presented or extracting disease information from biological data. They also undertake vital tasks such as optimizing the continuity of the system by ensuring that the bearings of an industrial machine are integrally disengaged or optimizing and making plans to ensure the availability of continuous and reliable electrical energy. These tasks find their field of application in social sciences and engineering (Dong et al., 2009; Chen et al., 2014; Grover et al., 2018).

Electrical engineering has been the most affected by the development of information and communication technologies. The first input of artificial intelligence technology to electrical power engineering was with smart meters. In the first studies, smart meter and sensor technologies and data collection systems were installed on energy users (Ongsakul & Vo, 2013; Alahakoon & Yu, 2015). Thus, the data collection from the traditional data distribution systems can be performed instantaneously. Information layers were added to traditional transmission and distribution lines for the immediately analysis of the data collected. The concept of intelligent networks is expanding: this essentially concerns data collection of the sensors and the evaluation of this collected data. In order to be smarter this requires initial energy transmission lines to be traceable at every point. As a result of monitoring and data collection, it is necessary to analyze the data collected including the characterization of the data to contribute to the healthy operation of the network. Although typical data analysis can produce meaningful results for general problems, detailed analysis of multi-point information in the system may be possible by using advanced algorithms. Geographical information requests and the addition of data from meteorological information systems to the analysis data in the energy transmission lines will make a significant contribution to the energy continuity of the existing power system. These contributions will improve customer satisfaction and social welfare by offering better quality, uninterrupted energy to users (Ongsakul & Vo, 2013; Klaimi et al., 2016).

This study focuses on the approaches of artificial intelligence and big data to electrical power systems engineering.

Artificial Intelligence (AI)

Artificial Intelligence (AI) can be defined as a technology that allows machines to produce, like humans, solutions to complex problems. Today this discipline, which was first proposed by John McCarthy in 1956 at the Dartmouth Conference, is a term used in every field of technology (McCarthy, 1992; Rajarman, 2014). In general, Artificial intelligence can be examined in two categories: software that addresses a narrower area designed for a specific task and more powerful software having human cognitive abilities. The latter has the ability to recommend the best solutions to problems. It is safe to say that the advanced artificial intelligence approach started with the Turing Test developed by Alan Turing in 1950 (Turing, 1950; Li et. al., 2018).

Initially, artificial intelligence software was developed so that experiences could not be used in future activities. With the development of Artificial Neural Network (ANN) models AI has become software, which can learn, make predictions, and analyze. This method is now widely used in smart car technology and autopilot systems. Artificial intelligence technology, which can understand the decisions and reactions of people and socially connect, will be possible in the future. Automatic machines, which have been developing since the Industrial Revolution, have evolved into robots that have artificial intelligence systems and can be programmed to perform high volume tasks. Today in the health sector, artificial intelligence is used to analyze the data obtained from the patients and to determine the most economical treatments methods to reduce costs. The AI in finance applications provides financial advice by collecting personal data. It also contributes to faster decision making and more accurate document matching in the practice of law (Turing, 1950; Mijwel, 2015; Bohanec, 2009).

Artificial intelligence (AI) is now used as a common name for many intelligent computer-based techniques. AI imitates the human brain with software techniques by investigating human behavior and rules and can make decisions and control operations very quickly. AI is inspired by the human brain. Artificial intelligence techniques comprise brain science; based on neurology, software and statistics sciences. AI is a system that simulates the behavior of the human brain and solves the problems in the fastest manner. Consequently, AI can make logical, reasoning theorems and proof that can execute fast programming, natural language processing problems can solve. However, the ability of self-development is increasing day-by-day (Li & Du, 2018).

Figure 1. The relationship between artificial intelligence and other disciplines

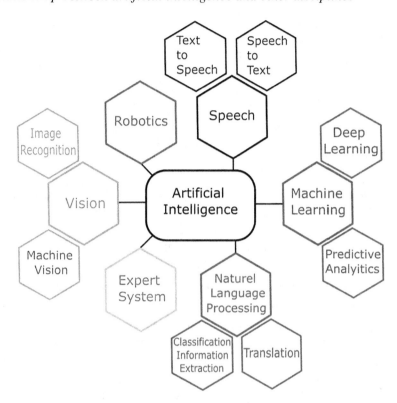

In the light of all these definitions and applications, with an ideal approach, artificial intelligence is an artificial operating system that is expected to exhibit high cognitive functions or autonomous behaviors, such as perception, learning, thinking, reasoning, problem solving, communicating, deducing and decision-making, all of which are specific to human intelligence. Figure 1 shows the relationship of artificial intelligence with other algorithms (Uliyar, 2017; Amazonaws, 2019; Brain, 2019; Ertel, 2017). Here, it is seen that AI bears a general meaning that involves machine learning and other learning algorithms.

The statistics of the artificial intelligence terms via Google are given in Figure 2 (Upwork, 2019). In-these statistics, AI, Deep Learning, ANN and Big Data were examined with data between 2004 and 2019. When data is analyzed, it is possible to understand the popularity of these terms by period. Artificial Neural Networks (ANN) was the most popular research subject in 2004 (it was known since the beginning of 1990's. In the early 2000's, Big Data and Deep Learning were limited, so the frequency of these terms is also limited. After 2012, Big Data and Deep Learning terminology-was popular in the Google search engine as well as in scientific studies. Nowadays, AI is less investigated because all these methods are grouped under the AI method (Google, 2019).

All artificial intelligence methods are closely related to the science of statistics. Moreover, the basis of these methods; machines learn from experience, adapt to new inputs, predict next steps, and perform many tasks as would human beings. Autonomous vehicles, chess-playing computers, and mobile phones can instantly translate into many examples. The real application of-artificial intelligence techniques that can make the right decisions on our behalf is industry. Artificial intelligence techniques are very practical and economical in heavy working conditions, which require intensive calculation as a result of decision-

Figure 2. Google statistics for investigating Artificial Intelligence methods (Google, 2019)

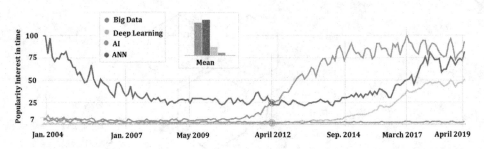

making mechanisms (Min, 2010; Witten et al., 2017). In areas where heavy industrial machines are used, industrial robots go beyond their software and apply the information they learned from previous experiences in different situations. AI can enable existing products to behave intelligently through software. Usually, these products are not sold as an individual application but they are being developed with the software that will provide AI capabilities to the product. Today, mobile phones are the best example for this definition. Perform programming of collected data using AI progressive learning algorithms. The algorithm here learns the structure of the data as a classifier or determinant. These learning algorithms reveal the ability of the AI structure to teach and teach. When it meets new data, it can adjust the model and make adjustments and adjustments in the model (Witten et al., 2017).

With its multi-layer or multi-layered structure (with a network of neural networks), AI provides more detailed analysis of data. Deep learning algorithms provide these. In an environment where massive amounts of data are received, learning systems can be created with direct data. The fact is that data that is applied to deep learning algorithms is more accurate. AI can analyze the data and bring them into intellectual property while making conclusions. With AI, the role of data is increasing, and it is the basis for the formation of new sectors and markets to increase competition in the sector.

Nowadays, the optimization and decision-making problems that arise in the stages of generation, delivery and marketing of energy in electric power systems are solved by artificial intelligence approaches (Zhang et al., 2018). These approaches and algorithms ensure that the system is more reliable, economical and efficient, and provide more comfortable facilities to users. Most of the smart grid approaches are artificial intelligence-oriented systems.

Machine Learning (ML)

Safety is the most important element in the electrical power systems and their functions. The need for deregulation of the power system and the increased need to operate systems closer to the operating limits require the use of more systematic approaches to safety to keep reliability at an acceptable level. It has been proven that a multiple control security application and evaluation can only be made with ML applications (Uliyar, 2017).

ML, a sub-branch of Artificial Intelligence, was developed from digital learning and model recognition studies. ML investigates the working systems of algorithms that can learn and make estimations using the data. These in contrast to static algorithms, have the ability to take decisions by making database estimates from model-based sample inputs. ML uses supervised learning and unsupervised learning methods to make inferences (Figure 3) (Mathworks, 2019; Sahu, 2018).

Figure 3. ML Learning Algorithm

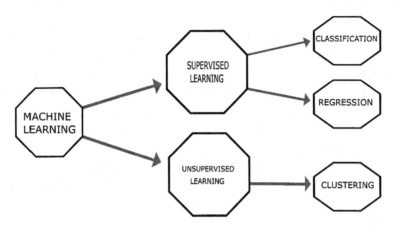

When scientific research revealed that the machines had to learn the data after a certain stage, the problems started to be solved by various symbolic approaches. In particular, this method has found use in automated medical diagnostic systems. These developments have allowed ML to be considered as a separate area and have been redeveloped since 1990. The aim of accepting ML as a separate field was to find solutions to practical problems with a new perspective. In this sense, the methods used by data mining are significantly overlapping. However, here; machine learning focuses on estimates made from data learned based on known features. Data mining focuses on revealing unknown (historical) features in the data. It can be described as feature extraction from these databases. In fact, data mining uses the general ML methods to extract information (Alpaydin, 2010).

The general relationship graph given in Figure 4 shows the association between the science of machine learning, data mining and statistics (Papageorgiou, 2019). In fact, when artificial intelligence and other learning algorithms are involved in this relationship analysis, a much more sociable and complex structure will emerge. ML in the near future is expected to develop by focusing on cyber physical systems, online learning, sensor integration methods, smart electricity networks, data acquisition, data integration and state imaging. Optimizing transmission and distribution networks in electrical power systems is a versatile field for ML applications. ML offers very practical approaches to solving optimization problems.

Figure 4. Machine Learning and other related areas

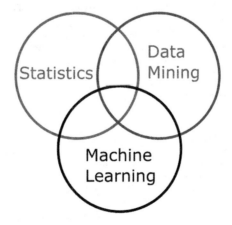

Artificial Neural Networks (ANN)

Artificial neural networks (ANN) are parallel computing devices that emerged with the idea of making a computer model of the brain. The goal of artificial neural networks is to develop a system for making complex computer operations faster than traditional systems: in other words, artificial neural networks. Inspired by the human brain, mathematical modeling of the learning process emerged. Therefore, the mathematical model of a neuron in artificial neural networks is an important step for artificial intelligence applications. Each element in the biological nerve cell is shown in Figure 6. However, the simplified version of biological systems is used for the mathematical model. In fact, a biological nerve cell is known to perform more complex biological and chemical processes. In the mathematical model structure, there are three sections in general terms. These are input, network structure and output. The network structure and hidden layers of the system can be composed of many layers depending on the type of problem (Figure 5.) (Haykin, 2009).

Effector transforms the electrical impulses produced by the brain into appropriate responses as organisms. Information in the central neural network is evaluated in the direction of forward and backward feeding between the receptor and response nerves to produce appropriate responses. In this respect, the biological nervous system is similar to the closed loop control system. The general usage area of artificial neural networks is decision making, classification and optimization but it is also used in pattern recognition and optimization problems. (Brain, 2019).

In the 1980s, the idea of a machine thinking like a person demonstrated a great improvement in the artificial neural networks technology in the 1990's. ANN is a subset of artificial intelligence. It generalizes the data and when introduced to unknown data and is capable of making a decision based on what it has previously learned. Today, because of its learning and generalization features, ANN has the ability to successfully solve complex problems in every field. The first computational model on which ANN is based was Pitts' and McCulloch's work (Pitts & McCulloch, 1947; Hertz et al., 2018; Shanmuganathan, & Samarasinghe, 2016). In 1954, Farley and Clark developed a model that can react and adapt to warnings within a network (Farley & Clark, 1954; Clark & Farley, 1955). The first programmable computer in the literature is Z3 developed by Zuse. Zuse also developed a first high-level programming language. (Giloi, 1997). Although the first digital computer was developed in 1941, the first neural computer was completed in 1950 (Paluszek & Thomas, 2017). While the work on the ANN models until the 1980s was very inefficient, in 1985 ANN finally obtained its well-known structure. Artificial neural networks have been applied in all fields, from economy to health and engineering, with high success rate in researches such as forecasting, estimation, failure analysis (Akinci et al., 2012; Nogay, 2016; Shahid et al., 2019; Yuce & Avci, 2017; Nogay et al., 2012). ANN is the approach that reflects the nervous system model in living organisms to the electronic environment with a programming discipline. In this sense, it has the skills of learning, remembering and updating the knowledge learned. In order to model the behaviors of

Figure 5. Block Diagram of Nervous System

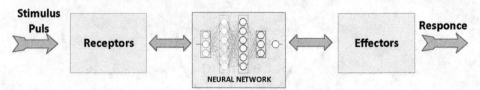

the nervous system, the model should be built in compliance with its structure. Accordingly, scientists have used the structure of the nervous system to model the ANN. The biological structure of a simple nerve cell is shown in Figure 6 (Brain, 2019) and the mathematical structure of a neuron is given in Figure 7. (Richard, 2018).

Dendrites are electrically passive arms that collect the signals from other cells and are the entrance to the system. Here, the axon is an electrically active body in which output pulses are generated and also the output of the system that provides one-way transmission. The synapse provides the connection of the axons of the cells with other dendrites. The myelin sheath acts as an insulating material that affects the spreading speed. The nucleus allows the periodic reproduction of the signals along the axon (Haykin, 2009; Richard, 2018; Akbal, 2018). The mathematical model of a human nerve cell is given in Figure 7.

$$S = w1.i1 + w2.i2 + \ldots\ldots + wn. \in -\theta = \sum_{i=1}^{n} wi.ii - \theta \tag{1}$$

Figure 6. Biological Representation of a Nerve Cell

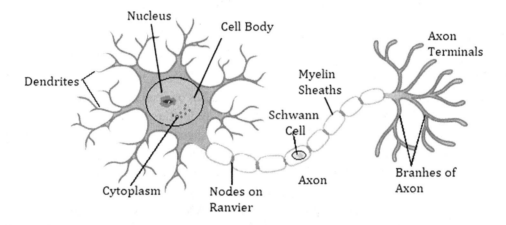

Figure 7. Mathematical Representation of a Nerve Cell

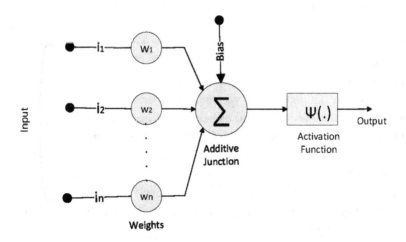

ANN is composed of various forms of artificial nerve cells connected to each other and is usually arranged in layers. As hardware, it can be implemented as software in electronic circuits or computers. In accordance with the brain's information processing method, ANN is a parallel-scattered processor capable of collecting information after a learning process, linking weights between cells, and storing and generalizing this information. The learning process involves learning algorithms that allow ANN's weights to be regenerated to achieve the desired goal. Figure 6. shows an artificial neural network model (Brain, 2019; Haykin, 2009; Richard, 2018). It can be understood that ANN is derived from its parallel scattered structure as well as its ability to learn and generalize. Generalization implies that ANN also produces appropriate responses for entries that are not encountered during learning.

In accordance with its structure, ANN can use many different algorithms such as Back-propagation Algorithm, Flexible Propagation Algorithm, Delta Algorithm, Rapid Propagation Algorithm and the Levenberg-Marquardt Method (Haykin, 2009; Richard, 2018). These algorithms are successful approaches that can predict the demands in the electrical power systems in advance.

In electrical power systems, load flow planning, anticipation of energy demands and taking the necessary action are very important issues. The current load on the energy transmission lines, which are fed by either hydroelectric power plants or by renewable power plants, needs to be transferred safely to needy users. In determining the immediate and the short-term need of this load, the artificial neural networks approach performs very well compared to traditional methods.

Deep Learning (DL)

Although there are solutions to a number of power control and decision-making problems in electrical power systems with different algorithm approaches, the studies on searching for the algorithm with the fastest and highest performance ratio are still in progress. While the ANN method has a high success rate, it has been observed that in terms of decision-making speed it has a very low performance when introduced to excessive visual data (Brownlee, 2013; Schmidhuber, 2015; Suk, 2017).

Although there are solutions to power control and decision-making problems in electrical power systems with different approach algorithms, the studies on seeking the algorithm with the fastest and highest performance ratio are still in progress. Although the ANN method has a high success rate, it has been observed that it has very low performance when faced with high visual data in terms of decision-making speed (Papageorgiou, 2019; Brownlee, 2013).

Nowadays as artificial intelligence, machine learning and deep learning are being compared with human intelligence, these terms are often confused with each other. Although artificial intelligence and machine learning are usually perceived as synonyms, they are quite different from each other (Magesh & Swarnalatha, 2017; Mocanu et al., 2016). In 1950, Alan Turing performed a study on whether machines could think or not. With the Turing Test he assembled, it was possible to determine if a machine was intelligent. If a person could not distinguish whether the interaction encountered was from a machine or a human, then that machine was considered as being intelligent. In 1956, John McCarthy premised a more in-depth study of artificial intelligence (McCarthy, 1992; Rajarman, 2014).

While ML is capable of processing on a single layer, DL has the ability to operate on several layers at the same time. Thus, DL can use a group of machine learning algorithms simultaneously and achieve a result in a single operation. DL algorithms are very rapid and successful in language processing and image recognition by adopting large output layers and using decision trees. The relationship between AI, ML and DL is given in Figure 8. (Brain, 2019).

Common faults in electrical power systems are load flow problems, energy supply problem, malfunctions occurring in transmission lines, faults occurring in generator and production plants. These faults can also be caused by a combination of several different faults. In other words, faults that have occurred it may also contain a lot of different failure. These are classified as difficult to understand and difficult to solve. The detection time of the fault and the repair of the fault is very important. Detection of a fault in a power grid consisting of long transmission lines is a very difficult task. Detection of the fault can only be made by analyzing a number of numerical signals and image data. If there are images between the data for the analysis of the fault, deep learning is mandatory.

In case such visual data processing takes time, the approaches made with deep learning algorithms offer much faster and more reliable solutions. This method provides brand-new approaches to the analysis of future needs in electrical power systems.

Big Data (BD)

The data produced during the transmission, distribution and marketing of electricity includes details that cannot be calculated by conventional methods. Storage and analysis of produced information are among the most important research topics (Begoli, & Horey, 2012; Zhang, et al., 2018). The amount of information produced and stored at the global level is too large to calculate. BD focuses on what can be done with the data rather than the data size. In this sense, BD provides the best decision by time and money optimization while using data from any source. With big data analysis, manufacturers can improve quality, reduce losses and obtain key roles in the financial market. It is also used in government institutions to extract information from many databases such as tax transactions, legal procedures, traffic congestion and medical applications (Chen et al., 2014; Samuel et al., 2015; Magesh & Swarnalatha 2017; Zhou et al., 2016).

Currently, as data is constantly produced by our mobile devices, software records, cameras, microphones, social media and all of our movements on the internet, it is transformed into stored data (Begoli & Honey, 2012). The continuous growth of data in terms of size, diversity and complexity has made it

Figure 8. Relationship between Artificial Intelligence and Deep Learning

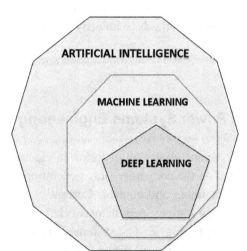

possible for the big data produced to become a focus of solutions with cloud computing. In addition, the internet of objects, big data and cloud computing provide rapid solutions to users. All these approaches are applied in other engineering fields as well as in the field of electrical power systems to analyze the data collected from consumers, meet the energy demands and solve the dynamic problems during the production of electrical energy. All data on the system is collected by Supervisory Control and Data Acquisition (SCADA) in electrical power systems. These data are collected in two groups. The first group is electrical data such as active power, reactive power, current, voltage, capacity changes. The second group consists of the mechanical and material data of all components on the energy transmission and distribution system. The size of these data is quite large. For this reason, the data collection time and the sampling time in the analysis are limited so that the collected data can be analyzed. These analyzes can only be done using big data methods. The method has limitations in terms of real-time processing of data and engineering operations in today's technology. In addition to all these electrical and mechanical variables, the data produced by renewable energy plants in energy systems are affected by meteorological factors. This complex situation allows too many parameters to be generated. This complexity is made possible by big data methods with the topology of the data and the strategy of processing the collected data with artificial intelligence methods.

METHODOLOGY

Today, due to non-linear loads, electrical networks are very complex. Faults on the network must be solved immediately. In this study's methodology it was found necessary to determine all the difficulties in the application of different electrical power systems of energy transmission lines. In the system center, a systematic combination of modern power electronic drivers for system theory, control engineering and optimization was prepared. Thus, advanced network integration was provided for interface elements for power electronics circuits. Modulation methods for medium and high-power lines in the network were developed and control and artificial intelligence algorithms formulated and implemented. In addition, flexible distributed operational strategies were created for next generation power systems. These strategies were based on artificial intelligence. Advanced software techniques were elaborated for power system analysis safety assessment, and preventive control. Finally, software-was designed to provide the economic supply demand balance for the network. In addition, a real energy transmission line model was created by the artificial intelligence methods to identify failure.

FINDINGS AND ANALYSIS

Applications of Electrical Power Systems Engineering

In recent years, data has become a very important criterion in electrical power systems engineering. Due to the variable (dynamic) load on the consumer side, generation, transmission and distribution of electrical energy requires instant planning and control. Currently, in addition to the classical energy generation methods, renewable energy sources feed the network. Large power plants, renewable energy sources and micro plants supply the power grid. The system also has storage units. The flow of energy in the network is monitored at every stage up until the end user. In energy distribution stations, SCADA systems are widely used to monitor the flow of energy (Manimuthu & Ramadoss, 2019). SCADA systems are uniquely used for observation. In the last ten years, many SCADA systems have been computer controlled. However, optimization and control cannot be done with the SCADA system. In order to control a system, it must have a data collection system. Collected data should be processed for analysis, control and planning. Data collected from the electricity network (production, transmission and distribution) are evaluated by various statistical analyses. The data collected here is considered as big data. Collecting and processing this data is done using special methods rather than classical methods (Wang et al., 2016). Mobile power stations are added or removed according to the demand. The development of renewable energy resources also increases investment incentives in this area. Many energy companies want to add a renewable energy source to the grid. Consumption also tends to increase on the user side. This dynamic change can only be controlled by artificial intelligence-based systems. Data management and real time monitoring systems for gas turbines are now being used in power plants (Frantz & Hunt, 2003). In most of these plants, real-time data analysis, data model results, optimization and dependency-analysis are performed.

The artificial neural networks method is widely-utilized to determine the location of the wind, power plant and planning, the production in concordance with energy markets as well as estimating the wind energy power source.

Figure 9 shows a real electric transmission line model. All operations on the power transmission line are controlled by the central control system which is also seen on the model. There are also sub-stations for communication in certain centers.-Fiber-optic transmission lines are utilized for communication and measurement at many points. The simulations can be simulated on-line. Short circuit failures for faults are investigated. In addition, the types of faults that occur can be classified by artificial intelligence methods.

Figure 10 shows the voltage fluctuation in the measuring stations of the three-phase normal system. This system has a voltage amplitude value of 380kV. However, the high amplitude region in the first second's results from the initial conditions. As it is known, in the initial conditions, the electrical machines can reach the amplitude values which are 3 to 5 times more than their nominal values. Figure 11 shows the analysis graphs for the normal voltage fluctuation given in Figure 10.

Figure 12. shows the voltage fluctuation of the three phase-ground short circuit faults of the three-phase system. Figure 13. shows the analysis graphs of the faulty voltage fluctuation given in Figure 12.

Figure 9. The real electric power network systems

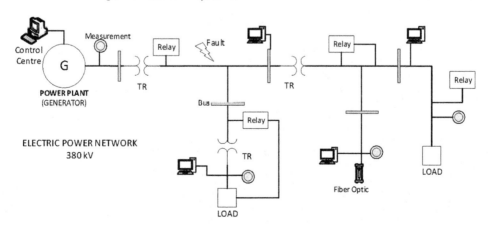

Figure 10. Normal voltage condition for measurement station 1

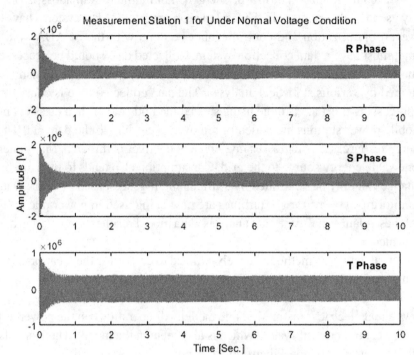

Figure 11. Normal voltage condition for measurement station 1

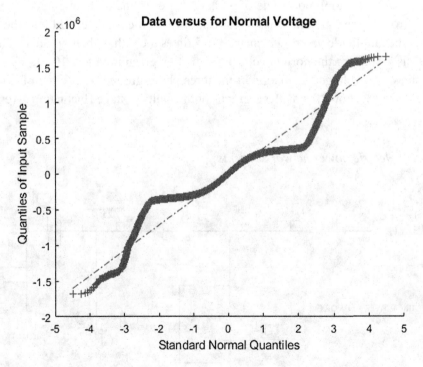

Figure 12. Fault voltage condition for measurement station 1

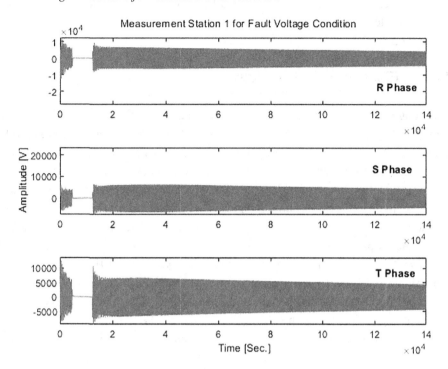

Figure 13. Data versus for normal voltage condition

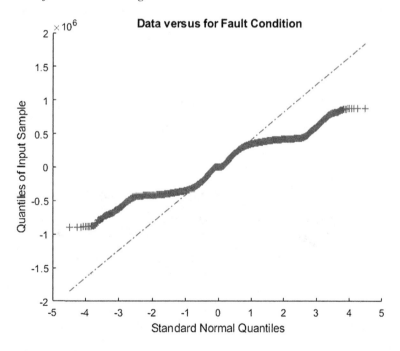

Data versus graphics normal and defective condition can be clearly identified. The control center opens the circuit to protect the system as soon as this graph reaches the data. The type and location of the fault can also be detected.

Optimization in Electric Power System

Electrical power engineering has the most comprehensive historical background in the basic electrical engineering science. Here, it is the main area of interest to deliver the energy from the production to the end-user. Therefore, it must use many different numerical optimization methods for the transmission and distribution of energy. By optimizing the system, the energy-generating plant reduces costs and provides enormous economic benefits. In optimization, the cost of fuel, energy security, continuity, quality criteria are calculated. Its main purpose is to minimize the various limitations of the enterprise and to minimize the cost of production in energy systems. It is also to increase the efficiency and efficiency of renewable energy sources in energy production. Figure 14 illustrates the relationship between artificial intelligence and management systems in electrical power systems (Zhao & Zhang, 2016).

Recently, artificial intelligence-based methods are widely used to solve optimization problems in electrical power systems (Paracha, 2009; Ongsakul & Vo, 2013). These methods have great advantages compared to the solution of problems that cannot be solved by the classical method. Advances in software and hardware technology have led to the development of artificial intelligence-based methods. The solution of problems such as control, planning, estimation, timing in electrical power systems is facilitated by AI. This concerns especially the problem of optimization which is considered the most

Figure 14. The relationship between artificial intelligence and management systems in electric power systems

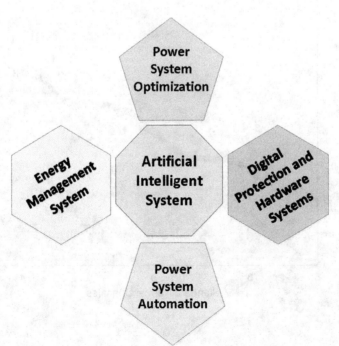

fundamental problem of electric power systems. The continuous change in the number of users and the energy consumed in the electrical system requires instant programming for optimal distribution of energy. AI methods provide excellent opportunities to system engineers in planning and analysis and even in predictive fault detection (Banu & Suja, 2012; Hessine & Saber, 2014; Zhang et al., 2018).

In the structure of a typical electrical power system, there are generators, AC and DC transmission systems, distribution systems load and Flexible AC Transmission Systems (FACTS) devices. The main purpose is to plan the low-cost expansion of the electrical power system. Therefore, supply and demand must be met at minimum cost. In order to achieve a variable supply-demand balance, components need to be continuously optimized by controlling artificial intelligence. In this planning, all parameters such as demand, energy transmission line, FACTS layout, environmental effects should be considered.

FUTURE RESEARCH DIRECTIONS

In the future, electric power systems will gain many different dimensions in both transmission and distribution stages. In this sense, it is thought that the components used in the transmission and distribution side in electric power systems will change greatly. The fiber-optics and components of the transmission lines provide the following benefits:

- Losses in energy transmission line is reduced.
- Collect quality data online.
- Reduces the danger of high voltage on power transmission lines.

However, fiber-optic technology has a high cost and requires large investment. Fiber-optics and technology imply the meaning of the integration of power systems engineering with communications engineering applications. In addition, research on superconductivity is not possible to apply the superconductivity technology to power transmission lines in the short and medium term. All these researches show that the transmission and distribution technology of electric energy will change dramatically in the future. The use of renewable energy sources will be used in the production of electricity. Renewable energy-based power plants will be installed in places where thermal based power plants will be reduced. Renewable power plants will be installed in smart micro-grid systems by establishing them close to local areas. All these electrical networks will be equipped with artificial intelligence methods and energy users will be provided with high quality and cheap electric energy. All electrical energy systems will be equipped with traccable and controllable systems over the cloud (Akinci et al., 2019).

CONCLUSION

In recent years artificial intelligence techniques and big data have found a field of application in every field. Electrical engineering and technologies are at the top of these fields. Many scholars consent that transmission, distribution, planning and economics of electrical energy require enormous expenses and involve very important engineering problems. The production, transmission and distribution of electrical energy, which together are called electrical power engineering are subjects that separately demand optimization and analysis. Estimation of the energy demands in electric power systems and estimation of future predictions are possible with artificial intelligence methods. Since the data produced by the electricity users are massive, processing and evaluation can only be analyzed with the big data approach. In this study, the ANN method and DL algorithms approach were investigated in order to meet the power flow and demands in the electric power systems. Also, the types of short circuit faults on electrical power network were determined by artificial intelligence methods.

REFERENCES

Akbal, B. (2018). Hybrid GSA-ANN methods to forecast sheath current of high voltage underground cable lines. *Journal of Computers*, *13*(4), 417–425.

Akinci, T. C. (2011). Short term speed forecasting with ANN in Batman, Turkey. *Elektronika ir Elektrotechnika*, *107*(1), 41–45.

Akinci, T. C., Korkmaz, U., Turkpence, D., & Seker, S. (2019). Big Data Base Energy Management System. *3rd International Symposium on Innovative Approaches in Scientific Studies-Engineering and Natural Sciences, (ISAS 2019)*, 48.

Akinci, T. C., Nogay, H. S., & Yilmaz, O. (2012). Application of artificial neural networks for defect detection in ceramic materials. *Archives of Acoustics*, *37*(3), 279–286. doi:10.2478/v10168-012-0036-1

Alahakoon, D., & Yu, X. (2015). Smart electricity meter data intelligence for future energy systems: A survey. *IEEE Transactions on Industrial Informatics*, *12*(1), 1–12.

Alpaydin, E. (2010). *Introduction to machine learning*. Cambridge, MA: The MIT Press.

Amazonaws. (2019). *Machine Learning*. Retrieved from https://wordstream-files-prod.s3.amazonaws.com/s3fs-public/machine-learning.png

Banu, G., & Suja, S. (2012). ANN based fault location technique using one end data for UHV lines. *European Journal of Scientific Research*, *77*(4), 549–559.

Begoli, E., & Horey, J. (2012). Design Principles for effective knowledge discovery from big data. In *Joint working conference on Software Architecture & 6th European Conference on Software Architecture*, (pp. 215-218). Helsinki, Finland: IEEE. 10.1109/WICSA-ECSA.212.32

Bogdan, Ž., Cehil, M., & Kopjar, D. (2009). Power system optimization. *Energy*, *32*(6), 955–960. doi:10.1016/j.energy.2007.01.004

Bohanec, M. (2009). Decising Making: A computer-science and information- technology viewpoint. *Interdisciplinary Description of Complex Systems*, *7*(2), 22–37.

Brain, C. (2019). *The artificial intelligence ecosystem*. Retrieved from https://medium.com/@b.james-curry/the-artificial-intelligence-ecosystem-f11107f7b306

Brownlee, J. (2013). *A tour of machine learning algorithms*. Retrieved from https://machinelearning-mastery.com/a-tour-of-machine-learning-algorithms/

Bryant, A. (2014). Thinking about the information age. *Informatics*, *1*(3), 190–195. doi:10.3390/informatics1030190

Chen, M., Mao, S., & Liu, Y. (2014). Big data: A survey. *Mobile Networks and Applications*, *19*(2), 171–209. doi:10.100711036-013-0489-0

Clark, W. A., & Farley, B. G. (1955). Generalization of Pattern Recognition in a Self-Organizing System. *Western Joint Computer Conference*, 86-91. 10.1145/1455292.1455309

Dong, Z., Zhang, P., Ma, J., Zhao, J., Ali, M., Meng, K., & Yin, X. (2009). *Emerging techniques in power system analysis*. Beijing: Springer, Higher Education Press.

Ertel, W. (2017). *Introduction to artificial intelligence*. Springer International Publishing. doi:10.1007/978-3-319-58487-4

Farley, B. G., & Clark, W. A. (1954). Simulation of self-organizing systems by digital computer. *Transaction of the IRE*, 76-84. 10.1109/TIT.1954.1057468

Fikri, M., Cheddadi, B., Sabri, O., Haidi, T., Abdelaziz, B., & Majdoub, M. (2018). *Power Flow Analysis by Numerical Techniques and Artificial Neural Networks. IEEE, Renewable Energies, Power Systems & Green Inclusive Economy*. Casablanca, Morocco: REPS-GIE.

Frantz, R. J., & Hunt, S. R. (2003). *System and method for monitoring gas turbine plants*. General Electric Company. Retrieved from https://patents.google.com/patent/US6542856B2/en

Giloi, W. K. (1997). Konrad Zuse's Plankalkül: The first high-level, "non von Neumann" programming language. *IEEE Annals of the History of Computing*, *19*(2), 17–24. doi:10.1109/85.586068

Google. (2019). Retrieved from https://ai.google/research

Grover, V., Chiang, R. H. L., Liang, T. P., & Zhang, D. (2018). Creating strategic business value from big data analytics: A research framework. *Journal of Management Information Systems*, *35*(2), 388–423. doi:10.1080/07421222.2018.1451951

Haykin, S. (2009). *Neural Networks and Learning Machines*. Pearson-Prentice Hall Publish.

Hertz, J., Krogh, A., & Palmer, R. G. (2018). *Introduction to the theory of neural computation*. CRC Press. doi:10.1201/9780429499661

Hessine, M. B., & Saber, S. B. (2014). Accurate fault classifier and locator for EHV transmission lines based on artificial neural networks. *Mathematical Problems in Engineering*, *2*, 1–19. doi:10.1155/2014/240565

Hidayatullah, N. A., Stojcevski, B., & Kalam, A. (2011). Analysis of distributed generation systems, smart grid technologies and future motivators influencing change in the electricity sector. *Smart Grid and Renewable Energy*, *2*(03), 216–229. doi:10.4236gre.2011.23025

Jiang, Y., Liu, C. C., & Xu, Y. (2016). Smart distribution systems. *Energies*, *9*(4), 297–317. doi:10.3390/en9040297

Kadar, P. (2013). Application of optimization techniques in the power system control. *Acta Polytechnica Hungarica*, *10*(5), 221–236.

Khan, N., Yaqoob, I., Haskem, I. A. T., Inayat, Z., Ali, W. K. M., Alam, M., ... Gani, A. (2014). Big data: Survey, technologies, opportunities, and challenges. *The Scientific World Journal*, 1–18. PMID:25136682

Klaimi, J., Rahim-Amoud, R., Merghem-Boulahia, L., & Jrad, A. (2016). Energy Management Algorithms in Smart Grids: State of the Art and Emerging Trends. *International Journal of Artificial Intelligence and Applications*, 7(4), 25–45. doi:10.5121/ijaia.2016.7403

Li, D., & Du, Y. (2018). *Artificial Intelligence with uncertainty*. CRC Press, Taylor & Francis.

Li, L., Zheng, N. N., & Wang, F. Y. (2018). On the crossroad of artificial intelligence: A revisit to Alan Turing and Norbert Wiener. *IEEE Transactions on Cybernetics. Early Access*, 1-9.

Magesh, G., & Swarnalatha, P. (2017). Big data and its applications: A survey. *Research Journal of Pharmaceutical, Biological and Chemical Sciences*, 8(2), 2346–2358.

Manimuthu, A., & Ramadoss, R. (2019). Absolute Energy Routing and Real-Time Power Monitoring for Grid-Connected Distribution Networks. *IEEE Design & Test, 36*(2). Retrieved from https://ieeexplore.ieee.org/abstract/document/8636507

Mathworks. (2019). *Machine learnin with Matlab*. Retrieved from https://www.mathworks.com/campaigns/offers/machine-learning-with-matlab.html

McCarthy, J. (1992). Reminiscences on the history of time sharing. *IEEE Annals of the History of Computing*, 14(1), 19–24.

Mijwel, M. M. (2015). *History of artificial intelligence*. Retrieved from https://www.researchgate.net/publication/322234922_History_of_Artificial_Intelligence

Min, H. (2010). Artificial intelligence in supply chain management: Theory and applications. *International Journal of Logistics: Research and Applications*, 13(1), 13–39. doi:10.1080/13675560902736537

Mocanu, E., Nguyen, P. H., Gibescu, M., & Kling, W. L. (2016). Deep learning for estimating building energy consumption. *Sustainable Energy. Grids and Networks*, 6, 91–99.

Nogay, H. S. (2016). Determination of leakage reactance in monophase transformers using by cascaded neural network. *Balkan Journal of Electrical & Computer Engineering*, 4(2), 89–96.

Nogay, H. S., Akinci, T. C., & Eidukeviciute, M. (2012). Application of artificial neural networks for short term wind speed forecasting in Mardin, Turkey. *Journal of Energy in Southern Africa*, 23(4), 2–7. doi:10.17159/2413-3051/2012/v23i4a3173

Ongsakul, W., & Vo, D. N. (2013). *Artificial Intelligence in Power System Optimization*. CRC Press.

Paluszek, M., & Tomas, S. (2017). *MATLAB machine learning*. Apress Pub. doi:10.1007/978-1-4842-2250-8

Papageorgiou, A. (2019). *Exploring the meaning of AI, data science and Machine Learning with the latest Wikipedia Clickstream*. Retrieved from https://towardsdatascience.com/exploring-the-meaning-of-ai-data-science-and-machine-learning-with-the-latest-wikipedia-5fea5f0a2d46

Paracha, Z. J., Kalam, A., & Ali, R. (2009). A novel approach of harmonic analysis in power distribution networks using artificial intelligence. In *International Conference on Information and Communication Technologies*, (157-160). Karachi, Pakistan: Academic Press. 10.1109/ICICT.2009.5267198

Pitts, W., & McCulloch, W. S. (1947). How we know universals, the perception of auditory and visual forms. *The Bulletin of Mathematical Biophysics, 9*(3), 127–147. doi:10.1007/BF02478291 PMID:20262674

Rajarman, V. (2014). John McCarthy-father of artificial intelligence. *Asia Pacific Mathematics Newsletter., 4*(3), 15–20.

Richard, N. (2018). *The differences between Artificial and Biological Neural Networks*. Retrieved from https://towardsdatascience.com/the-differences-between-artificial-and-biological-neural-networks-a8b46db828b7

Russell, S. J., & Norvig, P. (2016). *Artificial intelligence a modern approach*. Pearson Education Limited.

Sahu, D. K. (2018). *Supervised and unsupervised learning in data mining*. Retrieved from https://www.digitalvidya.com/blog/supervised-and-unsupervised-learning-in-data-mining/

Samuel, S. J., Koundinya, P. V. P., Sashidhar, K., & Bharathi, C. R. (2015). A Survey on big data and its research challenges. *Journal of Engineering and Applied Sciences (Asian Research Publishing Network), 10*(8), 3343–3347.

Schmidhuber, J. (2015). Deep Learning in Neural Networks: An Overview. *Neural Networks, 61*, 85–117. doi:10.1016/j.neunet.2014.09.003 PMID:25462637

Shahid, N., Rappon, T., & Berta, W. (2019). Applications of artificial neural networks in health care organizational decision-making: A scoping review. *PLoS One, 14*(2), 1–22. doi:10.1371/journal.pone.0212356 PMID:30779785

Shanmuganathan, S., & Samarasinghe, S. (2016). *Artificial neural network modelling*. Springer International Publishing. doi:10.1007/978-3-319-28495-8

Suk, H. I. (2017). An introduction to neural networks and deep learning. In *Deep Learning for Medical Image Analysis*. Academic Press.

Turing, A. M. (1950). Computing machinery and intelligence. *Mind, 4*(256), 433–460. doi:10.1093/mind/LIX.236.433

Uliyar, S. (2017). *A Primer: oracle intelligent bots - powered by artificial intelligence*. Oracle. Retrieved from http://www.oracle.com/us/technologies/mobile/chatbots-primer-3899595.pdf

Upwork. (2019). *Overview of artificial intelligence and natural language processing*. Retrieved from https://www.upwork.com/hiring/for-clients/artificial-intelligence-and-natural-language-processing-in-big-data/

Wang, P., Liu, B., & Hong, T. (2016). Electric load forecasting with recency effect: A big data approach. *International Journal of Forecasting, 32*(3), 585–597. doi:10.1016/j.ijforecast.2015.09.006

Witten, I. H., Frank, E., Hall, M. A., & Christopher, J. P. (2017). *Data mining: Practical machine learning tools and techniques*. Cambridge, MA: Morgan Kaufman-Elsevier.

Yuce, H., & Avci, K. (2017). Establishment of diagnosing faults and monitoring system with neural networks in air conditioning systems. *The Journal of Cognitive Systems*, 2(1), 63–69.

Zhang, Y. Z., Huang, T., & Bompard, E. F. (2018). Big data analytics in smart grids: A review. *Energy Informatics*, 1(8), 1–24.

Zhao, Z., & Zhang, X. (2016). Artificial intelligence applications in power system. Advances in intelligent systems research. In 2ⁿᵈ *International Conference on Artificial Intelligence and Industrial Engineering*, (133, 158-161). Atlantis Press.

Zhou, K., Fu, C., & Yang, S. (2016). Big data driven smart energy management: From big data to big insights. *Renewable & Sustainable Energy Reviews*, 56, 215–225. doi:10.1016/j.rser.2015.11.050

Zimmer, M., & Kurlanda, K. K. (2017). *Internet research ethics for the social age: New challenges, cases, and contexts*. Peter Lang International Academic Publishers. doi:10.3726/b11077

Chapter 10
Blockchain and Its Integration as a Disruptive Technology

Dhanalakshmi Senthilkumar
https://orcid.org/0000-0003-0363-5370
Malla Reddy Engineering College (Autonomous), India

ABSTRACT

Blockchain is the process of development in bitcoin. It's a digitized, decentralized, distributed ledger of cryptocurrency transactions. The central authorities secure that transaction with other users to validate transactions and record data, data is encrypted and immutable format with secured manner. The cryptography systems make use for securing the process of recording transactions in private and public key pair with ensuring secrecy and authenticity. Ensuring bitcoin transaction, to be processed in network, and ensuring transaction used for elliptic curve digital signature algorithm, all transactions are valid and in chronological order. The blockchain systems potential to transform financial and model of governance. In Blockchain, databases hold their information in an encrypted state, that only the private keys must be kept, so these AI algorithms are expected to increasingly be used, whether financial transactions are fraudulent, and should be blocked or investigated.

INTRODUCTION

Blockchain is a decentralized, distributed database used to maintain a continuously growing list of records, called blocks. Each block contains a timestamp and a link to a previous block. Blockchain serves as an open distributed ledger that can record transactions between two parties efficiently and in a verifiable and permanent way (Bruyn, 2017). Blockchain technology uses bitcoin; bitcoin is the first official cryptocurrency in the form of electronic cash. Bitcoin and its underlying Blockchain technology were first conceptualized by Nakamoto (2008) but implemented in 2009, as a core peer-to-peer version of an electronic cash system. These cash systems allow online payments to send directly from one party to another party without the central trusted authorities like bank systems or payment services (Nakamoto, 2008). Blockchain consists of a peer-to-peer network. It consists of a network of nodes that maintain a decentralized shared database of records. Records on the other hand, transfer the transactions of bitcoin

DOI: 10.4018/978-1-5225-9687-5.ch010

cryptocurrency between participating parties. Each party in the transaction has a public and private key pair of public key infrastructure (PKI). Transaction parties sign the transactions using their private key, verified by other parties using a public key. The transactions are broadcast to all the other nodes in the network (Shen & Pena-Mora, 2018). This permits bitcoin to be used like other assets in exchange for goods and services. Additionally, it is easily portable, divisible and irreversible. Bitcoin uses blockchain technology to maintain its public ledger of every single transaction ever made with Bitcoin.

In the understanding of blockchain technology, four kinds of keywords are essential. The four keywords are open, distributed or decentralized ledger, efficient, verifiable and permanent. The first keyword is open; whatever information you are putting inside the blockchain should be accessible to all; everyone will be able to observe and validate that information. The second keyword is a distributed ledger, where a copy of that public ledger to every individual party who is there in the platform as well as their communicating with each other is kept, so that the platform can be either distributed or decentralized based on a given application. The third keyword is efficiently. It is important to ensure the efficiency of the information and efficiency of the protocol. Furthermore, the protocol needs to be fast and scalable. The fourth keyword is verifiable. It is a crucial keyword that permits all on the network to check the validity of information. The final keyword is permanent, which indicates that all information registered in the blockchain remains persistent. It is sometimes referred to as tamper proof (Iansiti & Lakhani, 2017). Tamper proof signifies that once the information is registered into the blockchain, the information will not be able to be modified nor updated in future time. The blockchain technique thus ensures that all bitcoin transactions are recorded, organized and stored in cryptographically secured blocks, chained in a verifiable and persistent manner.

In the course of blockchain implementation, three basic capabilities are supported in the bitcoin network techniques; They are; a) hash chained storage where two fundamental building blocks can be used for hash chained storage capabilities, namely, hash pointer and merkle trees, b) digital signature: by using cryptographic algorithm, establishes the validity of data items, and c) commitment consensus: this technique secures the expansion of the global ledger and precludes malicious attacks (Zhang, Xue & Liu, 2019).

Blockchain can be classified into three categories: Public blockchain, Consortium blockchain, and Private blockchain.

1. A public blockchain enables anyone to read, send or receive transactions and allows any participant to join the consensus procedure of making the decision on which blocks contain correct transactions and get added to the blockchain.
2. A Consortium blockchain can read any participant in the network but write permissions only to a pre-selected set of participants in the network who control the consensus process, and the third blockchain type is,
3. In a private blockchain, write permissions are restricted strictly to a single participant, read permissions are open to the public or constrained to a subset of participants in the network, where only chosen players have the rights to join the network which then creates a closed loop environment (Zhang et al, 2019). Bitcoin is a public Blockchain, because it was designed completely open, decentralized, and permissionless, that means anyone can participate without establishing an identity and no central authority (Bojana, Elena & Anastas, 2017).

In this study, the author considers the recent surge in blockchain interest as an alternative to traditional centralized systems, and consider the emerging applications thereof.

In particular, key techniques required for blockchain implementation are assessed, offering a primer to guide research practitioners.

This is undertaken by firstly outlining the general blockchain framework and subsequently providing a detailed review of its component data and network structures.

Additionally, the breadth of applications to which blockchain has been applied is considered.

Finally, the various challenges to blockchain and its integration as a disruptive technology is assessed.

BACKGROUND

Blockchain technologies is not just one technique, containing cryptography, mathematics and algorithm but all the techniques combining peer-to-peer networks, using distributed consensus algorithm to solve distributed database synchronize problem (Lin & Liao, 2017). The Blockchain technologies derive the name from the essential data structure, a shared and distributed database and transactions which are there in chronological order. Whether public or private, transactions of blocks connected through a linked list manner; each element in the list point to the previous block, each pointer of a block contains the hash of the previous block, forming a chain of records in sequencing order of events, is blockchain (Andoni, Robu, Flynn, Abram, Geach, Jenkins, McCallum & Peacock, 2019). The Blockchain technology consists of different key element; the key elements are i) Decentralized – the data can be recorded by a Blockchain system and distributively updated, ii) Transparent – the data can be transparent to each node and also update the data, iii) Open Source – Blockchain system is open to everyone, publicly the record can be checked, iv) Autonomy – In Blockchain system, every node can transfer or modify the data safely, v) Immutable – records will be forever and can't be changed unless someone can take and vi) Anonymity – user know the blockchain address. Blockchain technology works with the process of sending a data node to the receiving data node via broadcasting networks. After receiving, the data node checks the message from those data and if the checked message was correct, then it will be stored with the use of Proof-of-Work (PoW) or Proof-of-Stake (PoS) consensus algorithm the node to be executed (Lin & Liao, 2019).

Blocks

Blockchain is a decentralized and trustful database that contains all records of events or transactions that have been executed and shared between participating parties (Shrier, Sharma & Pentland, 2016a). Blockchain came from the chained list of blocks in data structure. The chained blocks distribute over a peer to peer network where every node maintains the latest version, in which each block can contain information about transactions. The structure of block consists of two fields; every block has a header field and body field (Mahdi & Maaruf, 2018). The block has two components, block header and list of transactions or block body. The first components of a block header are; A block where in a blockchain is a collection of data recording the transactions and the second components are a list of transactions or block body, which represents the action triggered by the participant. The block body is composed of a transaction counters and transactions (Zheng, Xie, Dai & Wang, 2018).

The block header contains metadata to verify the validity of this block. It has a unique identifier called block header hash and inside the header are three main components;

- Previous Block Hash
- Timestamp, Difficulty and Nonce
- Merkle Tree Root

The metadata every block inherits, comes from previous block. Previous block's hash create the new block's hash and make them tamper proof. To generate the hash for Bitcoin mining and merkle tree is root of all transactions. Blockchain design structure is organized and used to construct the block hash. If there are changes in any transaction in the header, then the need to change all subsequent blocks hash and finally all transaction will result in changes at the block header. The headers of all subsequent blocks are connected in a chain. The Blockchain network maintains the local copy of Blockchain (Zheng, Xie, Dai & Wang, 2017). Figure 1 shows the structure of block header in Blockchain. The block header in typical metadata contains; block version, previous block header hash, merkle tree root hash, timestamp, nBits, Nonce and parent block hash.

- **Block Version:** current version of the block structure, indicates which set of block validation rule to follow
- **Previous Block Header Hash:** the reference of block parent block
- **Merkle Tree Root hash:** cryptographic hash value of all the transactions in the block
- **Timestamp**: current time as seconds in universal time, the time that this block was created
- **nBits**: target threshold of a valid block hash (current difficulty that was used to create this block)
- **Nonce**: it's a 4-byte field, which usually starts with 0 and increases for every hash calculation
- **Parent Block Hash**: 256-bit hashes value that point to the previous block

These fields constitute the block header while the rest of a block contains transactions and users create transactions and submit to the network. The block body is composed of a transaction counter and transactions. The maximum number of transactions that a block can contain depends on the block size and the size of each transaction. Blockchain uses an asymmetric cryptography mechanism to validate the authentication of transactions. Blocks are chained together with previous blocks header. The Blockchains are probabilistic systems as a block is created and set around the network, then each node processes the block and decides to fit in the current overarching Blockchain ledger (Zheng et al., 2018).

Layers of the Blockchain

Blockchain should be decomposed into separate layers. Each layer is more abstract than the lower layer. It allows for robust system design because each layer can be upgraded, patched or even completely swapped out without affecting other layers. A simplified summary on the internet is the OSI model; The physical layer - transports the bits, the network layer - manages addressing and routing of packets between different physical routers (IP), the transport layer - manages raw connection state (TCP & UDP), the session layer - manages higher level connection state (HTTP), and the application layer – manages the actual applications live, ex. Google search, Face book (David, 2017). Blockchain protocol specifies two conceptual parts; the low-level protocol – node to node communication, message format and high-level

Figure 1. Structure of Block Header in Blockchain (Rosic, 2017)

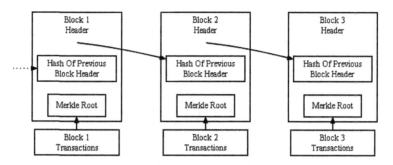

protocol – blockchain overlay network management and blockchain data exchange strategy (Alex, 2015). Figure 2. shows that OSI seven-layer models with blockchain protocol architecture. The lowest level on blockchain message exchange is like TLS which specifies handshake logic to right protocol and compatibility between nodes. The next layer above is the blockchain overlay network which contains higher level functionality, like the discovery of new and management of overlay networks related to blockchains.

Hashing Mechanisms

A Blockchain is organized using hash pointers. The hash pointers link to the data blocks. The data blocks can be used to check whether the data has been tampered or not. The hash of the data points to which location the data is stored. The hash pointer pointing to the predecessor block and each block indicates the address and which predecessor blocks the data must be stored. The user verifies the hash of the stored data and proves that stored data has not been tampered (Zhang, Xue & Liu, 2019). Example H(x)

Figure 2. OSI 7 layers in Blockchain Protocols (Oberhauser, 2015)

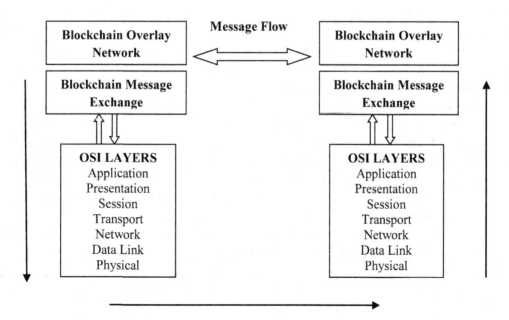

Figure 3. Sequence of Timestamp in Hash (Haber & Stornetta, 1991)

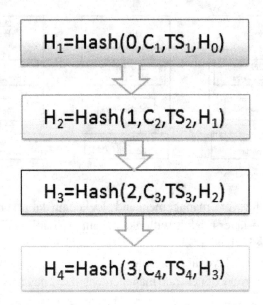

= x % n, where x and n are integers and % is the modular (remainder after division by n) operations. X can be of any arbitrary length, but H(x) is within the range [0, n-1]. X is message and H(X) is called the message digest. Figure 3. represents that sequence of a times tamp order in hash functions.

The cryptographically secured chain of blocks uses a time-stamp for a digital document. A sequence of timestamp denoting that (TS_1, TS_2 and so on) the documents are created or edited. Whenever a client accesses a document H=Hash ($0, C_1, TS_1, H_0$), it indicates that 0 for a construct block consisting of the sequence number of access, C1 – client ID, TS1 – timestamp value and H0 indicates that a hash value from the previous request; and the entire hash lists to connect it to the previous blocks. Bitcoin network uses hash functions like SHA-256, SHA-256; input message can be transforms into 256-bit message digest (Lee, 2015). The SHA-256 is a cryptographic hash function; the hash function takes as input its random size and produces the output as a fixed size. It is impossible to use the output of hash function to reconstruct it's given input. Bitcoin network utilize SHA-256 hash functions in two main ways; Mining and Creation of bitcoin addresses. The first way of hash function in a Bitcoin network is Mining, which is a method used to secure and verify the Bitcoin network and Bitcoin transactions; it's a peer-to-peer computer process. This process is used to describe adding transaction records to the bitcoin blockchain. They must first operate as mining node and successfully setting up mining node, the individual can construct candidate blocks. The blocks have six parameters; version, previous block hash, merkle root, timestamp, target and nonce. Based on the six parameters construct, the block is called block header. After successful construction of block, the miner can begin with the mining process. The second way of hash function is the creation of bitcoin addresses. The bitcoin address is 160 bits long of the public portion of public/private ECDSA key pair; and to produce the bitcoin address, select the random number as private key and multiply using an ECDSA to produce a public key. Where K = the public key and A = Bitcoin address: A = RIPEMD-160(SHA-256(K)). The creation of a Bitcoin address consists of two SHA-256 (Public key) and RIPEMD-160 (encrypting the result of public key) hashing algorithms (Bisola, 2018).

Merkle Tree

The blockchain may construct a binary search tree with its tree nodes linked to one another; the tree is constructed from bottom to top, replacing the regular pointers by hash pointers, to obtain a merkle tree. In merkle tree, every leaf node is labeled with hash of data blocks and the intermediate nodes contain the cumulative hash pointers to the respective sub trees in hierarchical fashion. The hash pointers are connected to the linked list of blocks. The hash pointer of the root node is the constant size for the whole tree (Top Hash) and in case of network hash pointer of the root node stored as a distributed fashion for every entity (Gandhi & Ramasastri, 2017). Merkle trees provide for efficient and secure verification of large amounts of data. Merkle tree uses; Peer- to-Peer Networks (Data blocks received in undamaged and unaltered) and bitcoin implementations (shared information are unaltered; no one can lie about a transaction). It is designed to ensure blocks of data can receive from other peers in a peer to peer network. The merkle tree construction algorithm for each time pair of lower level nodes is grouped into one at the parent level and contains the hash value of each. Figure 4. represents that Merkle Tree as a Binary Tree in Hash Pointers. This process is repeated until reaching the root of the tree, even if we change the hash value of its parent node, we need to change all nodes on the path of the bottom to top (Zhang et al., 2019). Merkle trees allow small and simple smart phones, laptop and even internet of things device to powerful computers to run a blockchain (Mattias, 2017).

Bitcoin Transaction

Bitcoin was the first application that introduced Blockchain technology; it's distributed, cryptographic digital currency was developed by an open source community, was proposed in 2008 by Nakamoto (2008). Bitcoin system is a decentralized environment for cryptocurrency; the participants can buy and exchange goods with digital money (Yli-Huumo, Ko, Choj, Park, & Smolander, 2016). The idea of the Bitcoin system is that earlier transaction history is verified in a cryptographic manner. In Bitcoin Blockchain, transfer transactions of Bitcoin cryptocurrency between participating parties has a public

Figure 4. Merkle Tree in Hash Pointers (Merkle, 1979)

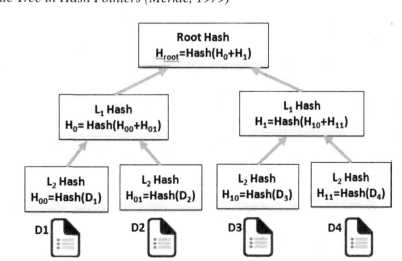

key infrastructure (PKI) mechanism (Shen et al., 2018). Bitcoin uses the public key infrastructure and the PKI user has a pair of private and public key. The address of the Bitcoin wallet is used for public key and authentication of users is the private key, so that transaction of the Bitcoin consists of sender and receiver of key pair based on the value to be transferred (Yli-Huumo et al., 2016). The transactions are broadcast to all peer nodes in the network, using consensus mechanism. Peer nodes check the valid transactions and those valid transactions are placed in a sequence manner, organized into blocks forming a linked chain (Shen et al., 2018). The entire history of Bitcoin transactions in the transactions systems is recorded in a time series of unspent transaction outputs model (UTXO). The benefits of the UTXO online transaction model; a) potentially high degree of privacy, b) potentially high degree of scalability, and c) potentially high degree of security (Zhang et al.,2019). The Bitcoin blockchain size is approximately 149 GB. The bitcoin transaction life cycle flows in the form of sender, network, miners and receiver. (Sender of Alice sends the message and Bitcoin Wallet to the Receiver of Bob. Between Alice to the Bob transaction, the life cycle of bitcoin performs in different sections). Figure 5. shows that the transaction life cycle in bitcoin technology. The following procedure represents that bitcoin transaction life cycle;

1. **The Sender (Alice):** Opens her bitcoin wallet then provides the Bob address and which amount to transfer to receiver
2. **The Network**: Wallet constructs the transaction, signs using Alice's private key and broadcast it to the network. Network nodes validate the transaction based on existing blockchain and propagate the transaction to the miners; miners include the transaction to the next block to be mined.
3. **The Miners**: It collects all the transactions for the time duration (10 minutes), then constructs a new block and tries to connect it with the existing blockchain through a cryptographic hash computation in the mining procedure; once the mining is over, the hash is obtained and the existing blockchain included in the blockchain, then the updated blockchain is propagated in the network
4. **The Receiver:** Bob opens his bitcoin wallet and refreshes whereby he gets the updated blockchain and the transaction reflects at Bob's wallet.

Figure 5. Bitcoin Transaction Life Cycle (Botjes, 2017)

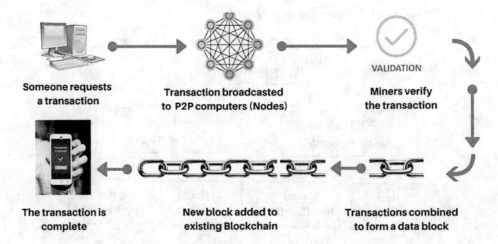

Smart Contracts

Smart Contract concept can be implemented with blockchain technology. Smart Contract is a second generation of Blockchains commonly known as Blockchain 2.0. Smart Contract is a computerized transaction protocol that executes the terms of contract (Szabo, 1997). Smart Contract can be made between entrusted, anonymous people. The execution of the contractual term is automatic and does not rely on third party for settling transactions and disagreements among pseudonymous participates. Smart Contracts use cryptographic mechanisms to enhance the security. The requirement of smart contracts techniques are; a) Blockchain and other distributed ledgers to maintain an immutable record of data, b) inherit the encryption pseudonymity of blockchains, c) able to prevent the unwanted monitoring and tracking, d) flexibility of programming languages and e) finally the potential to reduce risks, improve efficiency and save costs. (Hu, Liyanage, Manzoor, Thilakarathna, Jourjoin, Seneviratne & Yilanttila, 2018).

The Smart Contract code is stored on Blockchain and each contract is identified by unique addresses, users to operate it and they just send a transaction to this address. The correct execution of transaction address is enforced by blockchain consensus protocol (Reyna, Martin, Chen, Soler & Diaz, 2018). All the transaction information is present in smart contract and it executes automatically. Solidity is used for developing blockchain platforms in high level language. Smart contract is a blockchain technology that runs on executable code to facilitate, execute and enforce the terms of an agreement once the specified conditions are met, smart contract executes automatically in terms of an agreement. The Smart Contracts technique consists of four main tasks, the tasks are: Contractual Agreements, External Information Supply, Execution of Smart Contract and Settlement. Figure 6 represents that basic principle of Smart Contract systems which are contractual agreements, external information, execution and settlement (Hans, Zuber, Rizk & Steinmetz, 2017).

- **Contractual Agreements**: Contract terms are translated to executable code that can be evaluated automatically and independently
- **External Information**: Trustful data source to provide secure input information to the smart contracts
- **Execution**: Validate the execution of smart contracts, whether the predefined conditions are met
- **Settlement:** Settlements can be performed by tracking account modifications on the blockchain

Figure 6. Basic Principles of Smart Contract Systems (Botjes, 2017)

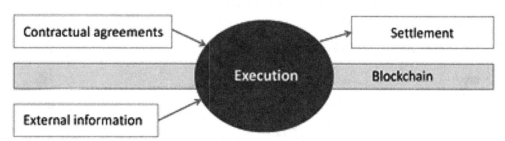

Digital Signature

Digital Signature is the digital identity and its equivalent to handwritten signatures but it's much more secure than a handwritten signature; it establishes the validity of a piece of data by using cryptographic algorithms. The cryptographic algorithm is associated with a digital object. These digital objects can verify but has not been altered. Digital signature formulates three core components; the first core component is key generation algorithms where each user creates two pair of keys for private key and public key. The private key is used to sign the transactions which is kept in confidential, digital signed transactions which are transmitted throughout the entire network called private key. The second core components are signing the algorithm where it produces a signature on the input message using a private key, and the third core component is verification of algorithms. This algorithm public key is an input, the input takes a signature and a message, using public key and validates the message and returns a Boolean value. Signatures are used to verify the user's identity and does not rely on a centralized party (Zhang et al., 2019). Figure 7. is an example of a digital signature used in Blockchain for two phases. For instance, a user Alice wants to send another user Bob a message. (1) In the signing phase, Alice encrypts her data with her private key and sends Bob the encrypted result and original data. (2) In the verification phase, Bob validates the value with Alice's public key. In the digital signature algorithms used in Blockchain is Elliptic Curve Digital Signature Algorithm (ECDSA). In ECDSA algorithms, private key and public key are generated in binary formats (generated public key is used as the users address and private key is used as password to spend the money). (Zheng et al., 2017)

Blockchain Consensus Algorithms

Important component of blockchain technology is consensus algorithms. Consensus means "general agreement", general agreement needed to record information, such as balance of every address and transactions on the blockchain. Consensus algorithms is open source, decentralized platform and decision-making process for a group, where individuals of the group construct and support decision for that works. Most of the decision need to support the individuals, whether they like it or not. Being a decentralized system, Blockchain systems do not need a third-party trusted authority (Li, Jiang, Chen, Luo & Wen, 2018). These algorithms are pieces of code that make sure there is concurrency over the distributed data-set and it's used to achieve agreement on a single data value. Blockchain consensus algorithms consists of different objectives; the objectives are agreement (the mechanism gathers all the agreements from the group), collaboration (every one of the group are in agreement), equal rights (every

Figure 7. Phases in Digital Signature (Zheng et al., 2017)

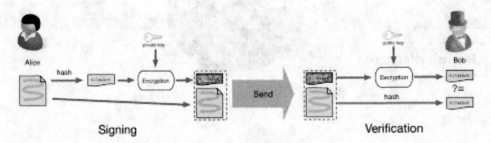

270

single participant has the same value in voting), co-operation (every individual will work as team), participation (everyone needs to participate in voting, no one will be left out or can stay out without a vote) and activity (every member of the group is equally active). The purpose of consensus algorithm is any one can be a node and remain anonymous. It's possible for a node to tamper with transactions and add them in a new block then the blockchain can end up with what we call fork. The main thing is to avoid such forks so that everyone agrees to a single version of the truth. Some possible attacks on blockchain is decentralization and immutable records. It's a distributed database on the network called nodes and is maintained in a shared manner. All nodes are complete ledgers and the ledgers have entire transactions on the blockchain called a distributed ledger network. The most commonly used consensus mechanisms for blockchain technologies are proof of work (PoW), Proof-of-Stake (PoS), Delegated Proof-of-Stake (DPoS) and Practical Byzantine fault Tolerance (PBFT).

Proof-of-Work (PoW)

PoW is Proof-of-Work which is the original and the very first implementation of a distributed and trustless consensus algorithm in blockchain network called PoW algorithm (Table 1). This algorithm is used to confirm transactions and produce new blocks to the chain. Figure 8 represents that PoW consensus mechanism. These mechanisms prove the credibility of the data to solve the solution of puzzle. The proving mechanisms is usually hard but easily verifies the problem for that puzzle. After resolving a PoW puzzle, it will broadcast to other nodes. The other nodes verify the puzzle then to achieve the consensus mechanisms (Li et al., 2018).

The PoW consensus mechanisms required to compute the puzzle are miners to solve complex cryptographic puzzles so they can add a block to the blockchain. Each block is added to the blockchain and you must follow the certain set of consensus rules and if the block does not follow consensus rules, it will be rejected by the network nodes. The synthesis of PoW consensus algorithms and consensus rules produce a reliable network and a shared state of blockchain can be achieved (Li et al., 2018).

Figure 8. Consensus Mechanisms in PoW (Li et al., 2018)

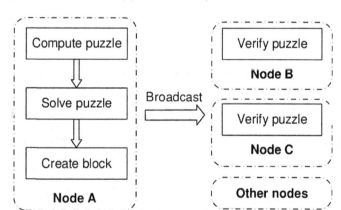

Proof-of-Stake (PoS)

PoS are proof of stake consensus algorithms used by cryptocurrencies to validate blocks (Table 1). The purpose is like proof of work, but a different way to verify and validate the transactions or blocks. Once the user is selected to validate the block and accurately verify all transactions in that block, then they are rewarded a certain amount of digital currency for their work. PoS replace computational work in the random selection process (Andoni et al., 2019). The first idea was proof of stake in bitcoin talk in 2011, but the first PoS based currency was PeerCoin in 2012 together with ShadowCash, Nxt, BlackCoin, NuShares/NuBits, Qors and Nav Coin. This approach works by calculating the weight of a node according to its currency staking or holding only, which means more power and electricity saving. These two approaches, proof of work and proof of stake can be used for any crypto currency. Proof-of-Work method will cause a lot of electricity power and computing power wasted, but PoS doesn't need expensive computing power as the resources are compared to Bitcoin and this method provides increased protection from malicious attack on the network (Lin & Liao, 2017).

Delegated Proof-of-Stake (DpoS)

Delegated Proof-of-Stake (DPoS) which is like PoS, miners get their priority to generate the blocks according to their stake (Table 1). Stakeholders elect their delegates to generate and validate a block. This process is termed as very reliable and most efficient consensus algorithms in blockchain networks and it's a method for validating transactions and adding shared ledger of systems in the blockchain network (Supriya & Kulkarni, 2017). DPoS often described as representative democratic, but PoS is described as direct democratic. Using reputation algorithms and real-time voting, the Stakeholders elect their delegates to generate and validate a block. With significantly fewer nodes to validate the block, the validated blocks are chosen in random order and if one says no; the choice goes to next randomly chosen delegate making the transactions to be confirmed quickly. For tuned parameters of the block size and block intervals, users do not need to worry about the dishonest delegates because the delegates could be voted out easily. DPoS use more efficient algorithm than PoS algorithms. DPoS delegated users quickly remove bad actors and sanction bad behavior. Delegated Proof-of-Stake (DPoS) cryptocurrencies are Bit Shares (BTS), EOS, STEEM, LSK, OXY and ARK. The backbone of Delegated Proof-of-Stake (DPoS) algorithm is Bitshares. Table.1. indicates the Different Consensus Algorithms and its goals.

Table 1. Comparison of Consensus Algorithms (Rosic, 2018)

Name of Consensus Algorithms	Goals	Domain	Implementations
Proof-of-Work (PoW)	Barrier to publishing blocks, difficult puzzle to solve and enable transactions between untrusted participants.	Permissionless Systems	Bitcoin, Ethereum
Proof-of-Stake (PoS)	Verify and validate the publishing blocks of transactions and enable less intensive barrier but still enable transactions between untrusted participants.	Permissionless Systems	Ethereum, Casper, krypton
Delegated Proof- of-Stake (DPoS)	Enable a more efficient consensus model thro liquid democracy, using cryptographically signed messages and elect and revoke the rights of delegates to validate and secure the blockchain.	Permissioned Systems	Bitshares, STEEM, EOS

Practical Byzantine Fault Tolerance (PBFT)

It is an algorithm for solving the problem of low efficiency in Original Byzantine Fault Tolerance algorithm and it's a message-based consistency method, to run on three phase protocol; pre-prepare, prepare and commit (Zheng et al., 2018). A client sends the request from one peer to other peers for pre-prepare messages in turn of broadcast channels; in the next stage of prepare, a prepare message is multicasted to other nodes; when replica receives prepare message and it matches with the pre-prepare message, once the message is matched, it changes the state to committed and executes in message operation. The message is executed, reply information is sent to the client. So, for PBFT method, before reaching agreement depends on three stages for exchanging a message (Zheng et al., 2017).

Types of Blockchains

Generally, types of Blockchains depend on managed data, availability of data and what actions can be performed by the user. Blockchain technologies can be categorized into three types. Public Blockchain, Private Blockchain and Consortium Blockchain (Zheng et al., 2017; Buterin, 2015).

Public Blockchain

Public Blockchain has no single owner, all the data in the Public Blockchain is accessible and visible to anyone; their consensus process is open to all. Public Blockchain can receive and send transactions from anybody in the world, but everyone can check and verify the transaction, and also participate in the process of getting consensus. In a Public Blockchain, anyone can join the Blockchain without any approval and can act as a simple node or as a miner (node), but they are not only decentralized but fully distributed. It is easy to detect fraud on the chain (Karim, Ubar & Rubina, 2018). These types of Blockchains are usually given an economic incentive, such as in cryptocurrency networks, but usually use proof of work (all the nodes on the network can solve cryptographic puzzles) or proof of stake (verification does not rely on excessive mechanisms) for consensus mechanisms (Nadir Abdelrahman, 2018). Examples of Public Blockchain are Bitcoin, Ethereum, Monero, Dash, and Litecoin

Private Blockchain

Node will be restricted and not every node can participate in this Blockchain. Only enables chosen nodes to join the network and data can be accessed on strict authority management. It's a distributed yet centralized network. They are managed by one organization in a trusted party then it's used for private purposes. Each transaction can be write and verified on private Blockchain completely faster and allows much greater efficiency. It uses privileges to control who can read from and write to the Blockchain and is also called permissioned (Karim et al., 2018). These Blockchains usually built organizations for their specific business need (Nadir Abdelrahman, 2018). Hyperledger fabric and ripple are examples of private Blockchain networks.

Consortium Blockchain

A consortium blockchain is part private, part public. Only selected group of nodes can participate in the consensus process, which had the authority and can be chosen in advance (Ganne, 2018). On consortium blockchain, a pre-selected group of nodes control the consensus process, but the other nodes may be allowed to participate in creating new transactions / reviewing it. Its hybrid between the 'low-trust' provided by public blockchains and 'single highly-trusted entity' model of private blockchains. It can be established in business for business, financial sector, insurance companies, government institutions. The data in blockchain can be open or private and can be seen as partly decentralized. Like Hyperledger and R3 (Banks), EWF (Energy), B3i (Insurance), Corda are both consortium blockchains. Public blockchains are when no central entity is available to verify a transaction and full decentralization is needed. Private and consortium blockchains allow defining different permissions on different users on the network. Table 2. represents the classification of blockchains. The advantages of private and consortium Blockchains are; Such as lower validation costs and shorter validation times (given the fact that, because of the smaller number of nodes, the mathematical problem can be simplified), reduced risk of attacks (since nodes that validate transactions are known) and increased privacy (as read permissions could be granted only to selected nodes).

Distributed Ledger Technology

Distributed ledger technology (DLT) enables the maintenance of a global, data structure by a set of mutually untrusted participants in a distributed environment. Distributed ledger is suitable for tracking the ownership of digital assets and Bitcoin network. Bitcoin network is the most prominent application in DLT Systems (Bencic & Zarko, 2018). DLT is a fast-evolving approach to recording and sharing the data in multiple data stores. It's an innovative method for storing and updating the data within and between organizations. Normally the distributed system has no central administrator or centralized data

Table 2. Classification of Blockchains (Zheng et al., 2017 & Zhang et al., 2019)

Properties	Public Blockchain	Private Blockchain	Consortium Blockchain
Efficiency	Low	High	High
Consensus Process	Permission-less	Permissioned	Permissioned
Describe	Anyone can participate and accessible worldwide	Controlled by an organization	Controlled by pre-selected nodes within consortium
Centralized	No	Yes	Partial
Access	Public	Restricted	Restricted
Determination	Miners	One Organization	Selected set of nodes
Network	Decentralized	Distributed	Traditional centralized systems
Example	Bitcoin, Etherum	Hyperledger Fabric, Ripple	Corda, R3, B3i
Speed	Slower	Faster	Medium
SoC (Speed of Consensus)	Slow	Fast	Slight Fast
#TA (Trust Authority)	0	1	≥ 1

storage, is an asset database that can be shared across a network of multiple sites requiring consensus of different users (Suvarna, 2018). The data can be shared and spread across multiple locations or nodes, all participants/user with in a network can have their own identical copy of the ledger, any changes to the ledger, within minutes or seconds can be reflected in all copies.

The Distributed Ledger Technology focuses on three key areas; these key areas are data coordination, cryptoeconomic internal incentive layers and integration into the digital commoditization of assets.

- **Data Coordination:** Within a system, how the information is distributed and allocated among the stakeholders
- **Crypto Economic Internal Incentive Layers:** To ensure the functionality of system, and different stake holders and users are motivated based on economic incentives
- **Integration into the Digital Commoditization of Assets:** How the systems can integrate into a digital goods (Rakesh, 2018).

Distributed Ledger Technology systems like IBM Fabric or R3 Corda are having similar functionalities as Blockchain systems, but it should be a separate subset of distributed ledgers. The distributed ledger technology has some functionality and use cases of these systems; the systems are; hyper ledger fabric, Corda R3. Main areas of these systems are; state, transactions and smart contract.

- **State:** It refers to the main unit of logic code, facilitate the representation of information in computing environment.
- **Transactions:** it can either initiate contracts or call upon pre-existing contracts, the generation of state or state transitions occur in development ecosystem.
- **Smart Contracts:** the individual units of code that execute actions within the platform ecosystem.

Hyperledger Fabric

In Hyperledger Fabric / IBM Fabric, replaces the key principles of Blockchain system, then maintaining the execution of all transactions in multichannel architecture to ensure high transaction throughput in a trusted environment. But IBM fabric is a distributed ledger technology, not a Blockchain (Revoredo, 2018). The technological features of Hyperledger Fabric systems are;

1. **State**, preserved on key/value pair's stores for the state. The key values are known as the world state, world state is indexed for view in to the transaction logs and then interaction between the chain code programs, and the database process that receives updates from the chain code APIs. Figure 9 shows that the IBM Fabric architecture; the state is nested within the database environment with traditional software development tools, the database is; Level DB (creates a key/value database) and Couch DB (would hold the document JSON database).
2. **Transactions** in IBM Fabric are all executed in fabric multichannel architecture, then the transactions are added to the share ledger to ensure high transaction throughput. This fabric allows for the read/write access environment. In hyperledger, fabric transactions deploy and invoke transactions; deploy transactions create new chain code and installs chain code in the software development environment and invoke transactions are previous chain code to be invoked for some functions. Functions that are successfully invoked; the chain code introduces changes to the state and that deploys and

Figure 9. IBM Fabric Architecture (Brent, 2018)

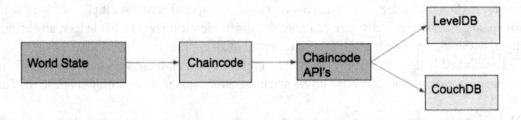

invokes transaction functions to activate chain code, and **iii) Smart Contract** in hyperledger fabric determine that code-based instructions to the system and the chain code is not deployed in smart contract, but run in secured Docker containers and code is written in Go or Node.js (Brent, 2018).

Corda R3

In R3 Corda architecture, the processing of shared data occurs in a partially reliable environment, this platform does not have the components of a Blockchain system, but Corda R3 systems just like Hyperledger Fabric, and it's a permissioned network. In R3 systems, participant of the network is selected in advance and access to the network is restricted (Revoredo, 2018). The technological features of Hyperledger Fabric systems are;

1. **State** is considered to include opaque data and this system is comparable to disk file not necessarily updated, then used to generate new successors. In R3 Corda of state consists of vault/database of consumed and unconsumed states,
2. **Transactions** in R3 Corda are considered to update the database vault and used for bitcoin ecosystem concepts, though the avoidance of double spending systems, and
3. **Smart Contract** in R3 Corda, smart contract classes implement in contract interface, written in Java/Kotlin and compiled through the Java Virtual Machine (JVM) where the main function of smart contract in R3 Corda is to verify function (Brent, 2018).

Permissioned and Permissionless Blockchain

Blockchain have two types such as Permissionless Blockchain (public) and Permissioned Blockchain (private).

Permissionless Blockchain

In Permissionless Blockchains, otherwise known as public Blockchains, the public Blockchains are open for all, which means anyone can join them to post transactions and to participate in the consensus process of adding a new block of a transactions system. Permissionless Blockchain is a decentralized ledger platform available to anyone publishing blocks, without permission from any authority. It's an open source downloaded software platform available to anyone to publish blocks. This Blockchains usually works on Proof-of-Work system, in which each block contains hash, a unique identifier of the data contained within the block. This system allows the timestamp server's implementation of P2P network; this network can read and write to the ledger (Nadir Abdelrahman, 2018). This permissionless Blockchain networks utilize the multiparty agreement or consensus system and creates smart contract, but it does not apply the public Blockchain Proof- of-Work (Dob, 2018). Bitcoin, Ethereum and Dash are some example of permissionless Blockchain.

The characteristics of permissionless models are;

- **Decentralization:** No central entity has the authority to edit the ledger, these models are only based on consensus protocols
- **Digital Assets:** Its presence of a financial system on the network
- **Anonymity:** The user does not submit the personal information
- **Permissionless Network**: the user identity is indirectly tied to the addresses
- **Transparency:** The transparent network needs to freely access all information of the users

Permissioned Blockchain

Permissioned Blockchains act as closed ecosystems. The users are not freely able to join the network, have special permission to access the chain, and check the record history or own transactions. It's preferred by centralized or decentralized organizations, where users publishing blocks must be authorized by some authority. Since only authorized users are maintaining the Blockchain, then its restricted read access and restricts who can issue transactions. They may restrict this access only to authorized individuals and instantiated using open source or closed source software. The permissioned Blockchains are also likely to be private Blockchains to securely record transactions and exchange information with one another. Permissioned Blockchains used by organizations that wish to work together, protect their Blockchains against how much they trust one another and invite business partners to record their transactions on a shared distributed ledger (Dob, 2018). Examples of permissioned Blockchain are Corda and Quorum.

The characteristics of permissionless models are;

- **Varying Decentralization**: For private Blockchain, it is entirely accepted if they are fully centralized or partially decentralized
- **Transparency and Anonymity**: Private Blockchains are not required to be transparent, but they can choose to operate freely, depending on the inner organizations of the businesses.
- **Governance:** In permissioned Blockchains, the governance is decided by members of the business network.

The resemblance between permissioned and permissionless Blockchain are; distributed through peer-to-peer network, unalterable for digitally signed ledgers, updated through consensus protocol and it's maintained immutable ledger. Table 2 indicates the Blockchain classifications. In public Blockchain, anyone can contribute and there is no need for third authority to grant permissions. Private Blockchain is a network where all the participants are known and trusted in the consensus process. The properties of public and private Blockchains are; platforms, actors, services, processes and data models (Table 3). The blockchain technology can be formed into four groups of Bitcoin, Multi Chain, Ethereum and Chain Core.

Table 3. Properties of Blockchain Classification (Dib, Brousmiche, Durand, Thea & Hamida, 2018)

Property	Public Blockchain (Permissionless Blockchain)	Private Blockchain (Permissioned Blockchain)
Consensus Type	**Consensus is managed by Public**	**Consensus is managed by a Single Owner**
Validation Transaction	Validated by any node (miner)	Validated by a list of authorized nodes (validators)
Consensus Algorithms	Without permission (PoW, PoS etc.,)	With permission (PBFT, PoA)
Network Scalability	High	Low to Medium
Transaction Throughput	Low (a few dozen of transactions validate per second)	High (hundred/thousands of transactions validate per second)
Infrastructure	Highly Decentralized	Distributed databases
Data Immutability	Rollback of blockchain is almost impossible	Rollback of blockchain is possible
With Smart Contracts	Ethereum	Chain Core
Transactions only	Bitcoin	Multicoin
Examples	Bitcoin, Ethereum, Dash, Lisk, Factom and Blockstream	Multichain, Blockstack, Quorum and Hyperledger

Certificate Authority

Certificate authority is a trusted entity, providing several certificate services to users of a blockchain, which issues digital certificates. It's a part of public key infrastructure (PKI) with Secure Sockets Layer and the layers authenticate content from web browsers to web servers. Certificate authorities classify that enrollment certificate authority (ECA) and transaction certificate authority (TCA). New users register with the blockchain network which enables registered users to request an enrollment certificate pair from the authority. These certificates are used for deploying chain code and invoking transaction from the chain code with the blockchain technology. Single transaction certificate can be used for multiple transactions. To secure their communication channels, users will need TLS certificate; these certificates can be requested from TLS certificate authority (TLSCA).

APPLICATIONS OF BLOCKCHAIN

Nakamoto (2008) published an invention which is Bitcoin in 2009. The Blockchain protocol was first designed for the development of Bitcoin cryptocurrency and has its ups and downs. The public ledger that records every Bitcoin transaction is known as a Blockchain. The Blockchain is now an exciting new alternative to traditional currency. Centralized banking and transaction methods are not only changing the way to handle financial transactions, but also offering alternative uses that will change the world. Blockchain is a distributed ledger that maintains a continuously-growing list of every transaction across every network distributed over tens of thousands of computers (Rampton, 2019). Blockchain application will directly compete with the market and other financial services institutions (Leandros, 2018). But the real potential of Blockchain goes far beyond the financial sector (Domingos, 2017).

Blockchain in Financial Services

The financial services are fundamentally about facilitating the trusted exchange of value between multiple, untrusting parties. Today financial service institutions are looking to Blockchain to enable more efficient cross-organizational collaboration, eliminate intermediaries and create disruptive business models. Using distributed ledger technology, Blockchain is a secure, shared, distributed ledger, and a new shared data structure, can record the transactions and work together to update the validates.

Cross Border Payments

Without the need of a third party, the transactions to allow for blockchain technology such as its transparency and anonymity of transactions, where every transaction is recorded on the blockchain, cannot be altered. It is unalterable because every transaction is recorded in blockchain so that risk of fraud is greatly reduced. Blockchain completely transforms how financial services work. It specifies the first production of the network of blockchain use cases cross border payments. Classic use case for bitcoin was created and perhaps of cryptocurrency which are really currencies in some way. In economics, a currency must follow the following criteria then it's accepted as a currency. Indian Rupee, US dollar or Euro all are currencies. In economics, the following criteria must be satisfied (Clayton, 2015).

- **Medium of Exchange**: Are merchants willing to accept the currency in exchange for goods and services
- **Unit of Account**: Unit of account is simply the unit by which the prices of all other items are quoted. It measures the real value of goods and services. The price of each item had to be quoted in terms of every other item
- **Store of Value:** When people are paid in money, they expect to be able to spend that money on a purchase (A mode of investment)

It almost never satisfies the unit of account to criteria and in some cases satisfies medium of exchange criteria. So, until it satisfies all these criteria, it is not going to be a currency, in the sense that economics economists think it. In the core of payments, the need for this kind of a currency that satisfies these properties is required to be able to do payables.

Stellar Consensus Protocol and Network

Stellar protocol is a decentralized hybrid blockchain platform with open membership. It provides a way to reach consensus without relying on a closed system to accurately record financial transactions. Its running on global networks where anyone can join this network, run a stellar node and they can be a participant in this network to execute transactions and store the state of payments or accounts in this network. The stellar distributed network can be used to track, hold and transfer any type of assets: Dollars, Euros, Bitcoin, Stocks, Gold and other tokens of value, any asset on the network can be traded and exchanged with any other. They have a cryptocurrency called Lumens (XLM) that is a native currency of the network and any account can issue assets on the stellar network which are called anchors. Anchors can run by individuals, small business, nonprofits, local communities and organizations etc. It is the protocol they have called Federated Byzantine Agreement (FBA) protocol, and it is an open membership where anybody can join in this protocol. When you submit a transaction for a payment on the stellar network, it will pass through in 2 to 5 seconds which is not bad for a blockchain platform. Network is formed by a set of anchors or they act as bridges that allow converting existing currency. These act as the currency converters and internally distributed exchange. If you are in India and want to transfer money to somebody in the US using dollars, then stellar network will do automatically within the network. Dollars will be converted at the lowest rate that is available. It's an interesting project and a good set of people are using that network.

Ripple Protocol and Network

Ripple is a decentralized, blockchain based on real-time gross settlement systems (RTGS), currency exchange and remittance network. It allows banking institutions to transfer large amounts of money freely and without delays on distributed networks. The Ripple project is older than Bitcoin itself. The original implementation was created by Ryan Fugger in 2004, who wanted to create a monetary system in a de-centralized manner and effectively create their own money and allow for individuals and communities (Buterin, 2017). It's the first open standard internet protocol-based technology for Payment Service Pro-viders (PSPs) to clear and settle transactions in distributed networks. Ripple protocol consensus algorithm is applied every few seconds by all nodes, to maintain the correctness and agreement of the network (Schwartz, Youngs & Britto, 2018). It confirms transactions in fewer than 4 seconds and it handles more than 10,000 transactions per second. Bitcoin is a Blockchain-based currency using mining and Ripple uses an iterative consensus ledger, validating servers' network along with XRP cryptocurrency tokens.

Blockchain in Compliance

Compliance is the use case of Blockchain in financial services with a process called KYC (Know Your Customer Process). It's a standard process to verify the identity of your business and then it protects your business from risks, designed to prevent fraud and is also frequently paired with Anti-Money Launder-ing (AML) verification. The KYC/AML process is implemented in financial institutions in blockchain technology. They play an integral role in client on boarding at financial services providers and are known as Know Your Customer (KYC) and Anti-Money Laundering (AML) processes. Blockchain's ability to securely record and store data in an immutable manner, share with permission to third parties, should come as no surprise using the distributed ledger technology. Know Your Customer process really is about collecting personal information. This information about your identity, when were you born, what is your address, where do you live, what do you do, where do you work maybe, whether you are a student are collected. And apart from that, they might collect other aspects on your financial standing for instance if you are opening a bank account. Once the KYC process is done, there is a lot of processes that hap-pen for AML. AML is an anti-money laundering and again there is a whole lot of due diligence that is done for every transaction that you perform. There is an anti-money laundering set of processes that happen to make sure you are not doing something suspicious or fraudulent right. The key participant in the network is a set of banks and customers could have identity on this network.

Blockchain in Mortgage

Mortgage use cases of blockchain in financial services. High-level overview and benefits of mortgage are used in blockchain to support mortgage processes. It can be viewed as a distributed database. A blockchain maintains a growing list of ordered records in blocks. The blocks have a timestamp and link to a previous block. The information's in the blocks are secured using public and private key cryptography. In blockchain, private key is used for digitally signed transactions and information for individual users. Cryptography makes these signatures impossible to forge or reproduce by other users. Additionally, each user has a public address, linked to his or her private key.

Blockchain in Supply Chain Management

Supply chain management in integrated planning with execution of different processes. The finished products or goods from one to another point are supply chain management. It's a network of individual entities, organizations, businesses, resources that technologies combine for manufacturing the product or service. The supply chain progresses delivering raw materials from supplier to manufacturer and ends by the final product to be delivered to the consumer. Blockchain technology in supply chain management can access the product information with the help of sensors and RFID (Radio Frequency Identification) tags. In RFID, a wireless technology uses transmitted radio signals to tag an item, track and trace its movement without human intervention. Each system of RFID contains an RF subsystem, consists of a reader (interrogator) and tag (transponder), its connected to a silicon chip with an antenna. The subsystem is supported through enterprise which is called middle-ware, networking services and analytic systems. A RFID tagged items and products can be tracked during the life cycle from the manufacturing final purchase of the finished goods. The subsystem utilizes that information stored over RFID tags. The supply chain management blockchain technology with usage of RFID systems, examined that to improve its effectiveness and mission performance.

DISRUPTIVE INNOVATION

Disruptive Innovation, a term by Clayton Christensen, describes a process by which a product or service takes root initially in simple applications at the bottom of a market and then relentlessly moves up market, eventually displacing established competitors. This has proved to be a powerful way of thinking about innovation driven growth (Christensen, 2019). Defined by Christensen, "Disruptive Innovation refers to a technological or market advance that fundamentally transforms or creates entirely new markets". The disruptive innovation techniques completely alter the existing techniques or entirely create the new market for products/services. Blockchain has the capability to operate both as a market disruptor and a new market creator (Leandros, 2018). Disruptive innovation describes a process by which a product or service takes root initially in simple applications at the bottom of a market and eventually displacing establishing competitors. There are four different types of innovation in Blockchain Disruption; a) Sustaining Innovation – does not significantly affect existing markets, b) Evolutionary Innovation – improves a product in an existing market, c) Revolutionary Innovation – unexpected but does not affect existing markets and d) Disruptive Innovation – creates a completely new market by providing a different set of values which ultimately overtakes an existing market (Christensen, 2019) - Table 4.

Table 4. Examples of Disruptive Innovation (Christensen, 2019)

Disruptor	Disruptee
Personal computers	Mainframe and mini computers
Mini mills	Integrated steel mills
Cellular phones	Fixed line telephony
Community colleges	Four-year colleges
Discount retailers	Full-service department stores
Retail medical clinics	Traditional doctor's offices

Table 5. Key Features of Blockchain Integration with AI (Salah et al., 2018)

Blockchain	AI
Decentralized	Centralized
Deterministic	Changing
Immutable	Probabilistic
Data Integrity	Volatile
Attacks Resilient	Data, Knowledge, and Decision-metric

AI AND BLOCKCHAIN: A DISRUPTIVE INTEGRATION

AI and Blockchain are among the most disruptive technologies. The Blockchain can connect data storage and data users from multiple domains altogether, and artificial intelligence helps to run on different kind of analytics application for the development of AI algorithms. Blockchain can power decentralized marketplaces and coordination platforms for various components of AI and AI is the key to providing users with confidentiality and privacy (Dinh & Thai, 2018). Blockchain and artificial intelligence are driving technological innovation, and both have profound implications for the future of business as well as our personal data. These two technologies are merged into a decentralized manner (Table 5). The Blockchain is a decentralized technology, which is a global network of computers and Artificial intelligence consisting of core techniques and technologies. Decentralized AI is basically a combination of AI and Blockchain. The decentralized AI enables processing and performs analytics or decision making on trusted, digitally signed, secured shared data which has been transacted and stored on the Blockchain. AI works with huge volumes of data and Blockchain is foreseen as a trusted platform to store such data.

Blockchain is all about keeping track of authentic records and AI helps in the decision-making process. Consolidation of AI and Blockchain can create secure, immutable, decentralized system for the highly sensitive information that AI driven systems must collect, store and utilize (Salah, Rehman, Nizamuddin & Al-Fuqaha, 2018). The Benefits of Blockchain for AI are:

- Enhanced Data Security
- Improved Trust on Robotic Decisions
- Collective Decision Making
- Decentralized Intelligence
- High Efficiency

The concept of Blockchain and Artificial Intelligence has been combined altogether. This technology develops for different use cases. The first use case is called SingularityNET. This SingularityNET provides a global AI network.

SingularityNET

SingularityNET is an open-source protocol and collection of smart contracts for a decentralized market of coordinated AI services. SingularityNET is backed by the SingularityNET foundation. SingularityNET is a Blockchain based framework designed to serve the needs of AI agents as they interact with each other with external customers. SingularityNET is a set of smart contract templates that AI agents can use to request AI work to be done, to exchange data, and to supply the results of AI work. It's widely accepted throughout the business worlds. Blockchain provides a powerful tool for managing transactions in a singularity era economy. Blockchain based framework is designed to serve the needs of AI agents as they interact with each other that encourages collaborations between these tools and decentralized sharing of information. It's essentially for distributed computing architecture with AI and machine learning concepts and to facilitate the market interactions with smart contracts technology.

SingularityNET is essentially a distributed computing architecture. The smart contracts are to facilitate market interactions with AI and machine learning tools. The design principles or features of SingularityNET are;

- Interoperability – The SingularityNET is a global AI network. It can interface with multiple Blockchains. There can be multiple stakeholders which can have multiple private Blockchains for running private applications.
- Data Sovereignty and Privacy – User data access is control based on privacy policy of individual stakeholders, and the access is securely validated to smart contracts. It validates and supports modular design application
- Modularity – Design the module for flexible network capabilities to make it create custom topologies, AI agent collaboration arrangements and failure recovery
- Scalability – Securely hosts both private and public contracts (Goertzel, 2017).

DeepBrain Chain

The second use case is DeepBrain Chain or DBC. This DBC provides AI computing platform with Blockchain. It was launched April 2017 and is the first national AI brain open platform. Artificial intelligence enterprises can reduce the hardware cost by 70% by buying data and processor power. DeepBrain Chain is based on the concept of artificial neural network; DeepBrain Chain provides a low-cost, private, flexible, secure and decentralized artificial intelligence computing platform with artificial intelligence products. The underlying algorithm design an application of AI operating system specify some principles; the principles are extended principle, stretching principle and privacy principle. The systems are;

- **Extended Principle:** Each module is loosely coupled, new modules are added easily and update any module should not need other module interface change,
- **Stretching Principle:** Flexible for customer access, if there is many users accessing a node it inevitably brings service breakdown to the node, so the container of the node itself should be automatically deployed.
- **Privacy Principle:** All participants of the DeepBrain Chain ecosystem, mining nodes, artificial intelligence manufacturers, and data providers, can get the privacy protection, according to their own need's. Participants can select the nodes. DeepBrain Chain focuses on low cost, security, pri-

vacy protection and flexible computing (Yong, Lee, & Wang, 2018). Examples of cryptocurrencies are BTC, ETH, BNB, KRW, and USDT. The DeepBrain Chain (DBC) supports the platform to run on Artificial Neural Network (ANN) in a decentralized way. The DBC and ANN indicates that;

- ANN algorithms require huge computing power
- A DBC provides a computing platform to support the creation of AI algorithms, based on ANN. DBC aims to reduce the computation cost by 70%.
- A decentralized neural network is used to describe a complex ANN based AI system for training applications

Numerai

The third use case of artificial intelligence and Blockchain is the SoHE application called Numerai. Numerai provides a Blockchain power AI coordinated hedge fund platform. Numerai has built a unique concept over Blockchain. Their concept allows data scientist to make predictions on more competitive markets in the stock market. The stock market has very slow progress in the fields of ML. In this progress, the data scientists must access the raw data, and it allows data scientists from all over the world to try to submit predictions based on machine learning models to improve the AI (Lopes & Alexandre, 2018). So, different data scientist can design different kinds of prediction models, and that kind of prediction model will run on top of a hedge fund to find out the probability of success for the investors. The data scientist will collaborate with the hedge fund and hedge fund platform, the data scientist wants to ensure certain requirements. Numerai hosts the weekly competitions for data scientist for submission of prediction models and they execute that prediction model on top of this Numerai network, which is again a peer to peer network powered by Blockchain. Based on the output of that prediction models, these data scientists are incentive high; so, that they can periodically participate in this Numerai platform.

Matrix AI Network

Matrix AI Network is an open source intelligent Blockchain platform. It supports smart contracts and machine learning services altogether, and artificial intelligence is used to automatically self-optimize the Blockchain network. So, that automates multiple processes inside the matrix Blockchain platform.

CONCLUSION

Blockchain Technology, is useful to understand the Blockchain in the context of blocks, hashing functions and merkle tree structures in bitcoin development. The Blockchain ecosystems need Bitcoin mechanisms and consensus algorithms, such as proof of work, proof of stake and delegated proof of stake. It's about maintaining distributed ledger technologies and disruptive innovation such as financial service, ripple network, supply chain management and stellar protocol network. For Blockchain technology to be fully utilized, educate the people on the nature of technology and actively engage policy makers.

This study furthermore serves as a guide to institutions to make blockchain adoption decisions more systematically.

REFERENCES

Alex, O. (2015). *Blockchain Protocol Series – Introduction.* Retrieved from https://medium.com/the-blockchain/blockchain-protocol-series-introduction-79d7d9ea899

Andoni, M., Robu, V., Flynn, D., Abram, S., Geach, D., Jenkins, D., ... Peacock, A. (2019). Blockchain technology in the energy sector: A systematic review of challenges and opportunities. *Renewable & Sustainable Energy Reviews, 100,* 143–174. doi:10.1016/j.rser.2018.10.014

Bencic, F. M., & Zarko, I. P. (2018). *Distributed Ledger Technology: Blockchain Compared to Directed Acyclic Graph.* 2018 IEEE 38th International Conference on Distributed Computing Systems (ICDCS), Vienna, Austria.

Bisola, A. (2018). *What is SHA-256 and how is it related to Bitcoin?* Retrieved from https://www.my-cryptopedia.com/sha-256-related-bitcoin/

Bojana, K., Elena, K., & Anastas, M. (2017). Blockchain Implementation Quality Challenges: A Literature Review. *Proceedings of the SQAMIA, 2017,* 6.

Brent, X. (2018). *Blockchain vs. Distributed Ledger Technologies Part 2: Governing Dynamics.* Retrieved May 23 2018, from https://media.consensys.net/blockchains-vs-distributed-ledger-technologies-part-2-governing-dynamics-a697848d5b82

Bruyn, A. S. (2017). *Blockchain an Introduction.* Research paper. VU University Amsterdam. Retrieved from https://beta.vu.nl/nl/Images/werkstuk-bruyn_tcm235-862258.pdf

Buterin, V. (2013). *Introducing Ripple.* Retrieved from https://bitcoinmagazine.com/articles/introducing-ripple/

Buterin, V. (2015). *On Public and Private Blockchains.* Retrieved August 6, 2015, from https://blog.ethereum.org/2015/08/07/on-public-and-private-blockchains/

Christensen, C. (2019). Disruptive Innovation. *Harvard Business School.* Retrieved 2019, from http://claytonchristensen.com/biography/

Clayton, S. (2015). *Money.* Dallas, TX: Everyday Economics Federal Reserve Bank of Dallas. Retrieved from https://www.dallasfed.org/~/media/documents/educate/everyday/money.pdf

David, X. (2017). *The Four Layers of the Blockchain.* Retrieved from https://medium.com/@coriacetic/the-four-layers-of-the-blockchain-dc1376efa10f

Dib, O., Brousmiche, K. L., Durand, A., Thea, E., & Hamida, E. B. (2018). Consortium Blockchains: Overview, Applications and Challenges. *International Journal on Advances in Telecommunications, 11*(1 & 2), 51–64.

Dinh, T. N., & Thai, M. T. (2018). AI and Blockchain: A Disruptive Integration. *IEEE Computer Society, 51*(9), 48-53. Retrieved from http://www.people.vcu.edu/~tndinh/papers/IEEEComp18_Blockchain+AI.pdf

Dob, D. (2018). *Permissioned vs. Permissionless Blockchains: Understanding the Differences*. Retrieved July 17, 2018, from https://blockonomi.com/permissioned-vs-permissionless-blockchains/

Domingos, T. (2017). *Blockchain: beyond payments*. Retrieved from https://atos.net/en/blog/blockchain-beyond-payments

Gandhi, R., & Ramasastri, A. S. (2017). *Applications of Blockchain Technology to Banking and Financial Sector in India*. IDRBT Publication.

Ganne, E. (2018). *Can Blockchain Revolutionize International Trade?* World Trade Organization. Retrieved from https://www.wto.org/english/res_e/booksp_e/blockchainrev18_e.pdf

Goertzel, B. (2017). *SingularityNET: A decentralized, open market and inter-network for AIs*. Thoughts, Theories & Studies on Artificial Intelligence (AI) Research.

Haber, S., & Stornetta, W. S. (1991). How to Time-Stamp a Digital Document. *Journal of Cryptology*, *3*(2), 99–111. doi:10.1007/BF00196791

Hans, R., Zuber, H., Rizk, A., & Steinmetz, R. (2017). *Blockchain and Smart Contracts: Disruptive Technologies for the Insurance Market*. Twenty-third Americas Conference on Information Systems (AMCIS), Boston, MA. Retrieved from ftp://www.kom.tu-darmstadt.de/papers/HZR+17-1.pdf

Hu, Y., Liyanage, M., Manzoor, A., Thilakarathna, K., Jourjoin, G., Seneviratne, A., & Yilanttila, M. (2018). The use of smart contracts and challenges. Computers and Society. *Cornell University*, *1*(1), 1–13.

Iansiti, M., & Lakhani, K. (2017). The Truth about Blockchain. *Harvard Business Review*. Retrieved from https://hbr.org/2017/01/the-truth-about-blockchain

Karim, S., Umar, R., & Rubina, L. (2018). *Conceptualizing Blockchains Characteristics & Applications*. 11th IADIS International Conference Information Systems 2018, Lisbon, Portugal.

Leandros, P. (2018). *Disruptive Innovation and Blockchain in Financial Services*. Retrieved from https://medium.com/datadriveninvestor/disruptive-innovation-and-blockchain-in-financial-services-911b102e785b

Lee, D. K. C. (2015). Handbook of Digital Currency. In Bitcoin Innovation, Financial Instruments, and Big Data (pp. 1-612). Elsevier.

Li, X., Jiang, P., Chen, T., Luo, X., & Wen, Q. (2018). A Survey on the Security of Blockchain Systems. *IT Security Conference*.

Lin, I., & Liao, T. (2017). A survey on Blockchain security issues and challenges. *International Journal of Network Security*, *19*(5), 653–659.

Lopes, V., & Alexandre, L. A. (2018). *An overview of Blockchain integration with Robotics and Artificial Intelligence*. Cornell University. Retrieved from https://arxiv.org/abs/1810.00329

Mahdi, H. M., & Maaruf, A. (2018). Applications of Blockchain Technology beyond Cryptocurrency. *Annals of Emerging Technologies in Computing*, *2*(1), 1–6. doi:10.33166/AETiC.2018.01.001

Mattias, S. (2017). *Performance and Scalability of Blockchain Networks and Smart Contracts*. Retrieved from https://umu.diva-portal.org/smash/get/diva2:1111497/FULLTEXT01.pdf

Merkle, R. C. (1979). Secrecy, authentication, and public key systems. ACM Digital Library.

Nadir Abdelrahman, A. (2018). Blockchain Technology: Classification, Opportunities, and Challenges. *International Research Journal of Engineering and Technology, 5*(5), 3423-3426. Retrieved from https://www.irjet.net/archives/V5/i5/IRJET-V5I5659.pdf

Nakamoto, S. (2008). *Bitcoin: A peer-to-peer electronic cash system*. Retrieved from https://bitcoin.org/bitcoin.pdf

Oberhauser, A. (2015). *Blockchain Protocol Series – Introduction*. Retrieved May 24, 2015, from https://medium.com/the-blockchain/blockchain-protocol-series-introduction-79d7d9ea899

Rakesh, K. (2018). *Blockchain vs. Distributed Ledger Technology*. Retrieved December 10, 2018, from https://medium.com/coinmonks/blockchain-vs-distributed-ledger-technology-b7b2e434093b

Rampton, J. (2018). *5 applications for Blockchain in your business*. Retrieved from https://execed.economist.com/blog/industry-trends/5-applications-blockchain-your-business

Revoredo, T. (2018). *Blockchains vs. DLTs*. Retrieved Jul 9, 2018, from https://medium.com/coinmonks/blockchains-vs-dlts-8fe03df39737

Reyna, A., Martin, C., Chen, J., Soler, E., & Diaz, M. (2018). On Blockchain and its Integration with IoT. Challenges and Opportunities. *Future Generation Computer Systems, 88*, 173–190. doi:10.1016/j.future.2018.05.046

Rosic, A. (2017). *What Is Hashing? Step-by-Step Guide-Under Hood Of Blockchain*. Retrieved from https://blockgeeks.com/guides/what-is-hashing/

Rosic, A. (2018). *Basic Primer: Blockchain Consensus Protocol*. Retrieved from https://blockgeeks.com/guides/blockchain-consensus/

Salah, K., Rehman, M. H., Nizamuddin, N., & Al-Fuqaha, A. (2018). Blockchain for AI: Review and Open Research Challenges. *IEEE Computer Society, 4*.

Schwartz, D., Youngs, N., & Britto, A. (2018). *The Ripple Protocol Consensus Algorithm*. Retrieved from https://ripple.com/files/ripple_consensus_whitepaper.pdf

Shen, C., & Pena-Mora, F. (2018). Blockchain for Cities – A Systematic Literature Review. *IEEE Access: Practical Innovations, Open Solutions, 1*, 1–33. Retrieved from https://www.researchgate.net/publication/328896113-Blockchain-for-Cities-A-Systematic-Literature-Review

Shrier, D., Sharma, D., & Pentland, A. (2016a). *Blockchain & Financial Services: The Fifth Horizon of Networked Innovation*. Retrieved from http://cdn.resources.getsmarter.ac/wp-content/uploads/2017/06/MIT_Blockchain_whitepaper_ PartOne.pdf

Supriya, T. A., & Kulkarani, V. (2017). Blockchain and Its Applications – A Detailed Survey. *International Journal of Computers and Applications, 180*(3), 29–35. doi:10.5120/ijca2017915994

Suvarna, K. (2018). *Review of distributed Ledgers: The technological Advances behind cryptocurrency.* Retrieved from https://www.researchgate.net/publication/323628539_Review_of_Distributed_Ledgers_The_technological_Advances_behind_cryptocurrency

Szabo, N. (1997). *The Ideas of Smart Contracts.* Retrieved from http://www.fon.hum.uva.nl/rob/Courses/InformationInSpeech/CDROM/Literature/LOTwinterschool2006/szabo.best.vwh.net/idea.html

Yli-Huumo, J., Ko, D., Choi, S., Park, S., & Smolander, K. (2016). Where is Current Research on Blockchain Technology?-A Systematic Review. *PLoS One, 11*(10). doi:10.1371/journal.pone.0163477 PMID:27695049

Yong, H., Lee, C., & Wang, D. (2018). *DeepBrain Chain: Artificial Intelligence Computing Platform Driven by Blockchain.* White Paper of DeepBrain Chain, Version 1.1.0.

Zhang, R., Xue, R., & Liu, L. (2019). Security and Privacy on Blockchain. *ACM Computing Surveys, 1*(1), 1-34. Retrieved from https://arxiv.org/pdf/1903.07602.pdf

Zheng, Z., Xie, S., Dai, H. N., & Wang, H. (2017). An Overview of Blockchain Technology: Architecture, Consensus, and Future Trends. *IEEE 6th International Congress on Big Data Congress.*

Zheng, Z., Xie, S., Dai, H. N., & Wang, H. (2018). Blockchain challenges and opportunities: A Survey. *International Journal of Web and Grid Services, 14*(4), 352–375. doi:10.1504/IJWGS.2018.095647

KEY TERMS AND DEFINITIONS

Artificial Intelligence: Simulation of human intelligence processes by machines; these processes include learning, reasoning, and self-correction. To perform tasks commonly associated with intelligent beings.

Bitcoin: Bitcoin is a complete decentralized peer to peer and permission less cryptocurrency, it's a form of electronic cash system with completely decentralized; cannot be any central party for ordering or recording or controlling your currency like bank or government.

Blockchain: Blockchain is a decentralized computation and information sharing platform which enables multiple authoritative domains that do not trust each other to cooperate coordinate and collaborate in a rational decision-making process.

Digital Signature: It is a mathematical scheme for presenting the authenticity of digital messages or documents, it uses public and private keys to encrypt and decrypt the data.

Distributed Ledger Technology: It's a decentralized technology to eliminate the need for a central authority or intermediary to process, validate or authenticate transactions or other types of data exchanges.

Hyderledger Fabric: Is responsible for ordering a sequence of transactions into blocks and it go it is going to deliver these sequence of totally ordered transactions to all the pears in the network.

Merkle Tree: Merkle tree is a tree structure, where the leaf nodes they will contain the hash of the document and every individual node or intermediate node, they will contain the hash of the combination of the left child under a right child.

Permissioned Blockchain: It works in open environment and large network of participants, so the participants do not need to reveal their own identity, the users do not need to reveal their own identity to other peers. It's something called private model. In a permissioned blockchain, only a restricted set of users have the rights to validate the block transactions.

Permissionless Blockchain: Here anyone can join the network, participate in the process of block verification to create consensus and also create smart contracts. Permissioned blockchains do not have to use the computing power-based mining to reach a consensus since all of the actors are known.

Smart Contracts: Its basically provide a decentralized platform, which can be utilized to avoid the intermediately in a contract. It's an automated computerized protocol used for digitally facilitating, verifying or enforcing the negotiation or performance of a legal contract by avoiding intermediates and directly validating the contract.

Supply Chain Management: the management of the flow of goods and services involves the movement and storage of raw materials, of work-in-process inventory, and of finished goods from point of origin to point of consumption. It's the network to be connected of individuals, organizations, activities, resources and technologies, sale of product or service.

APPENDIX: ACRONYMS

AML: Anti-money laundering.
ANN: Artificial neural network.
API: Application programming interface.
ARK: open-source cryptocurrency.
BTS: Bit Shares,
DB: Data base.
DBC: Deepbrain chain.
DpoS: Delegated proof-of-stake.
DLT: Distributed ledger technology.
ECA: Enrollment certificate authority.
ECDSA: Elliptic curve digital signature algorithm.
EOS: Open-source cryptocurrency.
FBA: Federated Byzantine agreement.
HTTP: Hyper text transfer protocol.
IP: Internet protocol.
JSON: Javascript object notation.
JVM: Java virtual machine.
KYC: Know your customer process.
LSK: Lisk cryptocurrency.
OSI: Open systems interconnection.
OXY: Oxycoin cryptocurrency.
PBFT: Practical Byzantine fault tolerance.
PKI: Public key infrastructure.
PoS: Proof-of-stake.
PoW: Proof-of-work.
PSPs: Payment service provider.
RFID: Radio frequency identification.
RTGS: Real-time gross settlement systems.
STEEM: Cryptocurrency.
TCA: Transaction certificate authority.
TCP: Transmission control protocol.
TLS: Transport layer security.
TLSCA: TLS certificate authority.
UDP: User datagram protocol.
UTXO: Unspent transaction outputs model.

Chapter 11
Cyber Secure Man-in-the-Middle Attack Intrusion Detection Using Machine Learning Algorithms

Jayapandian Natarajan
https://orcid.org/0000-0002-7054-0163
Christ University, India

ABSTRACT

The main objective of this chapter is to enhance security system in network communication by using machine learning algorithm. Cyber security network attack issues and possible machine learning solutions are also elaborated. The basic network communication component and working principle are also addressed. Cyber security and data analytics are two major pillars in modern technology. Data attackers try to attack network data in the name of man-in-the-middle attack. Machine learning algorithm is providing numerous solutions for this cyber-attack. Application of machine learning algorithm is also discussed in this chapter. The proposed method is to solve man-in-the-middle attack problem by using reinforcement machine learning algorithm. The reinforcement learning is to create virtual agent that should predict cyber-attack based on previous history. This proposed solution is to avoid future cyber middle man attack in network transmission.

INTRODUCTION

Data security and data analytics are a two major pillars of the modern business world. Cyber security is not only the association of data security and privacy, it also consists of a multiple of other components. Cyber security is a process that comprises data, network, storage and computing. The market growth of cyber security reached around 135 billion US dollar in the year 2017. The expected market growth during the period 2018 to 2022 is projected to be 200 billion US dollars (Steve Morgan, 2018). Almost all the utilization services are migrating to the cloud platform. These utilization services are storage,

DOI: 10.4018/978-1-5225-9687-5.ch011

network and infrastructure. The reason for this migration is easy accessibility and lower cost. The other positive aspect of this migration is reducing establishment and computational costs. Third party service providers are also facing serious data security problems. This security problem is termed cybersecurity and addresses data and network security. Cyber security issues and crimes are officially published in many documents in more than fifty countries (Gercke, 2012). The nature and scope of the problem is network security. Cyber security is similar to the banyan tree, where the leaf of this tree is to maintain security and risk management. This cybersecurity is a part of information security to manage various security tools (Kaufman, 2009). The Figure 1. illustrates different elements of cyber security (Schatz, Bashroush & Wall, 2017). The role of the roots is to provide higher security data.

Cyber security is the attainment of data organization and device computation. This device computation relates to various computers and deals with many traditional and non-traditional data. The objective of information security is a circle of three elements that is termed availability, integrity and confidentiality (Jouini & Rabai, 2019). Data availability refers to accessing data from server machines. The second element is integrity which deals with data accuracy and quality (Luo, Hong & Fang, 2018). The most important element is confidentiality which concerns handling data security mechanisms. Both cybersecurity and information security are communication security protocols (Von Solms & Van Niekerk, 2013).

The Table 1 illustrates the fundamental differences between cyber and information security systems (Luiijf, Besseling, Spoelstra & De Graaf, 2011). The primary difference between these two securities mechanisms refers to dealing with physical and digital data. The second difference is dealing with its own organization and public data. Public data means handling internet digital information. Cybersecurity signifies operating at a level above boundary level, which means handling cyber and physical attacks. Physical attack implies the physically theft of information (Pasqualetti, Dörfler & Bullo, 2013). This physical data protection consists of handling the information security protocols. Recently, apart from this physical data, all data is managed in digital format with the help of the cloud computing platform which is a technology that provides different levels of servers. Customers who utilize this technology

Figure 1. Cyber Security Elements

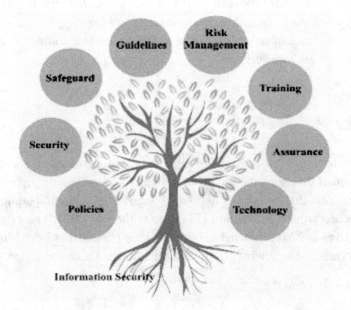

Table 1. A Comparison between Cyber Security and Information Security

Cyber Security	Information Security
Cyber security deals with data protection in digital format.	Information security providing physical and digital security.
Client to server communication protect unauthorized digital communication access.	Major focus on availability of data with data confidentiality.
It concerns handling advanced technology and it is named internet communication.	Information security is a basic security mechanism.
Data breach and data phishing is a primary goal.	Basic security protocols used for data security.
Third party security is needed.	Inter security mechanism is enough.

store their data and access it anywhere in the world. The backbone of this technology is internet communication. On the other hand, the main heart of cyber and information security is information and communication technology (ICT), where all data communication is interlinked. ICT is incorporated in cloud technology, which finds its importance even in the educational sector (Jayapandian, Pavithra & Revathi, 2017). ICT security also provides security protocol and mainly focuses on data accountability. It, in addition, provides reliability and data authenticity. Cyber security deals with different level of attackers. A cybercriminal is an indirect attacker where there is no direct-attack on the data (Lau, Xia & Ye, 2014). The control and the prevention of hacking server machines can be implemented by controlling the physical system by using software applications. Disgruntled employees are another major domain of cyber-attacks. The professional hacker is known to attempt hacking some confidential data from cloud servers whose data are misused during terrorist activities. These reflect some of the major cyber-attacks practiced in modern data base storage.

Machine learning is a sub-division of the artificial intelligence technique. Artificial intelligence is the process of automatically detecting the solution from prior experiences (Michalski, Carbonell & Mitchell, 2013). The major advantage of this methodology is that it permits making decisions without human intervention. The basic working mechanism of machine learning algorithms is working with existing history. General machine learning is an application program that is similar to data mining. It retrieves and subsequently extracts data from the database. The best working model of machine learning is the neural network whose working methodology is to connect nodes in a loosely coupled structure. This loosely coupled node operates in a similar manner to biological brain structure. Conceptually, the biological brain structure is preinstalled and used in network connection (Brugere, Gallagher & Berger-Wolf, 2018). In the 1990's, machine learning dealt with the neural network structure. This deep learning technology is the hottest topic in the machine learning field. The Bayesian network structure also deals with machine learning algorithms (Chaturvedi et al., 2018). The reason behind this is that normal databases evolved into the concept termed big data. Handling this data base is very critical in the modern world. Massive volumes of data are accessed in short period of time necessitating that machine learning algorithms also adopt deep learning concepts. Compared to normal programming, machine learning is a low-cost implementation. This gives rises to a feasible structure with more accuracy. The reinforcement method is one of the best approaches in machine learning, which provides a better solution to solve network security problems. The principal objective of this chapter is to compare and analyze traditional and reinforcement algorithms. Different parameters are relating to this comparison that is packet transmission time and delivery percentage. The other objective and hypothesis of the chapter is to solve network security issues by using machine learning algorithms.

MACHINE LEARNING METHODOLOGY

The basic working principle of machine learning is gathering results from the existing methods. As a rule, there are three major components used to analyze and gather results from machine learning. This concept is elaborated in Figure 2. The first component is a representation that means using formal programming language to classify the data-set. This classification set is named, hypothesis space, which is very useful for the learning component. The second component is evaluation; this component is utilized to find a worthy classifier in the regular data-set. The alternate name of this evaluation function is the objective function, which signifies finding good objectives from the bad data-sets. The major and final component of machine learning is optimization (Luo et al., 2019) whose goal is to increase efficiency from learners. This concept of self-optimization technique implies that there is no human intervention to automatically optimize the data from the data-set.

Machine learning representation is involved in several elements. Instance is a one of the important elements in the learning component. It – the instance - helps to find neighbor data elements and to support the vector machine. The extreme learning approach is one type of learning methodology utilized to predict data (Mahmoud, Dong & Ma, 2018). A decision tree is a reasonable application of the learning method essentially because the tree structure helps to take decisions in real-time applications (Park, Haghani, Samuel & Knodler, 2018). It creates a rule, whose task is to assist the logical program and finds the propositional solution. Neural networks and graphical models are used in network communication protocols; this structure is also incorporated in artificial intelligent mechanisms. The evaluation component is to provide accuracy and will also trace the error rate in the data-set. This component is used to obtain information at a lower cost. Optimization is a normal method; it has been previously used in basic data structure algorithms. Greed search is a search optimization method to find better solutions. There are two varieties of optimization employed, that is, assembled and continuous optimization. Greedy

Figure 2. Machine Learning Process

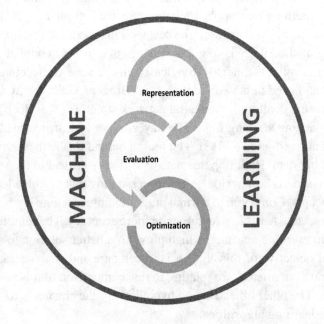

and beam search are example of assembled optimization methods (Fanjul-Peyro & Ruiz, 2010). In some cases, linear and quadratic programming is used during this optimization. Different types of machine learning algorithms are available and employed in a variety of applications. These algorithms are used in dynamic situations with a different types of predictive data-sets. This prediction instance is known as a label structure: also referred to as supervised method. Unlabeled structure implies an unsupervised algorithm (Fan, Su, Nien, Tsai, & Cheng, 2018). There are several problems faced while the implementing machine learning algorithm. The primary problem is data-set collection which is normally involved in field export structure. The most common method is the brute force technique implying a trial-and-error method; it finds the possible solution in a systematic way. This algorithm is used to find the best data-set in machine learning. Mining the data is also involved in the machine learning approach. This data mining is the process of data that is named data pre-processing, a technique that converts unreadable format data to normal raw data (Woo, Shin, Seo & Meilanitasari, 2018). There are a quantity of simulation tools available in data pre-processing methodology, among them are Weka, Rapid Miner and DataMelt (Rubab, Taqvi & Hassan, 2018). After data pre-processing, data cleaning is the toughest task. An example of this data cleaning of illegal values in a database. This family of problem is resolved by using mathematical variance and deviation. Misspellings is another commonly encountered issue which of sorting and finding the feature value in a database. Data sampling is a recent research option for data-set selection (Li, et. al., 2018). There are two sampling techniques normally used; the first one is random sampling, which is the selecting instance of sub-set values in a random structure. The second one is stratified sampling where are selected sample values in the most frequently accessed data-sets. Stratified sampling is sometimes used in imbalanced data-sets (Buda, Maki & Mazurowski, 2018). In the real-world data-set data values are incomplete; this type of data is unavoidable. In this situation existing data parameters are attributed to incomplete words and given some possible solution. The future sub-set selection process is categorized into three different methods that are termed, relevant data, irrelevant data and redundant data. The machine learning algorithm focuses uniquely on the method of future selection techniques which addresses optimal sub-sets based on previous data values. These are the general data prediction methods where the prediction accuracy is the most important factor (Ghahramani, 2015). It is not possible to archive this accuracy in normal data prediction algorithm necessitating the introduction of many special machine learning algorithms.

MACHINE LEARNING ALGORITHM

Machine learning is an algorithm that is categorized in many traditional structures. The methodology of this algorithm is gathering input from various sources and predicts the future based on experience. In general, the working structures of machine learning algorithms are based on predictive and mining approaches (Moeyersoms et al., 2015). The percentage of pattern matching is the most important factor. This model is compared with existing data patterns and provides the best result compared to existing structure.

Figure 3 elaborates the different machine learning algorithms.

Figure 3. Machine Learning Algorithm

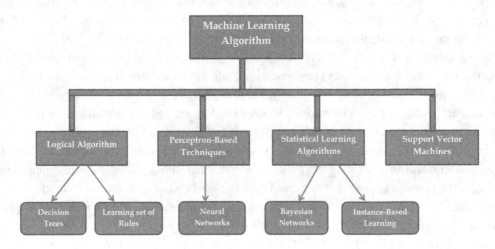

Logical Approach

The logical approach is the process of step-by-step procedure using logical conditions. This logical algorithm is used to solve logical problems in data processing systems (Wang & Pedrycz, 2017). It is the combination of conditional and mathematical operations. This approach solves the problem based on existing logical knowledge. Logical approaches are generally categorized into two major types, the first is a decision tree approach and another the rule or condition-based approach. The decision tree is the major part of a supervised machine learning algorithm. The primary aim of this decision tree method is to a create training data set that helps to predict the future result. This tree concept is used to predict the data elements by using nodes and leaves. Decision tree methods are implemented in different applications (Lino, Rocha, Macedo & Sizo, 2019). The working methodology of decision tree is a non-deterministic polynomial time (NP) complete problem, which is intended to find the best heuristic search to develop decision tree in the near-optimal structure (Boas, Santos & de Campos Merschmann, 2018). The best training data-set for this tree is root node and, based on this node decisions are subdivided and accessed. Finding the best data-set is to divide the tree structure from top to bottom. However, the true factor is never hundred percent correct. In order to find the best data-set value, individual data-set, together with group data-sets are compared with each other. This method works in a cyclic process until the best data set value in a database is detected. Thereafter, the training data-set is used to predict the target or class variable based on decision protocols. This protocol structure is to decide the result based on existing training set. Figure 4 is a working example of a decision tree structure; this diagram predicts the customer loan repayment history. Currently, most financial companies utilize this machine learning algorithm to analyze customer loan repayment history: based on this decision tree suggestion they decide to approve or disapprove the loan (Chen, Zhang & Ng, 2018). First step of the loan process is to check the customer credit score. If that score is above 800 it means that the customer loan application is automatically moved to the next approval process. Where the credit score is less than 700, indicating a poor credit history, the loan request is declined as suggested by the decision tree. For credit score between 700 and 800; the decision-making situation is critical. There is evidence of customer perfectly repaying their existing loan but due to some technical problems their loan history bears a negative remark. For example, sometimes electronic clearance system (ECS) or check clearance is activated due to the technical issue; this kind of problem indicates a negative score in the credit history. This type of problem is to be solved manually by the bank employee and the, loan application is checked manually based on previous repayment history.

Figure 4 is a model diagram of a decision tree for loan approvals. The decision tree creates separate rules for each root node and leaf node. This rule is to apply individual data-sets and, that data-set assigns a variety of rules. These rules are divided into two categories. First, the smallest rule-based algorithm which maintains consistent levels in the training data-set. Second, is largest rule-based algorithm which maintains the level of the data-set. The concept of the data structure is used to find the best rule, either the divide or the conquer algorithm. Here the algorithm is separated by two different modules and it finds the optimal solution. This method is used to learn more on rule-based in dynamic situations. The rule is retrospectively implemented in-depth in the root and leaf node. There are many rule-based algorithms available in machine learning. One of the best rule-based algorithm is the Repeated Incremental Pruning to Produce Error Reduction (RIPPER) (Xuan, 2018). This is pure rule-based approach and, the working principle of these algorithms is two-process, that is, growing (incremental) and pruning. The incremental method is a higher restrictive phase during the training data-set process and is utilized to analyze very close structures. The growing method is generally a closed loop structure in the root-to-leaf node. The second method is pruning process; this is a normal restrictive structure. This method is to reduce over-fitting during rule analysis.

This is a multi-class structure to maintain lower to higher training data-sets. Like that of fuzzy rule-based algorithm, it is used in machine learning (Wang, Liu, Pedrycz & Zhang, 2015). It works with the principle of true value structure, which assigns the value of zero or one. The condition is set to true or

Figure 4. Decision Tree Methodology

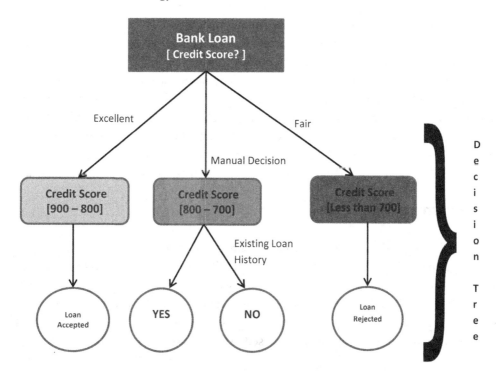

false model, based on that condition rule is assigned and data is predicted from the data set. An alternative decision algorithm, PART (Hall et al., 2009) is used to minimize the post processing during rule association. Most of these decision tree algorithm function with similar type of rule-based decisions by using tree concepts. Inductive logic programming (ILP) is another decision analysis technique in machine learning. This method is to predict the future, based on previous positive or negative data for certain situations.

Perceptron Algorithm

Machine learning is the concept of predictions taken from various parameters. Binary classification is also one of the prediction mechanisms: it is termed the "Perceptron algorithm" and utilizes the concept of function (Min & Jeong, 2009). This prime mode of the algorithm is to decide the suitability of an input parameter. The function formation is created with sub-class vector numbers. In general, this is a sub-classification of linear classification mechanisms.

As both algorithms use restricted linear functions as their hypotheses, Winnow and Perceptron can be considered similar. The principal advantage of the Winnow algorithm is its utilization of multi-dimensional data. This method provides higher accuracy in comparison to other methods. Moreover the Perceptron algorithm is directly implemented in neural networks (Pérez-Sánchez, Fontenla-Romero & Guijarro-Berdiñas, 2018). It is acknowledged that biological nervous systems form the platform for artificial neural networks. The former collaborates with the functional part of brain and data processing. This information processing unit relates to the nervous structure. This nervous is generally connected to many interconnected structures. The interconnected structure is to solve various unison problems. Similarly, neural networks solve various complex problems in network communication. Machine learning neural network operate at three different levels (Camacho, 2018). The first layer is the input layer which deals with input parameters based on trained data-set. These input values are gathered from different network resources. The second layer is the output layer; this is an outcome that provides the final result. In between these two layers, is an important layer which deals with the prediction process, that is termed the hidden layer. These hidden layers are like those of a central processing unit (CPU) whose hidden layer predicts the result based on existing data. The major challenge of this structure is ingrained in the hidden layer node calculation where it is required to establish the best result. Both structures underestimate and overestimate creation of the hidden layer. The neural count, based on fact that a dynamic hidden layer creation is required, had proved to be unpredictable. An alternate name of this method is the feedforward neural network and the main advantage of this method is that data is transferred and moved uniquely in one direction.

Statistical Learning Algorithms

A statistical learning algorithm is a functional algorithm, used to predict the data from various statistics (Miller, Nagy, Schlueter, 2018). The combination of statistics and the functional model is implemented to predict the final result. This algorithm is mostly implemented in the field of computer virtual vision and bio-metric systems. In this algorithm, the primary goal is learning, and the secondary goal is prediction. The method is to implement both supervised and unsupervised learning methods (Chapelle, Scholkopf & Zien, 2009). Occasionally online learning is also used to predict the result. The online learning is utilized in online shopping, where, based on customers previous history system, it automatically predicts future customer needs and provides unique offers. This statistical model works with the probability structure. There are many network models implemented in this statistical algorithm. A Bayesian network is one of the important methods of this learning algorithm. It is categorized as a relational and probabilistic graphical structured model (Neal, 2012). The computational probabilistic condition is used to predict the data-set. Bayesian networks are implemented in the conception of directed acyclic graphs (DAG) (Li & Yang, 2018). The DAG method is used in different variables that are designated "S" and "X". S is a structure node in a graph and X is a feature prediction. This graph structure creates a table and maintains the parameter variable during dynamic situations. In networking there are two conditions used who possess respectively a structured network and an unstructured network. The structured network is a normal network structure and consequentially there is no complexity associated with this model. The network transmission variable is to maintain a separate table, the conditional probability table, whose function is to maintain all the network transmission and node connection details. This table is used to calculate the network connection estimation time. This table provides appropriate time for data in transit from one node to another node. Compared to other method major benefits of this Bayesian network is the inherent limitation (Wang et al., 2018). This limitation structure is used to analyze difficulties of the unknown network model. The model is to provide best available network based on previous table values. In many prediction systems they use greedy search optimization models to provide better prediction. Search optimization is implemented in the network node connection where this model is used to reduce the complexity of network connection. Local search optimization is also a familiar method network solution (Jayapandian, Rahman & Gayathri, 2015). This model updates the data at regular intervals in a local search on a particular data element.

Instance-based learning is one of the familiar methods in the machine learning algorithm (Zhang, 2019). The technical label of this technique is, memory-based learning algorithm. The previous network connection visited details, based on their memory structures, are updated in the table and, predict the future data connection. This method analyzes new problems with existing data elements. The prediction is only valid for working with instance-based mechanisms. Furthermore, the methodology works with the structure of hypotheses construction from the training data-set. The complexity level of prediction as well as growing hypotheses is reduced. If the training data-set of hypotheses is n, the complexity level is to solve is $O(n)$, which is a new predicted instance based on previous learning. This notation provides advantage of not visiting previously failed data. The successful training data provides higher prediction mechanism in machine learning. The best example of this instance is the k-nearest neighbors search algorithm. This algorithm finds the nearest neighbors based on most visited data history. The previous data-set is stored and the training data-set value predicts the new instance data.

Support Vector Machine

The Support Vector Machine (SVM) is a category of supervised machine learning algorithm whose method is used to solve classification as well as time regression problems (Lorena et al., 2018). Classification problem segregate the training data-set during data prediction. The data-set is drawn into a separate plot in trained data elements. This plot creates a dimensional n-graph and creates a support line based on similarity index for data-sets. This method creates a hyper-plane and links the distance from one node to another. The support vector system is used to handle larger data-sets in a training instance.

The data label is important because supervised learning algorithms are uniquely utilized in labeled data. At the same time unlabeled data-sets are used for unsupervised learning methods. Furthermore, support vector machine only use labeled data-sets and are thus unsuitable for the unlabeled.

MACHINE LEARNING APPLICATION

Recently machine learning had found its utility in many technological applications. Figure 5 demonstrates different real-time applications of the machine learning approach. Machine learning is also implemented in Internet of Things' (IoT) devices (Cui et al., 2018). IoT is known to encounter different security issues in the healthcare sector, which is a major research area in the machine learning approach (Bitra, Jayapandian & Balachandran, 2018). The technology applications of these kinds of machine learning algorithms are used in modern gadgets who are invariably associated with this internet technology. The virtual personal assistant is an example of a recent machine learning application (Karov et al., 2017). Currently, there are numerous products available related to this technology, for example, smart phones, smart speakers and mobile phone applications. The principal aim of this virtual assistant is to assist without human intervention. This happens based on our personal data and that of existing friends. If, for instance, an individual uploads a group photo, social media software automatically detects the associated face and traces the origin of the page. This happens thanks to machine learning face detection software (Sun, Wu & Hoi, 2018). In the technological domain, letter communication is facilitated by the email transmission. In many countries, email communication is one of the authorized communications for some government offices. The problem encountered in this type of communication is spam mail and malware viruses. Manual detection is not possible because of interminable email communication happening within short time intervals. It is essentially for this reason that machine learning algorithms detect viruses and hence protect user systems. An example of this type of machine learning virus filter (uni-variate or multi-variate predictors) is the C4.5 decision tree induction system (Mishra, Yadav, Kumar & Jain, 2019).

Online customer support is a natural supporting system on internet web pages utilized by most online based services to provide customer support. This technology works with a current customer requirement based on existing data. If any customer needs assistance for a product related query it is searched in the database and a predefined solution is provided for that customer. At the end stage if the customer needs additional assistance then manual customer support will be provided from the company side. That is the reason many online shopping companies use this machine learning algorithm to predict customer feature products and send recommendation for customers (Guan, Wei & Chen, 2019). Sometime based on user browser history they provide product suggestions. The Internet search engine is the heart of the internet technology where users utilize search engines to find information. This data searching is an arduous task because large quantities of data are available on the internet server. To solve this problem, machine learning algorithms provide the best solution for search engine optimization.

COMPONENTS OF MODERN NETWORKS

Networking is the heart of modern communication technology. Modern technology needs node-to-node data transmission with the help of this networking. Data is transferred from one terminal to another termi-

Figure 5. Machine Learning Application

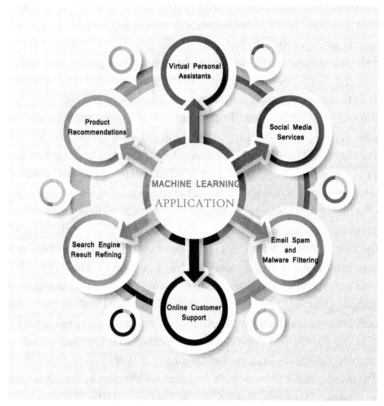

nal. There are six major components involved in this network communication (Zhu & Choulli, 2018). By way of this networking structure, computer systems are connected from one remote terminal to another. These components are used to connect systems with with similar protocol structures thus permitting a collection of rules to maintain and establish reliable communications. Firstly, the component network interface card - a hardware component - is generally integrated with the computer motherboard. This end-to-end terminal concept is used to connect the network cable in this interface card (Huang, Zhao, Zhou, & Xing, 2018). Secondly, the component serves as a hub and connects more than one computer device to the common port (Cai et al., 2018). This is a central device connecting multiple devices to a single hub. The working methodology of this hub is to send signals from one computer device to all other computer devices on a common platform. The network interface card sends and receives signals from this hub port (Majumder & Nath, 2019). Thirdly the component switch is considered to be the most important one in networking. Switches transfer signals from one device to another. These transfers are actuated by precise signals from one terminal to another based on the device ID. In the hub a common transmitting structure transfers signals from one particular device to all other devices disregarding the device ID (Borky & Bradley, 2019). These three components function within the local area network (LAN) connection where for out of the LAN connection some smart devices are required such as a router. The router's task to connect one LAN network to another. This is important for a larger network connections. Routers are regarded as smart technology whose main advantage is to update the networking information in a dynamic manner. Communication media plays a major role in network connection where it helps to establish a physical connection from one computer terminal to another (Henry, Adamchuk, Stanhope, Buddle & Rindlaub, 2019). The technical name of this communication media is the ethernet cable where the concept is a twisted pair wire. Modern cables are used in the latest technology such as in optical fiber cables. This fiber optical cable is used to transfer signals more rapidly than traditional methods. The latter normal utilize electrical signals with the main drawbacks being data loss and less transmit power. A wireless router uses the same communication structure and procedure (Monteiro, Souto, Pazzi, & Nogueira, 2019). All of these components are hardware, but without the software component is not possible to establish computer communication. Transmission Control Protocol/ Internet Protocol (TCP/IP) is the most important software component in establishing network connections (Edwards & Bramante, 2015). Utilizing server layers, TCP/IP protocol has a set of rules which permit data communication. The application layer is normally an end-user layer used to create end-to-end connections; an examples of this layer being Domain Name System (DNS) and Hyper Text Transfer Protocol (HTTP). This layer creates a process and is utilized in some applications. The presentation layer is a data representation and data encryption process dealing with a Secure Sockets Layer (SSL) certificate and image encryption (Zhang, Yang, Castiglione, Chen, & Li, 2019). This layer is the most important in security situations. The session layer is to establish session connection from one port to another. The technical name of this layer is inter-host connection establishment; an example of this layer being socket connection and Application Programming Interface (API). The transport layer is an end-to-end connection to maintain protocol that is termed Transmission Control Protocol (TCP) and User Data-gram Protocol (UDP). The network layer transfers the packet from one device to another. All the data is converted into packages - normally computer communication is in packet structure. It determines the path and finds the logical address. The data link layer creates frames with physical addressing concepts to establish the physical connection using the Ethernet (Graveto, Rosa, Cruz & Simões, 2019). The physical layer then converts the signal into binary format which is in binary transmission structure. These are the basic hardware and software components of a network connection establishment. This cyber physical system handles the modern technologies that is named Internet of Things (IoT) and Smart City. This Internet of Things is involved in many modern applications like smart homes, smart schools, smart industries and smart world (Bitra, Jayapandian & Balachandran, 2018). These smart devices are handled and controlled with the id of IoT (Whitmore, Agarwal & Da Xu, 2015). During the next ten years without this internet none of the device will be operative, because speed with IoT technology is implemented in all of the applications. The major advantage of this IoT technology is that it can controlled anywhere in the world.

CYBER ATTACKS IN NETWORK

Data security deals with many encryption methods and algorithms which are intended to provide better security for the server or client machine. Encryption is specifically consists of two methods, symmetric and asymmetric (Jayapandian, Rahman, Radhikadevi & Koushikaa, 2016). The major challenge is network security, especially because data is transferred from one particular network terminal to another. During this transmission some hackers attempt to get access to the data: whether data is encrypted or non-encrypted becomes a secondary secondary problem. This kind of network data attack is referred to as the Man-in-the-Middle (MITM) Attack. There are many cyber security attack problems faced in network data transmission. Denial of Service (DoS) is a common problem in data attack. In this attack, users try to access the information from the client machine to the server machine making data unavailable during that period. This type of attack - Distributed Denial of Service (DDoS) - is also directed at distributed network environments. The major difference between these two attacks is based on the character of the computer connectivity and data access. DoS deals with a single computer and internet connectivity while DDoS handles multiple computers connected to the internet network (Toklu & Şimşek, 2018). There are many methods to deal with this DDoS service; one of them being the User Datagram Protocol (UDP). Phishing attacks are presently considered as a normal events in internet world. In our contemporary world, a great deal of information is shared in email communication. This type of communication is not secure because of phishing attacks. This attack transfers the sensitive information like passwords and credit or debit card information (Krombholz, Hobel, Huber & Weippl, 2015). Spear phishing attacks are a sub-category of this attack. This attack only targets individual users or individual organizations. The attack is executed by installing malware which illegally accesses our personal data and transfers that data to hackers. Most of the time this type of attack happens via public network accessibility (Nithya & Gomathy, 2019). Drive-by overloads are a common method malware dispersal in some unauthorized websites. This kind of attack is executed by using HTTP or Hypertext Pre-processor (PHP) programs. Password hacking or cracking refers to the retrieved or to recover the password from the public computer system. There are two ways to retrieve the password, one is a stored password and another one is during transmission. This type of hacking happens because of less sensitive passwords. Sometimes users utilize their name or DoB in password combinations and is easily hacked. Stealing of data and interrupting the access control during data transmission is referred as an eavesdropping attack. This attack steals information from a server or network transmission. The Man-in-the-Middle attack (MITM) is the process of inserting or hijacking the information in the middle of data transmission from the client to server machine (Ahmad et. al., 2019). This type of attack is quite natural during the network data transmission. Sometimes session hijacking also happens during network communication. This session is the process of establishing a trusted connection between the clients to the server machine. IP spoofing is modifying IP header and unauthorized access to the server machine (Dayanandam, Rao, Babu & Durga, 2019). This is host-to-host communication to create a duplicate IP address and try to access the server machine. These are all the common attacks that occurred during network data transmission. Hybrid encryption also provides better data security in the cloud environment (Jayapandian & Rahman, 2017).

AVOIDING MAN-IN-THE-MIDDLE ATTACK USING REINFORCEMENT LEARNING

The previous section discussed various cyber security attacks in network communications. Network security is additionally a challenging task during online communication. This section examines another complex task; it concerns the Man-in-the-Middle attack. The operative mode of this type attack is some hacker tries to attack the data during network communications. Network communication happens both with internet and without internet. This attack mostly occurs with the internet connection. Operationally, the internet connects several computers and shares information from one remote location to another. The problem with this category of communication is the use of public networks where it very difficult to detect attackers. Furthermore, another challenge is that most of the users always depend on third-party servers. This is because maintaining individual servers is a burdensome task and that is expensive. This section will propose solution to solve this Man-in-the-Middle attack by using reinforcement learning. Before proceeding with to the proposed method, first discuss the reinforcement learning method.

The process of reinforcement learning is congruent with machine learning and is a sub-division of the artificial intelligence methodology. The main aim of reinforcement learning is to reduce the rewards in a perdition situation. It also provides special permission of finding the best prediction and behavior in computer machine and some third-party agents. This mechanism creates an additional component; the agent, which decides the next movement. In supervised learning, decisions are taken in the form of training data-sets. This reinforcement method is used to take decisions based on agent mechanisms based on previous experience. Supervised learning algorithms are used only for some test cases; it is not suitable for dynamic situations. The working principle of reinforcement methods is learning from experience. The user input level is maintained and starts the process. There is a different type of solution provided in a problem which is the unique advantage of this reinforcement method. Every individual solution is to provide some reward points where the points are based on the success ratio of that particular attempt. Machine learning is always keeping some training data-set; this training data-set is used to maintain all these solutions with reward points. The individual user will decide on the best solution; sometime the dynamic solution is also recommended for a problem. This is a sequential decision-making system where output is dependent on previous input in a problem. Each decision will have a unique label and based on that label, a decision is sequentially taken for a specific job. Reinforcement is categorized into two types, namely, positive and negative reinforcement. Positive reinforcement refers to the creation of higher strength in the behavior model frequency. It will create a positive approach in the behavior selection. The main advantage of positive reinforcement is to increase system performance. The long-term decision-making system is to sustain and provide a better solution. At the same time the drawback of this method is that the reinforcement state is damaged. The second type is negative reinforcement; this method is to provide a higher behavior model and avoid negative results. This structure is to increase model behavior during prediction. The proposed model uses the reinforcement learning concept during network data transmission to provide solutions for the Man-in-the-Middle attack. This architecture diagram in Figure 6. elaborates the functional components of Man-in-the-Middle attack with the reinforcement learning method.

The client machine communicates data to the server machine through some network over the internet. This server works with the concept of cloud technology; most of the service providers are now moving to cloud platforms because of the cost effectiveness. The MITM attack is a network cyber attacker. This attacker is a professional who focuses on network protocol s and hacks the network data. The entire network data is transferred in the form of packets having two fields where one is header field and other one is the data field. The MITM attacker always focuses on this data field. There are two kinds of attacks that

Figure 6. Proposed Reinforcement Learning Approach

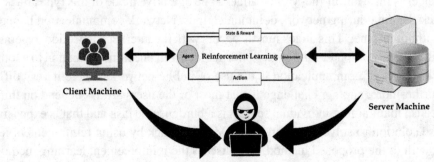

happen during network communication. The first one is data attack without modification which means the attacker does not modify any data during the data hack. This kind of hacking sometimes happens on social media by just hacking that data and disclosing it. This attack is faced by some government servers and terrorists that attack military information and disclose that information in public forums. Another kind of attack is data modification which means after the attack, the data hacker can try to modify that original data for a short duration of time. The major challenge of this task is to modify the data in some time interval, because networking protocol is working with time interval protocol. If the data is not reached within that time duration, that data is automatically disqualified and destroyed from the network terminal. This type of attack is also attempted in some banking and financial transactions. Most of the time this type of attack happens to individual server maintenance because individual server maintenance is more expensive so they do not concentrate on security. It is for this reason that many small industries depend on third party service providers which regrettably may not be trustworthy. The problem of third-party service providers is the lack of their privacy of their data. It also arrives that third party service providers misuse the client personal data for the purposes of marketing. So in either case, the security problem is permanently prevalent inciting the author to propose machine learning security protocol as a solution. The main objective of the network data transmission security is to deal with third eye or learning algorithm. In machine learning, there are many learning algorithms available, with reinforcement being the most successful. This proposed method is to create virtual agent during network transmission where the reinforcement decision is taken in the agent-based system. The purpose of this virtual agent is to avoid the middle man attack during data transmission. The MITM attack is involved at the mid-point between the client and server; similarly the agent mechanism is also installed in middle of the data communication. The working principle of this virtual agent system is to maintain a dynamic table and update all the network activity problems and solutions. Thus whenever a similar type of problem is encountered during network transmissions this virtual agent refers previous solutions and provide suggestions on how to avoid the current data attack. All this activity is happens in dynamic situations. The main advantage of this virtual method is that it is continuously attempting to find the best solution based on previous case histories. Reinforcement learning in two different terminals are referred to as an agent and environment. In this case environment is the network. There are three state conditions associated with the state mechanism. In the upper case, two states that are designated, state and reward; both states are connected to agents from the network environment. The reward state provides rewards for every individual solution in the virtual agent. This happens dynamically after the third or fourth cycle when that particular solution gets higher rewards. This virtual agent is automatically predicted the best solution based on this reward procedure. The solution give a warning to potential data attack activity to the client or service provider. The third state is the action state, which is the waiting signal from the agent component. This is engaged once the best solution is selected based on reward points. The network administrator always monitors the activities and subsequently provide higher security in the network terminal. The reinforcement learning algorithm functioning in dynamic situations thus provides situation-based solutions. Many other learning algorithms are also available under the machine learning concept, with the approach and uniqueness of this reinforcement algorithm being dynamism and its situational prediction mechanism.

RESULTS AND DISCUSSION

These result and discussion parts are focused on the analysis of security enhancement for network communication. This chapter uses reinforcement algorithm to solve network communication attacks. This proposed method is simulated with the help of Java Network Simulator (JNS). This JNS is used to create a network simulator (NS2) network environment by using Java. Network protocol is designed and establishes the connection between one computer system to another terminal. This proposed method concept is to store all the network activities in a separate table. That table is accessed in dynamic situations and is likewise updated in dynamic environments.

These network activities are updated in the memory table; system unique media access control (MAC) ID, IP address and type of network attacks. The purpose of this table is to compare the existing network attacks. The resulting type of attack is matched with that current table to existing table data. If that is matched, it means that a reinforcement agent provides a better solution based on the previous prediction match percentages. This simulation takes various parameters like the network attack ratio, Transmission time and Quality of Service (QoS). QoS parameter measures the following metrics that is named as throughput, packet delivery percentage and average packet delay ratio.

Figure 7 illustrates the network packet attacker ratio comparison between the stable methods, and proposed reinforcement method where the latter who provide better results. That means that the attacker ratio is reduced; in this case by 3.8%. The traditional method average attacker's ratio is 11.6% and at the same time, reinforcement algorithm method average attacker's ratio is 7.8%.

Figure 8 discusses network transmission time which is the overall transmission time used to calculate transmission of packets from one remote terminal to another. The average transmission time is almost 5.2% less in the proposed reinforcement method. Traditional method requires 0.5282 ms for five differ-

Figure 7. Packet Attack Ratio Analysis

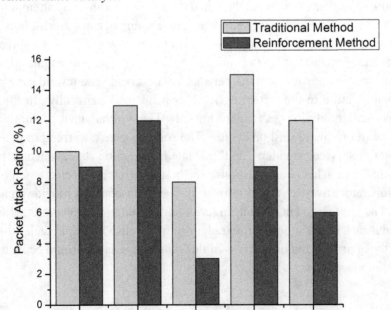

Figure 8. Packet Transmission Time Analysis

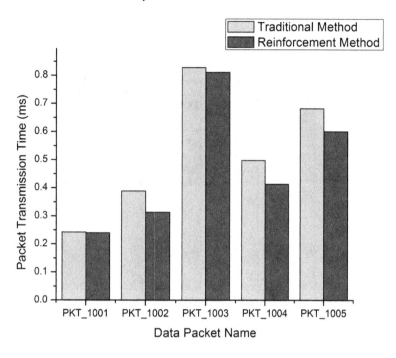

Figure 9. Throughput Analysis

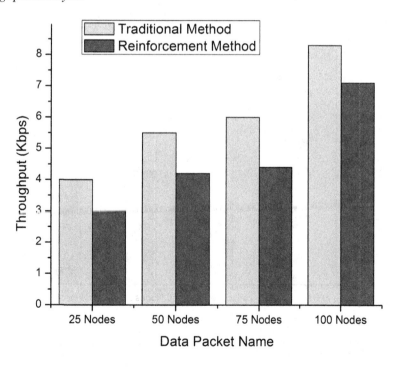

Figure 10. Packet Delivery Percentage Analysis

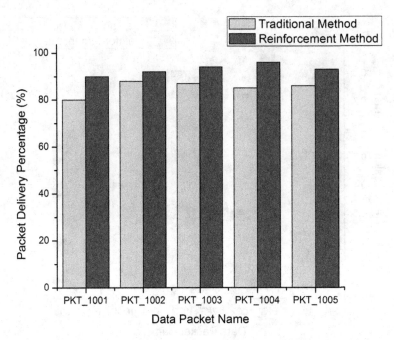

Figure 11. Packet Delay Time Analysis

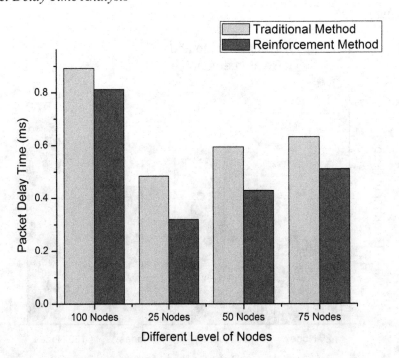

ent data packet transmissions, while the reinforcement method takes only 0.4762 ms for the same data packets. It shows that the proposed method is taking less transmission time.

Figure 9 analyzes network throughput where the general number of user nodes increase implies a decrease in the network throughput. The average throughput rate is reduced by 0.01275%. The four-difference network node is accessed is categorized by the difference in range from node 25 to 100.

Figure 10 shows the data packet delivery percentage; here average packet delivery percentage is increased by 7.8%. The traditional method delivers 85.2% of the packets while for the same scenario, after implementing the reinforcement method packet, delivery percentage increases to 93%.

Figure 11 demonstrates the network packet delay ratio; this indicates direct network quality. Compared to the traditional method, reinforcement method delay ratio is gradually reduced by 13.265%. The above result indicates that proposed reinforcement method provide better security in network data transmission.

Based on the simulation result, the network attacker ratio, transmission time, throughput and packet delay ratio are decreased in the proposed method. The packet delivery percentage is increased which indicates that the proposed QoS is increased compared to the existing method.

CONCLUSION

In our contemporary world modern technology is ineluctable. Network is a pillar of this modern and ever-expanding technology. Networking is not only utilized to connect computer devices, it is also used to connect modern gadgets. Machine learning and artificial intelligence algorithms are used in many traditional areas. These algorithms are used in the financial and banking sector to reduce the human manpower. This chapter reviewed machine learning algorithms and networking components. At the present, nearly ninety-nine percentage of modern devices are employed in IoT technology and all these devices are working with the help of network protocols. One of the major challenges in this technology are network data attacks. This chapter provided an innovative solution for this network security attack. In machine learning, algorithm are a prediction-based. Reinforcement learning apparently is one of the best machine learning algorithms. The proposed method is to use this reinforcement learning method to solve this network security problem. This method is provided a better and more rapid solution for network hacking. The main aim of this proposed method is to avoid middle man attack by using this reinforcement learning approach. This proposed idea is to increase the network transmission quality and provide higher security protocol in data transmission. Based on the simulation result, it concludes that the reinforcement algorithm provides better result compared to traditional methods.

REFERENCES

Ahmad, M., Younis, T., Habib, M. A., Ashraf, R., & Ahmed, S. H. (2019). A Review of Current Security Issues in Internet of Things. In *Recent Trends and Advances in Wireless and IoT-enabled Networks* (pp. 11–23). Cham: Springer. doi:10.1007/978-3-319-99966-1_2

Bitra, V. S., Jayapandian, N., & Balachandran, K. (2018). Internet of Things Security and Privacy Issues in Healthcare Industry. In *International Conference on Intelligent Data Communication Technologies and Internet of Things* (pp. 967-973). Springer.

Boas, M. G. V., Santos, H. G., & de Campos Merschmann, L. H. (2018). Optimal Decision Trees for Feature Based Parameter Tuning: Integer Programming Model and VNS Heuristic. *Electronic Notes in Discrete Mathematics*, *66*, 223–230. doi:10.1016/j.endm.2018.03.029

Borky, J. M., & Bradley, T. H. (2019). Developing the Network Dimension. In *Effective Model-Based Systems Engineering* (pp. 327–344). Cham: Springer. doi:10.1007/978-3-319-95669-5_9

Brugere, I., Gallagher, B., & Berger-Wolf, T. Y. (2018). Network structure inference, a survey: Motivations, methods, and applications. *ACM Computing Surveys*, *51*(2), 24. doi:10.1145/3154524

Buda, M., Maki, A., & Mazurowski, M. A. (2018). A systematic study of the class imbalance problem in convolutional neural networks. *Neural Networks*, *106*, 249–259. doi:10.1016/j.neunet.2018.07.011 PMID:30092410

Cai, J., Wang, Y., Liu, Y., Luo, J. Z., Wei, W., & Xu, X. (2018). Enhancing network capacity by weakening community structure in scale-free network. *Future Generation Computer Systems*, *87*, 765–771. doi:10.1016/j.future.2017.08.014

Camacho, D. M., Collins, K. M., Powers, R. K., Costello, J. C., & Collins, J. J. (2018). Next-generation machine learning for biological networks. *Cell*, *173*(7), 1581–1592. doi:10.1016/j.cell.2018.05.015 PMID:29887378

Chapelle, O., Scholkopf, B., & Zien, A. (2009). Semi-supervised learning. *IEEE Transactions on Neural Networks*, *20*(3), 542–542. doi:10.1109/TNN.2009.2015974

Chaturvedi, I., Ragusa, E., Gastaldo, P., Zunino, R., & Cambria, E. (2018). Bayesian network based extreme learning machine for subjectivity detection. *Journal of the Franklin Institute*, *355*(4), 1780–1797. doi:10.1016/j.jfranklin.2017.06.007

Chen, Y. Q., Zhang, J., & Ng, W. W. (2018). Loan Default Prediction Using Diversified Sensitivity Undersampling. In *2018 International Conference on Machine Learning and Cybernetics (ICMLC)* (Vol. 1, pp. 240-245). IEEE. 10.1109/ICMLC.2018.8526936

Cui, L., Yang, S., Chen, F., Ming, Z., Lu, N., & Qin, J. (2018). A survey on application of machine learning for Internet of Things. *International Journal of Machine Learning and Cybernetics*, 1–19.

Dayanandam, G., Rao, T. V., Babu, D. B., & Durga, S. N. (2019). DDoS Attacks—Analysis and Prevention. In *Innovations in Computer Science and Engineering* (pp. 1–10). Singapore: Springer. doi:10.1007/978-981-10-8201-6_1

Edwards, J., & Bramante, R. (2015). *Networking self-teaching guide: OSI, TCP/IP, LANs, MANs, WANs, implementation, management, and maintenance*. John Wiley & Sons.

Fan, S. K. S., Su, C. J., Nien, H. T., Tsai, P. F., & Cheng, C. Y. (2018). Using machine learning and big data approaches to predict travel time based on historical and real-time data from Taiwan electronic toll collection. *Soft Computing*, *22*(17), 5707–5718. doi:10.100700500-017-2610-y

Fanjul-Peyro, L., & Ruiz, R. (2010). Iterated greedy local search methods for unrelated parallel machine scheduling. *European Journal of Operational Research*, *207*(1), 55–69. doi:10.1016/j.ejor.2010.03.030

Gercke, M. (2012). *Understanding cybercrime: phenomena, challenges and legal response.* Academic Press.

Ghahramani, Z. (2015). Probabilistic machine learning and artificial intelligence. *Nature, 521*(7553), 452–459. doi:10.1038/nature14541 PMID:26017444

Graveto, V., Rosa, L., Cruz, T., & Simões, P. (2019). A stealth monitoring mechanism for cyber- physical systems. *International Journal of Critical Infrastructure Protection, 24,* 126–143. doi:10.1016/j.ijcip.2018.10.006

Guan, Y., Wei, Q., & Chen, G. (2019). Deep learning based personalized recommendation with multiview information integration. *Decision Support Systems, 118,* 58–69. doi:10.1016/j.dss.2019.01.003

Hall, M., Frank, E., Holmes, G., Pfahringer, B., Reutemann, P., & Witten, I. (2009). *The Weka Data Mining Software.* Retrieved from https://dl.acm.org/citation.cfm?Id=1656278

Henry, E., Adamchuk, V., Stanhope, T., Buddle, C., & Rindlaub, N. (2019). Precision apiculture: Development of a wireless sensor network for honeybee hives. *Computers and Electronics in Agriculture, 156,* 138–144. doi:10.1016/j.compag.2018.11.001

Huang, J., Zhao, M., Zhou, Y., & Xing, C. C. (2018). In-vehicle networking: Protocols, challenges, and solutions. *IEEE Network,* (99): 1–7.

Jayapandian, N., & Md Zubair Rahman, A. M. J. (2018). Secure Deduplication for Cloud Storage Using Interactive Message-Locked Encryption with Convergent Encryption, To Reduce Storage Space. *Brazilian Archives of Biology and Technology, 61*(0), 1–13. doi:10.1590/1678-4324-2017160609

Jayapandian, N., Pavithra, S., & Revathi, B. (2017). Effective usage of online cloud computing in different scenario of education sector. In *Innovations in Information, Embedded and Communication Systems (ICIIECS), 2017 International Conference on* (pp. 1-4). IEEE. 10.1109/ICIIECS.2017.8275970

Jayapandian, N., & Rahman, A. M. Z. (2017). Secure and efficient online data storage and sharing over cloud environment using probabilistic with homomorphic encryption. *Cluster Computing, 20*(2), 1561–1573. doi:10.100710586-017-0809-4

Jayapandian, N., Rahman, A. M. Z., & Gayathri, J. (2015). The online control framework on computational optimization of resource provisioning in cloud environment. *Indian Journal of Science and Technology, 8*(23), 1–13. doi:10.17485/ijst/2015/v8i23/79313

Jayapandian, N., Rahman, A. M. Z., Radhikadevi, S., & Koushikaa, M. (2016). Enhanced cloud security framework to confirm data security on asymmetric and symmetric key encryption. In *Futuristic Trends in Research and Innovation for Social Welfare (Startup Conclave), World Conference on* (pp. 1-4). IEEE. 10.1109/STARTUP.2016.7583904

Jouini, M., & Rabai, L. B. A. (2019). A security framework for secure cloud computing environments. In Cloud Security: Concepts, Methodologies, Tools, and Applications (pp. 249-263). IGI Global. doi:10.4018/978-1-5225-8176-5.ch011

Karov, Y., Breakstone, M., Shilon, R., Keller, O., & Shellef, E. (2017). *U.S. Patent No. 9,772,994*. Washington, DC: U.S. Patent and Trademark Office.

Kaufman, L. M. (2009). Data security in the world of cloud computing. *IEEE Security and Privacy*, *7*(4), 61–64. doi:10.1109/MSP.2009.87

Kotsiantis, S. B., Zaharakis, I., & Pintelas, P. (2007). Supervised machine learning: A review of classification techniques. Emerging artificial intelligence applications in computer engineering, 160, 3-24.

Krombholz, K., Hobel, H., Huber, M., & Weippl, E. (2015). Advanced social engineering attacks. *Journal of Information Security and Applications*, *22*, 113-122.

Lau, R. Y., Xia, Y., & Ye, Y. (2014). A probabilistic generative model for mining cybercriminal networks from online social media. *IEEE Computational Intelligence Magazine*, *9*(1), 31–43. doi:10.1109/MCI.2013.2291689

Li, B., & Yang, Y. (2018). Complexity of concept classes induced by discrete Markov networks and Bayesian networks. *Pattern Recognition*, *82*, 31–37. doi:10.1016/j.patcog.2018.04.026

Li, J., Cheng, K., Wang, S., Morstatter, F., Trevino, R. P., Tang, J., & Liu, H. (2018). Feature selection: A data perspective. *ACM Computing Surveys*, *50*(6), 94.

Lino, A., Rocha, Á., Macedo, L., & Sizo, A. (2019). Application of clustering-based decision tree approach in SQL query error database. *Future Generation Computer Systems*, *93*, 392–406. doi:10.1016/j.future.2018.10.038

Lorena, A. C., Maciel, A. I., de Miranda, P. B., Costa, I. G., & Prudêncio, R. B. (2018). Data complexity meta-features for regression problems. *Machine Learning*, *107*(1), 209–246. doi:10.100710994-017-5681-1

Luiijf, H. A. M., Besseling, K., Spoelstra, M., & De Graaf, P. (2011). Ten national cyber security strategies: A comparison. In *International Workshop on Critical Information Infrastructures Security* (pp. 1-17). Springer.

Luo, J., Hong, T., & Fang, S. C. (2018). Benchmarking robustness of load forecasting models under data integrity attacks. *International Journal of Forecasting*, *34*(1), 89–104. doi:10.1016/j.ijforecast.2017.08.004

Luo, X., Jiang, C., Wang, W., Xu, Y., Wang, J. H., & Zhao, W. (2019). User behavior prediction in social networks using weighted extreme learning machine with distribution optimization. *Future Generation Computer Systems*, *93*, 1023–1035. doi:10.1016/j.future.2018.04.085

Mahmoud, T., Dong, Z. Y., & Ma, J. (2018). An advanced approach for optimal wind power generation prediction intervals by using self-adaptive evolutionary extreme learning machine. *Renewable Energy*, *126*, 254–269. doi:10.1016/j.renene.2018.03.035

Majumder, A., & Nath, S. (2019). Classification of Handoff Schemes in a Wi-Fi-Based Network. In *Enabling Technologies and Architectures for Next-Generation Networking Capabilities* (pp. 300–332). IGI Global. doi:10.4018/978-1-5225-6023-4.ch014

Michalski, R. S., Carbonell, J. G., & Mitchell, T. M. (Eds.). (2013). *Machine learning: An artificial intelligence approach. Springer Science* Business Media.

Min, J. H., & Jeong, C. (2009). A binary classification method for bankruptcy prediction. *Expert Systems with Applications*, *36*(3), 5256–5263. doi:10.1016/j.eswa.2008.06.073

Mishra, A. K., Yadav, D. K., Kumar, Y., & Jain, N. (2019). Improving reliability and reducing cost of task execution on preemptible VM instances using machine learning approach. *The Journal of Supercomputing*, *75*(4), 2149–2180. doi:10.100711227-018-2717-7

Moeyersoms, J., de Fortuny, E. J., Dejaeger, K., Baesens, B., & Martens, D. (2015). Comprehensible software fault and effort prediction: A data mining approach. *Journal of Systems and Software*, *100*, 80–90. doi:10.1016/j.jss.2014.10.032

Monteiro, A., Souto, E., Pazzi, R., & Nogueira, M. (2019). Context-aware network selection in heterogeneous wireless networks. *Computer Communications*, *135*, 1–15. doi:10.1016/j.comcom.2018.11.006

Neal, R. M. (2012). *Bayesian learning for neural networks* (Vol. 118). Springer Science & Business Media.

Nithya, S., & Gomathy, C. (2019). Smaclad: Secure Mobile Agent Based Cross Layer Attack Detection and Mitigation in Wireless Network. *Mobile Networks and Applications*, 1–12.

Park, H., Haghani, A., Samuel, S., & Knodler, M. A. (2018). Real-time prediction and avoidance of secondary crashes under unexpected traffic congestion. *Accident; Analysis and Prevention*, *112*, 39–49. doi:10.1016/j.aap.2017.11.025 PMID:29306687

Pasqualetti, F., Dörfler, F., & Bullo, F. (2013). Attack detection and identification in cyber-physical systems. *IEEE Transactions on Automatic Control*, *58*(11), 2715–2729. doi:10.1109/TAC.2013.2266831

Pérez-Sánchez, B., Fontenla-Romero, O., & Guijarro-Berdiñas, B. (2018). A review of adaptive online learning for artificial neural networks. *Artificial Intelligence Review*, *49*(2), 281–299. doi:10.100710462-016-9526-2

Quinlan, J. R. (2014). *C4. 5: programs for machine learning*. Elsevier.

Rubab, S., Taqvi, S. A., & Hassan, M. F. (2018). Realizing the Value of Big Data in Process Monitoring and Control. In *Current Issues and Opportunities. In International Conference of Reliable Information and Communication Technology* (pp. 128-138). Springer.

Schatz, D., Bashroush, R., & Wall, J. (2017). Towards a more representative definition of cyber security. *Journal of Digital Forensics. Security and Law*, *12*(2), 8.

Shon, T., & Moon, J. (2007). A hybrid machine learning approach to network anomaly detection. *Information Sciences*, *177*(18), 3799–3821. doi:10.1016/j.ins.2007.03.025

Steve Morgan. (2018). *2018 Cybersecurity Market Report*. Retrieved from https://cybersecurityventures.com/cybersecurity-market-report/

Sun, X., Wu, P., & Hoi, S. C. (2018). Face detection using deep learning: An improved faster RCNN approach. *Neurocomputing*, *299*, 42–50. doi:10.1016/j.neucom.2018.03.030

Toklu, S., & Şimşek, M. (2018). Two-Layer Approach for Mixed High-Rate and Low-Rate Distributed Denial of Service (DDoS) Attack Detection and Filtering. *Arabian Journal for Science and Engineering*, *43*(12), 7923–7931. doi:10.100713369-018-3236-9

Tsai, C. F., Hsu, Y. F., Lin, C. Y., & Lin, W. Y. (2009). Intrusion detection by machine learning: A review. *Expert Systems with Applications*, *36*(10), 11994–12000. doi:10.1016/j.eswa.2009.05.029

Von Solms, R., & Van Niekerk, J. (2013). From information security to cyber security. *Computers & Security*, *38*, 97-102.

Wang, H., Xu, Z., & Pedrycz, W. (2017). An overview on the roles of fuzzy set techniques in big data processing: Trends, challenges and opportunities. *Knowledge-Based Systems*, *118*, 15–30. doi:10.1016/j.knosys.2016.11.008

Wang, M., Cui, Y., Wang, X., Xiao, S., & Jiang, J. (2018). Machine learning for networking: Workflow, advances and opportunities. *IEEE Network*, *32*(2), 92–99. doi:10.1109/MNET.2017.1700200

Wang, X., Liu, X., Pedrycz, W., & Zhang, L. (2015). Fuzzy rule based decision trees. *Pattern Recognition*, *48*(1), 50–59. doi:10.1016/j.patcog.2014.08.001 PMID:25395692

Whitmore, A., Agarwal, A., & Da Xu, L. (2015). The Internet of Things-A survey of topics and trends. *Information Systems Frontiers*, *17*(2), 261–274. doi:10.100710796-014-9489-2

Witten, I. H., Frank, E., Hall, M. A., & Pal, C. J. (2016). *Data Mining: Practical machine learning tools and techniques*. Morgan Kaufmann.

Woo, J., Shin, S. J., Seo, W., & Meilanitasari, P. (2018). Developing a big data analytics platform for manufacturing systems: Architecture, method, and implementation. *International Journal of Advanced Manufacturing Technology*, *99*(9-12), 2193–2217. doi:10.100700170-018-2416-9

Xuan, S., Man, D., Yang, W., Wang, W., Zhao, J., & Yu, M. (2018). Identification of unknown operating system type of Internet of Things terminal device based on RIPPER. *International Journal of Distributed Sensor Networks*, *14*(10). doi:10.1177/1550147718806707

Zhang, Q., Yang, L. T., Castiglione, A., Chen, Z., & Li, P. (2019). Secure weighted possibilistic c-means algorithm on cloud for clustering big data. *Information Sciences*, *479*, 515–525. doi:10.1016/j.ins.2018.02.013

Zhang, X., Li, R., Zhang, B., Yang, Y., Guo, J., & Ji, X. (2019). An instance-based learning recommendation algorithm of imbalance handling methods. *Applied Mathematics and Computation*, *351*, 204–218. doi:10.1016/j.amc.2018.12.020

Zhu, X., & Choulli, E. (2018). Acquisition and communication system for condition data of transmission line of smart distribution network. *Journal of Intelligent & Fuzzy Systems*, 1-14.

KEY TERMS AND DEFINITIONS

Cyber Security: Cyber security is the category of information security system during internet communication data attack is happen avoiding this attack is named as cyber security.

Machine Learning: Machine learning is the process of predicting future result based on previous history.

Man-in-the-Middle Attack (MITM): Man-in-the-middle attack means hacking the data during network transmission in public network.

Network Security: Network security provides better security system during network data transmission.

Reinforcement Learning: Reinforcement learning method is to create virtual agent for the purpose of taking dynamic decision.

APPENDIX: ACRONYMS

API: Application programming interface.
DAG: Directed acyclic graphs.
DDoS: Distributed denial of service.
DNS: Domain name system.
DoB: Date of birth.
DoS: Denial of service.
ECS: Electronic clearance system.
HTTP: Hypertext transfer protocol.
ILP: Inductive logic programming.
IoT: Internet of things.
IP: Internet protocol.
JNS: Java network simulator.
LAN: Local area network.
MAC: Media access control.
MITM: Man-in-the-middle attack.
NP: Non-deterministic polynomial.
PART: Partial decision tree (RWeka).
PHP: Hypertext pre-processor.
QoS: Quality of service.
RIPPER: Repeated incremental pruning to produce error reduction.
SSL: Secure sockets layer.
SVM: Support vector machine.
TCP: Transmission control protocol.
TCP/IP: Transmission control protocol/internet protocol.
UDP: User data-gram protocol.

Chapter 12
The Intersection of Data Analytics and Data–Driven Innovation

Marcus Tanque
Independent Researcher, USA

Harry J. Foxwell
George Mason University, USA

ABSTRACT

This chapter discusses businesses, key technology implementations, case studies, limitations, and trends. It also presents recommendations to improve data analysis, data-driven innovation, and big data project implementation. Small-to-large-scale project inefficiencies present unique challenges to both public and private sector institutions and their management. Data analytics management, data-driven innovation, and related project initiatives have grown in scope, scale, and frequency. This evolution is due to continued technological advances in analytical methods and computing technologies. Most public and private sector organizations do not deliver on project benefits and results. Many organizational and managerial practices emphasize these technical limitations. Specialized human and technical resources are essential for an organization's effective project completion. Functional and practical areas affecting analytics domain and ability requirements, stakeholder expectations, solution infrastructure choices, legal and ethical concerns will also be discussed in this chapter.

INTRODUCTION

There has been an increase in the integration of the technology labor force, autonomous machines, and computing devices in the last decade. The intersection of technology and labor force is due to enhanced technological innovations that continues to be observed in recent years (Organization, 2015). This chapter examines data analytics management and data-driven innovation (Desjardins, 2017). The juncture of technology and human capital spans disruptive technology evolution and challenges as well as trends affecting data analytics management and data-driven innovation (OECD, 2015). It also emphasizes

DOI: 10.4018/978-1-5225-9687-5.ch012

big data analytics, data-driven innovations, organizational and technical challenges. This technology trend also addresses domain ability requirements, stakeholder expectations, legal & ethical concerns, and solution infrastructure decision methods and techniques. This chapter discusses underlying data analytics evolution, data-driven innovations, and trends (Desjardins, 2017). The next section discusses the background of the chapter.

BACKGROUND

Applying the right knowledge to questions and challenges is vital for producing actionable data collection results. Data analytics is a process of parsing, purging, cleansing, moving, and visualizing information (OECD, 2015; Desjardins, 2017). The process further builds on the goal of assembling valuable and actionable data (OECD, 2015; Rankin, 2013; Maydon, 2017). The chapter also emphasizes data analytics methods and procedures used in the modern-day technology landscape. It underlines regulations, analytics methods, and the level of management oversight decision makers need to deliver successful project outcomes (Maydon, 2017). Data analysis is a term that describes the end-to-end domain process. Hence, this method applies to analysts and data scientists who have valuable capabilities to acquire, process, visualize, and interpret small and large datasets. The following segment focuses on data analytics' evolution, challenges, techniques, opportunities, and trends.

Data-Driven Innovation

Data-Driven Innovation (DDI) is a process of generating innovative outputs or amount of produced data. DDI encompasses advanced applications deriving from data analytics. For many years, this method has increased the production rate in the digital age. This growth is due to continued progress made in information and communications technology, data science, and research and development domains. This innovation would have never been possible without enormous technology contributions that Amazon, Apple, Google, Microsoft, Facebook, IBM, Dell Inc., Tweeter, and others have made in recent years. These multinational corporations have made considerable progress in data-driven analytics and related domains of ability. These technology indicators and trends have led to new data analytics' discoveries spanning various industries: healthcare, banking, cyber, defense, social media, engineering, aviation, telecom, automobile, entertainment, research & development, education, oil and gas, real estate, supply chain, forensics, law enforcement, aerospace, and others. DDI processes and effects consist of:

- **Current Data**
  - Internal Data Sets
  - External Data Sets
  - Government Data
  - Open Data
  - Fee-only Data Sets
- **Data Processing and Analysis**
  - Computers
  - Keyboards/Consoles
  - Tables

- **Innovation**
  - *Processes*
    - Increasing process efficiency
    - Increasing supply chain productivity
    - Improving data-based decision-making process
  - *Marketing*
    - Finding and defining the target pool of audiences
    - Exclusively tailored advertising processes and marketing channel optimization
    - Individually customized products or merchandises
    - Launching new merchandise and service lines
  - *Business*
    - Data-based innovation services
    - Data-based systems and tools
    - Rendering data-based innovation services and products

Many researchers argue that DDI has paved the way for more technological innovations in the 21st century. This progress includes new corporate solutions: methods, policies, procedures, tools, and practices for data science, big data, and business intelligence. The following processes describe the data value categorization:

- Knowledge Base
- Decision-Marking
  - Value-added Proposition, Growth, and Well-being
- Data Collection and Datafication
- Big Data
- Data Analytics, such as software and skill-based capabilities

DDI spans several technologies, business methods, and practices. The need for organizations to adapt to increasing data analytics innovations is key to data-driven solutions. These trends focus on the ability for an organization to modernize its traditional IT solutions, to define, extract, measure, capture, and process large data volumes to produce desired outputs. Aside from continued progress, both public and private sector organizations have made, there are still significant challenges affecting DDI processes. In recent years, these challenges have affected several industries. Limited investment or deployment of mobile broadband in rural and remote areas has been one of the significant technological setbacks' governments continue to deal with in the 21st century. Vendors should develop proven solutions to support these initiatives. Despite these constraints, governments and hi-tech firms continue to work together to find solutions to these technological challenges affecting rural areas.

Use Case

In 2013 McKinsey Global Institute researched data-driven innovation. It showed that each year, open source data produces nearly three-trillion dollars to the global economy (Mandinach, 2012; Rankin, 2013). In its findings, McKinsey also argued that DDI is a process that focuses on economic growth and infrastructural resources, which can be available for public use and unlimited purposes. In the supply chain, data is used to collect and analyze data sets needed for processing services and goods (Mandinach, 2012). Such process, however, proves that large volumes of data outside of public and private IT sectors can be collected, analyzed, and processed from the following communities: public administration, educational, and health services (Rankin, 2013). These industries also rely on data for scientific medical discoveries, researching on chronical and non-chronical diseases, to improve people's lives and adaptive learning environments. Governments use data to support their citizenry needs through accountability, transparency and increase openness. These processes allow that government to serve their communities as well as assist in rebuilding public trust in their officials (Mandinach, 2012; Rankin, 2013). Another potential area of data-driven innovation is the development of smart cities. "Smart cities' portals offer a great amount of data that can be used by the private and public entities to create new services. These data are also a valuable source for the deployment of big data businesses" (Abella et al., 2017), and provide a model for innovation in three stages:

- The first one shapes the release of data by the smart city
- The second stage analyses the mechanisms to create innovative products and services
- The last stage explains how these products and services impact their society.

Analytics Evolution, Opportunities, and Trends

The ecosystem can decide essential data-driven innovations. For many years data strategy has played an integral role in a data analytics environment. Hence, more processes are being improved to support continued demand for data-driven analytics. This process involves decision-makers' ability to make informed decisions for their daily organizational requirements. These technological efforts involve systematizing an organizational infrastructure that offers guidance and balanced visions needed for business success. How organizations adhere to their corporate leadership visions differs from industry-to-industry (Maydon, 2017). A categorization of developmental processes can find critical results of these technology advances, such as solution adoption and storage management. Most organizations can invest in technology solutions to support commercial, scientific, and social media applications (OECD, 2015). In theory, a single minute of Internet traffic may involve more than three-million Google searches, sixteen million text messages, hundred fifty-six million e-mails, and four-million YouTube videos. For many years, thousands of petabytes of data have been transferred and stored on organizational and individual systems (Desjardins, 2017).

Decision makers view analytics as a concept for operationalizing devices, systems, applications, and driving business growth. These solutions have increased organizational business and operational capabilities. Many researchers and practitioners define this analytics method as an orderly process offering decision makers practical tools to support their decision actions (OECD, 2015). These trends are factors needed to provide an organization with enhanced analytics capabilities, for instance, "*right-time and real-time*." The proposed analytical model has played an integral role in the modern-day business and technology industries (Desjardins, 2017). Business and technologies leaders depend on analytics tools, to make the right market and operational decisions while driving organizations' competitive advantages and annual profit margins. Thus, executive leaders define data analytics a technology platform that organizations to make business dealings through effectiveness.

This method is tailored to organizations whose business focus on generating product revenues by using data analytics as a platform for supporting product marketability (Rankin, 2013). These trends are determined by a detected parity in volume and frequency, particularly data increase and results. These outcomes are showing measures that managers should rely on to make informed decisions. These analytical activities can be implemented automatically through the reliance of devices, systems, databases, applications others (OECD, 2015). Through unified analytics capabilities, organizations can increase interaction with customers while satisfying their business requirements. This process guarantees that decision makers have the toolkit to monitor, market, and make actionable decisions that can help customer's business and technology requirements (Maydon, 2017). In data analytics, a large amount of data is needed for mining actionable and tracing content stored in various data repositories. This process generates statistical analysis, exploring, summarizing, and visualizing dataset types and volumes. These components range beyond the scope of numerical data presentation to specific levels of ability needed to manage data collection and analysis, e.g., curating data repositories and infrastructures (Maydon, 2017). The following section examines organizational and technical analytics.

Organizational and Technical Analytics

For many organizations, analysts and managers have played a critical role in improving day-to-day business growth and success. These professionals often perform as technical subject matter experts in the organization's engineering teams. While the analysts are responsible for understanding holistic/involvedness of operational dimensions relating to data residing in the organization. A data-driven organization typically encompasses many teams consisting of technical experts (OECD, 2015). These professionals consist of analysts, engineers, scientists, quants, statisticians, accountants, auditors, and others, whose role is to offer the ultimate ability to support organizational business efficiencies. These roles often intersect at the organizational, business, and technical levels. Driving productivity and allowing analysts to strive is every organization's business goal (Shingles et al., 2017).

In engineering, analytics is a business concept that involves multi-layer operations within an organization. An individual having this skill set can engage in data-driven activities within an organization. Merging technical and organizational data analytics structures is an involving process that often yields positive impacts on human elements. Data analytics and data-driven innovation are topics that include technical and organizational analytics structures. These schemes encompass eminence staff augmentation and organizational ability. Vendors are committed to developing innovative analytics toolsets. These capabilities are developed to support data agility, reliability, aptness, and scalability (OECD, 2015; Shingles et al., 2017). Organizations should hire and keep the most exceptional data analytics personnel to serve in crucial various business roles or dimensions. In an organization, the roles of data analysts and engineers are to get, cleanse, munge, and alter data to serve a requirement. These experts' responsibilities encompass processing data, to where analysts can evaluate/analyze the daily production and visible project results (Shingles et al., 2017). Furthermore, analysts handle adopting or implementing business intelligence toolsets, often used in the analytics data infrastructure.

Data analytics teams include machine learning, big data, data visualization, mathematics/operations research, statistics, development, and operations, also known as DevOps, business, and programming. Each of these areas consists of experts with varied backgrounds in information technology, software development/engineering, system administration, and others. Through data business, people can categorize some of these industrious analytics sectors, data researchers, data developers, and data analysts. Data analytics' roles stem from individual and group skill set/talents. These resources or capacities involve (Rankin, 2013; OECD, 2015): numerating, detail-oriented & methodical, aptly unconvinced, poised, inquiring, excellent communicators and narrators, patients, data lovers, life-learners, practical and business-savvy. Each of these sectors merely defines the role of data analysts and scientists (Shingles et al., 2017). Analytics process and challenges are discussed in the next section.

Analytics Processes and Challenges

Technology industries have worked toward helping customers achieve large-scale project goals. These industry experts' bearable contributions to implement agile data analytics capabilities, to support an organization's productivity and making analytical data estimations is vital to an organization's growth. Decision-makers handle refining data analytics in aid of large-scale project missions (OECD, 2015). Besides, small, medium, and large-scale organizations continue to seize the opportunity to make parallel and develop better-quality data analytics capabilities to augment effectiveness, advance policymaking procedures, and gaining competitive benefits. Using analytics capabilities and methods organizations can evaluate, use, and define any variations between product losses and gains in the pipeline. Mapping these analytical solutions can be complicated and challenging to implement. Decision-makers opt to define and apply determinate policies focused on supporting the organization's day-to-day projects and results (Rankin, 2013).

Lacking confidence in analytical data processing, visualization, protection, and storing can be detrimental to organizations and their decision makers. Despite technological and technical shortcomings; many industries have made concerted advances to ensure that some of these technical limitations affecting data, can be addressed, and overcome. Such process stems from best practices and technologies being developed or adopted to minimize many challenges (OECD, 2015). Continuous progress in data analytics focuses on enhancing an organization's capability to systematically operationalize analytical capabilities and enhance decision makers' overall operational visions. If well planned, executed, and implemented these advances can reward many organizations and their customers (OECD, 2015). The next section supplies in-depth analytics methods and associated technologies.

Analytics Methods and Technologies

How organizations often work with their customers is essential to business success. Many organizations often receive help from using integrated data analytics in business systems (Halper, 2015, p. 7). This process allows decision-makers to make actionable business and operational decision. Any analytical capabilities without well-defined actions are useless to organizations program missions. Decision makers rely on business and operationalized organizational guidelines to make a mutual decision. These activities can perform quasi-automatically or automatically process. This process also focuses on aiding organizations with preventive maintenance routine services: monitoring infrastructure resources experiencing anomalies. If these rules can be appropriately implemented and configured, they can enhance infrastructure performance and increase production (Halper, 2015, p. 7).

Delivering integrated data analytics solutions is an essential process that organizations often pay attention to within business and operational landscapes. Distributing process starts from the dashboards core layer to devices, the database system(s), then to integrated analytics applications (Halper, 2015, p. 8). Various business depends on consoles to strategize, unify, and operationalize analytics capabilities. This procedure focuses analytics on being fused into applications software. While in analytics operationalization, businesses are entrusted with ensuring that a similar method can directly be converted into business methods and benchmarks. More importantly, 'analytic models' are deployed to database systems, to process requested information that may be provisioned through various catalogs. Data scientists handle retention model. The model can be united data analytics in a database system to supply real-time results (Halper, 2015, p. 8).

Data scientists use *'interactive'* as an essential concept for unified data analytics. Through this process, end-users may retrieve single or multiple alerts living on a functioning system in real-time. How various organizations normally operationalize their analytics systems is essential for the actionable/informed decision-making process. This method also ensures that analytics as a concept can be integrated or adopted efficiently once deployed. As a result, decision-makers argue that organizational processes are the most crucial components needed for operationalizing data analytics (Halper, 2015, p. 14). In the analytics building, modern technologies and processes can improve production time and minimizes business underperformances. Such a method can be promoted through long-established organizational policies, system algorithms, processes, and techniques. Additionally, organizations use methodical and custom-tailored solutions to ensure that these activities effectively done.

The need for organizations to match customer's product development requirements through aligning their project capabilities with instituted terms and conditions is vital for the success of all parties involved business transactions. Continued customer's participation in wide-ranging technology dealings and measures needs ability and the right tools to process acquired, visualized, and processed data (Boyd & Crawford, 2011). It includes collecting actionable data through an iterative process calling for tenable managers' to seamlessly make decisions for adapting and improving operational capabilities. These methods involve adapting and supporting an organization's mission and vision. The next section discusses both legal and ethical analytics essentials.

Legal and Ethical Analytics Essentials

Most data analytics project failures often occur due to technical and management issues. These analytics catastrophes often result from practitioners' poor understanding of legal and ethical issues, some of which involve side effects of biased data gathering methods and the resulting detrimental decisions based on those biases. The PredPol system piloted in Kent, England, was adopted to predict crimes in that city (O'Neil, 2016). This process of incorporating data on various crimes occurred in the British inner cities. At the outset, PredPol model was adopted to ensure that the police sensibly introduced this concept, when arresting and imprisoning underprivileged citizens that is those, who looked to prevent the spreading of significant crimes. Data analytics failures are due to lack of attention, which often results in unexpected consequences for modeling algorithms. The right to use public data, such as social media postings have raised ethical questions in data analytics community (Ceraolo, 2017; Moody, 2015; O'Neil, 2016). The next segment focusses on data analytics and its involved transformational solutions.

Data Analytics

In recent years digital and technological innovations have transformed businesses, automation, and operationalization landscapes. Data analytics is a method of examining, purging, cleaning, altering, and visualizing data to gather valuable information and actionable data (Shingles, Phipps, Davenport, & Iglesias, 2017). This method supplies relevant information to help decision-makers make informed decisions. Data analytics makes up methods and procedures, listings of organizational and scientific capabilities, and social knowledge. In consequence of these data analytics, more organizations have introduced new technical concepts, such as systematic operationalization of analytics capabilities, to increasing and modernizing business processes (Maydon, 2017). Using data analytical tools involves a specialized skill set. This level of ability is needed when deploying new hardware and software solutions. Managing these systems is critical to the implementation of data analytics project and processes (Maydon, 2017).

According to Maydon, these projects incorporate (OECD, 2015; Maydon, 2017):

- **Descriptive Analytics**: An initial phase of data manipulation focusing on historical material, to produce noticeable results. This process includes preparing data for future analytical staging in producing data on past and present events or conditions (Ahlm & Litan, 2016)
- **User Behavior Analytics**: A cybersecurity method that is developed to detect insider threats, targeted incidences, and financial fraud activities. This method includes human behavior patterns that can be viewed by functional processes or numerical examination. Many organizations rely on user behavior analytics, to distinguished and meaningful variances. These include generated random patterns that may involve adverse irregularities that can lead to active and future risks or threats. Big data products include Apache Hadoop solution that has vital toolsets for increasing user behavioral analytics functionality. These tools aim to analyze petabytes of data for detecting insider and unconventional persistent risk or threat levels (Ahlm & Litan, 2016)
- **Predictive Analytics**: Statistical techniques, predictive modeling, to machine learning and data mining. This concept finds and measures the concrete results needed to measure conclusions. It allows these constraints and suggestions to support decision options. It describes forecasts and tendencies on pending analytical events (Ahlm & Litan, 2016)
- **Prescriptive Analytics**: Prescriptive analytics supplies top marks to event reports. It mitigates growing risks and threats affecting both public and private sector organizations. This process also focuses on evaluating essential historical proofs for predicting future threats and supplying measures for unknown risks. It serves as a crucial tool for addressing several security risks (Ahlm & Litan, 2016)
- **Diagnostic Analytics**: A method of intensely analyzing data, to actionable visualization of causes, behaviors, and events associated (Ahlm & Litan, 2016)

Descriptive analytics uses statistical methods to display data set characteristics and predictive analytics. This process, however, focuses on predicted events and trends. Prescriptive analytics supplies recommendations for managers to make concerted decisions (Cao, 2017). Many data analytics projects do not generate credible results. Critical reasons for these failures include, but are not limited to (OECD, 2015; Maydon, 2017):

- Projects lacking goals and coherence
- Decision makers ability to make comprehensible business cases
- Organizational leadership's failure to define project benchmarks and milestones; and
- Organization's inability to keep proper human capital and the right skillset for the assigned responsibilities

Despite the sustained implementation of data analytics, most data analytics projects do not produce plausible results. These failures often result from project staging, though most of the assigned tasks have never come to fruition. The observed failures are due to the lack of traceable return-on-investments (Gartner, 2016). In large-scale organizations, there are non-technical and technical limitations affecting data analytics practices. These technical limitations range from (NIST, 2017):

- Lacking stakeholder's clear project requirements definitions and product agreements;
- Budget and exclusive product licensing
- The inadequacy of processes that is, the proof-of-concept (POC) to production systems;
- Compliance with privacy and regulations
- The inconsistency of metadata standards
- Silos involving data and access restrictions
- Shifting from incorporated stewardship to distributed and granular models
- Legacy access methods that focus on greater compliance and integration issues
- Proprietary, patented access procedures that pose a barrier to building connectors
- Organizational maturity
- Shortage of data analytics experts, to the ability to handle complex and technical software issues

Data analysts and data scientists should be equipped with the technical capabilities needed to support the everyday organization's requirements. More business skills are needed to sustain organizational operations (Wiechecki, 2017). Having business knowledge, aligned with project development and evaluation, project management; distributing business processes, change management, to people, and leadership skills can be beneficial (Hilbert, 2015). Data-driven innovation will be discussed in detail in the next section.

Data-Driven Entrepreneurship

In recent years, the migration of social and economic activities to the Internet has increased exponentially (Rankin, 2013). These indices are due to continuing data collection, storage system, and processing cost, which has decreased significantly. In modern-day, data is often provisioned from the Internet of Things intelligent devices, i.e., self-directed machine-to-machine infrastructures. Internet firms such as Facebook, Google, Twitter, and others continue to receive help from data-driven innovation (Rankin, 2013; Wiechecki, 2017). These advances are due to more companies entering the global information and communications technology sector, mostly in the following domains: artificial intelligence, big data, data analytics, data driven innovation, predictive analytics, data science, machine learning to name a few (Rankin, 2013; Wiechecki, 2017).

As new business opportunities continue to arise, tech giants are keen to provide consumers with proven leading-edge technological solutions aimed at supporting and meeting their business/market requirements. In the short run, these unveiled technology indexes have helped the following business/ market segments (OECD, 2015; Rankin, 2013): start-ups, mid-to-large companies, public administration, health services, and education. The need for such disruptive data analytics and data-driven innovations is to ensure that the customer has agile technology solutions needed to satisfy the client's day-to-day operational requirements (Rankin, 2013; Wiechecki, 2017).

Case Study

In 2013 McKinsey Global Institute (MGI) researched data-driven innovation. MGI also showed that each year, open source data produces nearly three-trillion dollars to the global economy (MGI, 2011; Mandinach, 2012; Rankin, 2013). MGI also states that DDI is a process that focuses on economic growth and infrastructural resources, which can be available for public use and unlimited purposes. In a supply chain, data is used to collect and analyze datasets needed for processing services and goods (Mandinach, 2012). In most cases, large volumes of data outside of public and private IT sectors can be collected, analyzed, and processed from the following communities (MGI, 2011; Merelli & Rasetti, 2013): public administration, educational, and health services (Rankin, 2013). These industries rely on data for scientific medical discoveries, researching on chronical/non-chronical ailments, to improve people is lives and adaptive learning environments (OECD, 2015).

Many governments around the world often use data to support their citizenry needs, through accountability, transparency, and increase openness. These processes allow those governments to serve their communities and aid in rebuilding public trust in their officials (Mandinach, 2012; Rankin, 2013). Sears Corporation's Master Data Management describes the case of involving data analytics project failures (Guess, 2012). The lack of integration and standards of data sharing, the internal sale units can encumber the procurement of products. This process has collapsed, which has resulted in customers' lack of continuous interaction with various Sears' departments. This system had multiple 'error-prone' manual processes. This was due to the lack of inventory management and financial reconciliation processes by implementing data analytics was not a practical decision (Guess, 2012).

Major companies, such as Merck, Bechtel, and Monsanto, have effectively implemented large data analytics projects. In contrast, many corporations had adopted a similar concept, which yielded significant values. Shared threads and data-informed decision-making, data-based decision-making, and data-driven planning processes, which will be discussed in the next section. Some of the areas to be emphasized are as follows (Guess, 2012): The next section will discuss data-informed decision-making

- How to develop centralized data repositories, access tools, and procedures
- How to train personnel and develop POC projects
- How to make the most of machine learning techniques

Data-Informed Decision-Making

Adopting best practices to analytically and strategically collect and parse data can be an essential process for decision makers. Data-Informed Decision-Making (DIDM) is a process of collecting and analyzing data (Delort, 2012; OECD, 2015). It is a method that supports managers when making informed decisions, to improve their organizations' day-to-day operational management (USDEO, 2009). Decision-making is a process that was Dantzig coined in the 1950s, despite the foundation that was set forth by both Charles and Cooper after that (OECD, 2015; Rankin, 2013). DIDM has been widely used in many communities: business, education, marketing, researching, engineering, and scientific sectors. This process allows data to be collected, analyzed, processed, and visualized to support assigned tasks/missions (Delort, 2012). For example, in business and educational sectors, DIDM helps business owners, educators, and students through advancing their curricula and setting up agile processes needed to make informed business decisions (Rankin, 2013; Mandinach, 2012).

Data-driven decision-making involves several metrics (Rankin, 2013; Wiechecki, 2017). Decision makers rely on these methods to assess their organizational goals, results, and critical events affecting their organizations' day-to-day operations. By applying key performance indicators, aka KPIs, managers can make better decisions. This process includes, but it is not limited to proving determinate goals that can have a direct impact on an organization's operational activities and business results (Delort, 2012; OECD, 2015) — for example, finding key audiences, who rely on collected data to support business requirements. This process includes selecting KPIs by beginning to categorize performance metrics with financial, mission-driven activities, and business decision-making processes (Delort, 2012). In an organization setting, decision makers regularly work together, improving their ability to make informed decisions, as part of their organizational and business operations (OECD, 2015). This process often can be illustrated by managers' ability to build, incorporate, and make use of agile processes to help improve data-based analytics functionality (OECD, 2015). In recent years, many companies have invested their financial capital in helping build agile, responsive, reliable, and sustainable data-driven solutions (Delort, 2012).

Data-Driven Planning

Data-Driven Planning (DDP) is an essential process in the data-driven decision process (Rankin, 2013; OECD, 2015). Often decision makers struggle to set up a right model that can effectively support this process. Every organization should adopt the following DDP analytical and strategic steps (Delort, 2012): establishing strategic plan, forecast and trend data, conduct a capacity analysis, articulate and measure indicators of accomplishments, prioritize action plans needed to satisfy the strategic goals, align resources and capabilities with organization's business model, scope, and mission, implement results-attained assessment program evaluation (Delort, 2012; Rankin, 2013).

Managers should be able to examine these processes systematically or patently to substantiate, authenticate, and confirm the desired mission outcomes (OECD, 2015). The ability for managers to identify, budge needed resources to refine each goal is critical for any project success. Meriting these processes often helps organizations show or implement a useful DDD model (Delort, 2012; OECD, 2015).

Data-Based Decision-Making

- In recent years, DIDM has been applied in education and global development sectors to resolve educational and citizenry's issues (OECD, 2015; Rankin, 2013). Also, this system provides its users with the ability to assess data system's activities and performance (Delort, 2012; Rankin, 2013). Policymakers continue to develop policies and procedures to support DIDM solutions. Leading technology companies are focusing on modernized tools, strategies, and processes to support Data-Based Decision-making (DBDM) methods and practices. DBDM is a synonymous term for DIDM process (Rankin, 2013; Mandinach, 2012). In many industries, business and education terms are used to ensure that every manager, educator, and student have adequate tools to make an informed decision without basing their decision on quantitative data. In most colleges, universities, and K-12 programs educators depend on data system, to collect and analyze students' data, i.e., embedding labels, supplemental documentation, content decision, making key posts and presentations (Delort, 2012). To refining educators' and students' academic/pedagogical performance at school, this process often involves DIDM methods and techniques to help solve complex problems (Rankin, 2013). The following are four data types being used in most business and education communities (OECD, 2015; Rankin, 2013):

- Demographic Data
- Perception Data
- Student Learning/Employee Data
- School/Organizational Processes

These processes include governments, business owners, and educators, who handle collecting and analyzing distinct types of datasets. This process gives professionals a set of tools necessary to make adequate and informed decisions (Mandinach, 2012; Rankin, 2013). The next section will address data acquisition and management methods as well as relevant scenarios.

Data Acquisition and Management

In recent years, research institutions have received help from producing revenues while keeping complete ownership of their technology solution findings. Similar studies have proved that federally research institutions colleges and universities have complete proprietorship of their data. In various research institutions, principal investigators handle approving and managing access and retrieve to data. These principal investigators may be assigned as overseers of the data being collected, recorded, stored, kept, and recycled if the information is no longer in need (Boulton, 2017).

In the case of publishing data, delegated principal investigators will oversee negotiating and approving the copyright agreements of the material into consideration. Data can be described as a knowledge cornerstone and examination. Researchers describe data a driving force that business has relied on for many years, to extract, analyze, process, and visualize volumes of datasets. The research organizations lean on a series of unified factors to determine, gather, keep/store, and share data among selected or responsible parties. Organizations responsible for managing and supplying data should do so in such a manner. The process will deter/prevent any security breaches from affecting current or future data collection results and processing. For example, higher education institutions are interested in embarking on forward-thinking and relevant research projects involving these areas of interest—whether by developing patents that have exclusive licensing rights, for example, keeping ownership of their product 'patent' after being published. The next section addresses big data analytics, project failures, and successes, and the seven V's of the big data model as well as each involved method and process.

Big Data Analytics

Big data analytics is a term that was coined by John Mashey in the 1990s. The term 'big data' describes the first three big data V's: volume, velocity, and variety. It focuses on supporting the processing, managing, purging, and storing of data (Boyd & Crawford, 2011). Big data analytics describes composite and massive datasets—this process substitutes for the legacy "data processing application software." This method includes the following key areas: data capturing, storing, distributing, transferring, visualizing, enquiring, updating, and privacy (Boyd & Crawford, 2011; Marr, 2015). The operationalization of analytics capabilities includes management's decision, a variety of business, and operational insights needed to support the organization's requirements. This vision incorporates decision makers' ability to implementing exclusive business models/processes to daily organizational operation's capabilities/processes.

Analytics capabilities were adopted or integrated into dashboards, applications software, autonomous systems, devices, and catalogs in recent years. For many years, integrating analytics processes into other applications, systems, and devices was viewed as an evolving process (Cao, 2017). Likewise, this practice gives organizations the capabilities necessary to build on data volumes and frequencies. Systematizing business and operational decisions are essential to support decision-making processes (Deloitte & GMA, 2017e). If effectively implemented, these processes can produce competitive advantages for analytics operationalization (Marr, 2015). Boyd & Crawford (2011) cites, "there is little uncertainty that the quantities of data now available are indeed large, but that is not the most relevant characteristic of this new data ecosystem." Predictive, user behavior analytics and associated substitute methods encompass big data analytics. These processes aim to an abstract numerical value(s) from acquired or supplied data (Marr, 2015).

Project Failures and Successes

This subsection illustrates some of the effects of big data encompassing project failures or successes. Walsh (2014) said that the Google Flu Trends project is one of the cited case studies of data analytics project failures. This research was conducted between 2008 through 2014. The study also emphasized that influenza widespread outbreaks. It was also based on a large amount of data collected from half a billion user searches consisting of research on flu-related terms. It also focusses on critical areas of the existence and spread of this ailment. Not obtaining and exploit these epidemiological results have affected Google's predictive algorithms. This deterrence "overestimate the prevalence of flu in the 2012-2013 and 2011-2012 seasons by more than 50%;" but covered "over-predict the prevalence of the flue in 100 out 108 weeks from August 2011 to September 2013"; the approach predicted many flu cases that the United States Center for Disease Control had predicted prior to Google's survey. There is a parallel between Google's influenza's investigation project and data analytics research. Using material technology and data collection ability can lead to project failures. It further illustrates the big data—seven V's, data acquisition, and management characteristics.

Seven V's of the Big Data Model

Apart from these evolutions, managers should not overlook other problems affecting data analytics projects. In this chapter, we also discuss the seven relevant "V characteristics" of big data, although some authors (Shafer, 2017) have "discovered" as many as 42 such characteristics.

These big data seven *V's* methods consist of (Borne, 2014; Laney, 2001; Hilbert, 2015):

- **Volume**: An amount of data that can be displayed and processed to produce the required result(s). Today data volume is either measured in Gigabytes, Zettabytes, and Yottabytes. Big data numerical scale often poses challenges to data volume. This numerical analysis can be determined by a domain-specific; on the other hand, the authors agree that this measured value can be determined by a "ton of bytes." Due to the exponential growth of the Internet of Things devices, yet the measurement of data can be done through a high quality of data. Studies show that in the upcoming years, data volume will increase. This surge will need better technology to manage and process large datasets

- **Velocity**: A speed that allows for viewing and accessing data. This includes a large amount of data extracted and provisioned by the system in real-time. This term involves the rapidity and speediness the data travels. The speed makes it difficult for analysts, scientists, and managers to determine how data can be extrapolated into many data sets to produce the desired results

- **Variety**: A process that often describes challenges affecting big data analytics. It describes thousands of numerical values, e.g., features per data entry. Moreover, these elements include video, text, SMS, images, Extensible Markup Language, audio, etc. Variety describes the categorization of dimensionality, combinatorial analytics explosion, data types, and formats. These days organizations and individual citizens can choose toolsets, for example, type of search engines or medium needed for sending and receiving data over the Internet

- **Variability**: There is a difference between variability and variety. This process incorporates dynamic, spatiotemporal data, time-series, seasonal, and non-static behaviors. It also includes data sources, customers, objects of a survey

- **Veracity**: Integrates essential and enough datasets needed for testing various hypotheses. How authentic the data is often can be proportional to the relevance of data. Mostly, this is obtained from the processed data. It also describes 'data authenticity and accuracy.' Veracity prevents corrupted data from building up in the system. This method includes training material for examining sampled micro-scales, model-building, and validation; micro-grained "truth," for instance, objects that are a part of the data collection process. This includes the process of implementing whole-population analytics

- **Visualization**: A process of combining diagrams and graphs to visualize enormous quantities of complex data. Visualization is also a careful process of interpreting meaningful data entered and viewed in the form of tables and reports. In general, data is exhibited in numbers and methods allowing scientists and analysists to view the extrapolated or presented information in real-time; and

- **Value**: Describes return-on-investment, business value, and related extents of big data. This method transforms an organization's business processes, procedures using a 'top-to-bottom' or 'essential point approach.' This process needs a significant amount of time, effort, and resources to produce effective results

These characteristics aim to introduce and articulate specific data management challenges involving an organization's business operations (Hilbert, 2015). The following characteristics related issues do lack the standards and terminologies needed for gathering and communicating functioning datasets (GMA, 2017). Most organizational data analytics projects often do not provide acceptable results. Over the years, there have been significant benefits in types of studies (Boulton, 2017). The next section underscores customer relationship management analytics and essential Customer Relationship Management (CRM) analytics.

Customer Relationship Management Analytics

Customer Relationship Management (CRM) analytics is a process for gathering, organizing, and producing customer data. This method helps organizations in resolving their business issues. The concept includes functions as well as a set of tools, explicitly dashboards, portals, and other procedures. It includes the process of integrating mixed programming toolsets and other software capabilities. CRM analytics consists of the business concept and online analytical processes for delivering data mining (Boyd &

Crawford, 2011). These methods include online tools and websites. This toolkit offers decision makers and their organizations a set of capabilities needed for analyzing data in real-time.

An organization CRM analytics procedures and standards incorporate quality and efficiency, to total cost reduction, results, sustainability, enterprise agility, customer attention, and increase on profitability margins. This approach includes parsing clients' requirements and supplying actionable results. It also helps organizations, and their corresponding executive leadership makes informed decisions. Viewing an organization's aptitude in collecting and analyzing actionable data can be a practical asset for many enterprise project requirements (GMA, 2017). Many technology companies have developed agile tools needed for supporting analytical business requirements. For that reason, decision-makers are considered for handling and aligning CRM analytics with vertical organization's capabilities. This concept, however, can be achieved by adopting essential measures necessary to sustain suitable business results. These processes consist of: (Boyd & Crawford, 2011)

- **Direct client's response**: A central process that organizations need for engaging their customers. This stems from obtaining feedback on project milestone(s), division(s), and touchpoint(s). This procedure encompasses data mining, extrapolating processes, and techniques
- **Wide-ranging view of client's business understandings**: A practical method that decision makers need to adapt and supply realistic client responses, despite any level of communication or agreements that often is established between an organization and the customer. Today, organizations can determine/measure their customers' satisfaction by using analytics tools to support their findings. This ranges from helping customers aligning and defining the type of capabilities needed for storing, processing, and visualizing data
- **Measuring project milestones and production processes**: Through this practice, organizations and decision-makers can adopt suitable applications, such as software requirements that are crucial for determining and supporting any client's projects and milestones. For decades, this process has helped customers determining the level of incremental revenue profitability margins and desired analytical results
- **Determining and tracing project growth, benchmarks, trends, and milestones**: Decision makers rely on the first call resolution to enhance customer project efficiency. This process is implemented through e-mails, chats, phone calls, to name a few. Besides, decision-makers should consider the following: customer satisfaction and problem resolution. These managerial and technical areas can be achieved by employing data analytics toolsets. The tools help managers, analysts, and scientists when computing, regulating, and monitoring product growth and fixing the following issues affecting customer's productivity. Decision makers should articulate and define project requirements, by using various analytics toolkit, to predict, benchmark, and forecast any general project preliminaries as well as results
- **Ability to scale and check the client's value**: This process should aid organizations in measuring their annual growth margins. This includes lifetime value giving decision-makers the ability to regulate and track their practicability margins/equities. Hence, profitable customers work toward a sustainable growth strategy applicable for determining business efficiency and returns. This involves actual customer value effort and related contributions, which can be determined by profitable gains. These products often do make through the product pipeline. CRM analytics is a proven analytical tool for determining key performance indicators

Many organizations around the world often deliver more significant benefits by applying CRM analytics model (Boyd & Crawford, 2011). This toolkit offers terms of sales and related marketing activities. This includes the ability to excel/improve on current and future supply chain management capabilities. Consequently, key performance indicators are used to support organizations' metrics that decision makers, the need for determining and making informed decisions (Boyd & Crawford, 2011). Thus, CRM analytics has offered consumers a segmentation of federated functions and analytical solutions' capabilities, to support varied business environments Boyd & Crawford, 2011). These analytical measures also have provided decision makers with the tools needed to make informed decisions. The advent of CRM analytics has given businesses the analytical tool much needed to collect, process, personalize, monitor, and visualize data in real-time. This method also determines that supporting the customer's business requirements is paramount. Thus, managers should be introducing "*what-if scenarios*" when making informed decisions. This includes complementing other areas, that is, customer-to-customer that makes up short-lived/long-term business models (Boyd & Crawford, 2011). The next section will discuss systemic infrastructure data integration.

Systemic Infrastructure Data Integration

This process involves integrated measures that focus on data stored in various repositories. The process includes user's provisions consenting for a reliable exposition of data. How the information can be retrieved, processed, visualized, and manipulated leaves from distinct types of data (Lane, 2006). These practicalities span commercial and scientific assumptions that can be extracted from an assemblage of bioinformatics sources and domains. Data integration can be determined by the advent of amplified rate recurrence that regularly can be seen by a host of data volumes (Lane, 2006). Incorporating data sources is a process involving material silos—this technique includes a single query interface.

In 1991, the University of Minnesota fused the first collective public use microdata series (IPUMS). More importantly, the IPUMS' method relied on data warehousing schemes. This method further involves abstractions, transformations, and stacking colossal data from a different source, allowing for a single view classification. This was supported out by extracting and allotting several sources through a compatible process. The next subsection will emphasize technology and marketing challenges.

Technology and Marketing Challenges

Many organizations have endured decision-making and technological challenges involving large-scale big data resource consumptions. Despite these issues, managers continue to have difficulties when dealing with a large volume of data. These concerns are the results of several variations affecting data structures. Lacking these capabilities makes it hard for organizations to process, visualize, and manipulate massive data sources (Boyd & Crawford, 2011). However, best practices and integrated solutions should be introduced to ensure that data can be easily retrieved if the right processes are implemented and accessible for customer usage. This process gives decision makers the tools needed for making informed conclusions. Collecting capabilities to mining data are fundamental to an organization's business continuity (Boyd & Crawford, 2011).

These organizations have developed a list of processes and procedures to ensure for successful project implementation. Gartner (2015) suggested the following best practices for data analytics projects:

- The ability to find and make informed decisions, which can yield valuable data analytics product results to sustain an organization's day-to-day processes. This process ensures that continuous support of involved sponsors, such as project owners are entirely ranked
- Deciding if it is practical to develop in-house skill-based capabilities; these project efforts, should be cautiously mapped, to align with existing or future customer technical and domain knowledge-based requirements. These capabilities include planning, attaining, developing, and keeping internal labor force, instead of outsourcing competency-based skill set

Soderlund (2017) cited that Amazon Web Services (AWS) is one of the global leading data infrastructure management technology companies. Soderlund points out that AWS focuses on massive data management managerial and technical issues. This includes measuring and recommending suitable toolsets needed to implement, scale, and support future projects. In recent years, many organizations opt for AWS solutions as a preferred platform to support their everyday operations. These capabilities are uniquely applicable and include the following procedures (Boulton, 2017):

- Creating decoupled systems for managing data collection and processing
- Implementing tools to support organizational project requirements. These assets range from deploying integrated hardware components and applications software
- Having a coherent project profoundness—this includes selecting batch, interactive and real-time analytics processes
- Management's ability to find and clarify customer's requirements, internal business, and technology efficiencies
- Technology providers should be aware of data-related costs and low-cost customer ownership, before embarking on any project implementation

Decision-makers argue that these technology deterrents are due to the lack of enhanced analytical capabilities, data-driven innovation, and best practices needed to support ecosystem efforts and continued technological innovation (Desjardins, 2017). This process needs sustainable business assets, such as capital resource management needed to develop ultramodern data analytics capabilities (Shingles et al., 2017). Decision makers are still optimistic about improving their business models. This includes the following areas needing immediate attention (Halper, 2016): The next section explains the platforms, models, and tools.

- Lacking skilled domain and subject matter experts
- Insufficient attention to data governance and quality
- Lacking executive, stakeholder, and financial support
- Insufficient understanding of the potential return on investment for these projects

Platforms, Models, and Tools

In recent years increased growth in data analytics sources has proven that many organizations are excited about implementing multitenant software technologies in their environments. Despite continued advances in many analytics projects, vendors are developing these pioneering cross-platform software solutions, for example, R, Python, MapReduce, Hadoop, Vertica, Flume, Hive, Pig, Spark, and others.

In addition, vendors have developed predictive/prescriptive analytics software expressly Rapid Miner Studio and Server, MATLAB, Predixion Insight, STATISTICA, Google Cloud Prediction API, Anaconda, AdvancedMiner, GMDH Shell, Angoss, Emcien, KNIME Analytics, Dataiku DSS, Minitab, GraphLab Create, Ayata, NGData, discovery Software Suite, CMSR Data Miner Suite, Vanguard Business Analytics Suite, DataRPM, Skytree, QIWare, Grapheur, SPM, Rapid Insight Veera, DMWay, Lavastorm, TIMi Suite, TIBCO Spotfire, Atleryx Analytics/designer, LIONoso, Profitect, and others. These advanced analytics software tools were built to support customer's business, development, and infrastructure requirements (Halper, 2015).

Multi-tenant software capabilities are built to interact with and complemented by other machine learning free solutions. These capabilities include open source and commercial-off-the-shelf (COTS) products. These applications have perfected/unified to functioning/scaling in a self-governing or shared infrastructure environment and fusion-sourced style. Besides, these products provide the customer with the provisioning, elasticity, and flexible capabilities focused on selecting between readily available (enterprise open source and COTS applications). More to the point, customer excitement demands among small, medium, to large-scale organizations for adopting application solutions. Such requirements continuing to acquire similar analytics products are on the rise.

Despite these increases, software/technology vendors should handle educating their customers on primary functions associated with these analytics' software products. This multitenant applications software expressly integrity, the three I's that often symbolize data integrity and business intelligence (Deloitte & GMA, 2017c). Deloitte organization has granted permission to adapt and cite this model or its respective components (Deloitte & GMA, 2017d). Each of these proven stages signifying an organization's current industry business and operational performance posture involving the consumer-packaged goods (CPG). Figure 1 explains a five-stage analytical capability that most organizations have employed over the past decade (Deloitte & GMA, 2017a). It further explains how CPG firms mostly compete, analyze, aspire, localize, identify, and determine each level of analytical capabilities, to ensure that these elements are aligned with both the capital expenditures (CAPEX) and operational expenditures (OPEX).

Decision makers rely on CAPEX and OPEX outliers to determine and forecast on profits & losses and several predictions about an organization's production and financial margins. This chapter recom-

Figure 1. Analytical capability levels
Source: Deloitte GMA Industry Research

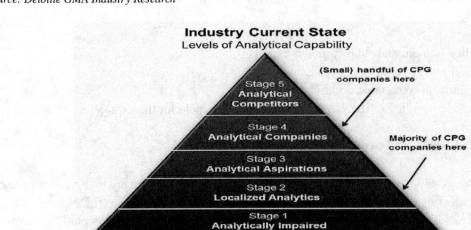

mends that one of the CPG businesses continues to lack the necessary level of analytical maturity and initial diagnostic capabilities. These internal resources include, but are not limited to: enterprise resource planning, total productive maintenance, electronic data exchange, the point of sale/point of purchase and syndicated/partnership (Deloitte & GMA, 2017a). Often, CAPEX and OPEX can be used to determine operational determinants an organization could be experiencing and define suitable corrective measures (Deloitte & GMA, 2017a). Stage two in the above diagram, concludes that aside from organizations' inability to offset their quarterly & annual performance and product margins, many CPG companies have profited off the illustrated analytical models.

Decision-makers argue that this illustrated list of analytical capability is needed to enhance organizational business and operation functions. Phase two of the analytical maturity model shows five-stage business functions. Based on each of these functions, decision-makers can map and assess customer's requirements and make informed business decisions concerning historic and collected data. This process provides identified analytics, trade promotion, and grouping management as well as a request for predicting analytics (Deloitte & GMA, 2017a). Apart from prevailing technical and operational issues affecting everyday organization's operations, vendors are not convinced that data analytics tools and systems pose a significant risk/threat to the holistic technology ecosystem. Such analytical progress, many organizations have been implementing analytics products; most of the decision makers are still uncertain about adopting this technology in their respective organizations.

Lacking data management maturity methods to support adopted analytical capabilities have been one of the core reasons for decision makers' skepticism in deploying analytics product resources in their businesses (Deloitte & GMA, 2017a). The below five-stage analytical maturity model describes a high-level summary that describes the step-by-step organization's business/operation's developments. These methods (stage one-five) are intended to support decision makers should be capable of explaining the variance between least developed analytical organizations, including stages one-two. Stages four-five emphasized advanced and outstanding analytical maturity capabilities an organization can adapt to support its day-to-day business or operation's performance. This five-stage process encompasses tools & techniques, information and insights, decision-making processes, talent and organization, leadership, and culture. Also, the organization granted permission to adapt/change and cite the below figure and its corresponding reference materials, to line up with the theme of this topic. Figure 2 describes the analytical maturity model and its conforming five iterative stages and processes (Deloitte & GMA, 2017a).

For many years, CPG firms adopted and deployed several computer applications needed to support their customers (Halper, 2015). This list of custom-based applications and vendor's capabilities includes, but it is not limited to:

- **WebFOCUS**: A self-service analytics developed for balancing or orchestrating tools and applications. This application consists of a subset of analytical tools that support businesses, power users, and designers' mixed methods and platforms. The application improves the processing time that users often need to publish and share their insights as "InfoApps." This tool has helped users, who have limited technical ability, to promote their printed materials
- **Omni-Gen**: A platform for modeling data incorporation and agile development. This platform manages data quality and inherent solutions
- **Omni-Payer/Omni-Patient**: Data management applications developed for health indemnification and markets. This solution provides a holistic perspective of each patient or health insurance subscriber and multiple business environments

Figure 2. Analytical maturity model
Source: Deloitte Consulting LLP

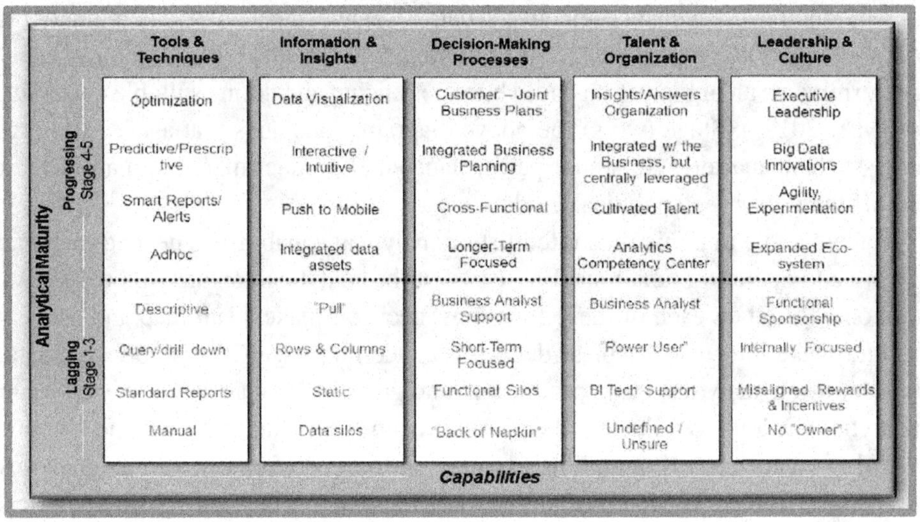

	Tools & Techniques	Information & Insights	Decision-Making Processes	Talent & Organization	Leadership & Culture
Progressing Stage 4-5	Optimization	Data Visualization	Customer – Joint Business Plans	Insights/Answers Organization	Executive Leadership
	Predictive/Prescriptive	Interactive / Intuitive	Integrated Business Planning	Integrated w/ the Business, but centrally leveraged	Big Data Innovations
	Smart Reports/ Alerts	Push to Mobile	Cross-Functional	Cultivated Talent	Agility, Experimentation
	Adhoc	Integrated data assets	Longer-Term Focused	Analytics Competency Center	Expanded Eco-system
Lagging Stage 1-3	Descriptive	"Pull"	Business Analyst Support	Business Analyst	Functional Sponsorship
	Query/drill down	Rows & Columns	Short-Term Focused	"Power User"	Internally Focused
	Standard Reports	Static	Functional Silos	BI Tech Support	Misaligned Rewards & Incentives
	Manual	Data silos	"Back of Napkin"	Undefined / Unsure	No "Owner"

Analytical Maturity (vertical label on left)

Capabilities

- **iWay**: A collection of toolsets showing a pillar for integrating business/technology capabilities that are integrated into Omni and WebFOCUS artifact lines. This method focuses on multifaceted and synchronized merger subjects
- **OpenText Corporation**: In business intelligence ecosystem, OpenText Information Hub or iHub is an organization that empowers businesses to strategize, mature, and integrate/process reports. This method aims to visualize and integrate dashboards into application and device's displays. Further, this integrated analytics supplies granulated security capabilities distributes both on-site and in the cloud technology capabilities. This application introduces present-day application programming interfaces (APIs) in varied iHub environments. Similarly, the firm specializes in the use of programming applications, that is application programming interface (API) and representational state transfer API. These solutions produce transport material such as data from iHub to mobile and web applications
- **OpenText Analytics Designer**: A mature unified development environment aiming to link to sources explicitly content and data objects. The application also aims to produce extensible, markup language projects to support iHub solutions. This application can be offered as an open source solution for many users worldwide. In the new analytics environment, iHub continues to play an integral role through advanced/integrated capabilities built-in within its product stack. This technology company has developed analytical products to support its brands of "enterprise content management" and data exchange/collections in recent years
- **Pentaho**: A Hitachi Group solution built for offering useful open source business analytics applications. The product can merge information access, incorporation, and analytics into unified capabilities. Pentaho emphasizes big data composition and integrated analytics to sustain time-value. Its data integration capabilities are the core of Pentaho ecosystem. It delivers an in-built painterly understanding to help formulate, merge, and dispense/provision data at the measure. The system gives data analysts and designers the ability to contribute, blend, and process information from

data warehouse platforms, Hadoop, NoSQL, analytical databases, and information-driven tendering software solutions. Thus, programming applications: R and Weka have been contributory to the predictive and operationalization of actionable models shown in the information pipeline. Continuous technology studies prove that Pentaho is one of the crucial analytical platforms built to date. Its software and technology capabilities are deployed on the cloud as flexible solutions for embedding analytics solutions into a large-scale application platform. Such a method includes Pentaho analytics needed for interacting with other systems as an add-on application, i.e., user interface and user experience for designing ecosystem. The entire analytical technology practice is possible through Pentaho's open architecture solution and enhanced JavaScript API. JavaScript is proven to be one of the industry's leading programming software that provides integrated original equipment manufacturer's solution distribution

- **SAP**: Integrates software solutions that support business intelligence and analytics environments. In the operationalized and integrated environments, systems, applications, products, data processing, and SAP HANA solution are equipped with a synchronized and interactive capability to share data in real-time. SAP gives clients an integrated capability for connecting and integrating its advanced solutions with those deployed into business intelligence/data warehousing settings. Inherently, customers have the way to custom-tailor these unified solutions to make parallel with time-value solutions

- **HANA**: Open source method that retrieves data and interacts with other assigned/targeted applications, that is, Java database connectivity/open database connectivity, SQL, and multiple dimensional eXpressions. HANA can support the topmost business intelligence capabilities for SAP and non-SAP ecosystems

- **SAP HANA**: Consists of graphical data flow showing and related programming toolset. In this ecosystem, data schemes are stored as source objects, then to be reprocessed for a myriad of application solutions. Database platforms use SAP HANA as an integrated application that supports the management of data. This platform offers data collections/catalogs needed to support advanced analytics products. Such a method includes predictive analytics and interactive process with R, typescript, and geospatial analytics

- **SAS**: A platform deployed to support technical aspects of business analytics lifecycle. It encompasses data organization, determining and showing of analytics capabilities to support an organization's day-to-day business and operation. SAS is a leading and innovative advanced analytics firm/platform. Its technology solutions cover machine learning and data mining. It provides customers with a suite of prearranged and unstructured solutions, and the ability to process data analysis applications, to support business and operation's requirements. For several years data scientists and business analysts have relied on SAS capabilities to support their ecosystems and related program missions. SAS is a user-friendly application and can be easily deployed on any technology environments, to support the operationalization and integration of analytical capabilities into business usability

- **Space-Time Insight**: A platform that provides real-time circumstantial actionable solution capabilities to support customer's business requirements. Public and private sector organizations have preferred space-time insight as an analytical tool for determining situational intelligence. Recommended tool provides calculations of unidentified business/technical outlooks that can impede decision makers' ability to make collective business decisions. This application can be used by multiple organizations for scheduling and planning business milestones and associated project requirements. This tool's capability forms business operational and intelligence capability for gathering actionable data that an organization will need to make an informed decision. This capability is developed as COTS solutions. It processes, correlates, analyzes, and visualizes data in real-time

- **Tableau**: It gives high-level visualization of the entire ecosystem. Despite its ability to democratizing analytical capabilities (vertically or horizontally), this tool is user-friendly and environmentally responsive. The solution supports several mobile businesses intelligent solutions--when deployed, it can analyze evolutions, and trends involving big data analytical capabilities. It gives decision makers, line managers, and the ordinary people the ability to access, view, and process actionable data and to make an informed decision based on their business areas of responsibility. Its desktop version is customized with features allowing users to analyze data without the need to code through an interactive dashboard solution environment. This application functions in unison with the Tableau server platform. The application allows users to view web pages, blogs, web applications, and other social media resources available via intranet portals. The deployment of Tableau JavaScript APIs can carry out this process, IFrame, often injected using user's non-proprietary programming scripts. This solution can be deployed as an add-on to sustain project missions scaled by Canvas, a Salesforce business tool for joining open source applications. Tableau provides customers with other cloud solution capabilities

- **Talend**: An integrated freely-based software supporting organizations data-driven initiatives. Many organizations use Talend Data to link, exchange data and related application-based solutions. This integration consists of on-premises, to multitenancy cloud architecture environments, such as enterprise data warehouses, Hadoop, and associated infrastructures. These include sensors embedded into enterprise devices, to name a few. In recent decade many organizations have relied on this technology to combine solutions, which is data integration, big data analytics, cloud/master data management solutions. These capabilities can be singly deployed to support customer's project missions. Its application software capabilities. These solutions include Apache Spark, Lambda, and NoSQL. These solutions can be deployed as an add-on, to deliver a batch application to support the inherent functionality of database systems.

Many organizations are responsible for setting project goals that are essential to their business values. Decision makers are accountable for developing and forecasting on coherent/cohesive business strategies for supporting data, applications, and analytics capabilities. These processes need human capital solutions and business toolsets to functioning within integrated environments. Managers oversee crafting/developing a common language essential for sustaining an organizationn's operational goals and business results (Deloitte & GMA, 2017a). The next segment provides the conclusion of the chapter and provides recommendations for future research direction.

FINDINGS AND ANALYSIS

This chapter concludes with solutions and recommendations for small, medium, and large-scale data analytics projects. Given the rapid advances in technology and issues involving the grey literature, the authors propose that future study be considered to address some of the constraints, setbacks, and forecasts involving methods, technologies, and techniques (Desjardins, 2017). Many organizations have fully adopted analytics and data-driven solutions to support on-premises operations and client's requirements. Lacking analytical maturity models to define and measure an organization's business and operational performances can pose a significant concern to decision makers. More technology resources must guarantee that organizations have the tools and capabilities to support their customers' business requirements (Deloitte & GMA, 2017b). Continuous successes and failures of significant data analytics projects often can pose significant challenges to decision makers. It is evident that these managerial challenges are not limited to technological, organizational, and business decisions.

FUTURE RESEARCH DIRECTIONS

Adopting these challenges in a comprehensible approach can help public and private sector organizations and governments effectively address issues, affecting productivity and technology advances that are needed to support data analytics capabilities. Such issues range from managers' vision and decision to assign and work on small analytics projects, to effactually evaluate daily project milestones and successes. A similar strategy can be a useful tool for decision makers to have a better understanding of their mission failures and successes. Expert training involving all aspects of analytical processes often can be critical toward producing data analytics development performances, data-driven solutions, and reliable results. It involves creating and managing useful and operational data repositories, e.g., evolving process.

Organizational data generating methods and results can be needed and useful tool to satisfy customer's analytics and data-driven project requirements and forecasts. This process also includes data types needed to support enterprise analytics projects and milestones. For instance, *"buying-in"* from stakeholders often can be achieved through proper channels of communication stemming from time-value organizations need to align their data analytics while gaining the reputation of their customers. Measuring these metrics growth and compliance factors often depend on the other vital areas, productive trends, and growth margins. Decision makers should consider as *"you cannot manage what you do not measure"* process to improve and support their organizations' data analytics project successes (Lavinsky, 2017).

REFERENCES

Ahlm, E., & Litan, A. (2016). Market Trends: User and Entity Behavior Analytics Expand Their Market Reach. *Gartner*. Retrieved from https://www.gartner.com/doc/reprints?id=1-370BP2V&ct=160518&st=sb

Bernhardt, V. (2013). Data analysis for continuous school improvement. Routledge.

Borne, K. (2014). *Top 10 Big Data Challenges – A Serious Look at 10 Big Data V's*. Retrieved from https://mapr.com/blog/top-10-big-data-challenges-serious-look-10-big-data-vs/

Boulton, C. (2017). *6 data analytics success stories: An inside look*. Retrieved from https://www.cio.com/article/3221621/analytics/6-data-analytics-success-stories-an-inside-look.html

Boyd, D., & Crawford, K. (2011). Six Provocations for Big Data. *Social Science Research Network: A Decade in Internet Time: Symposium on the Dynamics of the Internet and Society*. doi:10.2139srn.1926431

Cao, L. (2017). Data science: A comprehensive overview. *ACM Computing Surveys*, *50*(3), 43. doi:10.1145/3076253

Ceraolo, M. (2017). *Ethics and the misuse of social media data*. Retrieved from http://blog.ukdataservice. ac.uk/ethics-and-the-alleged-misuse-of-social-media-data/

Chang, W. (2017). *NIST Big Data Reference Architecture for Analytics and Beyond*. Retrieved from https://bigdatawg.nist.gov/Day2_15_NBDRA_Anaytics_and_Beyond_WoChang.pdf

Deloitte Consulting in partnership with Grocery Manufacturers Association. (2017a). *Recommendation and Conclusion 1*. Retrieved from http://www.gmaonline.org/issues-policy/collaborating-with-retailers/ big-data-analytics/recommendation-and-conclusion-1/

Deloitte Consulting in partnership with Grocery Manufacturers Association. (2017b). *Overview*. Retrieved from http://www.gmaonline.org/issues-policy/collaborating-with-retailers/big-data-analytics/ introduction#2by2

Deloitte Consulting in partnership with Grocery Manufacturers Association. (2017c). *Tools and Techniques*. Retrieved from http://www.gmaonline.org/issues-policy/collaborating-with-retailers/big-data-analytics/recommendation-and-conclusion-1/tools-and-techniques

Deloitte Consulting in partnership with Grocery Manufacturers Association. (2017d). *Information and insight*. Retrieved from http://www.gmaonline.org/issues-policy/collaborating-with-retailers/big-data-analytics/recommendation-and-conclusion-1/information-insights

Deloitte Consulting in partnership with Grocery Manufacturers Association. (2017e). *Analytical Maturity: Talent and Organization*. Retrieved from http://www.gmaonline.org/issues-policy/collaborating-with-retailers/big-data-analytics/recommendation-and-conclusion-1/talent-and-organization

Deloitte Consulting in partnership with Grocery Manufacturers Association. (2017f). *Leadership and Culture*. Retrieved from http://www.gmaonline.org/issues-policy/collaborating-with-retailers/big-data-analytics/recommendation-and-conclusion-1/leadership-and-culture

Delort, P. (2012). ICCP Technology Foresight Forum - "Harnessing data as a new source of growth: Big data analytics and policies." OECD.

Desjardins, J. (2017). *What happened in an Internet Minute in 2017*. Retrieved from http://www.visualcapitalist.com/happens-internet-minute-2017/

Gartner. (2015). *Gartner Says Business Intelligence and Analytics Leaders Must Focus on Mindsets and Culture to Kick Start Advanced Analytics*. Retrieved from https://www.gartner.com/newsroom/id/3130017

Gartner. (2016). *Gartner Survey Reveals Investment in Big Data Is Up but Fewer Organizations Plan to Invest*. Retrieved from https://www.gartner.com/newsroom/id/3466117

Grocery Manufacturers Association. (2017). *A formula for Growth: Innovation, Big Data & Analytics*. Retrieved from http://www.gmaonline.org/issues-policy/collaborating-with-retailers/big-data-analytics/

Guess, A. R. (2012). *Case Study: A Failure of Data Management at Sears*. Retrieved from http://www.dataversity.net/case-study-a-failure-of-data-management-at-sears/

Halper, F. (2015). Operationalizing and Embedding Analytics for Action. *TDWI*. Retrieved from https://www.sas.com/content/dam/SAS/en_us/doc/whitepaper2/tdwi-operationalizing-embedding-analytics-for-action-108112.pdf

Harvey, C. (2017). *Big Data Challenges*. Retrieved from https://www.datamation.com/big-data/big-data-challenges.html

Hilbert, M. (2015). *Big data for development. A review of promises and challenges.* Development Review. Retrieved from http://www.martinhilbert.net/big-data-for-development/

Lane, F. (2006). *IDC: World Created 161 Billion Gigs of Data in 2006.* Academic Press.

Laney, D. (2001). *3D Data Management: Controlling Data Volume, Velocity, and Variety.* Retrieved from https://blogs.gartner.com/doug-laney/files/2012/01/ad949-3D-Data-Management-Controlling-Data-Volume-Velocity-and-Variety.pdf

Lavinsky, D. (2017). *The Two Most Important Quotes in Business.* Retrieved from: https://www.growthink.com/content/two-most-important-quotes-business

Mandinach, E. (2012). A perfect time for data use. *Educational Psychologist*, *47*(2), 2. doi:10.1080/00461520.2012.667064

Marr, B. (2015). *Where big data project fail.* Retrieved from https://www.datacentral.com/profile/blogs/where-big-data-projects-fail

Maydon, T. (2017). *The 4 Types of data analytics.* Retrieved from https://www.datasciencecentral.com/profiles/blogs/the-4-types-of-data-analytics

McKinsey Global Institute. (2011). Big data: The next frontier for innovation, competition and productivity. *McKinsey & Company*. Retrieved from: http://www.mckinsey.com/~/media/McKinsey/dotcom/Insights%20and%20pubs/MGI/Research/Technology%20and%20Innovation/Big%20Data/MGI_big_data_full_report.ashx

Merelli, E., & Rasetti, M. (2013). Non-locality, topology, formal languages: New global tools to handle large datasets. *Procedia Computer Science*, *18*, 90–99. doi:10.1016/j.procs.2013.05.172

Moody, V. (2015). *Using big data responsibly: the ethics of big data.* Retrieved from http://blog.ukdataservice.ac.uk/using-big-data-responsibly-the-ethics-of-big-data/

OECD. (2015). Organization for Economic Cooperation and Development. *Data-Driven Innovation for Growth and Well-Being, OECD.*

Rankin, J. (2013). *How data Systems & reports can either fight or propagate the data analysis error epidemic, and how educator leaders can help. Presentation conducted by the Technology Information Center for Administrative Leadership (TICAL).* School Leadership Summit.

Shingles, M., Page, O., Phipps, J., Davenport, T., & Iglesias, J. (2017). *Acknowledgments and Key Contacts.* Retrieved from http://www.gmaonline.org/issues-policy/collaborating-with-retailers/big-data-analytics/appendix

Soderlund, J. (2017). *Big Data Architecture Patterns and Best Practices*. Retrieved from https://www.slideshare.net/AmazonWebServices/big-data-architectural-patterns-and-best-practices-75668557

USDEO Department of Education Office of Planning, Evaluation and Policy Development. (2009). Implementing data-informed decision making in schools: Teacher access, supports, and use. United States Department of Education. (ERIC Document Reproduction Service No. ED504191)

Villanova University. (2017). *When Big Data Does Not Work*. Retrieved from https://taxandbusinessonline.villanova.edu/resources-business/infographic-business/when-big-data-doesnt-work.html

Walsh, B. (2014). *Google's flue project shows the failings of big data*. Retrieved from http://time.com/23782/google-flu-trends-big-data-problems/

Weichecki, A. (2017). *Data Science Skill Set*. Retrieved from http://datasc.pl/2017/08/13/data-sciensitwg-philippa-diesinger/

APPENDIX: ABBREVIATIONS

AWS: Amazon web services.
CRM: Customer relationship management.
DBDM: Data-based decision making.
DDI: Data-driven innovation.
DDP: Data-driven planning.
DevOps: Development and operations.
MGI: McKinsey Global Institute.
NoSQL: Not only SQL database.

Compilation of References

Adams, K. M., Corrigan, J. M., & Greiner, A. C. (2004). *1st Annual Crossing the Quality Chasm Summit: A Focus on Communities*. Washington, DC: National Academies Press.

Adamu, A. A., & Shehu, A. (2018). Development of an Automatic Tomato Sorting Machine Based on Color Sensor. *International Journal of Recent Engineering Research and Development, 3*(11).

Adkins, M. (2013). *The Idea of the Sciences in the French Enlightenment: A Reinterpretation*. University of Delaware Press.

Afshar, V. (2017). *Blockchain Will Disrupt Every Industry*. Retrieved from https://www.huffingtonpost.com/entry/blockchain-will-disrupt-every-industry-us-5963868ce4b08f5c97d06b55

Aghighi, H., Azadbakht, M., Ashourloo, D., Shahrabi, H. S., & Radiom, S. (2018). Machine Learning Regression Techniques for the Silage Maize Yield Prediction Using Time-Series Images of Landsat 8 OLI. *IEEE Journal of Selected Topics in Applied Earth Observations and Remote Sensing*, 1–15. doi:10.1109/JSTARS.2018.2823361

Ahlm, E., & Litan, A. (2016). Market Trends: User and Entity Behavior Analytics Expand Their Market Reach. *Gartner*. Retrieved from https://www.gartner.com/doc/reprints?id=1-370BP2V&ct=160518&st=sb

Ahmadi, M., O'Neil, M., Fragala-Pinkham, M., Lennon, N., & Trost, S. (2018). Machine learning algorithms for activity recognition in ambulant children and adolescents with cerebral palsy. *Journal of Neuroengineering and Rehabilitation, 15*(1), 105. doi:10.1186/s12984-018-0456-x PubMed

Ahmad, M., Younis, T., Habib, M. A., Ashraf, R., & Ahmed, S. H. (2019). A Review of Current Security Issues in Internet of Things. In *Recent Trends and Advances in Wireless and IoT-enabled Networks* (pp. 11–23). Cham: Springer. doi:10.1007/978-3-319-99966-1_2

Ahmed, F., Al-Mamun, H. A., Bari, A. S. M. H., Hossain, E., & Kwan, P. (2012). Classification of crops and weeds from digital images: A support vector machine approach. *Crop Protection, 40*, 98–104. doi:10.1016/j.cropro.2012.04.024

Ajdadi, F. R., Gilandeh, Y. A., Mollazade, K., & Hasanzadeh, R. P. R. (2016). Application of machine vision for classification of soil aggregate size. *Soil & Tillage Research, 162*, 8–17. doi:10.1016/j.still.2016.04.012

Akbal, B. (2018). Hybrid GSA-ANN methods to forecast sheath current of high voltage underground cable lines. *Journal of Computers, 13*(4), 417–425.

Akbarzadeh, S., Paap, A., Ahderom, S., Apopei, B., & Alameh, K. (2018). Plant discrimination by Support Vector Machine classifier based on spectral reflectance. *Computers and Electronics in Agriculture, 148*, 250–258. doi:10.1016/j.compag.2018.03.026

Akinci, T. C. (2011). Short term speed forecasting with ANN in Batman, Turkey. *Elektronika ir Elektrotechnika, 107*(1), 41–45.

Akinci, T. C., Korkmaz, U., Turkpence, D., & Seker, S. (2019). Big Data Base Energy Management System. *3rd International Symposium on Innovative Approaches in Scientific Studies-Engineering and Natural Sciences, (ISAS 2019)*, 48.

Akinci, T. C., Nogay, H. S., & Yilmaz, O. (2012). Application of artificial neural networks for defect detection in ceramic materials. *Archives of Acoustics*, *37*(3), 279–286. doi:10.2478/v10168-012-0036-1

Al-Abdulqader, O., & Mohan, V. (2018). Learning by Demonstration with Baxter Humanoid. In K. Arai, S. Kapoor, & R. Bhatia (Eds.), *Intelligent Systems and Applications. IntelliSys 2018. Advances in Intelligent Systems and Computing* (Vol. 868). Cham: Springer.

Alahakoon, D., & Yu, X. (2015). Smart electricity meter data intelligence for future energy systems: A survey. *IEEE Transactions on Industrial Informatics*, *12*(1), 1–12.

Alberti-Alhtaybat, L., Al-Htaybat, K., & Hutaibat, K. (2019). A knowledge management and sharing business model for dealing with disruption: The case of Aramex. *Journal of Business Research*, *94*, 400–407. doi:10.1016/j.jbusres.2017.11.037

Alcaraz, J. C., Moghaddamnia, S., Poschadel, N., & Peissig, J. (2018). Machine Learning as Digital Therapy Assessment for Mobile Gait Rehabilitation. 2018 IEEE 28th International Workshop on Machine Learning for Signal Processing (MLSP), 1–6. 10.1109/MLSP.2018.8517005

Aleke, B., Ojiako, U., & Wainwright, D. W. (2011). ICT adoption in developing countries: Perspectives from small-scale agribusinesses. *Journal of Enterprise Information Management*, *24*(1), 68–84. doi:10.1108/17410391111097438

Alelaiwi, A., Hassan, M. M., & Bhuiyan, M. Z. A. (2017). A Secure and Dependable Connected Smart Home System for Elderly. 2017 IEEE 15th Intl Conf on Dependable, Autonomic and Secure Computing, 15th Intl Conf on Pervasive Intelligence and Computing, 3rd Intl Conf on Big Data Intelligence and Computing and Cyber Science and Technology Congress(DASC/PiCom/DataCom/CyberSciTech), 722–727. 10.1109/DASC-PICom-DataCom-CyberSciTec.2017.126

Alex, O. (2015). *Blockchain Protocol Series – Introduction*. Retrieved from https://medium.com/the-blockchain/blockchain-protocol-series-introduction-79d7d9ea899

Algergawy, A., Babalou, S., Kargar, M. J., & Hashem Davarpanah, S. (2015). Seecont: A new seeding-based clustering approach for ontology matching. In Lecture Notes in Computer Science (including subseries Lecture Notes in Artificial Intelligence and Lecture Notes in Bioinformatics) (Vol. 9282, pp. 245–258). Springer International Publishing. doi:10.1007/978-3-319-23135-8_17

Algergawy, A., Massmann, S., & Rahm, E. (2011). A clustering-based approach for large-scale ontology matching. In Lecture Notes in Computer Science (including subseries Lecture Notes in Artificial Intelligence and Lecture Notes in Bioinformatics) (Vol. 6909 LNCS, pp. 415–428). Springer. doi:10.1007/978-3-642-23737-9_30

Ali, A., Qadir, J., Rasool, R., Sathiaseelan, A., Zwitter, A., & Crowcroft, J. (2016). Big data for development: Applications and techniques. *Big Data Analytics*, *1*(1), 2. doi:10.118641044-016-0002-4

Ali, M., Deo, R. C., Downs, N. J., & Maraseni, T. (2018). Multi-stage committee based extreme learning machine model incorporating the influence of climate parameters and seasonality on drought forecasting. *Computers and Electronics in Agriculture*, *152*, 149–165. doi:10.1016/j.compag.2018.07.013

Alipio, M. I., Dela Cruz, A. E. M., Doria, J. D. A., & Fruto, R. M. S. (2017). A smart hydroponics farming system using exact inference in Bayesian network. *2017 IEEE 6th Global Conference on Consumer Electronics (GCCE)*, 1–5. 10.1109/GCCE.2017.8229470

Alpaydin, E. (2010). *Introduction to machine learning*. Cambridge, MA: The MIT Press.

Alqaraawi, A., Alwosheel, A., & Alasaad, A. (2016). Heart rate variability estimation in photoplethysmography signals using Bayesian learning approach. Healthcare Technology Letters, 3(2), 136–142. doi:10.1049/htl.2016.0006 PubMed

Al-Ramahi, M. A., Liu, J., & El-Gayar, O. F. (2017). Discovering Design Principles for Health Behavioral Change Support Systems: A Text Mining Approach. *ACM Transactions on Management Information Systems, 8*(2–3), 1–24. doi:10.1145/3055534

Altuntas, N., Imal, E., Emanet, N., & Ozturk, C. N. (2016). Reinforcement learning-based mobile robot navigation. *Turkish Journal of Electrical Engineering and Computer Sciences, 24*(3), 1747–1767. doi:10.3906/elk-1311-129

Amazonaws. (2019). *Machine Learning.* Retrieved from https://wordstream-files-prod.s3.amazonaws.com/s3fs-public/machine-learning.png

Amft, O. (2018). How Wearable Computing Is Shaping Digital Health. *IEEE Pervasive Computing, 17*(1), 92–98. doi:10.1109/MPRV.2018.011591067

Amin, M. B., Khan, W. A., Lee, S., & Kang, B. H. (2015). Performance-based ontology matching. *Applied Intelligence, 43*(2), 356–385. doi:10.100710489-015-0648-z

Ancin-Murguzur, F. J., Barbero-Lopez, A., Kontunen-Soppela, S., & Haapala, A. (2018). Automated image analysis tool to measure microbial growth on solid cultures. *Computers and Electronics in Agriculture, 151*, 426–430. doi:10.1016/j.compag.2018.06.031

Andoni, M., Robu, V., Flynn, D., Abram, S., Geach, D., Jenkins, D., ... Peacock, A. (2019). Blockchain technology in the energy sector: A systematic review of challenges and opportunities. *Renewable & Sustainable Energy Reviews, 100*, 143–174. doi:10.1016/j.rser.2018.10.014

Anjum, A., Sporny, M., & Sill, A. (2017). Blockchain standards for compliance and trust. *IEEE Cloud Computing, 4*(4), 84–90. doi:10.1109/MCC.2017.3791019

Aparanji, V. M., Wali, U. V., & Aparna, R. (2017). Robotic Motion Control using Machine Learning Techniques. In *International Conference on Communication and Signal Processing* (pp. 6–8). Academic Press. 10.1109/ICCSP.2017.8286579

Apte, P. R., Shah, H., & Mann, D. (2001). 5W's and an H of TRIZ Innovation. *The TRIZ Journal.* Retrieved from https://triz-journal.com/5ws-h-triz-innovation/

Aquino, C. F., Chamhum Salomao, L. C., & Azevedo, A. M. (2016). High-efficiency phenotyping for vitamin A in banana using artificial neural networks and colorimetric data. *Bragantia, 75*(3), 268–274. doi:10.1590/1678-4499.467

Arima, S., Kondo, N., Yagi, Y., Monta, M., & Yoshida, Y. (2001). Harvesting robot for strawberry grown on table top culture, 1: Harvesting robot using 5 DOF manipulator. *Journal of Society of High Technology in Agriculture.* Retrieved from http://agris.fao.org/agris-search/search.do?recordID=JP2001006293

Arima, S., Kondo, N., Shibano, Y., Fujiura, T., Yamashita, J., & Nakamura, H. (1994). Studies on Cucumber Harvesting Robot (Part 2). *Journal of the Japanese Society of Agricultural Machinery, 56*(6), 69–76. doi:10.11357/jsam1937.56.6_69

Arteta, T. A., Hameg, R., Landin, M., Gallego, P. P., & Barreal, M. E. (2018). Neural networks models as decision-making tool for in vitro proliferation of hardy kiwi. *European Journal of Horticultural Science, 83*(4), 259–265. doi:10.17660/eJHS.2018/83.4.6

Artificiallawyer. (2018). Hype Killer – Only 1% of Companies Are Using Blockchain. *Gartner Reports.* Retrieved from: https://www.artificiallawyer.com/2018/05/04/hype-killer-only-1-of-companies-are-using-blockchain-gartner-reports

Aslani, S., & Mahdavi-Nasab, H. (2013). Optical Flow Based Moving Object Detection and Tracking for Traffic Surveillance. *International Journal of Electrical, Computer, Energetic, Electronic and Communication Eng., 7*(9), 963–967. Retrieved from https://www.semanticscholar.org/paper/Optical-Flow-Based-Moving-Object-Detection-and-for-Aslani-Mahdavi-Nasab/a3248c45cdc41417ed2ea236829a99c9783bb3ca

Auer, S., Bizer, C., Kobilarov, G., Lehmann, J., Cyganiak, R., & Ives, Z. (2007). DBpedia: A nucleus for a Web of open data. In Lecture Notes in Computer Science (including subseries Lecture Notes in Artificial Intelligence and Lecture Notes in Bioinformatics) (Vol. 4825, pp. 722–735). Springer. doi:10.1007/978-3-540-76298-0_52

Ayankoya, K., Calitz, A. P., & Greyling, J. H. (2016). Real-Time Grain Commodities Price Predictions In South Africa: A Big Data And Neural Networks Approach. *Agrekon, 55*(4), 483–508. doi:10.1080/03031853.2016.1243060

Babazadeh, S., Moghaddam, P. A., Sabatyan, A., & Sharifian, F. (2016). Classification of potato tubers based on solanine toxicant using laser induced light backscattering imaging. *Computers and Electronics in Agriculture, 129*, 1–8. doi:10.1016/j.compag.2016.09.009

Badawy, R., Raykov, Y. P., Evers, L. J. W., Bloem, B. R., Faber, M. J., Zhan, A., ... Little, M. A. (2018). Automated Quality Control for Sensor Based Symptom Measurement Performed Outside the Lab. *Sensors (Basel), 18*(4), 1215. doi:10.3390/s18041215 PubMed

Badue, C., Guidolini, R., Carneiro, R. V., Azevedo, P., Cardoso, V. B., Forechi, A., & De Souza, A. F. (2019). *Self-Driving Cars: A Survey*. Cornell University.

Baig, M., & Mirza, F. GholamHosseini, H., Gutierrez, J., & Ullah, E. (2018). Clinical Decision Support for Early Detection of Prediabetes and Type 2 Diabetes Mellitus Using Wearable Technology. 2018 40th Annual International Conference of the IEEE Engineering in Medicine and Biology Society (EMBC), 4456–4459. 10.1109/EMBC.2018.8513343

Baker, J. D. (2016). The Purpose, Process, and Methods of Writing a Literature Review. *AORN Journal, 103*(3), 265–269. doi:10.1016/j.aorn.2016.01.016 PMID:26924364

Bakhshi, M., Pourtaheri, M., & Eftekhari, A. R. (2016). Developing a Model to Predict Success of Agricultural Production Enterprises Based on Their Capitals. *Journal of Agricultural Science and Technology, 18*(6), 1443–1454.

Bakhshipour, A., & Jafari, A. (2018). Evaluation of support vector machine and artificial neural networks in weed detection using shape features. *Computers and Electronics in Agriculture, 145*, 153–160. doi:10.1016/j.compag.2017.12.032

Bakhshipour, A., Jafari, A., Nassiri, S. M., & Zare, D. (2017). Weed segmentation using texture features extracted from wavelet sub-images. *Biosystems Engineering, 157*, 1–12. doi:10.1016/j.biosystemseng.2017.02.002

Baktir, A. C., Tunca, C., Ozgovde, A., Salur, G., & Ersoy, C. (2018). SDN-Based Multi-Tier Computing and Communication Architecture for Pervasive Healthcare. *IEEE Access : Practical Innovations, Open Solutions, 6*, 56765–56781. doi:10.1109/ACCESS.2018.2873907

Balog, M., Gaunt, A. L., Brockschmidt, M., Nowozin, S., & Tarlow, D. (2016). *DeepCoder: Learning to Write Programs. Computer Science. Machine Learning*. Cornell University.

Banerjee, A., Bandyopadhyay, T., & Acharya, P. (2013). Data Analytics: Hyped Up Aspirations or True Potential? *Vikalpa, 38*(4), 1–12. doi:10.1177/0256090920130401

Bansal, P., & Kockelman, K. M. (2018). Are we ready to embrace connected and self-driving vehicles? A case study of Texans. *Transportation, 45*(2), 641–675. doi:10.100711116-016-9745-z

Banu, G., & Suja, S. (2012). ANN based fault location technique using one end data for UHV lines. *European Journal of Scientific Research, 77*(4), 549–559.

Barber, C. B., Dobkin, D. P., & Huhdanpaa, H. (1996). The quickhull algorithm for convex hulls. *ACM Trans. Math. Softw.*, *22*(4), 469–483. Retrieved from doi:10.1145/235815.235821

Battaglia, P. W., Pascanu, R., Lai, M., Rezende, D., & Kavukcuoglu, K. (2016). *Interaction Networks for Learning about Objects, Relations and Physics*. Retrieved from http://arxiv.org/abs/1612.00222

Begoli, E., & Horey, J. (2012). Design Principles for effective knowledge discovery from big data. In *Joint working conference on Software Architecture & 6th European Conference on Software Architecture*, (pp. 215-218). Helsinki, Finland: IEEE. 10.1109/WICSA-ECSA.212.32

Bellare, M., & Rogaway, P. (2005). *Introduction*. Introduction to Modern Cryptography.

Bello-Orgaz, G., Jung, J. J., & Camacho, D. (2016). Social big data: Recent achievements and new challenges. *Information Fusion*, *28*, 45–59. doi:10.1016/j.inffus.2015.08.005

Bencic, F. M., & Zarko, I. P. (2018). *Distributed Ledger Technology: Blockchain Compared to Directed Acyclic Graph*. 2018 IEEE 38th International Conference on Distributed Computing Systems (ICDCS), Vienna, Austria.

Benckendorff, P. J., Sheldon, P. J., & Fesenmaier, D. R. (2014). *Tourism information technology* (2nd ed.). Tourism Information Technology. doi:10.1079/9781780641850.0000

Bengio, Y. (2009). Learning Deep Architectures for AI. *Foundations and Trends® in Machine Learning, 2*(1), 1–127. doi:10.1561/2200000006

Bengio, Y. (2013). Deep Learning of Representations: Looking Forward. In A.-H. Dediu, C. Martín-Vide, R. Mitkov, & B. Truthe (Eds.), Statistical Language and Speech Processing (Vol. 7978, pp. 1–37). Academic Press. doi:10.1007/978-3-642-39593-2_1

Ben, X., Meng, W., Wang, K., & Yan, R. (2016). An adaptive neural networks formulation for the two-dimensional principal component analysis.(Report). *Neural Computing & Applications*, 27.

Beranek, B., Simon, H. A., & McCorduck, P. (1977). History Of Artificial Intelligence. *IJCAI (United States)*, *2*, 4.

Bernhardt, V. (2013). Data analysis for continuous school improvement. Routledge.

Bharadwaj, S. (2018). *Blockchain based solutions to everyday problems*. Retrieved from https://medium.com/swlh/blockchain-based-solutions-to-everyday-problems-7c0bb3cb83dc

Bhise, S. (2018, October 11). Forces Driving the Growth of Wearable Medical Device Market. Retrieved January 24, 2019, from Health Works Collective website: https://www.healthworkscollective.com/forces-driving-the-growth-of-wearable-medical-device-market/

Bieger, T., & Laesser, C. (2004). Information sources for travel decisions: Toward a source process model. *Journal of Travel Research*, *42*(4), 357–371. doi:10.1177/0047287504263030

Biggio, B., & Roliab, F. (2018). Wild patterns: Ten years after the rise of adversarial machine learning. *Pattern Recognition*, *84*, 317–331. doi:10.1016/j.patcog.2018.07.023

Bishop, C. (1995). *Neural Networks for Pattern Recognition*. Oxford, UK: University Press.

Bisola, A. (2018). *What is SHA-256 and how is it related to Bitcoin?* Retrieved from https://www.mycryptopedia.com/sha-256-related-bitcoin/

Bitra, V. S., Jayapandian, N., & Balachandran, K. (2018). Internet of Things Security and Privacy Issues in Healthcare Industry. In *International Conference on Intelligent Data Communication Technologies and Internet of Things* (pp. 967-973). Springer.

Björk, P., & Kauppinen-Räisänen, H. (2015). Contemporary insights to the dynamic pre-trip information sourcing behaviour. *Tourism and Hospitality Research, 15*(1), 39–53. doi:10.1177/1467358414553871

Bley, A., & Martin, C. (2010). *SCITOS G5 –A mobile platform for research and industrial applications.* Retrieved from http://download.ros.org/data/Events/CoTeSys-ROS-School/metralabs.pdf

Boas, M. G. V., Santos, H. G., & de Campos Merschmann, L. H. (2018). Optimal Decision Trees for Feature Based Parameter Tuning: Integer Programming Model and VNS Heuristic. *Electronic Notes in Discrete Mathematics, 66*, 223–230. doi:10.1016/j.endm.2018.03.029

Boell, S. K., & Cecez-Kecmanovic, D. (2015). A Hermeneutic Approach for Conducting Literature Reviews and Literature Searches. *Communications of the Association for Information Systems, 34*, 12. Retrieved from http://aisel.aisnet.org/cais/vol34/iss1/12

Bogdan, Ž., Cehil, M., & Kopjar, D. (2009). Power system optimization. *Energy, 32*(6), 955–960. doi:10.1016/j.energy.2007.01.004

Bohanec, M. (2009). Decising Making: A computer-science and information- technology viewpoint. *Interdisciplinary Description of Complex Systems, 7*(2), 22–37.

Bojana, K., Elena, K., & Anastas, M. (2017). Blockchain Implementation Quality Challenges: A Literature Review. *Proceedings of the SQAMIA, 2017*, 6.

Bolesnikov, M., Popovic-Stijacic, M., Radisic, M., Takaci, A., Borocki, J., Bolesnikov, D., ... Dziendziora, J. (2018). Development of a Business Model by Introducing Sustainable and Tailor-Made Value Proposition for SME Clients. *Sustainability*. doi:10.1111/joms.12352

Borges, P. H. M., Mendoza, Z. M. S. H., Maia, J. C. S., Bianchini, A., & Fernandes, H. C. (2017). Estimation Of Fuel Consumption In Agricultural Mechanized Operations Using Artificial Neural Networks. *Engenharia Agrícola, 37*(1), 136–147. doi:10.1590/1809-4430-eng.agric.v37n1p136-147/2017

Borky, J. M., & Bradley, T. H. (2019). Developing the Network Dimension. In *Effective Model-Based Systems Engineering* (pp. 327–344). Cham: Springer. doi:10.1007/978-3-319-95669-5_9

Borne, K. (2014). *Top 10 Big Data Challenges – A Serious Look at 10 Big Data V's.* Retrieved from https://mapr.com/blog/top-10-big-data-challenges-serious-look-10-big-data-vs/

Borthakur, D., Dubey, H., Constant, N., Mahler, L., & Mankodiya, K. (2017). Smart fog: Fog computing framework for unsupervised clustering analytics in wearable Internet of Things. 2017 IEEE Global Conference on Signal and Information Processing (GlobalSIP), 472–476. doi:10.1109/GlobalSIP.2017.8308687

Boulton, C. (2017). *6 data analytics success stories: An inside look.* Retrieved from https://www.cio.com/article/3221621/analytics/6-data-analytics-success-stories-an-inside-look.html

Boyd, D., & Crawford, K. (2011). Six Provocations for Big Data. *Social Science Research Network: A Decade in Internet Time: Symposium on the Dynamics of the Internet and Society.* doi:10.2139srn.1926431

Brahmi, H., Ammar, B., & Alimi, A. M. (2013). Intelligent path planning algorithm for autonomous robot based on recurrent neural networks. *Advanced Logistics and Transport (ICALT), 2013 International Conference on*, 199–204.

Brain, C. (2019). *The artificial intelligence ecosystem*. Retrieved from https://medium.com/@b.jamescurry/the-artificial-intelligence-ecosystem-f11107f7b306

Brandes, U., Borgatti, S. P., & Freeman, L. C. (2016). Maintaining the duality of closeness and betweenness centrality. *Social Networks*, *44*, 153–159. doi:10.1016/j.socnet.2015.08.003

Brennan, D. (2018). *The Ultimate Disruptor: How Blockchain is Transforming Financial Services*. Retrieved from https://gowlingwlg.com/getmedia/ab5aecbb-8997-4bd2-96e7-58070b33e07a/how-blockchain-is-transforming-financial-services.pdf.xml?ext=.pdf

Brent, X. (2018). *Blockchain vs. Distributed Ledger Technologies Part 2: Governing Dynamics*. Retrieved May 23 2018, from https://media.consensys.net/blockchains-vs-distributed-ledger-technologies-part-2-governing-dynamics-a697848d5b82

BRESOV. (n. d.). *Breeding for Resilient, Efficient and Sustainable Organic Vegetable Production*. Retrieved from https://www.bresov.eu/about/tomatoes

Breton, P. (1995). *L'image de l'homme: du Golem aux créatures virtuelles. Open Library*. Paris: Seuil.

Brownlee, J. (2013). *A tour of machine learning algorithms*. Retrieved from https://machinelearningmastery.com/a-tour-of-machine-learning-algorithms/

Brugere, I., Gallagher, B., & Berger-Wolf, T. Y. (2018). Network structure inference, a survey: Motivations, methods, and applications. *ACM Computing Surveys*, *51*(2), 24. doi:10.1145/3154524

Bruyn, A. S. (2017). *Blockchain an Introduction*. Research paper. VU University Amsterdam. Retrieved from https://beta.vu.nl/nl/Images/werkstuk-bruyn_tcm235-862258.pdf

Bryant, A. (2014). Thinking about the information age. *Informatics*, *1*(3), 190–195. doi:10.3390/informatics1030190

Brynjolfsson, E., Hitt, L. M., & Kim, H. H. (2011). *Strength in Numbers: How Does Data-Driven Decision-Making Affect Firm Performance?* SSRN Electronic Journal. doi:10.2139srn.1819486

BTGA. (2018). *British Tomatoes Growers Association. Tomato Types*. Retrieved from http://www.britishtomatoes.co.uk/tomato-facts/tomato-types

Buda, M., Maki, A., & Mazurowski, M. A. (2018). A systematic study of the class imbalance problem in convolutional neural networks. *Neural Networks*, *106*, 249–259. doi:10.1016/j.neunet.2018.07.011 PMID:30092410

Buhalis, D., & Amaranggana, A. (2015). Smart Tourism Destinations Enhancing Tourism Experience Through Personalisation of Services. Information and Communication Technologies in Tourism 2015. doi:10.1007/978-3-319-14343-9_28

Bush, V. (1945). As we may think. A top US scientist foresees a possible future world in which man-made machines will start to think. Office of scientific research and development. *Atlantic Monthly*.

Buterin, V. (2013). *Introducing Ripple*. Retrieved from https://bitcoinmagazine.com/articles/introducing-ripple/

Buterin, V. (2014). *A next-generation smart contract and decentralized application platform*. White Paper, 2014. Retrieved from https://github.com/ethereum/wiki/wiki/White-Paper

Buterin, V. (2015). *On Public and Private Blockchains*. Retrieved August 6, 2015, from https://blog.ethereum.org/2015/08/07/on-public-and-private-blockchains/

Buterin, V. (2015). *On Public and Private Blockchains*. Retrieved from https://ethereum.org/

Byakatonda, J., Parida, B. P., Kenabatho, P. K., & Moalafhi, D. B. (2018). Influence of climate variability and length of rainy season on crop yields in semiarid Botswana. *Agricultural and Forest Meteorology, 248*, 130–144. doi:10.1016/j.agrformet.2017.09.016

Cai, J., Wang, Y., Liu, Y., Luo, J. Z., Wei, W., & Xu, X. (2018). Enhancing network capacity by weakening community structure in scale-free network. *Future Generation Computer Systems, 87*, 765–771. doi:10.1016/j.future.2017.08.014

Caires, E. F., & Guimaraes, A. M. (2018). A Novel Phosphogypsum Application Recommendation Method under Continuous No-Till Management in Brazil. *Agronomy Journal, 110*(5), 1987–1995. doi:10.2134/agronj2017.11.0642

Calderano, B. F., Polivanov, H., da Silva Chagas, C., de Carvalho, W. J., Barroso, E. V., Teixeira Guerra, A. J., & Calderano, S. B. (2014). Artificial Neural Networks Applied for Soil Class Prediction in Mountainous Landscape of The Serra Do Mar. *Revista Brasileira de Ciência do Solo, 38*(6), 1681–1693. doi:10.1590/S0100-06832014000600003

Camacho, D. M., Collins, K. M., Powers, R. K., Costello, J. C., & Collins, J. J. (2018). Next-generation machine learning for biological networks. *Cell, 173*(7), 1581–1592. doi:10.1016/j.cell.2018.05.015 PMID:29887378

Camargo, A., Molina, J. P., Cadena-Torres, J., Jimenez, N., & Kim, J. T. (2012). Intelligent systems for the assessment of crop disorders. *Computers and Electronics in Agriculture, 85*, 1–7. doi:10.1016/j.compag.2012.02.017

Cameron, M., Viviers, W., & Steenkamp, E. (2017). Breaking the "big data" barrier when selecting agricultural export markets: An innovative approach. *Agrekon, 56*(2), 139–157. doi:10.1080/03031853.2017.1298456

Cao, L. (2017). Data science: A comprehensive overview. *ACM Computing Surveys, 50*(3), 43. doi:10.1145/3076253

Castaneda-Miranda, A., & Castano, V. M. (2017). Smart frost control in greenhouses by neural networks models. *Computers and Electronics in Agriculture, 137*, 102–114. doi:10.1016/j.compag.2017.03.024

Castro, C. A. de O., Resende, R. T., Kuki, K. N., Carneiro, V. Q., Marcatti, G. E., Cruz, C. D., & Motoike, S. Y. (2017). High-performance prediction of macauba fruit biomass for agricultural and industrial purposes using Artificial Neural Networks. *Industrial Crops and Products, 108*, 806–813. doi:10.1016/j.indcrop.2017.07.031

Cave, S., & OhEigeartaigh, S. S. (2018). An AI Race for Strategic Advantage: Rhetoric and Risks. Association for the Advancement of Artificial Intelligence.

CEMA - European Agricultural Machinery. (2017, February 13). *Digital Farming: what does it really mean?* Retrieved September 24, 2018, from http://www.cema-agri.org/page/digital-farming-what-does-it-really-mean

Center for Disease Control (CDC). (2018, September 5). About Chronic Disease. Retrieved September 25, 2018, from https://www.cdc.gov/chronicdisease/about/index.htm

Ceraolo, M. (2017). *Ethics and the misuse of social media data*. Retrieved from http://blog.ukdataservice.ac.uk/ethics-and-the-alleged-misuse-of-social-media-data/

Cerbah, F. (2010). Learning ontologies with deep class hierarchies by mining the content of relational databases. In Studies in Computational Intelligence (Vol. 292, pp. 271–286). Academic Press. doi:10.1007/978-3-642-00580-0_16

Chan, C. M. L. (2013). From Open Data to Open Innovation Strategies: Creating E-Services Using Open Government Data. *Proceedings of the 46th Hawaii International Conference on System Sciences (HICSS-46)*, 1890–1899. 10.1109/HICSS.2013.236

Chandiok, A., & Chaturvedi, D. K. (2018). CIT: Integrated cognitive computing and cognitive agent technologies based cognitive architecture for human-like functionality in artificial systems. *Biologically Inspired Cognitive Architectures., 26*, 55–79. doi:10.1016/j.bica.2018.07.020

Chandra, N., Bipan, T., & Chiranjib, K. (2012). An Automated Machine Vision Based System for Fruit Grading and Sorting. *Six International Conference On Sensing Technology*. Retrieved from http://seat.massey.ac.nz/conferences/icst2012/files/icst2012program.pdf

Chang, W. (2017). *NIST Big Data Reference Architecture for Analytics and Beyond*. Retrieved from https://bigdatawg.nist.gov/Day2_15_NBDRA_Anaytics_and_Beyond_WoChang.pdf

Chan, M., Estève, D., Fourniols, J.-Y., Escriba, C., & Campo, E. (2012). Smart wearable systems: Current status and future challenges. *Artificial Intelligence in Medicine*, *56*(3), 137–156. doi:10.1016/j.artmed.2012.09.003 PubMed

Chantre, G. R., Vigna, M. R., Renzi, J. P., & Blanco, A. M. (2018). A flexible and practical approach for real-time weed emergence prediction based on Artificial Neural Networks. *Biosystems Engineering*, *170*, 51–60. doi:10.1016/j.biosystemseng.2018.03.014

Chapelle, O., Scholkopf, B., & Zien, A. (2009). Semi-supervised learning. *IEEE Transactions on Neural Networks*, *20*(3), 542–542. doi:10.1109/TNN.2009.2015974

Chapman, R., Cook, S., Donough, C., Lim, Y. L., Vun Vui Ho, P., Lo, K. W., & Oberthür, T. (2018). Using Bayesian networks to predict future yield functions with data from commercial oil palm plantations: A proof of concept analysis. *Computers and Electronics in Agriculture*, *151*, 338–348. doi:10.1016/j.compag.2018.06.006

Chaturvedi, I., Ragusa, E., Gastaldo, P., Zunino, R., & Cambria, E. (2018). Bayesian network based extreme learning machine for subjectivity detection. *Journal of the Franklin Institute*, *355*(4), 1780–1797. doi:10.1016/j.jfranklin.2017.06.007

Chaudhary, A., Kolhe, S., & Kamal, R. (2016). A hybrid ensemble for classification in multiclass datasets: An application to oilseed disease dataset. *Computers and Electronics in Agriculture*, *124*, 65–72. doi:10.1016/j.compag.2016.03.026

Chen, Z., & Liu, B. (2018). Lifelong Machine Learning (2nd ed.). Synthesis Lectures on Artificial Intelligence and Machine Learning. Morgan & Claypool Publishers.

Cheng, S., Tamil, L. S., & Levine, B. (2015). A Mobile Health System to Identify the Onset of Paroxysmal Atrial Fibrillation. 2015 International Conference on Healthcare Informatics, 189–192. doi:10.1109/ICHI.2015.29

Cheng, Y., Chen, P., Yang, C., & Samani, H. (2016). IMU based activity detection for post mini-stroke healthcare. 2016 International Conference on System Science and Engineering (ICSSE), 1–4. doi:10.1109/ICSSE.2016.7551611

Cheng, Y., Jiang, P., & Peng, Y. (2014). Increasing big data front end processing efficiency via locality sensitive Bloom filter for elderly healthcare. 2014 IEEE Symposium on Computational Intelligence in Big Data (CIBD), 1–8. doi:10.1109/CIBD.2014.7011524

Cheng, X., Zhang, Y., Chen, Y., Wu, Y., & Yue, Y. (2017). Pest identification via deep residual learning in complex background. *Computers and Electronics in Agriculture*, *141*, 351–356. doi:10.1016/j.compag.2017.08.005

Chen, M., Mao, S., & Liu, Y. (2014). Big data: A survey. *Mobile Networks and Applications*, *19*(2), 171–209. doi:10.100711036-013-0489-0

Chen, M., Yang, J., Zhou, J., Hao, Y., Zhang, J., & Youn, C. (2018). 5G-Smart Diabetes: Toward Personalized Diabetes Diagnosis with Healthcare Big Data Clouds. *IEEE Communications Magazine*, *56*(4), 16–23. doi:10.1109/MCOM.2018.1700788

Chen, X., Kong, Y., Fang, X., & Wu, Q. (2013). A fast two-stage ACO algorithm for robotic path planning. *Neural Computing & Applications*, *22*(2), 313–319. doi:10.100700521-011-0682-7

Chen, Y. Q., Zhang, J., & Ng, W. W. (2018). Loan Default Prediction Using Diversified Sensitivity Undersampling. In *2018 International Conference on Machine Learning and Cybernetics (ICMLC)* (Vol. 1, pp. 240-245). IEEE. 10.1109/ICMLC.2018.8526936

Chi, M., Plaza, A., Benediktsson, J. A., Sun, Z., Shen, J., & Zhu, Y. (2016). Big Data for Remote Sensing: Challenges and Opportunities. *Proceedings of the IEEE, 104*(11), 2207–2219. doi:10.1109/JPROC.2016.2598228

Cho, G. (2018). The Australian digital farmer: challenges and opportunities. *IOP Conference Series: Earth and Environmental Science, 185*, 012036. 10.1088/1755-1315/185/1/012036

Choi, R., Kang, W., & Son, C. (2017). Explainable sleep quality evaluation model using machine learning approach. 2017 IEEE International Conference on Multisensor Fusion and Integration for Intelligent Systems (MFI), 542–546. doi:10.1109/MFI.2017.8170377

Choi, S., Lehto, X. Y., & Oleary, J. T. (2007). What does the consumer want from a DMO website? A study of US and Canadian tourists' perspectives. *International Journal of Tourism Research, 9*(2), 59–72. doi:10.1002/jtr.594

Cho, J., Kim, J. T., Kim, J., Park, S., & Kim, K. (2016). Simple Walking Strategies for Hydraulically Driven Quadruped Robot over Uneven Terrain. *Journal of Electrical Engineering & Technology, 11*(5), 1433–1440. doi:10.5370/JEET.2016.11.5.1433

Christensen, C. (2019). Disruptive Innovation. *Harvard Business School.* Retrieved 2019, from http://claytonchristensen.com/biography/

Christensen, C. (1997). *The Innovator's Dilemma: When New Technologies Cause Great Firms to Fail.* Harvard Business School Press.

Christensen, C. (1997). *The Revolutionary Book that Will Change the Way You Do Business (Collins Business Essentials).* New York: Harper Paperbacks.

Chuen, L. K. D. (Ed.). (2015). *Handbook of Digital Currency.* Elsevier. Available: http://EconPapers.repec.org/RePEc:eee:monogr:9780128021170

Chung, C.-L., Huang, K.-J., Chen, S.-Y., Lai, M.-H., Chen, Y.-C., & Kuo, Y.-F. (2016). Detecting Bakanae disease in rice seedlings by machine vision. *Computers and Electronics in Agriculture, 121*, 404–411. doi:10.1016/j.compag.2016.01.008

Cilluffo, F. J., & Cardash, S. L. (2018). What's wrong with Huawei, and why are countries banning the Chinese telecommunications firm? *The Conversation.* Retrieved from https://theconversation.com/whats-wrong-with-huawei-and-why-are-countries-banning-the-chinese-telecommunications-firm-109036

Clark, W. A., & Farley, B. G. (1955). Generalization of Pattern Recognition in a Self-Organizing System. *Western Joint Computer Conference*, 86-91. 10.1145/1455292.1455309

Clavera, I., Rothfuss, J., Schulman, J., Fujita, Y., Asfour, T., & Abbeel, P. (2018). *Model-Based Reinforcement Learning via Meta-Policy Optimization. Computer Science. Machine Learning.* Cornell University.

Clayton, C. (2015, January 27). The future of healthcare analytics is prescriptive. Retrieved January 30, 2019, from Healthcare IT News website: https://www.healthcareitnews.com/blog/future-healthcare-analytics-prescriptive

Clayton, S. (2015). *Money.* Dallas, TX: Everyday Economics Federal Reserve Bank of Dallas. Retrieved from https://www.dallasfed.org/~/media/documents/educate/everyday/money.pdf

Clercq, M. D., Vats, A., & Biel, A. (2018). Agriculture 4.0: The Future of Farming Technology. *World Government Summit*, 30.

Coble, K. H., Mishra, A. K., Ferrell, S., & Griffin, T. (2018). Big Data in Agriculture: A Challenge for the Future. *Applied Economic Perspectives and Policy*, *40*(1), 79–96. doi:10.1093/aepp/ppx056

Cockburn, I., Henderson, R., & Stern, S. (2018). *The Impact of Artificial Intelligence on Innovation* (No. w24449). doi:10.3386/w24449

Coeckelbergh, M. (2018). *Should we ban fully autonomous weapons? Uni:View Magazin.* Centre for Digital Philosophy.

Cohen, I., & Medioni, G. (1999). *Detecting and Tracking Moving Objects for Video Surveillance* (Vol. 2). IEEE Proc. Comput. Vis. Pattern Recognit. Retrieved from http://citeseerx.ist.psu.edu/viewdoc/download?doi=10.1.1.20.7779&rep=rep1&type=pdf doi:10.1109/CVPR.1999.784651

Coinbase. (2019). Available at https://en.wikipedia.org/wiki/Coinbaset

Columbus, L. (2016). *McKinsey's 2016 Analytics Study Defines The Future Of Machine Learning.* Retrieved November 14, 2018, from https://www.forbes.com/sites/louiscolumbus/2016/12/18/mckinseys-2016-analytics-study-defines-the-future-machine-learning/#3da708d214eb

Corcho, O., Fernández-López, M., & Gómez-Pérez, A. (2003). Methodologies, tools and languages for building ontologies. Where is their meeting point? *Data & Knowledge Engineering*, *46*(1), 41–64. doi:10.1016/S0169-023X(02)00195-7

Correa, F. E., Oliveira, M. D. B., Gama, J., Correa, P. L. P., & Rady, J. (2016). Analyzing the behavior dynamics of grain price indexes using Tucker tensor decomposition and spatio-temporal trajectories. *Computers and Electronics in Agriculture*, *120*, 72–78. doi:10.1016/j.compag.2015.11.011

Costa, A. G., Pinto, F., Motoike, S. Y., Braga Júnior, R. A., & Gracia, L. M. N. (2018). Classification of Macaw Palm Fruits from Colorimetric Properties for Determining the Harvest Moment. *Engenharia Agrícola*, *38*(4), 634–641. doi:10.1590/1809-4430-eng.agric.v38n4p634-641/2018

Costill, A. (2013, August 7). Top 10 Places that Have Banned Google Glass. Retrieved January 24, 2019, from Search Engine Journal website: https://www.searchenginejournal.com/top-10-places-that-have-banned-google-glass/66585/

Coughlin, J. (2017). *The Longevity Economy: Unlocking the World's Fastest-Growing, Most Misunderstood Market.* Hachette Book Group.

Coverage Book. (n.d.). *Answer the Public.* Retrieved from https://answerthepublic.com/

Crunchbase. (2018). *Crunchbase: Discover innovative companies and the people behind them.* Retrieved December 17, 2018, from Crunchbase website: https://www.crunchbase.com

Csató, L. (2016). Measuring centrality by a generalization of degree. *Central European Journal of Operations Research*, 1–20. doi:10.100710100-016-0439-6

Cui, L., Yang, S., Chen, F., Ming, Z., Lu, N., & Qin, J. (2018). A survey on application of machine learning for Internet of Things. *International Journal of Machine Learning and Cybernetics*, 1–19.

Cuzzocrea, A., Song, I.-Y., & Davis, K. C. (2011). *Analytics over large-scale multidimensional data: the big data revolution!* Academic Press.

Cytoscape. (n.d.). *Cytoscape.* Retrieved from: http://www.cytoscape.org/

da Silva, C. A. Junior, Nanni, M. R., Teodoro, P. E., & Capristo Silva, G. F. (2017). Vegetation Indices for Discrimination of Soybean Areas: A New Approach. *Agronomy Journal*, *109*(4), 1331–1343. doi:10.2134/agronj2017.01.0003

Dalakov, G. (2018). *The History of Computers.* Retrieved from http://history-computer.com/index.html

Daraio, C., Lenzerini, M., Leporelli, C., Naggar, P., Bonaccorsi, A., & Bartolucci, A. (2016). The advantages of an Ontology-Based Data Management approach: Openness, interoperability and data quality. *Scientometrics*, *108*(1), 441–455. doi:10.100711192-016-1913-6

Das, S. (2017). *100%: Dubai Will Put Entire Land Registry on a Blockchain*. Retrieved from https://www.ccn.com/100-dubai-put-entire-land-registry-blockchain/

Dash, T., Nayak, T., & Swain, R. R. (2015). Controlling Wall Following Robot Navigation Based on Gravitational Search and Feed Forward Neural Network. *Proceedings of the 2nd International Conference on Perception and Machine Intelligence*, 196–200. 10.1145/2708463.2709070

DataGovIn. (2014). *Literacy Rates in India*. Retrieved from https://data.gov.in/catalog/literacy-rate-india-nsso-and-rgi

David, X. (2017). *The Four Layers of the Blockchain*. Retrieved from https://medium.com/@coriacetic/the-four-layers-of-the-blockchain-dc1376efa10f

Dayanandam, G., Rao, T. V., Babu, D. B., & Durga, S. N. (2019). DDoS Attacks—Analysis and Prevention. In *Innovations in Computer Science and Engineering* (pp. 1–10). Singapore: Springer. doi:10.1007/978-981-10-8201-6_1

de Barros, M. M., da Silva, F. M., Costa, A. G., Ferraz, G. A. e S., & da Silva, F. C. (2018). Use of classifier to determine coffee harvest time by detachment force. *Revista Brasileira de Engenharia Agrícola e Ambiental*, *22*(5), 366–370. doi:10.1590/1807-1929/agriambi.v22n5p366-370

De Castro, L. R., Cortez, L. A. B., & Vigneault, C. (2005). Effect of sorting, refrigeration and packaging on tomato shelf life. *World Food Journal*. Retrieved from https://www.researchgate.net/publication/266370688_Effect_of_sorting_refrigeration_and_packaging_on_tomato_shelf_life

Delgado, G., Aranda, V., Calero, J., Sanchez-Maranon, M., Serrano, J. M., Sanchez, D., & Vila, M. A. (2008). Building a fuzzy logic information network and a decision-support system for olive cultivation in Andalusia. *Spanish Journal of Agricultural Research*, *6*(2), 252–263. doi:10.5424jar/2008062-316

Deloitte Consulting in partnership with Grocery Manufacturers Association. (2017a). *Recommendation and Conclusion 1*. Retrieved from http://www.gmaonline.org/issues-policy/collaborating-with-retailers/big-data-analytics/recommendation-and-conclusion-1/

Deloitte Consulting in partnership with Grocery Manufacturers Association. (2017b). *Overview*. Retrieved from http://www.gmaonline.org/issues-policy/collaborating-with-retailers/big-data-analytics/introduction#2by2

Deloitte Consulting in partnership with Grocery Manufacturers Association. (2017c). *Tools and Techniques*. Retrieved from http://www.gmaonline.org/issues-policy/collaborating-with-retailers/big-data-analytics/recommendation-and-conclusion-1/tools-and-techniques

Deloitte Consulting in partnership with Grocery Manufacturers Association. (2017d). *Information and insight*. Retrieved from http://www.gmaonline.org/issues-policy/collaborating-with-retailers/big-data-analytics/recommendation-and-conclusion-1/information-insights

Deloitte Consulting in partnership with Grocery Manufacturers Association. (2017e). *Analytical Maturity: Talent and Organization*. Retrieved from http://www.gmaonline.org/issues-policy/collaborating-with-retailers/big-data-analytics/recommendation-and-conclusion-1/talent-and-organization

Deloitte Consulting in partnership with Grocery Manufacturers Association. (2017f). *Leadership and Culture*. Retrieved from http://www.gmaonline.org/issues-policy/collaborating-with-retailers/big-data-analytics/recommendation-and-conclusion-1/leadership-and-culture

Delort, P. (2012). ICCP Technology Foresight Forum - "Harnessing data as a new source of growth: Big data analytics and policies." OECD.

Demir, B. (2018). Application of data mining and adaptive neuro-fuzzy structure to predict color parameters of walnuts (Juglans regia L.). *Turkish Journal of Agriculture and Forestry*, *42*(3), 216–225. doi:10.3906/tar-1801-78

Demir, B., Gurbuz, F., Eski, I., Kus, Z. A., Yilmaz, K. U., & Ercisli, S. (2018). Possible Use of Data Mining for Analysis and Prediction of Apple Physical Properties. *Erwerbs-Obstbau*, *60*(1), 1–7. doi:10.100710341-017-0330-1

Desjardins, J. (2017). *What happened in an Internet Minute in 2017*. Retrieved from http://www.visualcapitalist.com/happens-internet-minute-2017/

DFID. (n.d.). *Growth: Building Jobs and Prosperity in Developing Countries*. Retrieved from https://www.oecd.org/derec/unitedkingdom/40700982.pdf

Dhaliwal, S. (2016). *IBM to Launch One of the Largest Blockchain Implementations in the World*. Retrieved from https://cointelegraph.com/news/ibm-to-launch-one-of-the-largest-blockchain-implementations-in-the-world

Diaz, I., Mazza, S. M., Combarro, E. F., Gimenez, L. I., & Gaiad, J. E. (2017). Machine learning applied to the prediction of citrus production. *Spanish Journal of Agricultural Research*, *15*(2), e0205. doi:10.5424jar/2017152-9090

Dib, O., Brousmiche, K. L., Durand, A., Thea, E., & Hamida, E. B. (2018). Consortium Blockchains: Overview, Applications and Challenges. *International Journal on Advances in Telecommunications*, *11*(1 & 2), 51–64.

Dickens, C. (1859). *A Tale of Two Cities*. CreateSpace Independent Publishing Platform.

Dieleman, J. L., Squires, E., Bui, A. L., Campbell, M., Chapin, A., Hamavid, H., ... Murray, C. J. L. (2017). Factors Associated with Increases in US Health Care Spending, 1996-2013. *Journal of the American Medical Association*, *318*(17), 1668–1678. doi:10.1001/jama.2017.15927 PubMed

Dietterich, T. G. (2015). *Benefits and Risks of Artificial Intelligence*. The Association for the Advancement of Artificial Intelligence (AAAI). Retrieved from https://medium.com/@tdietterich/benefits-and-risks-of-artificial-intelligence-460d288cccf3

Dimililer, K., & Zarrouk, S. (2017). ICSPI: Intelligent Classification System of Pest Insects Based on Image Processing and Neural Arbitration. *Applied Engineering in Agriculture*, *33*(4), 453–460. doi:10.13031/aea.12161

Ding, W., & Taylor, G. (2016). Automatic moth detection from trap images for pest management. *Computers and Electronics in Agriculture*, *123*, 17–28. doi:10.1016/j.compag.2016.02.003

Dinh, T. N., & Thai, M. T. (2018). AI and Blockchain: A Disruptive Integration. *IEEE Computer Society, 51*(9), 48-53. Retrieved from http://www.people.vcu.edu/~tndinh/papers/IEEEComp18_Blockchain+AI.pdf

Djeddi, W. E., & Khadir, M. T. (2014). A novel approach using context-based measure for matching large scale ontologies. In Lecture Notes in Computer Science (including subseries Lecture Notes in Artificial Intelligence and Lecture Notes in Bioinformatics) (Vol. 8646, pp. 320–331). Springer. doi:10.1007/978-3-319-10160-6_29

Doan, A., Halevy, A., & Ives, Z. (2012). Principles of Data Integration. *Principles of Data Integration*, 95–119. doi:10.1016/B978-0-12-416044-6.00004-1

Dob, D. (2018). *Permissioned vs. Permissionless Blockchains: Understanding the Differences*. Retrieved July 17, 2018, from https://blockonomi.com/permissioned-vs-permissionless-blockchains/

Domingos, T. (2017). *Blockchain: beyond payments*. Retrieved from https://atos.net/en/blog/blockchain-beyond-payments

Dong, X. L., & Srivastava, D. (2013). Big data integration. *2013 IEEE 29th International Conference on Data Engineering (ICDE)*, 1245–1248. 10.1109/ICDE.2013.6544914

Dong, Z., Zhang, P., Ma, J., Zhao, J., Ali, M., Meng, K., & Yin, X. (2009). *Emerging techniques in power system analysis*. Beijing: Springer, Higher Education Press.

Dornelles, E. F., Kraisig, A. R., da Silva, J. A. G., Sawicki, S., Roos-Frantz, F., & Carbonera, R. (2018). Artificial intelligence in seeding density optimization and yield simulation for oat. *Revista Brasileira de Engenharia Agrícola e Ambiental*, *22*(3), 183–188. doi:10.1590/1807-1929/agriambi.v22n3p183-188

Dos Santos Ferreira, A., Matte Freitas, D., Gonçalves da Silva, G., Pistori, H., & Folhes, M. T. (2017). Weed detection in soybean crops using ConvNets. *Computers and Electronics in Agriculture*, *143*, 314–324. doi:10.1016/j.compag.2017.10.027

Du, J., Li, W., & Huang, H. (2011). *A Study of Man-in-the-Middle Attack Based on SSL Certificate Interaction*. Retrieved from https://ieeexplore.ieee.org/abstract/document/6154142/authors#authors

Dubé, L., Du, P., McRae, C., Sharma, N., Jayaraman, S., & Nie, J. (2018). Convergent Innovation in Food through Big Data and Artificial Intelligence for Societal-Scale Inclusive Growth. The Technology Innovation Management Review.

Du, K., Sun, Z., Li, Y., Zheng, F., Chu, J., & Su, Y. (2016). Diagnostic Model For Wheat Leaf Conditions Using Image Features And A Support Vector Machine. *Transactions of the ASABE*, *59*(5), 1041–1052. doi:10.13031/trans.59.11434

Du, X., Zhai, J., & Lv, K. (2016). Algorithm Trading using Q-Learning and Recurrent Reinforcement Learning. *Positions*, 1–7.

Ebrahimi, E., & Mollazade, K. (2010). Integrating fuzzy data mining and impulse acoustic techniques for almond nuts sorting. *Australian Journal of Crop Science*, *4*(5), 353–358.

Eckartz, S., Van den Broek, T., & Ooms, M. (2016). Open data innovation capabilities: Towards a framework of how to innovate with open data. In Lecture Notes in Computer Science (including subseries Lecture Notes in Artificial Intelligence and Lecture Notes in Bioinformatics) (Vol. 9820, pp. 47–60). Springer. doi:10.1007/978-3-319-44421-5_4

Economist. (2015). *The great chain of being sure about things*. Retrieved from https://www.economist.com/briefing/2015/10/31/the-great-chain-of-being-sure-about-things

Edwards, J., & Bramante, R. (2015). *Networking self-teaching guide: OSI, TCP/IP, LANs, MANs, WANs, implementation, management, and maintenance*. John Wiley & Sons.

Eha, B. P. (2017). *Bank of England completes cross-border payment proof of concept 2017*. Retrieved from https://www.americanbanker.com/news/bank-of-england-completes-cross-border-payment-proof-of-concept

Ejupi, A., & Menon, C. (2018). Detection of Talking in Respiratory Signals: A Feasibility Study Using Machine Learning and Wearable Textile-Based Sensors. *Sensors (Basel)*, *18*(8), 2474. doi:10.3390/s18082474 PubMed

El Idrissi Esserhrouchni, O., Frikh, B., & Ouhbi, B. (2015). Learning non-taxonomic relationships of financial ontology. In *Proceedings of the 7th International Joint Conference on Knowledge Discovery, Knowledge Engineering and Knowledge Management* (pp. 479–489). Academic Press. 10.5220/0005590704790489

El Idrissi Esserhrouchni, O., Frikh, B., Ouhbi, B., & Ibrahim, I. K. (2017). Learning domain taxonomies: The TaxoLine approach. *International Journal of Web Information Systems*, *13*(3), 281–301. doi:10.1108/IJWIS-04-2017-0024

El-Bendary, N., El-Hariri, E., Hassanien, A. E., & Badr, A. (2015). *Using Machine Learning Techniques for Evaluating Tomato Ripeness*. Retrieved from https://www.academia.edu/23217224/Using_machine_learning_techniques_for_evaluating_tomato_ripeness

El-Gayar, O., Nasralah, T., & Noshokaty, A. E. (2018). IT for diabetes self-management - What are the patientsâ€™ expectations? AMCIS 2018 Proceedings. Retrieved from https://aisel.aisnet.org/amcis2018/DataScience/Presentations/18

El-Gayar, O., Nasralah, T., & Elnoshokaty, A. (2019). *Wearable devices for health and wellbeing: Design Insights from Twitter. In 52nd Hawaii International Conference on Systems Sciences (HICSS-52'19)*. Maui, HI: IEEE Computer Society; doi:10.24251/HICSS.2019.467.

Elliott, D., Keen, W., & Miao, L. (2019). *Recent advances in connected and automated vehicles. Journal of Traffic and Transportation Engineering*. doi:10.1016/j.jtte.2018.09.005

Ellis, D. A., & Piwek, L. (2017). When wearable devices fail: Towards an improved understanding of what makes a successful wearable intervention. 1st GetAMoveOn Annual Symposium.

Ellis, D. A., & Piwek, L. (2018). Failing to encourage physical activity with wearable technology: What next? Journal of the Royal Society of Medicine. doi: PubMed doi:10.1177/0141076818788856

ElSaadany, Y., Majumder, A. J. A., & Ucci, D. R. (2017). A Wireless Early Prediction System of Cardiac Arrest through IoT. 2017 IEEE 41st Annual Computer Software and Applications Conference (COMPSAC), 2, 690–695. 10.1109/COMPSAC.2017.40

Emamgholizadeh, S., Parsaeian, M., & Baradaran, M. (2015). Seed yield prediction of sesame using artificial neural network. *European Journal of Agronomy, 68*, 89–96. doi:10.1016/j.eja.2015.04.010

Ertel, W. (2017). *Introduction to artificial intelligence*. Springer International Publishing. doi:10.1007/978-3-319-58487-4

Espejo-Garcia, B., Martinez-Guanter, J., Perez-Ruiz, M., Lopez-Pellicer, F. J., & Javier Zarazaga-Soria, F. (2018). Machine learning for automatic rule classification of agricultural regulations: A case study in Spain. *Computers and Electronics in Agriculture, 150*, 343–352. doi:10.1016/j.compag.2018.05.007

Espinoza, K., Valera, D. L., Torres, J. A., Lopez, A., & Molina-Aiz, F. D. (2016). Combination of image processing and artificial neural networks as a novel approach for the identification of Bemisia tabaci and Frankliniella occidentalis on sticky traps in greenhouse agriculture. *Computers and Electronics in Agriculture, 127*, 495–505. doi:10.1016/j.compag.2016.07.008

Essayeh, A., & Abed, M. (2015). Towards ontology matching based system through terminological, structural and semantic level. Procedia Computer Science, 60, 403–412. doi:10.1016/j.procs.2015.08.154

European Commission. (2018). *Employment and social developments in Europe 2011*. Publications Office of the European Union. doi:10.2767/44905

Euzenat, J., & Shvaiko, P. (2007a). Ontology matching. Heidelberg, Germany: Springer. doi:10.1007/978-3-540-49612-0

Everingham, Y. L., Smyth, C. W., & Inman-Bamber, N. G. (2009). Ensemble data mining approaches to forecast regional sugarcane crop production. *Agricultural and Forest Meteorology, 149*(3–4), 689–696. doi:10.1016/j.agrformet.2008.10.018

Everingham, Y., Sexton, J., Skocaj, D., & Inman-Bamber, G. (2016). Accurate prediction of sugarcane yield using a random forest algorithm. *Agronomy for Sustainable Development, 36*(2), 27. doi:10.100713593-016-0364-z

Fanjul-Peyro, L., & Ruiz, R. (2010). Iterated greedy local search methods for unrelated parallel machine scheduling. *European Journal of Operational Research, 207*(1), 55–69. doi:10.1016/j.ejor.2010.03.030

Fan, S. K. S., Su, C. J., Nien, H. T., Tsai, P. F., & Cheng, C. Y. (2018). Using machine learning and big data approaches to predict travel time based on historical and real-time data from Taiwan electronic toll collection. *Soft Computing, 22*(17), 5707–5718. doi:10.100700500-017-2610-y

Fan, W., & Bifet, A. (2013). Mining big data: Current status and forecast to the future. *ACM SIGKDD Explorations Newsletter*, *14*(2), 1. doi:10.1145/2481244.2481246

FAO. (2016). *Food Agric. Organ.* Retrieved from http://faostat.fao.org

FAO. (2018). *Food and Agricultural Organisation of the United Nations, Statistics Division.* Available at http://www. fao.org/faostat/en/#data/QC/visualize

Farjam, A., Omid, M., Akram, A., & Niari, Z. F. (2014). A Neural Network Based Modeling and Sensitivity Analysis of Energy Inputs for Predicting Seed and Grain Corn Yields. *Journal of Agricultural Science and Technology*, *16*(4), 767–778.

Farley, B. G., & Clark, W. A. (1954). Simulation of self-organizing systems by digital computer. *Transaction of the IRE*, 76-84. 10.1109/TIT.1954.1057468

Feldstein, S. (2019). The Road to Digital Unfreedom: How Artificial Intelligence is Reshaping Repression. *Journal of Democracy*, *30*(1), 40–52. doi:10.1353/jod.2019.0003

Feng, Y., Peng, Y., Cui, N., Gong, D., & Zhang, K. (2017). Modeling reference evapotranspiration using extreme learning machine and generalized regression neural network only with temperature data. *Computers and Electronics in Agriculture*, *136*, 71–78. doi:10.1016/j.compag.2017.01.027

Ferentinos, K. P. (2018). Deep learning models for plant disease detection and diagnosis. *Computers and Electronics in Agriculture*, *145*, 311–318. doi:10.1016/j.compag.2018.01.009

Fernandes, J. L., Rocha, J. V., & Camargo Lamparelli, R. A. (2011). Sugarcane yield estimates using time series analysis of spot vegetation images. *Scientia Agrícola*, *68*(2), 139–146. doi:10.1590/S0103-90162011000200002

Fernandez, R., Montes, H., Surdilovic, J., Surdilovic, D., Gonzalez-De-Santos, P., & Armada, M. (2018). Automatic Detection of Field-Grown Cucumbers for Robotic Harvesting. *IEEE Access: Practical Innovations, Open Solutions*, *6*, 35512–35527. doi:10.1109/ACCESS.2018.2851376

Ferraro, D. O., Ghersa, C. M., & Rivero, D. E. (2012). Weed Vegetation of Sugarcane Cropping Systems of Northern Argentina: Data-Mining Methods for Assessing the Environmental and Management Effects on Species Composition. *Weed Science*, *60*(1), 27–33. doi:10.1614/WS-D-11-00023.1

Ferreira, G. B. S., Vargas, P. A., & Oliveira, G. M. B. (2014). An Improved Cellular Automata-Based Model for Robot Path-Planning. In *Towards Autonomous Robotic Systems* (pp. 25–36). Cham: Springer.

Fesenmaier, D. R., & Jeng, J.-M. (2000). Assessing Structure in the Pleasure Trip Planning Process. *Tourism Analysis*.

Feurer, M., Eggensperger, K., Falkner, S., Lindauer, M., & Hutter, F. (2018). Practical automated machine learning for the automl challenge 2018. *International Workshop on Automatic Machine Learning at ICML*.

Fikri, M., Cheddadi, B., Sabri, O., Haidi, T., Abdelaziz, B., & Majdoub, M. (2018). *Power Flow Analysis by Numerical Techniques and Artificial Neural Networks. IEEE, Renewable Energies, Power Systems & Green Inclusive Economy.* Casablanca, Morocco: REPS-GIE.

Finextra. (2017). *SWIFT Blockchain POC: enhanced cross-border payments.* Retrieved from https://www.finextra.com/blogposting/14321/swift-blockchain-poc-enhanced-cross-border-payments

Finucane, C., Jing, G., & Kress-Gazit, H. (2010). LTLMoP : Experimenting with Language, Temporal Logic and Robot Control. *Intelligent Robots and Systems (IROS), 2010 IEEE/RSJ International Conference on*, 1988–1993. 10.1109/IROS.2010.5650371

Firstpost. (2017). *Andhra Pradesh To Become First State To Deploy Blockchain Technology Across The Administration.* Retrieved from https://www.firstpost.com/tech/news-analysis/andhra-pradesh-to-become-first-state-to-deploy-blockchain-technology-across-the-administration-4125897.html

Food and Agriculture Organization of the United Nations. (Ed.). (2017). *The future of food and agriculture: trends and challenges.* Rome: Food and Agriculture Organization of the United Nations.

Food Security Information Network. (2018). *Global Report on Food Crises 2018.* Retrieved from http://www.fsincop.net/fileadmin/user_upload/fsin/docs/global_report/2018/GRFC_2018_Full_report_EN_Low_resolution.pdf

Forkan, A. R. M., Khalil, I., & Atiquzzaman, M. (2017). ViSiBiD: A learning model for early discovery and real-time prediction of severe clinical events using vital signs as big data. *Computer Networks, 113,* 244–257. doi:10.1016/j.comnet.2016.12.019

Fortin, J. G., Anctil, F., Parent, L.-E., & Bolinder, M. A. (2011). Site-specific early season potato yield forecast by neural network in Eastern Canada. *Precision Agriculture, 12*(6), 905–923. doi:10.100711119-011-9233-6

Frantz, R. J., & Hunt, S. R. (2003). *System and method for monitoring gas turbine plants.* General Electric Company. Retrieved from https://patents.google.com/patent/US6542856B2/en

Frost & Sullivan. (2018). *Frost & Sullivan's 10 Healthcare Predictions for 2018.* Retrieved from https://ww2.frost.com/frost-perspectives/frost-sullivans-10-healthcare-predictions-2018/

Fukuda, S., Spreer, W., Yasunaga, E., Yuge, K., Sardsud, V., & Mueller, J. (2013). Random Forests modelling for the estimation of mango (Mangifera indica L. cv. Chok Anan) fruit yields under different irrigation regimes. *Agricultural Water Management, 116,* 142–150. doi:10.1016/j.agwat.2012.07.003

Galceran, E., & Carreras, M. (2013). A Survey on Coverage Path Planning for Robotics. *Robotics and Autonomous Systems, 61*(12), 1258–1276. doi:10.1016/j.robot.2013.09.004

Galston, W. A., & Hendrickson, C. (2018). *A policy at peace with itself: Antitrust remedies for our concentrated, uncompetitive economy.* Washington, DC: Brookings Institution.

Gampfer, F. (2018). Managing Complexity of Digital Transformation with Enterprise Architecture. 31st Bled e-Conference: Digital Transformation: Meeting the Challenges.

Gandhi, R., & Ramasastri, A. S. (2017). *Applications of Blockchain Technology to Banking and Financial Sector in India.* IDRBT Publication.

Ganne, E. (2018). *Can Blockchain Revolutionize International Trade?* World Trade Organization. Retrieved from https://www.wto.org/english/res_e/booksp_e/blockchainrev18_e.pdf

Gans, D., Kralewski, J., Hammons, T., & Dowd, B. (2005). *Medical Groups' Adoption Of Electronic Health Records And Information Systems.* Retrieved from https://www.healthaffairs.org/doi/full/10.1377/hlthaff.24.5.1323

García, A., Linaza, M. T., Franco, J., & Juaristi, M. (2015). Methodology for the Publication of Linked Open Data from Small and Medium Size DMOs. Information and Communication Technologies in Tourism 2015. doi:10.1007/978-3-319-14343-9_14

Garcia-Santillan, I. D., & Pajares, G. (2018). On-line crop/weed discrimination through the Mahalanobis distance from images in maize fields. *Biosystems Engineering, 166,* 28–43. doi:10.1016/j.biosystemseng.2017.11.003

Garneto, B., Kandaswamy R., Lovelock, J.D., & Reynolds, M. (2017). *Forecast: Blockchain.* Academic Press.

Gartner. (2015). *Gartner Says Business Intelligence and Analytics Leaders Must Focus on Mindsets and Culture to Kick Start Advanced Analytics*. Retrieved from https://www.gartner.com/newsroom/id/3130017

Gartner. (2016). *Gartner Survey Reveals Investment in Big Data Is Up but Fewer Organizations Plan to Invest*. Retrieved from https://www.gartner.com/newsroom/id/3466117

Gartner. (2018). *Gartner IT Glossary*. Retrieved November 14, 2018, from https://www.gartner.com/it-glossary/

Gayathri, R., & Uma, V. (2018). Ontology based knowledge representation technique, domain modeling languages and planners for robotic path planning : A survey. *ICT Express*, *4*(2), 69–74. doi:10.1016/j.icte.2018.04.008

Gayathri, R., & Uma, V. (2019). A Review of Description Logic-Based Techniques for Robot Task Planning. *Integrated Intelligent Computing. Communication and Security*, *771*, 101–107. doi:10.1007/978-981-10-8797-4

Gercke, M. (2012). *Understanding cybercrime: phenomena, challenges and legal response*. Academic Press.

Ghahramani, Z. (2015). Probabilistic machine learning and artificial intelligence. *Nature*, *521*(7553), 452–459. doi:10.1038/nature14541 PMID:26017444

Gholami, M., Ejupi, A., Rezaei, A., Ferrone, A., & Menon, C. (2018). Estimation of Knee Joint Angle Using a Fabric-Based Strain Sensor and Machine Learning: A Preliminary Investigation. 2018 7th IEEE International Conference on Biomedical Robotics and Biomechatronics (Biorob), 589–594. 10.1109/BIOROB.2018.8487199

Ghosh, K., & Nath, A. (2016). Big Data: Security Issues, Challenges and Future Scope. *International Journal of Research Studies in Computer Science and Engineering*, *3*(3). doi:10.20431/2349-4859.0303001

Giloi, W. K. (1997). Konrad Zuse's Plankalkül: The first high-level, "non von Neumann" programming language. *IEEE Annals of the History of Computing*, *19*(2), 17–24. doi:10.1109/85.586068

Gisler, C., Ridi, A., Fauquex, M., Genoud, D., & Hennebert, J. (2014). Towards glaucoma detection using intraocular pressure monitoring. 2014 6th International Conference of Soft Computing and Pattern Recognition (SoCPaR), 255–260. 10.1109/SOCPAR.2014.7008015

Goertzel, B. (2017). *SingularityNET: A decentralized, open market and inter-network for AIs*. Thoughts, Theories & Studies on Artificial Intelligence (AI) Research.

Goldstein, A., Fink, L., Meitin, A., Bohadana, S., Lutenberg, O., & Ravid, G. (2018). Applying machine learning on sensor data for irrigation recommendations: Revealing the agronomist's tacit knowledge. *Precision Agriculture*, *19*(3), 421–444. doi:10.100711119-017-9527-4

Gómez-Casero, M. T., Castillejo-Gonzalez, I. L., Garcia-Ferrer, A., Peña-Barragán, J. M., Jurado-Expósito, M., Garcia-Torres, L., & López-Granados, F. (2010). Spectral discrimination of wild oat and canary grass in wheat fields for less herbicide application. *Agronomy for Sustainable Development*, *30*(3), 689–699. doi:10.1051/agro/2009052

Gonzalez-Sanchez, A., Frausto-Solis, J., & Ojeda-Bustamante, W. (2014). Predictive ability of machine learning methods for massive crop yield prediction. *Spanish Journal of Agricultural Research*, *12*(2), 313–328. doi:10.5424jar/2014122-4439

Goodfellow, I., Pouget-Abadie, J., Mirza, M., Xu, B., Warde-Farley, D., Ozair, S., . . . Bengio, Y. (2014). Generative Adversarial Nets. *Advances in Neural Information Processing Systems 27 (NIPS 2014)*. Retrieved from http://papers.nips.cc/paper/5423-generative-adversarial-nets

Goodwin, T. (2018). *Digital Darwinism: Survival of the Fittest in the Age of Business Disruption*. Kogan Page.

Google. (2019). Retrieved from https://ai.google/research

Gopalapillai, R., Gupta, D., & Tsb, S. (2017). Pattern Identification of Robotic Environments using Machine Learning Techniques. *Procedia Computer Science*, *115*, 63–71. doi:10.1016/j.procs.2017.09.077

Goswami, A. K., Gakkar, S., & Kaur, H. (2014). Automatic Object Recognition from Satellite Images using Artificial Neural Network. *International Journal of Computer Applications*, *95*(10). Retrieved from https://www.ijcaonline.org/archives/volume95/number10/16633-6502

Goumopoulos, C., O'Flynn, B., & Kameas, A. (2014). Automated zone-specific irrigation with wireless sensor/actuator network and adaptable decision support. *Computers and Electronics in Agriculture*, *105*, 20–33. doi:10.1016/j.compag.2014.03.012

Granville, K. (2018). Facebook and Cambridge Analytica: What You Need to Know as Fallout Widens. *New York Times*. Retrieved from https://www.nytimes.com/2018/03/19/technology/facebook-cambridge-analytica-explained.html

Gratton, L., & Scott, A. (2016). *The 100-Year Life: Living and Working in an Age of Longevity*. Bloomsbury Institute.

Graveto, V., Rosa, L., Cruz, T., & Simões, P. (2019). A stealth monitoring mechanism for cyber- physical systems. *International Journal of Critical Infrastructure Protection*, *24*, 126–143. doi:10.1016/j.ijcip.2018.10.006

Gray, M. (2017). *Introducing Project "Bletchley"*. Retrieved from https://github.com/Azure/azure-blockchain-projects/blob/master/bletchley/bletchley-whitepaper.md

Gretzel, U., Fesenmaier, D. R., & O'Leary, J. T. (2005). The transformation of consumer behaviour. In *Tourism Business Frontiers*. Consumers, Products and Industry. doi:10.4324/9780080455914

Gretzel, U., Sigala, M., Xiang, Z., & Koo, C. (2015). Smart tourism: Foundations and developments. *Electronic Markets*, *25*(3), 179–188. doi:10.100712525-015-0196-8

Gretzel, U., Werthner, H., Koo, C., & Lamsfus, C. (2015). Conceptual foundations for understanding smart tourism ecosystems. *Computers in Human Behavior*, *50*, 558–563. doi:10.1016/j.chb.2015.03.043

Gribel, L. (2018). Drivers of Wearable Computing Adoption: An Empirical Study of Success Factors Including IT Security and Consumer Behaviour-Related Aspects. University of Plymouth. Retrieved from https://pearl.plymouth.ac.uk/bitstream/handle/10026.1/11662/2018gribel10508825.pdf?sequence=1&isAllowed=y

Griffel, L. M., Delparte, D., & Edwards, J. (2018). Using Support Vector Machines classification to differentiate spectral signatures of potato plants infected with Potato Virus Y. *Computers and Electronics in Agriculture*, *153*, 318–324. doi:10.1016/j.compag.2018.08.027

Grocery Manufacturers Association. (2017). *A formula for Growth: Innovation, Big Data & Analytics*. Retrieved from http://www.gmaonline.org/issues-policy/collaborating-with-retailers/big-data-analytics/

Gross, A., Hartung, M., Kirsten, T., & Rahm, E. (2010). On matching large life science ontologies in parallel. In *International Conference on Data Integration in the Life Sciences* (pp. 35–49). Springer Berlin Heidelberg. 10.1007/978-3-642-15120-0_4

Grover, V., Chiang, R. H. L., Liang, T. P., & Zhang, D. (2018). Creating strategic business value from big data analytics: A research framework. *Journal of Management Information Systems*, *35*(2), 388–423. doi:10.1080/07421222.2018.1451951

Guan, Y., Wei, Q., & Chen, G. (2019). Deep learning based personalized recommendation with multi-view information integration. *Decision Support Systems*, *118*, 58–69. doi:10.1016/j.dss.2019.01.003

Guess, A. R. (2012). *Case Study: A Failure of Data Management at Sears*. Retrieved from http://www.dataversity.net/case-study-a-failure-of-data-management-at-sears/

Guha, R. V., Brickley, D., & MacBeth, S. (2015). Schema.org: Evolution of Structured Data on the Web. *Queue, 13*(9), 10–37. doi:10.1145/2857274.2857276

Guine, R. P. F., Matos, S., Goncalves, F. J., Costa, D., & Mendes, M. (2018). Evaluation of phenolic compounds and antioxidant activity of blueberries and modelization by artificial neural networks. *International Journal of Fruit Science, 18*(2), 199–214. doi:10.1080/15538362.2018.1425653

Guo, Y., & Liang, C. (2016). *Blockchain application and outlook in the banking industry.* Retrieved from https://jfin-swufe.springeropen.com/articles/10.1186/s40854-016-0034-9

Gutierrez, M., Alegret, S., Caceres, R., Casadesus, J., Marfa, O., & del Valle, M. (2008). Nutrient solution monitoring in greenhouse cultivation employing a potentiometric electronic tongue. *Journal of Agricultural and Food Chemistry, 56*(6), 1810–1817. doi:10.1021/jf073438s PMID:18303814

Gutierrez, P. A., López-Granados, F., Peña-Barragán, J. M., Jurado-Expósito, M., Gómez-Casero, M. T., & Hervas-Martinez, C. (2008). Mapping sunflower yield as affected by Ridolfia segetum patches and elevation by applying evolutionary product unit neural networks to remote sensed data. *Computers and Electronics in Agriculture, 60*(2), 122–132. doi:10.1016/j.compag.2007.07.011

Gutierrez, P. A., López-Granados, F., Peña-Barragán, J. M., Jurado-Expósito, M., & Hervas-Martinez, C. (2008). Logistic regression product-unit neural networks for mapping Ridolfia segetum infestations in sunflower crop using multitemporal remote sensed data. *Computers and Electronics in Agriculture, 64*(2), 293–306. doi:10.1016/j.compag.2008.06.001

Haber, S., & Stornetta, W. S. (1991). How to Time-Stamp a Digital Document. *Journal of Cryptology, 3*(2), 99–111. doi:10.1007/BF00196791

Haenssle, H. A., Fink, C., Schneiderbauer, R., Toberer, F., Buhl, T., Blum, A., ... Uhlmann, L. (2018). Man against machine: Diagnostic performance of a deep learning convolutional neural network for dermoscopic melanoma recognition in comparison to 58 dermatologists. *Annals of Oncology : Official Journal of the European Society for Medical Oncology, 29*(8), 1836–1842. doi:10.1093/annonc/mdy166

Haifa. (2018). *Crop Guide: Tomato.* Retrieved from https://www.haifa-group.com/tomato-fertilizer/crop-guide-tomato

Hale, S. (2018). *Death by Algorithm; Public opinion and the lethal autonomous weapons debate. European Master's Degree in Human Rights and Democratisation. A.Y. 2017/2018.* Aristotle University of Thessaloniki.

Hall, M., Frank, E., Holmes, G., Pfahringer, B., Reutemann, P., & Witten, I. (2009). *The Weka Data Mining Software.* Retrieved from https://dl.acm.org/citation.cfm?Id=1656278

Halper, F. (2015). Operationalizing and Embedding Analytics for Action. *TDWI.* Retrieved from https://www.sas.com/content/dam/SAS/en us/doc/whitepaper2/tdwi-operationalizing-embedding-analytics-for-action-108112.pdf

Hamedani, S. R., Liaqat, M., Shamshirband, S., Al-Razgan, O. S., Al-Shammari, E. T., & Petkovic, D. (2015). Comparative Study of Soft Computing Methodologies for Energy Input-Output Analysis to Predict Potato Production. *American Journal of Potato Research, 92*(3), 426–434. doi:10.100712230-015-9453-9

Hans, R., Zuber, H., Rizk, A., & Steinmetz, R. (2017). *Blockchain and Smart Contracts: Disruptive Technologies for the Insurance Market.* Twenty-third Americas Conference on Information Systems (AMCIS), Boston, MA. Retrieved from ftp://ftp.kom.tu-darmstadt.de/papers/HZR+17-1.pdf

Harada, M., Tominaga, T., Hiramatsu, K., & Marui, A. (2013). Real-Time Prediction Of Chlorophyll-A Time Series In A Eutrophic Agricultural Reservoir In A Coastal Zone Using Recurrent Neural Networks With Periodic Chaos Neurons. *Irrigation and Drainage, 62*(1), 36–43. doi:10.1002/ird.1757

Haralick, R. M., Shanmugam, K., & Dinstein, I. (1973). Textural Features for Image Classification. *IEEE Transactions on Systems, Man, and Cybernetics*, *SMC-3*(6), 610–621. doi:10.1109/TSMC.1973.4309314

Harris, A., True, H., Hu, Z., Cho, J., Fell, N., & Sartipi, M. (2016). Fall recognition using wearable technologies and machine learning algorithms. 2016 IEEE International Conference on Big Data (Big Data), 3974–3976. doi:10.1109/BigData.2016.7841080

Harshavardhan, N. G., Sanjeev, S. S., & Vijay, R. S., & R, A. (2012). Fruits Sorting and Grading using. *International Journal of Advanced Research in Computer Engineering & Technology*, *I*(6), 117–122.

Harvey, C. (2017). *Big Data Challenges*. Retrieved from https://www.datamation.com/big-data/big-data-challenges.html

Hassanien, A. E., Gaber, T., Mokhtar, U., & Hefny, H. (2017). An improved moth flame optimization algorithm based on rough sets for tomato diseases detection. *Computers and Electronics in Agriculture*, *136*, 86–96. doi:10.1016/j.compag.2017.02.026

Hausenblas, M. (2009). Exploiting linked data to build web applications. *IEEE Internet Computing*, *13*(4), 68–73. doi:10.1109/MIC.2009.79

Hawas, M. A. (2017). Are We Intentionally Limiting Urban Planning and Intelligence? A Causal Evaluative Review and Methodical Redirection for Intelligence Systems. *IEEE Access : Practical Innovations, Open Solutions*, *5*, 13253–13259. doi:10.1109/ACCESS.2017.2725138

Hayashi, S., Ganno, K., Ishii, Y., & Tanaka, I. (2002). Robotic Harvesting System for Eggplants. *Japan Agricultural Research Quarterly: JARQ*, *36*(3), 163–168. doi:10.6090/jarq.36.163

Haykin, S. (2009). *Neural Networks and Learning Machines*. Pearson-Prentice Hall Publish.

Heim, R. H. J., Wright, I. J., Chang, H.-C., Carnegie, A. J., Pegg, G. S., Lancaster, E. K., ... Oldeland, J. (2018). Detecting myrtle rust (Austropuccinia psidii) on lemon myrtle trees using spectral signatures and machine learning. *Plant Pathology*, *67*(5), 1114–1121. doi:10.1111/ppa.12830

Heirloom Organics Non HighBrid Seeds. (2017). Retrieved from https://www.non-hybrid-seeds.com/sp/tomato-lovers-s-pack-59.html

Helbing, D. (2018). Societal, Economic, Ethical and Legal Challenges of the Digital Revolution: From Big Data to Deep Learning, Artificial Intelligence, and Manipulative Technologies. In D. Helbing (Ed.), *Towards Digital Enlightenment*. Cham: Springer.

Henry, E., Adamchuk, V., Stanhope, T., Buddle, C., & Rindlaub, N. (2019). Precision apiculture: Development of a wireless sensor network for honeybee hives. *Computers and Electronics in Agriculture*, *156*, 138–144. doi:10.1016/j.compag.2018.11.001

Hernandez-Hernandez, J. L., García-Mateos, G., González-Esquiva, J. M., Escarabajal-Henarejos, D., Ruiz-Canales, A., & Molina-Martínez, J. M. (2016). Optimal color space selection method for plant/soil segmentation in agriculture. *Computers and Electronics in Agriculture*, *122*, 124–132. doi:10.1016/j.compag.2016.01.020

Heron, J. R., & Zachariah, G. L. (1974). *Automatic Sorting of Processing Tomatoes*. The American Society of Agricultural and Biological Engineers. Retrieved from doi: . doi:10.13031/2013.37013

Hertz, J., Krogh, A., & Palmer, R. G. (2018). *Introduction to the theory of neural computation*. CRC Press. doi:10.1201/9780429499661

Hessine, M. B., & Saber, S. B. (2014). Accurate fault classifier and locator for EHV transmission lines based on artificial neural networks. *Mathematical Problems in Engineering*, *2*, 1–19. doi:10.1155/2014/240565

Hidayat, W., Tambunan, T. D., & Budiawan, R. (2018). Empowering Wearable Sensor Generated Data to Predict Changes in Individual's Sleep Quality. 2018 6th International Conference on Information and Communication Technology (ICoICT), 447–452. 10.1109/ICoICT.2018.8528750

Hidayatullah, N. A., Stojcevski, B., & Kalam, A. (2011). Analysis of distributed generation systems, smart grid technologies and future motivators influencing change in the electricity sector. *Smart Grid and Renewable Energy*, *2*(03), 216–229. doi:10.4236gre.2011.23025

Hilbert, M. (2015). *Big data for development. A review of promises and challenges*. Development Review. Retrieved from http://www.martinhilbert.net/big-data-for-development/

Hill, T. (2016). *Visa Introduces Blockchain-based solution for Payment Services*. Retrieved from https://bitcoinist.com/visa-blockchain-solution-payments/

Hill, B. D., Kalischuk, M., Waterer, D. R., Bizimungu, B., Howard, R., & Kawchuk, L. M. (2011). An Environmental Model Predicting Bacterial Ring Rot Symptom Expression. *American Journal of Potato Research*, *88*(3), 294–301. doi:10.100712230-011-9193-4

Hjalager, A. M., & Nordin, S. (2011). User-driven Innovation in Tourism-A Review of Methodologies. *Journal of Quality Assurance in Hospitality & Tourism*, *12*(4), 289–315. doi:10.1080/1528008X.2011.541837

Holcomb, S. D., Porter, W. K., Ault, S. V., Mao, G., & Wang, J. (2018). Overview on DeepMind and Its AlphaGo Zero AI. *ICBDE '18 Proceedings of the 2018 International Conference on Big Data and Education*, 67-71. 10.1145/3206157.3206174

Horton, M., Jean, N., & Burke, M. (2016). *Stanford scientists combine satellite data, machine learning to map poverty*. Retrieved from https://news.stanford.edu/2016/08/18/combining-satellite-data-machine-learning-to-map-poverty/

Hua, A. V., & Notland, J. S. (2016). *Blockchain enabled Trust & Transparency in supply chains*. Retrieved from https://www.academia.edu/38200535/Blockchain_enabled_Trust_and_Transparency_in_supply_chains

Huang, C. D., Goo, J., Nam, K., & Yoo, C. W. (2017). Smart tourism technologies in travel planning: The role of exploration and exploitation. *Information & Management*, *54*(6), 757–770. doi:10.1016/j.im.2016.11.010

Huang, J., Zhao, M., Zhou, Y., & Xing, C. C. (2018). In-vehicle networking: Protocols, challenges, and solutions. *IEEE Network*, (99): 1–7.

Husin, Z., Shakaff, A. Y. M., Aziz, A. H. A., Farook, R. S. M., Jaafar, M. N., Hashim, U., & Harun, A. (2012). Embedded portable device for herb leaves recognition using image processing techniques and neural network algorithm. *Computers and Electronics in Agriculture*, *89*, 18–29. doi:10.1016/j.compag.2012.07.009

Hu, W., Qu, Y., & Cheng, G. (2008). Matching large ontologies: A divide-and-conquer approach. *Data & Knowledge Engineering*, *67*(1), 140–160. doi:10.1016/j.datak.2008.06.003

Hu, Y., Liyanage, M., Manzoor, A., Thilakarathna, K., Jourjoin, G., Seneviratne, A., & Yilanttila, M. (2018). The use of smart contracts and challenges. Computers and Society. *Cornell University*, *1*(1), 1–13.

Iansiti, M., & Lakhani, K. (2017). The Truth about Blockchain. *Harvard Business Review*. Retrieved from https://hbr.org/2017/01/the-truth-about-blockchain

Iapichino, A., De Rosa, A., & Liberace, P. (2018). Smart Organizations, New Skills, and Smart Working to Manage Companies' Digital Transformation. In L. Pupillo, E. Noam, & L. Waverman (Eds.), *Digitized Labor. Palgrave Macmillan.* Cham: Springer Link. doi:10.1007/978-3-319-78420-5_13

IBM. (2013). *Descriptive, predictive, prescriptive: Transforming asset and facilities management with analytics.* IBM.

IBM. (2015, September 11). *The new AI innovation equation.* Retrieved April 16, 2019, from IBM Cognitive - What's next for AI website: http://www.ibm.com/watson/advantage-reports/future-of-artificial-intelligence/ai-innovation-equation.html

IBM. (2018, October 30). *Big Data Analytics.* Retrieved January 9, 2019, from https://www.ibm.com/analytics/hadoop/big-data-analytics

ICAO. (2011). *International Civil Aviation Organization (ICAO, 2011). Unmanned Aircraft Systems (UAS).* Retrieved from https://www.icao.int/Meetings/UAS/Documents/Circular%20328_en.pdf

Ilic, M., Ilic, S., Jovic, S., & Panic, S. (2018). Early cherry fruit pathogen disease detection based on data mining prediction. *Computers and Electronics in Agriculture, 150,* 418–425. doi:10.1016/j.compag.2018.05.008

Iqbal, Z., Ilyas, R., Shahzad, W., & Inayat, I. (2018). A comparative study of machine learning techniques used in non-clinical systems for continuous healthcare of independent livings. 2018 IEEE Symposium on Computer Applications Industrial Electronics (ISCAIE), 406–411. doi:10.1109/ISCAIE.2018.8405507

Irudeen, R., & Samaraweera, S. (2013). Big data solution for Sri Lankan development: A case study from travel and tourism. *2013 International Conference on Advances in ICT for Emerging Regions (ICTer).* 10.1109/ICTer.2013.6761180

Isaac, K. A. (2015). An overview of post-harvest challenges facing tomato production in Africa. Dunedin: African Studies Association of Australasia and the Pacific (AFSAAP).

Ishiguro, H., & Nishio, S. (2018). Building Artificial Humans to Understand Humans. In H. Ishiguro & F. Dalla Libera (Eds.), *Geminoid Studies.* Singapore: Springer. doi:10.1007/978-981-10-8702-8_2

Ito, J., Narula, N., & Ali, R. (2017). *The Blockchain Will Do to the Financial System What the Internet Did to Media.* Retrieved from https://hbr.org/2017/03/the-blockchain-will-do-to-banks-and-law-firms-what-the-internet-did-to-media

Ivars-Baidal, J. A., Celdrán-Bernabeu, M. A., Mazón, J. N., & Perles-Ivars, Á. F. (2017). Smart destinations and the evolution of ICTs: A new scenario for destination management? *Current Issues in Tourism.* doi:10.1080/13683500.2017.1388771

Jackman, P., & Sun, D.-W. (2013). Recent advances in image processing using image texture features for food quality assessment. *Trends in Food Science & Technology, 29*(1), 35–43. doi:10.1016/j.tifs.2012.08.008

Jadhav, J. J. (2014). Moving Object Detection and Tracking for Video Surveillance. *IJERGS, 2*(4), 372–378. Retrieved from https://pdfs.semanticscholar.org/a58e/281e2d25fd8ba4e54860bf6c9b8781dd8e4a.pdf

Jafari, J. (2001). The scientification of tourism. Hosts AND guests revisited: tourism issues of the 21st century.

Janssen, M., Charalabidis, Y., & Zuiderwijk, A. (2012). Benefits, Adoption Barriers and Myths of Open Data and Open Government. *Information Systems Management, 29*(4), 258–268. doi:10.1080/10580530.2012.716740

Jayapandian, N., Pavithra, S., & Revathi, B. (2017). Effective usage of online cloud computing in different scenario of education sector. In *Innovations in Information, Embedded and Communication Systems (ICIIECS), 2017 International Conference on* (pp. 1-4). IEEE. 10.1109/ICIIECS.2017.8275970

Jayapandian, N., Rahman, A. M. Z., Radhikadevi, S., & Koushikaa, M. (2016). Enhanced cloud security framework to confirm data security on asymmetric and symmetric key encryption. In *Futuristic Trends in Research and Innovation for Social Welfare (Startup Conclave), World Conference on* (pp. 1-4). IEEE. 10.1109/STARTUP.2016.7583904

Jayapandian, N., & Md Zubair Rahman, A. M. J. (2018). Secure Deduplication for Cloud Storage Using Interactive Message-Locked Encryption with Convergent Encryption, To Reduce Storage Space. *Brazilian Archives of Biology and Technology*, *61*(0), 1–13. doi:10.1590/1678-4324-2017160609

Jayapandian, N., & Rahman, A. M. Z. (2017). Secure and efficient online data storage and sharing over cloud environment using probabilistic with homomorphic encryption. *Cluster Computing*, *20*(2), 1561–1573. doi:10.100710586-017-0809-4

Jayapandian, N., Rahman, A. M. Z., & Gayathri, J. (2015). The online control framework on computational optimization of resource provisioning in cloud environment. *Indian Journal of Science and Technology*, *8*(23), 1–13. doi:10.17485/ijst/2015/v8i23/79313

Jelen, L., & Milana, V. (2018). Building and nurturing trust among members in virtual project teams. Strategic Management, 23(3), 10-16.

Jiang, P., Winkley, J., Zhao, C., Munnoch, R., Min, G., & Yang, L. T. (2016). An Intelligent Information Forwarder for Healthcare Big Data Systems with Distributed Wearable Sensors. *IEEE Systems Journal*, *10*(3), 1147–1159. doi:10.1109/JSYST.2014.2308324

Jiang, Y., Liu, C. C., & Xu, Y. (2016). Smart distribution systems. *Energies*, *9*(4), 297–317. doi:10.3390/en9040297

Jimenez, D., Cock, J., Jarvis, A., Garcia, J., Satizabal, H. F., Van Damme, P., ... Barreto-Sanz, M. A. (2011). Interpretation of commercial production information: A case study of lulo (Solanum quitoense), an under-researched Andean fruit. *Agricultural Systems*, *104*(3), 258–270. doi:10.1016/j.agsy.2010.10.004

Jimenez, D., Cock, J., Satizabal, H. F., Barreto, M. A., Perez-Uribe, A., Jarvis, A., & Van Damme, P. (2009). Analysis of Andean blackberry (Rubus glaucus) production models obtained by means of artificial neural networks exploiting information collected by small-scale growers in Colombia and publicly available meteorological data. *Computers and Electronics in Agriculture*, *69*(2), 198–208. doi:10.1016/j.compag.2009.08.008

Johnson, D., Menezes, A., & Vanstone, S. (2001). The elliptic curve digital signature algorithm (ecdsa). *International Journal of Information Security*, *1*(1), 36–63. doi:10.1007102070100002

Jokic, A., Zavargo, Z., Gyura, J., Radivojevic, S., & Seres, Z. (2010). An Artificial Neural Network Approach to Prediction of Sugar Beet Yield and Quality in Serbia. In *Sugar Beet Crops: Growth, Fertilization & Yield* (pp. 153–166). Academic Press. Retrieved from https://www.researchgate.net/publication/281874792_An_artificial_neural_network_approach_to_prediction_of_sugar_beet_yield_and_quality_in_Serbia

Jones, J. B. (2017). *Growing Tomatoes*. Retrieved from http://www.growtomatoes.com/tomato-world-production-statistics/

Jouini, M., & Rabai, L. B. A. (2019). A security framework for secure cloud computing environments. In Cloud Security: Concepts, Methodologies, Tools, and Applications (pp. 249-263). IGI Global. doi:10.4018/978-1-5225-8176-5.ch011

Kadar, P. (2013). Application of optimization techniques in the power system control. *Acta Polytechnica Hungarica*, *10*(5), 221–236.

Kakavand, H., Kost De Sevres, N., & Chilton, B. (2017). *The blockchain revolution: An analysis of regulation and technology related to distributed ledger technologies*. Academic Press.

Kalampokis, E., Tambouris, E., & Tarabanis, K. (2011). A classification scheme for open government data: Towards linking decentralised data. *International Journal of Web Engineering and Technology*, *6*(3), 266. doi:10.1504/IJWET.2011.040725

Kalantari, M. (2017). Consumers adoption of wearable technologies: Literature review, synthesis, and future research agenda. *International Journal of Technology Marketing*, *12*(1), 1. doi:10.1504/IJTMKT.2017.10008634

Kambatla, K., Kollias, G., Kumar, V., & Grama, A. (2014). Trends in big data analytics. *Journal of Parallel and Distributed Computing*, *74*(7), 2561–2573. doi:10.1016/j.jpdc.2014.01.003

Kamilaris, A., Kartakoullis, A., & Prenafeta-Boldú, F. X. (2017). A review on the practice of big data analysis in agriculture. *Computers and Electronics in Agriculture*, *143*, 23–37. doi:10.1016/j.compag.2017.09.037

Kamilaris, A., & Prenafeta-Boldú, F. X. (2018). Deep learning in agriculture: A survey. *Computers and Electronics in Agriculture*, *147*, 70–90. doi:10.1016/j.compag.2018.02.016

Kaplinsky, R., & Morris, M. (2001). *A Handbook for Value Chain Research*. 113.

Kariappa, B. (2015). *Block Chain 2.0: The Renaissance of Money*. Retrieved from https://www.wired.com/insights/2015/01/block-chain-2-0/

Karim, S., Umar, R., & Rubina, L. (2018). *Conceptualizing Blockchains Characteristics & Applications*. 11th IADIS International Conference Information Systems 2018, Lisbon, Portugal.

Karov, Y., Breakstone, M., Shilon, R., Keller, O., & Shellef, E. (2017). *U.S. Patent No. 9,772,994*. Washington, DC: U.S. Patent and Trademark Office.

Karuppanan, K., Vairasundaram, A. S., & Sigamani, M. (2012, August 3). A comprehensive machine learning approach to prognose pulmonary disease from home. Academic Press. doi:10.1145/2345396.2345482

Kasturirangan, K., Aravamudan, R., Deekshatulu, B. L., Joseph, G., & Chandrasekhar, M. H. (1996). *Indian Remote Sensing satellite IRS-1C—the beginning of a new era*. Retrieved from http://repository.ias.ac.in/88316/1/88316.pdf

Katiyar, S. K., & Arun, P. V. (2014). *An enhanced neural network based approach towards object extraction*. Retrieved from https://arxiv.org/abs/1405.6137

Kaufman, L. M. (2009). Data security in the world of cloud computing. *IEEE Security and Privacy*, *7*(4), 61–64. doi:10.1109/MSP.2009.87

Kawamura, N., Namikawa, K., Fujiura, T., & Ura, M. (1984). Study on Agricultural Robot (Part 1). *Journal of the Japanese Society of Agricultural Machinery*, *46*(3), 353–358. doi:10.11357/jsam1937.46.3_353

Kelly, K. (2016). *The Inevitable: Understanding the 12 Technological Forces That Will Shape Our Future*. Viking Press.

Keyword Tool. (n.d.). Retrieved from https://keywordtool.io/

Khan, W. A., Idris, M., Ali, T., Ali, R., Hussain, S., & Hussain, M. ... Kang, B. H. (2015). Correlating health and wellness analytics for personalized decision making. 2015 17th International Conference on E-Health Networking, Application Services (HealthCom), 256–261. 10.1109/HealthCom.2015.7454508

Khanal, S., Fulton, J., Klopfenstein, A., Douridas, N., & Shearer, S. (2018). Integration of high resolution remotely sensed data and machine learning techniques for spatial prediction of soil properties and corn yield. *Computers and Electronics in Agriculture*, *153*, 213–225. doi:10.1016/j.compag.2018.07.016

Khan, M. J., Khan, H. S., Yousaf, A., Khurshid, K., & Abbas, A. (2018). Modern Trends in Hyperspectral Image Analysis: A Review. *IEEE Access: Practical Innovations, Open Solutions*, *6*, 14118–14129. doi:10.1109/ACCESS.2018.2812999

Khan, M. S. S., & Khan, A. S. (2017). A Brief Survey on Robotics. *International Journal of Computer Science and Mobile Computing*, *6*(9), 38–45.

Khan, N., Yaqoob, I., Haskem, I. A. T., Inayat, Z., Ali, W. K. M., Alam, M., ... Gani, A. (2014). Big data: Survey, technologies, opportunities, and challenges. *The Scientific World Journal*, 1–18. PMID:25136682

Khashei-Siuki, A., Kouchakzadeh, M., & Ghahraman, B. (2011). Predicting Dryland Wheat Yield from Meteorological Data Using Expert System, Khorasan Province, Iran. *Journal of Agricultural Science and Technology*, *13*(4), 627–640.

Khazaei, J., Naghavi, M. R., Jahansouz, M. R., & Salimi-Khorshidi, G. (2008). Yield estimation and clustering of chickpea genotypes using soft computing techniques. *Agronomy Journal*, *100*(4), 1077–1087. doi:10.2134/agronj2006.0244

Khoshnevisan, B., Rafiee, S., & Mousazadeh, H. (2013). Environmental impact assessment of open field and greenhouse strawberry production. *European Journal of Agronomy*, *50*, 29–37. doi:10.1016/j.eja.2013.05.003

Khoshnevisan, B., Rafiee, S., Omid, M., Mousazadeh, H., & Rajaeifar, M. A. (2014). Application of artificial neural networks for prediction of output energy and GHG emissions in potato production in Iran. *Agricultural Systems*, *123*, 120–127. doi:10.1016/j.agsy.2013.10.003

Kim, S., Lee, K., Lee, J., & Jeon, J. Y. (2016). EPOC aware energy expenditure estimation with machine learning. *2016 IEEE International Conference on Systems, Man, and Cybernetics (SMC)*, 001585–001590. 10.1109/SMC.2016.7844465

Kirsten, T., Kolb, L., Hartung, M., Groß, A., Köpcke, H., & Rahm, E. (2010). Data Partitioning for Parallel Entity Matching. *Strategies (La Jolla, Calif.)*, *3*(2), 11. Retrieved from http://arxiv.org/abs/1006.5309

Klaimi, J., Rahim-Amoud, R., Merghem-Boulahia, L., & Jrad, A. (2016). Energy Management Algorithms in Smart Grids: State of the Art and Emerging Trends. *International Journal of Artificial Intelligence and Applications*, *7*(4), 25–45. doi:10.5121/ijaia.2016.7403

Klein, D. J. (2010). Centrality measure in graphs. *Journal of Mathematical Chemistry*, *4*(47), 1209–1223. doi:10.100710910-009-9635-0

K-Meta. (n.d.). *Keyword Research Tool*. Retrieved from https://k-meta.com/suggestions

Knapp, S. (2006). *Artificial Intelligence: Past, Present, and Future. Vox of Dartmouth*. The Newspaper for the Dartmouth Faculty and Staff.

Kondo, N., Nishitsuji, Y., Ling, P. P., & Ting, K. C. (1996). Visual Feedback Guided Robotic Cherry Tomato Harvesting. *Transactions of the ASAE. American Society of Agricultural Engineers*, *39*(6), 2331–2338. doi:10.13031/2013.27744

Kotsiantis, S. B., Zaharakis, I., & Pintelas, P. (2007). Supervised machine learning: A review of classification techniques. Emerging artificial intelligence applications in computer engineering, 160, 3-24.

Kouadio, L., Deo, R. C., Byrareddy, V., Adamowski, J. F., Mushtaq, S., & Nguyen, V. P. (2018). Artificial intelligence approach for the prediction of Robusta coffee yield using soil fertility properties. *Computers and Electronics in Agriculture*, *155*, 324–338. doi:10.1016/j.compag.2018.10.014

Krishnan, P., & Surya Rao, A. V. (2005). Effects of genotype and environment on seed yield and quality of rice. *The Journal of Agricultural Science*, *143*(04), 283–292. doi:10.1017/S0021859605005496

Krombholz, K., Hobel, H., Huber, M., & Weippl, E. (2015). Advanced social engineering attacks. *Journal of Information Security and Applications*, *22*, 113-122.

Krotkov, E., Hackett, D., Jackel, L., Perschbacher, M., Pippine, J., Strauss, J., & Orlowski, C. (2018). The DARPA Robotics Challenge Finals: Results and Perspectives. In M. Spenko, S. Buerger, & K. Iagnemma (Eds.), *The DARPA Robotics Challenge Finals: Humanoid Robots To The Rescue. Springer Tracts in Advanced Robotics* (Vol. 121). Cham: Springer. doi:10.1007/978-3-319-74666-1_1

Kshetri, N. (2014). The emerging role of Big Data in key development issues: Opportunities, challenges, and concerns. *Big Data & Society, 1*(2). doi:10.1177/2053951714564227

Kulage, K. M., & Larson, E. L. (2016). Implementation and Outcomes of a Faculty-Based, Peer Review Manuscript Writing Workshop. *Journal of Professional Nursing, 32*(4), 262–270. doi:10.1016/j.profnurs.2016.01.008

Kumar, D., Adamowski, J., Suresh, R., & Ozga-Zielinski, B. (2016). Estimating Evapotranspiration Using an Extreme Learning Machine Model: Case Study in North Bihar, India. *Journal of Irrigation and Drainage Engineering, 142*(9), 04016032. doi:10.1061/(ASCE)IR.1943-4774.0001044

Kumari, S. (2014). Military Robots-A Survey. *International Journal of Advanced Research in Electrical. Electronics and Instrumentation Engineering, 3*(3), 77–80.

Kunze, L., Dolha, M. E., Guzman, E., & Beetz, M. (2011). Simulation-based Temporal Projection of Everyday Robot Object Manipulation. *International Foundation for Autonomous Agents and Multiagent Systems*, 107–114.

Kunze, L., Hawes, N., Duckett, T., Hanheide, M., & Krajnik, T. (2018). Artificial Intelligence for Long-Term Robot Autonomy : A Survey. *IEEE Robotics and Automation Letters, 3*(4), 4023–4030. doi:10.1109/LRA.2018.2860628

Kus, Z. A., Demir, B., Eski, I., Gurbuz, F., & Ercisli, S. (2017). Estimation of the Colour Properties of Apples Varieties Using Neural Network. *Erwerbs-Obstbau, 59*(4), 291–299. doi:10.100710341-017-0324-z

Lamsal, A., Welch, S. M., Jones, J. W., Boote, K. J., Asebedo, A., Crain, J., ... Arachchige, P. G. (2017). Efficient crop model parameter estimation and site characterization using large breeding trial data sets. *Agricultural Systems, 157*, 170–184. doi:10.1016/j.agsy.2017.07.016

Landherr, A., Friedl, B., & Heidemann, J. (2010). A Critical Review of Centrality Measures in Social Networks. *Business & Information Systems Engineering, 2*(6), 371–385. doi:10.100712599-010-0127-3

Lane, F. (2006). *IDC: World Created 161 Billion Gigs of Data in 2006.* Academic Press.

Laney, D. (2001). 3D data management: Controlling data volume, velocity and variety. META Group Research Note, 6(70), 1.

Laney, D. (2001). *3D Data Management: Controlling Data Volume, Velocity, and Variety* [Technical Report]. Retrieved from https://blogs.gartner.com/doug-laney/files/2012/01/ad949-3D-Data-Management-Controlling-Data-Volume-Velocity-and-Variety.pdf

Laney, D. (2001). *3D Data Management: Controlling Data Volume, Velocity, and Variety.* Retrieved from https://blogs.gartner.com/doug-laney/files/2012/01/ad949-3D-Data-Management-Controlling-Data-Volume-Velocity-and-Variety.pdf

Lantmateriet. (2019). Retrieved from https://www.lantmateriet.se/

Larson, M. (2018). Reality, requirements, regulation: Points of intersection with the machine-learning pipeline. In Assessing the impact of machine intelligence on human behaviour: an interdisciplinary endeavour. In *Proceedings of 1st HUMAINT workshop.* Barcelona, Spain: European Commission.

Lau, R. Y., Xia, Y., & Ye, Y. (2014). A probabilistic generative model for mining cybercriminal networks from online social media. *IEEE Computational Intelligence Magazine, 9*(1), 31–43. doi:10.1109/MCI.2013.2291689

Lavinsky, D. (2017). *The Two Most Important Quotes in Business.* Retrieved from: https://www.growthink.com/content/two-most-important-quotes-business

Leandros, P. (2018). *Disruptive Innovation and Blockchain in Financial Services.* Retrieved from https://medium.com/datadriveninvestor/disruptive-innovation-and-blockchain-in-financial-services-911b102e785b

Lee, D. K. C. (2015). Handbook of Digital Currency. In Bitcoin Innovation, Financial Instruments, and Big Data (pp. 1-612). Elsevier.

Lenzerini, M. (2002, June). Data integration: A theoretical perspective. *Pods*, 233–246. doi:10.1145/543613.543644

Leonard, S. (2017). *The Internet of Value: What It Means and How It Benefits Everyone*. Retrieved from https://ripple.com/insights/the-internet-of-value-what-it-means-and-how-it-benefits-everyone/

Li, L., Zheng, N. N., & Wang, F. Y. (2018). On the crossroad of artificial intelligence: A revisit to Alan Turing and Norbert Wiener. *IEEE Transactions on Cybernetics. Early Access*, 1-9.

Liao, M.-S., Chuang, C.-L., Lin, T.-S., Chen, C.-P., Zheng, X.-Y., Chen, P.-T., ... Jiang, J.-A. (2012). Development of an autonomous early warning system for Bactrocera dorsalis (Hendel) outbreaks in remote fruit orchards. *Computers and Electronics in Agriculture, 88*, 1–12. doi:10.1016/j.compag.2012.06.008

Li, B., & Yang, Y. (2018). Complexity of concept classes induced by discrete Markov networks and Bayesian networks. *Pattern Recognition, 82*, 31–37. doi:10.1016/j.patcog.2018.04.026

Liberati, A., Altman, D. G., Tetzlaff, J., Mulrow, C., Gøtzsche, P. C., Ioannidis, J. P., ... Moher, D. (2009). The PRISMA statement for reporting systematic reviews and meta-analyses of studies that evaluate health care interventions: Explanation and elaboration. *PLoS Medicine, 6*(7), e1000100. doi:10.1371/journal.pmed.1000100 PubMed

Li, D., & Du, Y. (2018). *Artificial Intelligence with uncertainty*. CRC Press, Taylor & Francis.

Liddell, H. G., Scott, R., Jones, H. S., & McKenzie, R. (1984). *A Greek-English Lexicon*. Oxford University Press.

Li, G., Law, R., Vu, H. Q., Rong, J., & Zhao, X. (2015a). Identifying emerging hotel preferences using Emerging Pattern Mining technique. *Tourism Management, 46*, 311–321. doi:10.1016/j.tourman.2014.06.015

Li, J., Cheng, K., Wang, S., Morstatter, F., Trevino, R. P., Tang, J., & Liu, H. (2018). Feature selection: A data perspective. *ACM Computing Surveys, 50*(6), 94.

Li, M., Porter, A. L., & Suominen, A. (2018). Insights into relationships between disruptive technology/innovation and emerging technology: A bibliometric perspective. *Technological Forecasting and Social Change, 129*, 285–296. doi:10.1016/j.techfore.2017.09.032

Lim, C.-G., Kim, Z. M., & Choi, H.-J. (2017). Context-based healthy lifestyle recommendation for enhancing user's wellness. *2017 IEEE International Conference on Big Data and Smart Computing (BigComp)*, 418–421. 10.1109/BIGCOMP.2017.7881748

Lingg, E., Leone, G., Spaulding, K., & B'Far, R. (2014). Cardea: Cloud based employee health and wellness integrated wellness application with a wearable device and the HCM data store. 2014 IEEE World Forum on Internet of Things (WF-IoT), 265–270. doi:10.1109/WF-IoT.2014.6803170

Lin, I., & Liao, T. (2017). A survey on Blockchain security issues and challenges. *International Journal of Network Security, 19*(5), 653–659.

Lino, A., Rocha, Á., Macedo, L., & Sizo, A. (2019). Application of clustering-based decision tree approach in SQL query error database. *Future Generation Computer Systems, 93*, 392–406. doi:10.1016/j.future.2018.10.038

Li, S. A., Feng, H. M., Huang, S. P., & Chu, C. Y. (2018). Fuzzy Self-Adaptive Soccer Robot Behavior Decision System Design through ROS. *The Journal of Imaging Science and Technology, 62*(3). doi:10.2352/J.ImagingSci.Technol.2018.62.3.030401

Liu, B., Dai, X., Gong, H., Guo, Z., Liu, N., Wang, X., & Liu, M. (2018). Deep Learning versus Professional Healthcare Equipment: A Fine-Grained Breathing Rate Monitoring Model. Mobile Information Systems, 2018, 1–9. doi:10.1155/2018/1904636

Liu, C., Zheng, B., Wang, C., Zhao, Y., Fu, S., & Li, H. (2017). CNN-Based Vision Model for Obstacle Avoidance of Mobile Robot. *MATEC Web of Conferences, 139*, 4–7. 10.1051/matecconf/201713900007

Liu, X., Rajan, S., Ramasarma, N., Bonato, P., & Lee, S. I. (2018). The Use of a Finger-Worn Accelerometer for Monitoring of Hand Use in Ambulatory Settings. IEEE Journal of Biomedical and Health Informatics. doi: PubMed doi:10.1109/JBHI.2018.2821136

Liu, K.-C., & Chan, C.-T. (2017). Significant Change Spotting for Periodic Human Motion Segmentation of Cleaning Tasks Using Wearable Sensors. *Sensors (Basel), 17*(1). doi: PubMed doi:10.339017010187

Liu, K., Hsieh, C., Hsu, S. J., & Chan, C. (2018). Impact of Sampling Rate on Wearable-Based Fall Detection Systems Based on Machine Learning Models. *IEEE Sensors Journal, 18*(23), 9882–9890. doi:10.1109/JSEN.2018.2872835

Liu, Z.-Y., Wu, H.-F., & Huang, J.-F. (2010). Application of neural networks to discriminate fungal infection levels in rice panicles using hyperspectral reflectance and principal components analysis. *Computers and Electronics in Agriculture, 72*(2), 99–106. doi:10.1016/j.compag.2010.03.003

Li, X., Jiang, P., Chen, T., Luo, X., & Wen, Q. (2018). A Survey on the Security of Blockchain Systems. *IT Security Conference*.

Logan, T. M., McLeod, S., & Guikema, S. (2016). Predictive models in horticulture: A case study with Royal Gala apples. *Scientia Horticulturae, 209*, 201–213. doi:10.1016/j.scienta.2016.06.033

Londhe, D., Nalawade, S., Pawar, S., Atkari, V., & Wandkar, S. (2013). Grader: A review of different methods of grading for fruits and vegetables. *Agricultural Engineering International: CIGR Journal, 15*(3). Retrieved from https://pdfs.semanticscholar.org/7c76/5a16c27361db3a8a041742406bf9e90eb7e9.pdf

Longhi, C., Titz, J. B., & Viallis, L. (2014). Open data: Challenges and opportunities for the tourism industry. In Tourism Management, Marketing, and Development: Volume I: The Importance of Networks and ICTs. Academic Press. doi:10.1057/9781137354358

Lopes, V., & Alexandre, L. A. (2018). *An overview of Blockchain integration with Robotics and Artificial Intelligence.* Cornell University. Retrieved from https://arxiv.org/abs/1810.00329

Lopez, G., Quesada, L., & Guerrero, L. A. (2017). Alexa vs. Siri vs. Cortana vs. Google Assistant: A Comparison of Speech-Based Natural User Interfaces. *Advances in Human Factors and Systems Interaction. International Conference on Applied Human Factors and Ergonomics.* Springer Link.

López-Granados, F., Gómez-Casero, M. T., Peña-Barragán, J. M., Jurado-Expósito, M., & García-Torres, L. (2010). Classifying Irrigated Crops as Affected by Phenological Stage Using Discriminant Analysis and Neural Networks. *Journal of the American Society for Horticultural Science, 135*(5), 465–473. doi:10.21273/JASHS.135.5.465

Lorena, A. C., Maciel, A. I., de Miranda, P. B., Costa, I. G., & Prudêncio, R. B. (2018). Data complexity meta-features for regression problems. *Machine Learning, 107*(1), 209–246. doi:10.100710994-017-5681-1

Lu, J. M., & Hsu, Y. L. (2015). Telepresence robots for medical and homecare applications. In M. Weijnen (Ed.), Contemporary Issues in Systems Science and Engineering (pp. 725–735). Academic Press. doi:10.1002/9781119036821.ch21

Lu, J., Hu, J., Zhao, G., Mei, F., & Zhang, C. (2017). An in-field automatic wheat disease diagnosis system. *Computers and Electronics in Agriculture, 142*(A), 369–379. doi:10.1016/j.compag.2017.09.012

Luiijf, H. A. M., Besseling, K., Spoelstra, M., & De Graaf, P. (2011). Ten national cyber security strategies: A comparison. In *International Workshop on Critical Information Infrastructures Security* (pp. 1-17). Springer.

Luo, J., Hong, T., & Fang, S. C. (2018). Benchmarking robustness of load forecasting models under data integrity attacks. *International Journal of Forecasting, 34*(1), 89–104. doi:10.1016/j.ijforecast.2017.08.004

Luo, X., Jiang, C., Wang, W., Xu, Y., Wang, J. H., & Zhao, W. (2019). User behavior prediction in social networks using weighted extreme learning machine with distribution optimization. *Future Generation Computer Systems, 93*, 1023–1035. doi:10.1016/j.future.2018.04.085

Lustrek, M., Gjoreski, H., Vega, N. G., Kozina, S., Cvetkovic, B., Mirchevska, V., & Gams, M. (2015). Fall Detection Using Location Sensors and Accelerometers. *IEEE Pervasive Computing, 14*(4), 72–79. doi:10.1109/MPRV.2015.84

Lv, C., Xing, Y., Lu, C., Liu, Y., Guo, H., Gao, H., & Cao, D. (2018). Hybrid-learning-based classification and quantitative inference of driver braking intensity of an electrified vehicle. *IEEE Transactions on Vehicular Technology*.

Lykou, G., Iakovakis, G., & Gritzalis, D. (2019). *Aviation Cybersecurity and Cyber-Resilience: Assessing Risk in Air Traffic Management* (pp. 245–260). Critical Infrastructure Security and Resilience. Retrieved from https://link.springer.com/bookseries/5540

Maccani, G., Donnellan, B., & Helfert, M. (2015). Open Data Diffusion for Service Innovation: An Inductive Case Study on Cultural Open Data Services. *Proceedings of the 19th Pacific-Asian Conference on Information Systems (PACIS 2015)*, 1–17.

Maedche, A., & Staab, S. (2001). Ontology Learning for the Semantic Web. *IEEE Intelligent Systems, 16*(2), 72–79. doi:10.1109/5254.920602

Magesh, G., & Swarnalatha, P. (2017). Big data and its applications: A survey. *Research Journal of Pharmaceutical, Biological and Chemical Sciences, 8*(2), 2346–2358.

Mahdi, H. M., & Maaruf, A. (2018). Applications of Blockchain Technology beyond Cryptocurrency. *Annals of Emerging Technologies in Computing, 2*(1), 1–6. doi:10.33166/AETiC.2018.01.001

Mahendran, R., Jayashree, G. C., & Alugusandaram, K. (2012). *Application of Computer Vision Technique on Sorting and Grading of of Fruits and Vegetables. Journal of Food Process Technol S1-001*. doi:10.4172/2157-7110.S1-001

Mahmoud, T., Dong, Z. Y., & Ma, J. (2018). Advanced method for short-term wind power prediction with multiple observation points using extreme learning machines. *The Journal of Engineering, 2018*(1), 29–38. doi:10.1049/joe.2017.0338

Mahmoud, T., Dong, Z. Y., & Ma, J. (2018). An advanced approach for optimal wind power generation prediction intervals by using self-adaptive evolutionary extreme learning machine. *Renewable Energy, 126*, 254–269. doi:10.1016/j.renene.2018.03.035

Majumder, A., & Nath, S. (2019). Classification of Handoff Schemes in a Wi-Fi-Based Network. In *Enabling Technologies and Architectures for Next-Generation Networking Capabilities* (pp. 300–332). IGI Global. doi:10.4018/978-1-5225-6023-4.ch014

Makhija, Y., & Sharma, R. S. (2019). *Face Recognition: Novel Comparison of Various Feature Extraction Technique. Harmony Search and Nature Inspired Optimization Algorithms. Advances in Intelligent Systems and Computing*. Springer.

Mallede, W., Marir, F., & Vassilev, V. (2013). Algorithms for Mapping RDB Schema to RDF for Facilitating Access to Deep Web. *Proceedings of the First International Conference on Building and Exploring Web Based Environments*.

Mandinach, E. (2012). A perfect time for data use. *Educational Psychologist, 47*(2), 2. doi:10.1080/00461520.2012.667064

Manimuthu, A., & Ramadoss, R. (2019). Absolute Energy Routing and Real-Time Power Monitoring for Grid-Connected Distribution Networks. *IEEE Design & Test, 36*(2). Retrieved from https://ieeexplore.ieee.org/abstract/document/8636507

Manyika, J., Chui, M., Groves, P., Farrell, D., Van Kuiken, S., & Doshi, E. A. (2013). *Open Data: Unlocking Innovation and Performance with Liquid Information.* McKinsey.

Mariani, M. M., Buhalis, D., Longhi, C., & Vitouladiti, O. (2014). Managing change in tourism destinations: Key issues and current trends. *Journal of Destination Marketing & Management, 2*(4), 269–272. doi:10.1016/j.jdmm.2013.11.003

Marine-Roig, E., & Anton Clavé, S. (2015). Tourism analytics with massive user-generated content: A case study of Barcelona. *Journal of Destination Marketing & Management, 4*(3), 162–172. doi:10.1016/j.jdmm.2015.06.004

Mario, C. (2018). Disruptive Firms and Technological Change. *Quaderni IRCrES-CNR, 3*(1), 3-18. Retrieved from https://ssrn.com/abstract=3103008

Mark, T. B., Griffin, T. W., & Whitacre, B. E. (2016). The Role of Wireless Broadband Connectivity on `Big Data' and the Agricultural Industry in the United States and Australia. *International Food and Agribusiness Management Review, 19*(A), 43–56.

Markkanen, A. (2015). IoT Analytics Today and in 2020. *ABI Research*, 10.

Marr, B. (2015). *Where big data project fail.* Retrieved from https://www.datacentral.com/profile/blogs/where-big-data-projects-fail

Marr, B. (2018). *Here Are 10 Industries Blockchain Is Likely To Disrupt.* Retrieved from https://www.forbes.com/sites/bernardmarr/2018/07/16/here-are-10-industries-blockchain-is-likely-to-disrupt/#3001d98bb5a2

Martin, K. (2016, April 27). How will artificial intelligence affect legal practice? *Thomson Reuters.* Retrieved April 1, 2019, from Answers On website: https://blogs.thomsonreuters.com/answerson/artificial-intelligence-legal-practice/

Marwala, T., & Hurwitz, E. (2017). *Introduction to man and machines. In Artificial intelligence and economic theory: skynet in the market* (pp. 1–14). Springer International Publishing AG.

Masley, S. (2019). *Tomato Varieties and Types of Tomatoes.* Retrieved from https://www.grow-it-organically.com/tomato-varieties.html

Mathur, N., Paul, G., Irvine, J., Abuhelala, M., Buis, A., & Glesk, I. (2016). A Practical Design and Implementation of a Low Cost Platform for Remote Monitoring of Lower Limb Health of Amputees in the Developing World. *IEEE Access : Practical Innovations, Open Solutions, 4*, 7440–7451. doi:10.1109/ACCESS.2016.2622163

Mathworks. (2019). *Machine learnin with Matlab.* Retrieved from https://www.mathworks.com/campaigns/offers/machine-learning-with-matlab.html

Mattar, M. A., El-Marazky, M. S., & Ahmed, K. A. (2017). Modeling sprinkler irrigation infiltration based on a fuzzy-logic approach. *Spanish Journal of Agricultural Research, 15*(1), e1201. doi:10.5424jar/2017151-9179

Mattias, S. (2017). *Performance and Scalability of Blockchain Networks and Smart Contracts.* Retrieved from https://umu.diva-portal.org/smash/get/diva2:1111497/FULLTEXT01.pdf

Matusov, E., Wilken, P., Bahar, P., Schamper, J., Golik, P., Zeyer, A., ... Peter, J. (2018). Neural Speech Translation at AppTek. *Proceedings of the 15th International Workshop on Spoken Language Translation.*

Maydon, T. (2017). *The 4 Types of data analytics.* Retrieved from https://www.datasciencecentral.com/profiles/blogs/the-4-types-of-data-analytics

Mazilu, S., Hardegger, M., Zhu, Z., Roggen, D., Tröster, G., Plotnik, M., & Hausdorff, J. M. (2012). Online detection of freezing of gait with smartphones and machine learning techniques. 2012 6th International Conference on Pervasive Computing Technologies for Healthcare (PervasiveHealth) and Workshops, 123–130. 10.4108/icst.pervasivehealth.2012.248680

McCarthy, J. (1992). Reminiscences on the history of time sharing. *IEEE Annals of the History of Computing, 14*(1), 19–24.

McClarkin, E. (2018). *MEPs back plan to revolutionise international trade*. Retrieved from http://conservativeeurope. com/news/meps-back-plan-to-revolutionise-international-trade

McCraw, T. K. (2010). *Prophet of Innovation: Joseph Schumpeter and Creative Destruction.* Cambridge, MA: Harvard University Press. doi:10.4159/9780674040779

McKinsey Global Institute. (2011). Big data: The next frontier for innovation, competition and productivity. *McKinsey & Company*. Retrieved from: http://www.mckinsey.com/~/media/McKinsey/dotcom/Insights%20and%20pubs/MGI/ Research/Technology%20and%20Innovation/Big%20Data/MGI_big_data_full_report.ashx

McLean A. (2018). Retrieved from https://www.zdnet.com/article/uob-claims-regions-first-real-time-cross-border-mobile-number-funds-transfer/

McLeod, S. (2018). *5 Principles of Blockchain CPAs Should Know.* Retrieved from: https://www.accountingweb.com/ technology/trends/5-principles-of-blockchain-cpas-should-know

Mei, Z., Gu, X., Chen, H., & Chen, W. (2018). Automatic Atrial Fibrillation Detection Based on Heart Rate Variability and Spectral Features. *IEEE Access : Practical Innovations, Open Solutions, 6*, 53566–53575. doi:10.1109/ ACCESS.2018.2871220

Mellander, C., Lobo, J., Stolarick, K., & Matheson, Z. (2015). *Night-Time Light Data: A Good Proxy Measure for Economic Activity?* Retrieved from https://www.researchgate.net/publication/283263879-Night-Time-Light-Data-A-Good-Proxy-Measure-for_Economic_Activity

Mendel, J. M. (2017). *Uncertain Rule-Based Fuzzy Systems. Introduction and new directions.* Springer International Publishing. doi:10.1007/978-3-319-51370-6

Mendoza, F., & Aguilera, J. M. (2004). Application of image analysis for classification of ripening bananas. *Journal of Food Science, 69*(9), 471–477. doi:10.1111/j.1365-2621.2004.tb09932.x

Menezes, A. J., van Oorschot, P. C., & Vanstone, S. A. (1997). Handbook of Applied Cryptography. Academic Press.

Mercer, K., Li, M., Giangregorio, L., Burns, C., & Grindrod, K. (2016). Behavior Change Techniques Present in Wearable Activity Trackers: A Critical Analysis. *JMIR mHealth and uHealth, 4*(2), e40. doi:10.2196/mhealth.4461 PubMed

Merelli, E., & Rasetti, M. (2013). Non-locality, topology, formal languages: New global tools to handle large datasets. *Procedia Computer Science, 18*, 90–99. doi:10.1016/j.procs.2013.05.172

Merkle, R. C. (1979). Secrecy, authentication, and public key systems. ACM Digital Library.

Mesh, J. (2018, August 2). How Wearables are Changing the Healthcare Industry. Retrieved September 27, 2018, from Healthcare IT Leaders website: https://www.healthcareitleaders.com/blog/how-wearables-are-changing-the-healthcare-industry/

MHA. (n.d.). *History of Census in India.* Retrieved from http://censusindia.gov.in/Ad_Campaign/drop_in_articles/05-History_of_Census_in_India.pdf

MHFW. (2015). *The Ministry of Health and Family Welfare. Government of India Ministry of Health and Family Welfare Statistics. Division Rural Health Statistics.* Retrieved from https://wcd.nic.in/sites/default/files/RHS_1.pdf

Michalski, R. S., Carbonell, J. G., & Mitchell, T. M. (Eds.). (2013). *Machine learning: An artificial intelligence approach. Springer Science* Business Media.

Michopoulou, E., & Buhalis, D. (2013). Information provision for challenging markets: The case of the accessibility requiring market in the context of tourism. *Information & Management, 50*(5), 229–239. doi:10.1016/j.im.2013.04.001

Microsoft, Inc. (2018). *Digital Agriculture: Farmers in India are using AI to increase crop yields.* Retrieved September 25, 2018, from Microsoft News Center India website: https://news.microsoft.com/en-in/features/ai-agriculture-icrisat-upl-india/

Mijwel, M. M. (2015). *History of artificial intelligence.* Retrieved from https://www.researchgate.net/publication/322234922_History_of_Artificial_Intelligence

Milhorance, F., & Singer, J. (2018). Media Trust and Use among Urban News Consumers in Brazil. *Ethical Space: The International Journal of Communication Ethics, 15*(3/4). Retrieved from http://www.communicationethics.net/espace/

Milicic, M. (2008). *Action.* Time and Space in Description Logics.

Mills, M. (2015). *Artificial Intelligence in Law – The State of Play in 2015?* Retrieved April 29, 2019, from Legal IT Insider website: https://www.legaltechnology.com/latest-news/artificial-intelligence-in-law-the-state-of-play-in-2015/

Min, H. (2010). Artificial intelligence in supply chain management: Theory and applications. *International Journal of Logistics: Research and Applications, 13*(1), 13–39. doi:10.1080/13675560902736537

Min, J. H., & Jeong, C. (2009). A binary classification method for bankruptcy prediction. *Expert Systems with Applications, 36*(3), 5256–5263. doi:10.1016/j.eswa.2008.06.073

Misaki, E., Apiola, M., Gaiani, S., & Tedre, M. (2018). Challenges facing sub-Saharan small-scale farmers in accessing farming information through mobile phones: A systematic literature review. *The Electronic Journal on Information Systems in Developing Countries, 84*(4), e12034. doi:10.1002/isd2.12034

Mishra, A. K., Yadav, D. K., Kumar, Y., & Jain, N. (2019). Improving reliability and reducing cost of task execution on preemptible VM instances using machine learning approach. *The Journal of Supercomputing, 75*(4), 2149–2180. doi:10.100711227-018-2717-7

Moawed, S., Algergawy, A., Sarhan, A., Eldosouky, A., & Saake, G. (2014). A latent semantic indexing-based approach to determine similar clusters in large-scale schema matching. In *New Trends in Databases and Information Systems* (pp. 267–276). Springer International Publishing. doi:10.1007/978-3-319-01863-8_29

Mocanu, E., Nguyen, P. H., Gibescu, M., & Kling, W. L. (2016). Deep learning for estimating building energy consumption. *Sustainable Energy. Grids and Networks, 6*, 91–99.

Moeyersoms, J., de Fortuny, E. J., Dejaeger, K., Baesens, B., & Martens, D. (2015). Comprehensible software fault and effort prediction: A data mining approach. *Journal of Systems and Software, 100*, 80–90. doi:10.1016/j.jss.2014.10.032

Moller, A. B., Beucher, A., Iversen, B. V., & Greve, M. H. (2018). Predicting artificially drained areas by means of a selective model ensemble. *Geoderma, 320*, 30–42. doi:10.1016/j.geoderma.2018.01.018

Montanati, A., Cigognini, I. M., & Cifarelli, A. (2017). *Agri and food waste valorisation co-ops based on flexible multi-feedstocks biorefinery processing technologies for newhigh added value applications.* Retrieved from http://agrimax-project.eu/files/2017/11/AGRIMAX-D.1.2_Mapping-of-AFPW-and-their-characteristics.pdf

Monteiro, A., Souto, E., Pazzi, R., & Nogueira, M. (2019). Context-aware network selection in heterogeneous wireless networks. *Computer Communications, 135*, 1–15. doi:10.1016/j.comcom.2018.11.006

Moody, V. (2015). *Using big data responsibly: the ethics of big data.* Retrieved from http://blog.ukdataservice.ac.uk/using-big-data-responsibly-the-ethics-of-big-data/

Mougayar, W. (2016). *The business blockchain: promise, practice, and application of the next Internet technology.* Wiley.

Mountasser, I., Ouhbi, B., & Frikh, B. (2015). From data to wisdom: A new multi-layer prototype for Big Data management process. In *International Conference on Intelligent Systems Design and Applications, ISDA* (pp. 104–109). IEEE. 10.1109/ISDA.2015.7489209

Mountasser, I., Ouhbi, B., & Frikh, B. (2016). Hybrid large-scale ontology matching strategy on big data environment. In *18th International Conference on Information Integration and Web-based Applications and Services* (pp. 282–287). ACM. 10.1145/3011141.3011185

Mukhopadhyay, S. C. (2015). Wearable Sensors for Human Activity Monitoring: A Review. *IEEE Sensors Journal, 15*(3), 1321–1330. doi:10.1109/JSEN.2014.2370945

Murphy, J. (2018). Artificial Intelligence, Rationality, and the World Wide Web. *IEEE Intelligent Systems, 33*(1).

Musto, C., Lops, P., de Gemmis, M., & Semeraro, G. (2017). Semantics-aware Recommender Systems exploiting Linked Open Data and graph-based features. *Knowledge-Based Systems, 136*, 1–14. doi:10.1016/j.knosys.2017.08.015

Nadir Abdelrahman, A. (2018). Blockchain Technology: Classification, Opportunities, and Challenges. *International Research Journal of Engineering and Technology, 5*(5), 3423-3426. Retrieved from https://www.irjet.net/archives/V5/i5/IRJET-V5I5659.pdf

Naids, A., Rossetti, D., Bond, T., Huang, A., Deal, A., Fox, K., ... Mikatarian, R. (2018). *The Demonstration of a Robotic External Leak Locator on the International Space Station.* Orlando, FL: AIAA SPACE and Astronautics Forum and Exposition. doi:10.2514/6.2018-5118

Najafabadi, M. M., Villanustre, F., Khoshgoftaar, T. M., Seliya, N., Wald, R., & Muharemagic, E. (2015). Deep learning applications and challenges in big data analytics. *Journal of Big Data, 2*(1), 1. doi:10.118640537-014-0007-7

Nakamoto, S. (2008). *Bitcoin: A Peer-to-Peer Electronic Cash System.* Retrieved from https://www.bitcoin.com/bitcoin.pdf

Nakamoto, S. (2008). *Bitcoin: A peer-to-peer electronic cash system.* Retrieved from https://bitcoin.org/bitcoin.pdf

Nandyala, C. S., & Kim, H.-K. (2016). Big and Meta Data Management for U-Agriculture Mobile Services. *International Journal of Software Engineering and Its Applications, 10*(2), 257–270. doi:10.14257/ijseia.2016.10.2.21

NASA. (2015). *National Aeronautics and Space Administration (NASA, 2015). UTM: Air Traffic Management for Low-Altitude Drones.* Retrieved from https://www.nasa.gov/sites/default/files/atoms/files/utm-factsheet-11-05-15.pdf

Navarro-Hellin, H., Martinez-del-Rincon, J., Domingo-Miguel, R., Soto-Valles, F., & Torres-Sanchez, R. (2016). A decision support system for managing irrigation in agriculture. *Computers and Electronics in Agriculture, 124*, 121–131. doi:10.1016/j.compag.2016.04.003

Neal, R. M. (2012). *Bayesian learning for neural networks* (Vol. 118). Springer Science & Business Media.

Nemati, E., Liaqat, D., Rahman, M. M., & Kuang, J. (2017). A novel algorithm for activity state recognition using smartwatch data. 2017 IEEE Healthcare Innovations and Point of Care Technologies (HI-POCT), 18–21. doi:10.1109/HIC.2017.8227574

Ngo, M. V., La, Q. D., Leong, D., & Quek, T. Q. S. (2017). User behavior driven MAC scheduling for body sensor networks. 2017 IEEE 19th International Conference on E-Health Networking, Applications and Services (Healthcom), 1–6. 10.1109/HealthCom.2017.8210762

Ngonidzashe, C. (2015). *Retailers, supermarkets blasted over horticultural products import*. Harare: The Sunday News.

Nguyen, H., Eguchi, A., & Hooten, D. (2011). *Search of a Cost Effective Way to Develop Autonomous Floor Mapping Robots. In Robotic and Sensors Environments* (pp. 107–112). ROSE.

Nieto, J., Slawinski, E., Mut, V., & Wagner, B. (2010). Online Path Planning based on Rapidly-Exploring Random Trees. In *IEEE International Conference on Industrial Technology* (pp. 1451–1456). IEEE. 10.1109/ICIT.2010.5472492

Nii, M., Iwamoto, T., Okajima, S., & Tsuchida, Y. (2016). Hybridization of standard and fuzzified neural networks from MEMS-based human condition monitoring data for estimating heart rate. 2016 International Conference on Machine Learning and Cybernetics (ICMLC), 1, 1–6. doi:10.1109/ICMLC.2016.7860868

Nithya, S., & Gomathy, C. (2019). Smaclad: Secure Mobile Agent Based Cross Layer Attack Detection and Mitigation in Wireless Network. *Mobile Networks and Applications*, 1–12.

Nogay, H. S. (2016). Determination of leakage reactance in monophase transformers using by cascaded neural network. *Balkan Journal of Electrical & Computer Engineering*, *4*(2), 89–96.

Nogay, H. S., Akinci, T. C., & Eidukeviciute, M. (2012). Application of artificial neural networks for short term wind speed forecasting in Mardin, Turkey. *Journal of Energy in Southern Africa*, *23*(4), 2–7. doi:10.17159/2413-3051/2012/v23i4a3173

Nomura Research Institute. (2016). *Survey on blockchain technologies and related services*. Tech. Rep. 2015. Retrieved from http://www.meti.go.jp/english/press/ 2016/pdf/0531 01f.pdf

O'Donnell, T. (2016). *Blockchain: Looking Beyond the Hype of Bitcoins*. Retrieved from https://www.infosys.com/blockchain/PublishingImages/pdf/hype-bitcoin.pdf

Oberhauser, A. (2015). *Blockchain Protocol Series – Introduction*. Retrieved May 24, 2015, from https://medium.com/the-blockchain/blockchain-protocol-series-introduction-79d7d9ea899

Obozintsev, L. (2018). *From Skynet to Siri: an exploration of the nature and effects of media coverage of artificial intelligence*. University of Delaware, Department of Communication. Retrieved from http://udspace.udel.edu/handle/19716/24048

OECD. (2015). Organization for Economic Cooperation and Development. *Data-Driven Innovation for Growth and Well-Being, OECD*.

Offray, J. (1912). *L'Homme Machine, Man A Machine: Including Frederick the Great's "Eulogy" on La Mettrie and Extracts from La Mettrie's "The Natural History of the Soul"*. The Open Court Publishing Company.

Oh, S.-H., & Lee, Y. (1995). *A Modified Error Function to Improve the Error Back-Propagation Algorithm for Multi-Layer Perceptrons*. Retrieved from https://onlinelibrary.wiley.com/doi/abs/10.4218/etrij.95.0195.0012

Olaniyi, E., Оланїі, Е., & Олании, Э. (2018). Digital Agriculture: Mobile Phones, Internet & Agricultural Development in Africa. *Actual Problems of Economics*, 16.

Omidi-Arjenaki, O., Maghaddam, P., & Motlagh, A. M. (2012). Online tomato sorting based on shape, maturity, size, and surface defects using machine vision. *Turkish Journal of Agriculture and Forestry, 37*(1), 62-68. Retrieved from DOI: . doi:10.3906/tar-1201-10

Omrane, H., Masmoudi, M. S., & Masmoudi, M. (2016). Fuzzy Logic Based Control for Autonomous Mobile. *Computational Intelligence and Neuroscience*. PMID:27688748

Ongsakul, W., & Vo, D. N. (2013). *Artificial Intelligence in Power System Optimization*. CRC Press.

Oo, L. M., & Aung, N. Z. (2018). A simple and efficient method for automatic strawberry shape and size estimation and classification. *Biosystems Engineering, 170*, 96–107. doi:10.1016/j.biosystemseng.2018.04.004

Orge Pinheiro, C. A., & de Senna, V. (2017). Multivariate analysis and neural networks application to price forecasting in the Brazilian agricultural market. *Ciência Rural, 47*(1). doi:10.1590/0103-8478cr20160077

Oshunsanya, S. O. (2013). Crop Yields as Influenced by Land Preparation Methods Established Within Vetiver Grass Alleys for Sustainable Agriculture in Southwest Nigeria. *Agroecology and Sustainable Food Systems, 37*(5), 578–591. doi:10.1080/21683565.2012.762439

Osman, J., Inglada, J., & Dejoux, J.-F. (2015). Assessment of a Markov logic model of crop rotations for early crop mapping. *Computers and Electronics in Agriculture, 113*, 234–243. doi:10.1016/j.compag.2015.02.015

Ospoval, D. (1995). *The Development of Darwin's Theory: Natural History, Natural Theology and Natural Selection, 1838-1859*. Cambridge University Press.

Otero-Cerdeira, L., Rodríguez-Martínez, F. J., & Gómez-Rodríguez, A. (2015). Ontology matching: A literature review. *Expert Systems with Applications, 42*(2), 949–971. doi:10.1016/j.eswa.2014.08.032

Otte, M. W. (2008). A Survey of Machine Learning Approaches to Robotic. *The International Journal of Robotics Research, 5*(1), 90–98.

Oung, Q. W., Hariharan, M., Lee, H. L., Basah, S. N., Sarillee, M., & Lee, C. H. (2015). Wearable multimodal sensors for evaluation of patients with Parkinson disease. 2015 IEEE International Conference on Control System, Computing and Engineering (ICCSCE), 269–274. doi:10.1109/ICCSCE.2015.7482196

OWASP (The Open Web Application Security Project). (2016). *Top 20 OWASP Vulnerabilities And How To Fix Them*. Retrieved from https://www.upguard.com/articles/top-20-owasp-vulnerabilities-and-how-to-fix-them

Özdemir, V., & Hekim, N. (2018). Birth of Industry 5.0: Making Sense of Big Data with Artificial Intelligence, "The Internet of Things" and Next-Generation Technology Policy. *OMICS: A Journal of Integrative Biology, 22*(1), 65–76. doi:10.1089/omi.2017.0194 PMID:29293405

Pallottino, F., Menesatti, P., Figorilli, S., Antonucci, F., Tomasone, R., Colantoni, A., & Costa, C. (2018). Machine Vision Retrofit System for Mechanical Weed Control in Precision Agriculture Applications. *Sustainability, 10*(7), 2209. doi:10.3390u10072209

Paluszek, M., & Tomas, S. (2017). *MATLAB machine learning*. Apress Pub. doi:10.1007/978-1-4842-2250-8

Panarello, A., Tapas, N., Merlino, G., Longo, F., & Puliafito, A. (2018). Blockchain and iot integration: A systematic survey. *Sensors (Basel), 18*(8), 2575. doi:10.339018082575 PMID:30082633

Panda, A. K. (2018). *A Case of Mutli-Label Image Classification: Build a state of the Art Multi-label Image Classifier*. Retrieved from https://towardsdatascience.com/fast-ai-season-1-episode-3-a-case-of-multi-label-classification-a4a90672a889

Pandorfi, H., Bezerra, A. C., Atarassi, R. T., Vieira, F. M. C., Barbosa Filho, J. A. D., & Guiselini, C. (2016). Artificial neural networks employment in the prediction of evapotranspiration of greenhouse-grown sweet pepper. *Revista Brasileira de Engenharia Agrícola e Ambiental, 20*(6), 507–512. doi:10.1590/1807-1929/agriambi.v20n6p507-512

Pantano, E., Priporas, C. V., & Stylos, N. (2017). 'You will like it!' using open data to predict tourists' response to a tourist attraction. *Tourism Management, 60*, 430–438. doi:10.1016/j.tourman.2016.12.020

Pantazi, X. E., Tamouridou, A. A., Alexandridis, T. K., Lagopodi, A. L., Kashefi, J., & Moshou, D. (2017). Evaluation of hierarchical self-organising maps for weed mapping using UAS multispectral imagery. *Computers and Electronics in Agriculture*, *139*, 224–230. doi:10.1016/j.compag.2017.05.026

Papageorgiou, A. (2019). *Exploring the meaning of AI, data science and Machine Learning with the latest Wikipedia Clickstream.* Retrieved from https://towardsdatascience.com/exploring-the-meaning-of-ai-data-science-and-machine-learning-with-the-latest-wikipedia-5fea5f0a2d46

Paracha, Z. J., Kalam, A., & Ali, R. (2009). A novel approach of harmonic analysis in power distribution networks using artificial intelligence. In *International Conference on Information and Communication Technologies*, (157-160). Karachi, Pakistan: Academic Press. 10.1109/ICICT.2009.5267198

Paraforos, D. S., Vassiliadis, V., Kortenbruck, D., Stamkopoulos, K., Ziogas, V., Sapounas, A. A., & Griepentrog, H. W. (2016). A Farm Management Information System Using Future Internet Technologies. *IFAC-PapersOnLine*, *49*(16), 324–329. doi:10.1016/j.ifacol.2016.10.060

Park, S., Chung, K., & Jayaraman, S. (2014). Wearables: Fundamentals, Advancements, and a Roadmap for the Future. In Wearable Sensors (pp. 1–23). Academic Press. doi:10.1016/B978-0-12-418662-0.00001-5

Park, E., Lee, S. I., Nam, H. S., Garst, J. H., Huang, A., Campion, A., ... Sarrafzadeh, M. (2017). Unobtrusive and Continuous Monitoring of Alcohol-impaired Gait Using Smart Shoes. *Methods of Information in Medicine*, *56*(1), 74–82. doi:10.3414/ME15-02-0008 PubMed

Park, H., Haghani, A., Samuel, S., & Knodler, M. A. (2018). Real-time prediction and avoidance of secondary crashes under unexpected traffic congestion. *Accident; Analysis and Prevention*, *112*, 39–49. doi:10.1016/j.aap.2017.11.025 PMID:29306687

Park, S. H., Kim, B., Kang, C. M., Chung, C. C., & Choi, J. W. (2018). Sequence-to-sequence prediction of vehicle trajectory via LSTM encoder-decoder architecture. *2018 IEEE Intelligent Vehicles Symposium (IV)*, 1672-1678. 10.1109/IVS.2018.8500658

Park, S., Im, J., Jang, E., & Rhee, J. (2016). Drought assessment and monitoring through blending of multi-sensor indices using machine learning approaches for different climate regions. *Agricultural and Forest Meteorology*, *216*, 157–169. doi:10.1016/j.agrformet.2015.10.011

Parsa, M., Panda, P., Sen, S., & Roy, K. (2017). Staged Inference using Conditional Deep Learning for energy efficient real-time smart diagnosis. Conference Proceedings: ... Annual International Conference of the IEEE Engineering in Medicine and Biology Society. IEEE Engineering in Medicine and Biology Society. Annual Conference, 2017, 78–81. doi:10.1109/EMBC.2017.8036767

Pasqualetti, F., Dörfler, F., & Bullo, F. (2013). Attack detection and identification in cyber-physical systems. *IEEE Transactions on Automatic Control*, *58*(11), 2715–2729. doi:10.1109/TAC.2013.2266831

Patel, M. S., Asch, D. A., & Volpp, K. G. (2015). Wearable Devices as Facilitators, Not Drivers, of Health Behavior Change. *Journal of the American Medical Association*, *313*(5), 459–460. doi:10.1001/jama.2014.14781 PubMed

Pathinarupothi, R. K., Prathap, J. D., Rangan, E. S., Gopalakrishnan, E. A., Vinaykumar, R., & Soman, K. P. (2017). Single Sensor Techniques for Sleep Apnea Diagnosis Using Deep Learning. 2017 IEEE International Conference on Healthcare Informatics (ICHI), 524–529. doi:10.1109/ICHI.2017.37

Pellerin, C. (2017). *Project Maven to Deploy Computer Algorithms to War Zone by Year's End.* DoD News, Defense Media Activity. Retrieved from https://dod.defense.gov/News/Article/Article/1254719/project-maven-to-deploy-computer-algorithms-to-war-zone-by-years-end/

Peloia, P. R., & Rodrigues, L. H. A. (2016). Identification Of Commercial Blocks Of Outstanding Performance Of Sugarcane Using Data Mining. *Engenharia Agrícola, 36*(5), 895–901. doi:10.1590/1809-4430-Eng.Agric.v36n5p895-901/2016

Peng, P., Tian, Y., Xiang, T., Wang, Y., Pontil, M., & Huang, T. (2018). Joint semantic and latent attribute modelling for cross-class transfer learning. *IEEE Transactions on Pattern Analysis and Machine Intelligence, 40*(7), 1625–1638. doi:10.1109/TPAMI.2017.2723882 PMID:28692964

Pentos, K., & Pieczarka, K. (2017). Applying an artificial neural network approach to the analysis of tractive properties in changing soil conditions. *Soil & Tillage Research, 165*, 113–120. doi:10.1016/j.still.2016.08.005

Perea, R. (2018). Prediction of applied irrigation depths at farm level using artificial intelligence techniques. *Agricultural Water Management, 206*, 229–240. doi:10.1016/j.agwat.2018.05.019

Perez, S. (2016). *Microsoft silences its new A.I. bot Tay, after Twitter users teach it racism.* Retrieved from https://techcrunch.com/2016/03/24/microsoft-silences-its-new-a-i-bot-tay-after-twitter-users-teach-it-racism/

Pérez-Sánchez, B., Fontenla-Romero, O., & Guijarro-Berdiñas, B. (2018). A review of adaptive online learning for artificial neural networks. *Artificial Intelligence Review, 49*(2), 281–299. doi:10.100710462-016-9526-2

Perumaal, S. S., & Jawahar, N. (2013). Automated Trajectory Planner of Industrial Robot for Pick-and-Place Task. *International Journal of Advanced Robotic Systems, 10*(2), 100. doi:10.5772/53940

Pesonen, J., & Lampi, M. (2016). Utilizing open data in tourism. Conference: Enter2016.

Pike, A. C., Mueller, T. G., Schoergendorfer, A., Shearer, S. A., & Karathanasis, A. D. (2009). Erosion Index Derived from Terrain Attributes using Logistic Regression and Neural Networks. *Agronomy Journal, 101*(5), 1068–1079. doi:10.2134/agronj2008.0207x

Pimentel, D. (2017). *US Navy to Integrate Blockchain to Manufacturing.* Retrieved from https://blocktribune.com/us-navy-integrate-blockchain-manufacturing/

Pineda, M., Pérez-Bueno, M. L., & Barón, M. (2018). Detection of Bacterial Infection in Melon Plants by Classification Methods Based on Imaging Data. *Frontiers in Plant Science, 9*, 164. doi:10.3389/fpls.2018.00164 PMID:29491881

Pitts, W., & McCulloch, W. S. (1947). How we know universals, the perception of auditory and visual forms. *The Bulletin of Mathematical Biophysics, 9*(3), 127–147. doi:10.1007/BF02478291 PMID:20262674

Popescu, C. (2018). Improvements in business operations and customer experience through data science and Artificial Intelligence. *Proceedings of the 12th International Conference on Business Excellence.* 10.2478/picbe-2018-0072

Poulsen, H. E., Andersen, J. T., Keiding, N., Schramm, T. K., Sorensen, R., Gislasson, G., … Torp-Pedersen, C. (2009). *Why Epidemiological and Clinical Intervention Studies Often Give Different or Diverging Results?* Retrieved from http://enghusen.dk/whyEpi.pdf

Pour, A. S., Chegini, G., Zarafshan, P., & Massah, J. (2018). Curvature-based pattern recognition for cultivar classification of Anthurium flowers. *Postharvest Biology and Technology, 139*, 67–74. doi:10.1016/j.postharvbio.2018.01.013

Prasad, R., Deo, R. C., Li, Y., & Maraseni, T. (2018). Soil moisture forecasting by a hybrid machine learning technique: ELM integrated with ensemble empirical mode decomposition. *Geoderma, 330*, 136–161. doi:10.1016/j.geoderma.2018.05.035

Priya, S. (2014). Rescue Robot-A Study ARE USED. *International Journal of Advanced Research in Electrical. Electronics and Instrumentation Engineering, 3*(3), 158–161.

Protopop, I., & Shanoyan, A. (2016). Big Data and Smallholder Farmers: Big Data Applications in the Agri-Food Supply Chain in Developing Countries. *International Food and Agribusiness Management Review, 19*(A, SI), 173–190.

Provost, F., & Fawcett, T. (2013). Data Science and its Relationship to Big Data and Data-Driven Decision Making. *Big Data, 1*(1), 51–59. doi:10.1089/big.2013.1508 PMID:27447038

Punagin, S., & Arya, A. (2015). Privacy in the age of Pervasive Internet and Big Data Analytics - Challenges and Opportunities - ProQuest. *International Journal of Modern Education and Computer Science, 7*(7), 36–47. doi:10.5815/ijmecs.2015.07.05

Qu, H., Xing, K., & Alexander, T. (2013). Neurocomputing An improved genetic algorithm with co-evolutionary strategy for global path planning of multiple mobile robots $. *Neurocomputing, 120,* 509–517. doi:10.1016/j.neucom.2013.04.020

Quinlan, J. R. (2014). *C4. 5: programs for machine learning.* Elsevier.

Quintal, V. A., Lee, J. A., & Soutar, G. N. (2010). Risk, uncertainty and the theory of planned behavior: A tourism example. *Tourism Management, 31*(6), 797–805. doi:10.1016/j.tourman.2009.08.006

Quora. (2017). *What does "Block Time" mean in crypto-currency?* Retrieved from: https://www.quora.com/What-does-Block-Time-mean-in-crypto-currency

Quraishi, M. Z., & Mouazen, A. M. (2013). Development of a methodology for in situ assessment of topsoil dry bulk density. *Soil & Tillage Research, 126,* 229–237. doi:10.1016/j.still.2012.08.009

Raconteur. (2016). *The future of blockchain in 8 charts.* Retrieved from: https://www.raconteur.net/business-innovation/the-future-of-blockchain-in-8-charts

Radhakrishnan, M. (2012). *Application of Computer Vision Technique on Sorting and Grading of Fruits and Vegetables.* Retrieved from https://www.researchgate.net/publication/308953918_Application_of_Computer_Vision_Technique_on_Sorting_and_Grading_of_Fruits_and_Vegetables

Rajapaksha, P., Farahbakhsh, R., Nathanail, E., & Crespi, N. (2017). ITrip, a framework to enhance urban mobility by leveraging various data sources. In Transportation Research Procedia (Vol. 24, pp. 113–122). Academic Press. doi:10.1016/j.trpro.2017.05.076

Rajarman, V. (2014). John McCarthy-father of artificial intelligence. *Asia Pacific Mathematics Newsletter., 4*(3), 15–20.

Rajeswari, S., Suthendran, K., & Rajakumar, K. (2017). A smart agricultural model by integrating IoT, mobile and cloud-based big data analytics. *2017 International Conference on Intelligent Computing and Control (I2C2),* 1–5. 10.1109/I2C2.2017.8321902

Rakesh, K. (2018). *Blockchain vs. Distributed Ledger Technology.* Retrieved December 10, 2018, from https://medium.com/coinmonks/blockchain-vs-distributed-ledger-technology-b7b2e434093b

Ramirez-Gil, J. G., Martinez, G. O. G., & Osorio, J. G. M. (2018). Design of electronic devices for monitoring climatic variables and development of an early warning system for the avocado wilt complex disease. *Computers and Electronics in Agriculture, 153,* 134–143. doi:10.1016/j.compag.2018.08.002

Rampton, J. (2018). *5 applications for Blockchain in your business.* Retrieved from https://execed.economist.com/blog/industry-trends/5-applications-blockchain-your-business

Randazzo, V., Pasero, E., & Navaretti, S. (2018). VITAL-ECG: A portable wearable hospital. *2018 IEEE Sensors Applications Symposium (SAS),* 1–6. 10.1109/SAS.2018.8336776

Rankin, J. (2013). *How data Systems & reports can either fight or propagate the data analysis error epidemic, and how educator leaders can help. Presentation conducted by the Technology Information Center for Administrative Leadership (TICAL).* School Leadership Summit.

Rantanen, M. T. (2011). *A Connectivity-Based Method for Enhancing Sampling in Probabilistic Roadmap Planners.* Academic Press. doi:10.100710846-010-9534-4

Rao, R. S. (2015). *Cleaning, Sorting and Grading of Tomatoes.* Retrieved from https://www.slideshare.net/SUDHA-KARARAOPARVATAN/tomato-grading

Rath, M., Pati, & Patanayak, B. K. (2019). An Overview on Social Networking: Design, Issues, Emerging Trends, and Security, "Social Network Analytics: Computational Research Methods and Techniques. Elsevier.

Rath, M. (2018). Big Data and IoT-Allied Challenges Associated With Healthcare Applications in Smart and Automated Systems. *International Journal of Strategic Information Technology and Applications, 9*(2), 18–34. doi:10.4018/IJSITA.2018040102

Rath, M., & Kumar, P. (2019). Security Protocol with IDS Framework Using Mobile Agent in Robotic MANET. *International Journal of Information Security and Privacy, 13*(1), 46–58. doi:10.4018/IJISP.2019010104

Rath, M., Pati, B., Panigraphi, C. R., & Sakar, J. L. (2019). QTM: A QoS Task Monitoring System for Mobile Ad hoc Networks. In P. Sa, S. Bakshi, I. Hatzilygeroudis, & M. Sahoo (Eds.), *Recent Findings in Intelligent Computing Techniques* (Vol. 707). Singapore: Springer. doi:10.1007/978-981-10-8639-7_57

Rathod, S., Singh, K. N., Patil, S. G., Naik, R. H., Ray, M., & Meena, V. S. (2018). Modeling and forecasting of oilseed production of India through artificial intelligence techniques. *Indian Journal of Agricultural Sciences, 88*(1), 22–27.

Ravi, D., Wong, C., Lo, B., & Yang, G. (2016). Deep learning for human activity recognition: A resource efficient implementation on low-power devices. 2016 IEEE 13th International Conference on Wearable and Implantable Body Sensor Networks (BSN), 71–76. 10.1109/BSN.2016.7516235

Ravi, D., Wong, C., Lo, B., & Yang, G. (2017). A Deep Learning Approach to on-Node Sensor Data Analytics for Mobile or Wearable Devices. *IEEE Journal of Biomedical and Health Informatics, 21*(1), 56–64. doi:10.1109/JBHI.2016.2633287 PubMed

Raviprakash, N., Suresh, M., Rathis, A., Yadav, A., Devarla, D., & Nagaraj, G. S. (2015). Shot Segmentation for Content Based Video Retrieval. *Proceedings of the International Conference. Computational Systems for Health & Sustainability.*

Razum, D., Seketa, G., Vugrin, J., & Lackovic, I. (2018). Optimal threshold selection for threshold-based fall detection algorithms with multiple features. 2018 41st International Convention on Information and Communication Technology, Electronics and Microelectronics (MIPRO), 1513–1516. 10.23919/MIPRO.2018.8400272

Redman, J. (2016). *Disney Reveals Dragonchain, an Interoperable Ledger.* Retrieved from: https://news.bitcoin.com/disney-dragonchain-interoperable-ledger/

Reid, D. (2017). *Blockchain set to disrupt aviation within 2 years claims Accenture.* Retrieved from https://www.cnbc.com/2017/06/27/blockchain-set-to-disrupt-aviation-within-2-years-claims-accenture.html

Reiss, A., & Stricker, D. (2014). Aerobic activity monitoring: Towards a long-term approach. *Universal Access in the Information Society, 13*(1), 101–114. doi:10.1007/s10209-013-0292-5

Remarczyk, M., Narayanan, P., Mitrovic, S., & Black, M. (2018). Our New Handshake with the Robot: How Our Changing Relationship with Machines Can Make Us More Human. *Proceedings of SAI Intelligent Systems Conference IntelliSys 2018: Intelligent Systems and Applications,* 839-851.

Renaud-Gentie, C., Burgos, S., & Benoit, M. (2014). Choosing the most representative technical management routes within diverse management practices: Application to vineyards in the Loire Valley for environmental and quality assessment. *European Journal of Agronomy, 56,* 19–36. doi:10.1016/j.eja.2014.03.002

Revoredo, T. (2018). *Blockchains vs. DLTs*. Retrieved Jul 9, 2018, from https://medium.com/coinmonks/blockchains-vs-dlts-8fe03df39737

Reyna, A., Martin, C., Chen, J., Soler, E., & Diaz, M. (2018). On Blockchain and its Integration with IoT. Challenges and Opportunities. *Future Generation Computer Systems, 88*, 173–190. doi:10.1016/j.future.2018.05.046

Ribeiro, C. O., & Oliveira, S. M. (2011). A hybrid commodity price-forecasting model applied to the sugar-alcohol sector. *The Australian Journal of Agricultural and Resource Economics, 55*(2), 180–198. doi:10.1111/j.1467-8489.2011.00534.x

Richard, N. (2018). *The differences between Artificial and Biological Neural Networks*. Retrieved from https://towardsdatascience.com/the-differences-between-artificial-and-biological-neural-networks-a8b46db828b7

Rivest, R. L. (1990). Cryptography. In *Handbook of Theoretical Computer Science*. Elsevier. doi:10.1016/B978-0-444-88071-0.50018-7

Rizwan, A., Zoha, A., Zhang, R., Ahmad, W., Arshad, K., Ali, N. A., ... Abbasi, Q. H. (2018). A Review on the Role of Nano-Communication in Future Healthcare Systems: A Big Data Analytics Perspective. *IEEE Access : Practical Innovations, Open Solutions, 6*, 41903–41920. doi:10.1109/ACCESS.2018.2859340

Rizzo, D., Martin, L., & Wohlfahrt, J. (2014). Miscanthus spatial location as seen by farmers: A machine learning approach to model real criteria. *Biomass and Bioenergy, 66*, 348–363. doi:10.1016/j.biombioe.2014.02.035

Roberge, J., & Chantepie, P. (2017). The Promised Land of Comparative Digital Cultural Policy Studies. *The Journal of Arts Management, Law, and Society, 47*(5), 295–299.

Rohmer, E., Singh, S. P. N., & Freese, M. (2013). V-REP : a Versatile and Scalable Robot Simulation Framework. *Intelligent Robots and Systems (IROS), 2013 IEEE/RSJ International Conference on*, 1321-1326.

Rokunuzzaman, M., & Jayasuriya, H. (2013). Development of a low cost machine vision system for sorting of. *Aricultural Engineering International Journal, 15*(1), 173–180.

Romero, J. R., Roncallo, P. F., Akkiraju, P. C., Ponzoni, I., Echenique, V. C., & Carballido, J. A. (2013). Using classification algorithms for predicting durum wheat yield in the province of Buenos Aires. *Computers and Electronics in Agriculture, 96*, 173–179. doi:10.1016/j.compag.2013.05.006

Rooney, J. J., & VanDen Heuvel, L. N. (2004). Root cause Analysis for Beginners. *Quality Basics*. Retrieved from https://www.formsbirds.com/download-root-cause-analysis-for-beginners

Rosic, A. (2017). *What Is Hashing? Step-by-Step Guide-Under Hood Of Blockchain*. Retrieved from https://blockgeeks.com/guides/what-is-hashing/

Rosic, A. (2018). *Basic Primer: Blockchain Consensus Protocol*. Retrieved from https://blockgeeks.com/guides/blockchain-consensus/

Ross, J.-M. (2010). Roger Magoulas on Big Data. *O'Reilly Radar*. Retrieved from http://radar.oreilly.com/2010/01/roger-magoulas-on-big-data.html

Ross, I., & Bettman, J. R. (1979). An Information Processing Theory of Consumer Choice. *Journal of Marketing, 43*(3), 124. doi:10.2307/1250155

Rosso, R., Munaro, G., Salvetti, O., Colantonio, S., & Ciancitto, F. (2010). CHRONIOUS: An open, ubiquitous and adaptive chronic disease management platform for Chronic Obstructive Pulmonary Disease (COPD), Chronic Kidney Disease (CKD) and renal insufficiency. 2010 Annual International Conference of the IEEE Engineering in Medicine and Biology, 6850–6853. doi:10.1109/IEMBS.2010.5626451

Rubab, S., Taqvi, S. A., & Hassan, M. F. (2018). Realizing the Value of Big Data in Process Monitoring and Control. In *Current Issues and Opportunities. In International Conference of Reliable Information and Communication Technology* (pp. 128-138). Springer.

Ruiz, M. P. (2017). Task based agricultural mobile robots in arable farming : A review. *Spanish Journal of Agricultural Research, 15*(1), 1–19.

Russell, S. J., & Norvig, P. (2010). *Artificial intelligence: A Modern Approach* (3rd ed.). Upper Saddle River, NJ: Prentice Hall.

Russell, S. J., & Norvig, P. (2016). *Artificial intelligence a modern approach.* Pearson Education Limited.

Russell, S. J., & Norvig, P. (2016). *Artificial intelligence: a modern approach.* Pearson Education Limited.

Sabanci, K., & Aydin, C. (2017). Smart Robotic Weed Control System for Sugar Beet. *Journal of Agricultural Science and Technology, 19*(1), 73–83.

Safa, M., Samarasinghe, S., & Nejat, M. (2015). Prediction of Wheat Production Using Artificial Neural Networks and Investigating Indirect Factors Affecting It: Case Study in Canterbury Province, New Zealand. *Journal of Agricultural Science and Technology, 17*(4), 791–803.

Safavi, H. R., Mehrparvar, M., & Szidarovszky, F. (2016). Conjunctive Management of Surface and Ground Water Resources Using Conflict Resolution Approach. *Journal of Irrigation and Drainage Engineering, 142*(4), 05016001. doi:10.1061/(ASCE)IR.1943-4774.0000991

Saha, S. S., Rahman, S., Rasna, M. J., Zahid, T. B., Islam, A. K. M. M., & Ahad, M. A. R. (2018). Feature Extraction, Performance Analysis and System Design Using the DU Mobility Dataset. *IEEE Access : Practical Innovations, Open Solutions, 6*, 44776–44786. doi:10.1109/ACCESS.2018.2865093

Sahu, D. K. (2018). *Supervised and unsupervised learning in data mining.* Retrieved from https://www.digitalvidya.com/blog/supervised-and-unsupervised-learning-in-data-mining/

Sa, I., Ge, Z., Dayoub, F., Upcroft, B., Perez, T., & McCool, C. (2016). DeepFruits: A Fruit Detection System Using Deep Neural Networks. *Sensors (Basel), 16*(8), 1222. doi:10.339016081222 PMID:27527168

Saidu, A., Clarkson, A. M., Adamu, S. H., Mohammed, M., & Jibo, I. (2017). Application of ICT in Agriculture. *Opportunities and Challenges in Developing Countries., 3*, 11.

Salah, K., Rehman, M. H., Nizamuddin, N., & Al-Fuqaha, A. (2018). Blockchain for AI: Review and Open Research Challenges. *IEEE Computer Society, 4*.

Samek, W., Wiegand, T., & Müller, K.-R. (2017). *Explainable artificial intelligence: Understanding, visualizing and interpreting deep learning models.* Retrieved from https://arxiv.org/abs/1708.08296

Sampada, M. (n.d.). e-*Tool for Human Resources Management System.* Retrieved from https://ehrms.nic.in/Home/HomePageFeatures?ID=AboutUs

Samuel, S. J., Koundinya, P. V. P., Sashidhar, K., & Bharathi, C. R. (2015). A Survey on big data and its research challenges. *Journal of Engineering and Applied Sciences (Asian Research Publishing Network), 10*(8), 3343–3347.

Santipantakis, G., Kotis, K., & Vouros, G. A. (2017). OBDAIR: Ontology-Based Distributed framework for Accessing, Integrating and Reasoning with data in disparate data sources. *Expert Systems with Applications, 90*, 464–483. doi:10.1016/j.eswa.2017.08.031

Sathyanarayana, A., Joty, S., Fernandez-Luque, L., Ofli, F., Srivastava, J., Elmagarmid, A., ... Taheri, S. (2016). Sleep Quality Prediction from Wearable Data Using Deep Learning. *JMIR mHealth and uHealth*, *4*(4). doi:10.2196/mhealth.6562

Sathyanarayana, A., Srivastava, J., & Fernandez-Luque, L. (2017). The Science of Sweet Dreams: Predicting Sleep Efficiency from Wearable Device Data. *Computer*, *50*(3), 30–38. doi:10.1109/MC.2017.91

Schatz, D., Bashroush, R., & Wall, J. (2017). Towards a more representative definition of cyber security. *Journal of Digital Forensics. Security and Law*, *12*(2), 8.

Scheurer, C., & Zimmermann, U. E. (2011). Path Planning Method for Palletizing Tasks using Workspace Cell Decomposition. In *IEEE International Conference on Robotics and Automation* (pp. 1–4). IEEE. 10.1109/ICRA.2011.5980573

Schluse, M., Priggemeyer, M., Atorf, L., & Rossmann, J. (2018). Experiment table digital twins streamlining simulation-based systems engineering for industry 4.0. *IEEE Transactions on Industrial Informatics*, *14*(4), 1722–1731. doi:10.1109/TII.2018.2804917

Schmidhuber, J. (2015). Deep Learning in Neural Networks: An Overview. *Neural Networks*, *61*, 85–117. doi:10.1016/j.neunet.2014.09.003 PMID:25462637

Schuhmacher, M., & Ponzetto, S. (2014). Ranking Entities in a Large Semantic Network. In *European Semantic Web Conference* (pp. 254–258). Springer International Publishing. 10.1007/978-3-319-11955-7_30

Schwartz, D., Youngs, N., & Britto, A. (2018). *The Ripple Protocol Consensus Algorithm*. Retrieved from https://ripple.com/files/ripple_consensus_whitepaper.pdf

Schwemmer, M. A., Skomrock, N. D., Sederberg, P. B., Ting, J. E., Sharma, G., Bockbrader, M. A., & Friedenberg, D. A. (2018). Meeting brain–computer interface user performance expectations using a deep neural network decoding framework. *Nature Medicine*, *24*(11), 1669–1676. doi:10.103841591-018-0171-y PMID:30250141

Schwikowski, B. (2015). Cytoscape: Visualization and Analysis of omis data in interaction networks, Institut Pasteur. *Gnome Research*. Retrieved from https://research.pasteur.fr/en/software/cytoscape/

Self Care Forum. (2019). What do we mean by self care and why is it good for people? Retrieved April 9, 2019, from Self Care Forum website: http://www.selfcareforum.org/about-us/what-do-we-mean-by-self-care-and-why-is-good-for-people/

Sennaar, K. (2017, October 16). *AI in Agriculture - Present Applications and Impact*. Retrieved September 24, 2018, from TechEmergence website: https://www.techemergence.com/ai-agriculture-present-applications-impact/

Serenelli, L. (2014). *EU, Interpol fight epidemic of stolen, fake passports*. Retrieved from https://www.usatoday.com/story/news/world/2014/05/21/stolen-passports/9351329/

Shackelford, S. (2018). *Smart Factories, Dumb Policy? Managing Cybersecurity and Data Privacy Risks in the Industrial Internet of Things*. Kelley School of Business Research Paper No. 18-80. doi:10.2139srn.3252498

Shafie, A. A., Hafiz, F., & Ali, M. H. (2009). Motion detection techniques using optical flow. World Academy of Science, Engineering and Technology. *International Journal of Electrical and Computer Engineering*, *3*(8). Retrieved from http://waset.org/publications/8745

Shahbaz, N., & Phukan, A. (2018). *A Legal and Ethical Examination of Photorealistic Videos Created Using Artificial Neural Networks*. Retrieved from https://www.sccur.org/sccur/Fall_2018_Conference/Multidisc_Posters/19/

Shahid, N., Rappon, T., & Berta, W. (2019). Applications of artificial neural networks in health care organizational decision-making: A scoping review. *PLoS One*, *14*(2), 1–22. doi:10.1371/journal.pone.0212356 PMID:30779785

Shamsfakhr, F., & Sadeghibigham, B. (2017). A neural network approach to navigation of a mobile robot and obstacle. *Turkish Journal of Electrical Engineering and Computer Sciences*, 25(3), 1629–1642. doi:10.3906/elk-1603-75

Shamshiri, R. R., Weltzien, C., Hameed, I. A., Yule, I. J., Grift, T. E., Balasundram, S. K., ... Chowdhary, G. (2018). Research and development in agricultural robotics: A perspective of digital farming. *International Journal of Agricultural and Biological Engineering*, 11(4), 1–14. doi:10.25165/j.ijabe.20181104.4278

Shanmuganathan, S., & Samarasinghe, S. (2016). *Artificial neural network modelling*. Springer International Publishing. doi:10.1007/978-3-319-28495-8

Sharma, N., Sharma, P., Irwin, D., & Shenoy, P. (2011). Predicting solar generation from weather forecasts using machine learning. *2011 IEEE International Conference on Smart Grid Communications (SmartGridComm)*, 528–533. 10.1109/SmartGridComm.2011.6102379

Shekoofa, A., Emam, Y., Ebrahimi, M., & Ebrahimie, E. (2011). Application of supervised feature selection methods to define the most important traits affecting maximum kernel water content in maize. *Australian Journal of Crop Science*, 5(2), 162–168.

Shen, C., & Pena-Mora, F. (2018). Blockchain for Cities – A Systematic Literature Review. *IEEE Access: Practical Innovations, Open Solutions*, 1, 1–33. Retrieved from https://www.researchgate.net/publication/328896113-Blockchain-for-Cities-A-Systematic-Literature-Review

Shin, L. (2017). *The First Government To Secure Land Titles On The Bitcoin Blockchain Expands Project.* Retrieved from https://www.forbes.com/sites/laurashin/2017/02/07/the-first-government-to-secure-land-titles-on-the-bitcoin-blockchain-expands-project/#5b9dbeab4dcd

Shin, G., Jarrahi, M. H., Fei, Y., Karami, A., Gafinowitz, N., Byun, A., & Lu, X. (2019). Wearable activity trackers, accuracy, adoption, acceptance and health impact: A systematic literature review. *Journal of Biomedical Informatics*, 93, 103153. doi:10.1016/j.jbi.2019.103153 PubMed

Shingles, M., Page, O., Phipps, J., Davenport, T., & Iglesias, J. (2017). *Acknowledgments and Key Contacts.* Retrieved from http://www.gmaonline.org/issues-policy/collaborating-with-retailers/big-data-analytics/appendix

Shon, T., & Moon, J. (2007). A hybrid machine learning approach to network anomaly detection. *Information Sciences*, 177(18), 3799–3821. doi:10.1016/j.ins.2007.03.025

Shrier, D., Sharma, D., & Pentland, A. (2016a). *Blockchain & Financial Services: The Fifth Horizon of Networked Innovation.* Retrieved from http://cdn.resources.getsmarter.ac/wp-content/uploads/2017/06/MIT_Blockchain_whitepaper_PartOne.pdf

Shvaiko, P., & Euzenat, J. (2013). Ontology Matching: State of the Art and Future Challenges. *IEEE Transactions on Knowledge and Data Engineering*, 25(1), 158–176. doi:10.1109/TKDE.2011.253

Sideratos, G., & Hatziargyriou, N. D. (2012). Probabilistic Wind Power Forecasting Using Radial Basis Function Neural Networks. *IEEE Transactions on Power Systems*, 27(4), 1788–1796. doi:10.1109/TPWRS.2012.2187803

Silva, L. O. L. A., Koga, M. L., Cugnasca, C. E., & Costa, A. H. R. (2013). Comparative assessment of feature selection and classification techniques for visual inspection of pot plant seedlings. *Computers and Electronics in Agriculture*, 97, 47–55. doi:10.1016/j.compag.2013.07.001

Simon, M. (2018, September 25). The Creepy-Cute Robot that Picks Peppers With its Face. *Wired.* Retrieved from https://www.wired.com/story/the-creepy-cute-robot-that-picks-peppers/

Simon, M., Rodner, E., & Denzler, J. (2016). *ImageNet pre-trained models with batch normalization*. Cornell University. Retrieved from https://arxiv.org/abs/1612.01452

Singh, N. (2018). *Talking machines: democratizing the design of voice-based agents for the home*. Program in Media Arts and Sciences (Massachusetts Institute of Technology). Retrieved from http://hdl.handle.net/1721.1/119089

Singh, N., Delwiche, M. J., & Johnson, R. S. (1993). Image analysis methods for real-time color grading of stonefruit. *Computers and Electronics in Agriculture, 9*(1), 71–84. doi:10.1016/0168-1699(93)90030-5

Sirsat, M. S., Cernadas, E., Fernandez-Delgado, M., & Khan, R. (2017). Classification of agricultural soil parameters in India. *Computers and Electronics in Agriculture, 135*, 269–279. doi:10.1016/j.compag.2017.01.019

Sivathanu, B., & Pillai, R. (2018). Smart HR 4.0 – how industry 4.0 is disrupting HR. *Human Resource Management International Digest, 26*(4), 7–11. doi:10.1108/HRMID-04-2018-0059

Smith, B. A., Hoogenboom, G., & McClendon, R. W. (2009). Artificial neural networks for automated year-round temperature prediction. *Computers and Electronics in Agriculture, 68*(1), 52–61. doi:10.1016/j.compag.2009.04.003

Snihur, Y., & Wiklund, J. (2018). *Searching for innovation: Product, process, and business model innovations and search behavior in established firms, Long Range Planning*. Wiley.

Snow, V. O., & Lovattb, S. J. (2008). A general planner for agro-ecosystem models. *Computers and Electronics in Agriculture, 60*(2), 201–211. doi:10.1016/j.compag.2007.08.001

Soderlund, J. (2017). *Big Data Architecture Patterns and Best Practices*. Retrieved from https://www.slideshare.net/AmazonWebServices/big-data-architectural-patterns-and-best-practices-75668557

Soh, Y. W., Koo, C. H., Huang, Y. F., & Fung, K. F. (2018). Application of artificial intelligence models for the prediction of standardized precipitation evapotranspiration index (SPEI) at Langat River Basin, Malaysia. *Computers and Electronics in Agriculture, 144*, 164–173. doi:10.1016/j.compag.2017.12.002

Soltani, M., Omid, M., & Alimardani, R. (2015). Egg volume prediction using machine vision technique based on pappus theorem and artificial neural network. *Journal of Food Science and Technology, 52*(5), 3065–3071. doi:10.100713197-014-1350-6 PMID:25892810

Song, F., Zacharewicz, G., & Chen, D. (2013). An Analytic Aggregation-Based Ontology Alignment Approach with Multiple Matchers. In *Advanced Techniques for Knowledge Engineering and Innovative Applications* (pp. 143–159). Springer Berlin Heidelberg. doi:10.1007/978-3-642-42017-7_11

Sonobe, R., Tani, H., Wang, X., Kojima, Y., & Kobayashi, N. (2015). Extreme Learning Machine-based Crop Classification using ALOS/PALSAR Images. *Japan Agricultural Research Quarterly, 49*(4), 377–381. doi:10.6090/jarq.49.377

Soper, S. (2018). Amazon Prime Day Shopping Topped $4 Billion, Analyst Estimates. *Bloomberg*. Retrieved from https://www.bloomberg.com/news/articles/2018-07-19/amazon-prime-day-shopping-topped-4-billion-analyst-estimates

Sousa, M. J., & Rocha, A. (2019). Skills for disruptive digital business. *Journal of Business Research, 94*, 257–263. doi:10.1016/j.jbusres.2017.12.051

Spice, B. (2017). *Machine Learning Will Change Jobs*. Carnegie Mellon University.

Srivastava, S., Riano, L., & Abbeel, P. (2014). Combined Task and Motion Planning Through an Extensible Planner-Independent Interface Layer. *Robotics and Automation*, 639–646.

Staranowicz, A., & Mariottini, G. L. (2011). A Survey and Comparison of Commercial and Open-Source Robotic Simulator Software. In *Proceedings of the 4th International Conference on PErvasive Technologies Related to Assistive Environments* (p. 56). ACM. 10.1145/2141622.2141689

Stastny, J., Konecny, V., & Trenz, O. (2011). Agricultural data prediction by means of neural network. *Agricultural Economics-Zemedelska Ekonomika, 57*(7), 356–361. doi:10.17221/108/2011-AGRICECON

Statistica. (2017). *Global production of vegetables in 2017, by type (in million metric tons).* Retrieved from https://www.statista.com/statistics/264065/global-production-of-vegetables-by-type/

Stefanie, C. (2018, June 4). Wearable Tech is Here to Stay with a Robust Presence in the Future Healthcare Industry. Retrieved September 19, 2018, from Wearable Technologies website: https://www.wearable-technologies.com/2018/06/wearable-tech-is-here-to-stay-with-a-robust-presence-in-the-future-healthcare-industry/

Steffen, D., Bleser, G., Weber, M., Stricker, D., Fradet, L., & Marin, F. (2011). A personalized exercise trainer for elderly. 2011 5th International Conference on Pervasive Computing Technologies for Healthcare (PervasiveHealth) and Workshops, 24–31. 10.4108/icst.pervasivehealth.2011.245937

Steve Morgan. (2018). *2018 Cybersecurity Market Report.* Retrieved from https://cybersecurityventures.com/cybersecurity-market-report/

Steyskal, S., & Polleres, A. (2013). Mix'n'Match: An alternative approach for combining ontology matchers (short paper). In Lecture Notes in Computer Science (including subseries Lecture Notes in Artificial Intelligence and Lecture Notes in Bioinformatics) (Vol. 8185, pp. 555–563). Springer. doi:10.1007/978-3-642-41030-7_40

Stroud, F. (2015). *Blockchain: Webopedia Definition.* Available at: https://www.webopedia.com/TERM/B/blockchain.html

Strydom, M., & Buckley, S. (2018). The Big Data Research Ecosystem: an Analytical Literature Study. In M. Strydom & S. Buckley (Eds.), *Big Data Governance and Perspectives in Knowledge Management.* IGI Global.

Subash, S. P., Kumar, R. R., & Aditya, K. S. (2019). Satellite data and machine learning tools for predicting poverty in rural India. *Agricultural Economics Research Review, 31*(2), 231–240. doi:10.5958/0974-0279.2018.00040.X

Sucan, I. A., Moll, M., & Kavraki, L. E. (2012). The open motion planning library. *IEEE Robotics & Automation Magazine, 19*(4), 72–82. doi:10.1109/MRA.2012.2205651

Sugumaran, V., & Storey, V. C. (2002). Ontologies for conceptual modeling: Their creation, use, and management. In Data and Knowledge Engineering (Vol. 42, pp. 251–271). Academic Press. doi:10.1016/S0169-023X(02)00048-4

Suh, H. K., Ijsselmuiden, J., Hofstee, J. W., & van Henten, E. J. (2018). Transfer learning for the classification of sugar beet and volunteer potato under field conditions. *Biosystems Engineering, 174,* 50–65. doi:10.1016/j.biosystemseng.2018.06.017

Sujaritha, M., Annadurai, S., Satheeshkumar, J., Sharan, S. K., & Mahesh, L. (2017). Weed detecting robot in sugarcane fields using fuzzy real time classifier. *Computers and Electronics in Agriculture, 134,* 160–171. doi:10.1016/j.compag.2017.01.008

Suk, H. I. (2017). An introduction to neural networks and deep learning. In *Deep Learning for Medical Image Analysis.* Academic Press.

Sulistyo, S. B., Woo, W. L., & Dlay, S. S. (2017). Regularized Neural Networks Fusion and Genetic Algorithm Based On-Field Nitrogen Status Estimation of Wheat Plants. *IEEE Transactions on Industrial Informatics, 13*(1), 103–114. doi:10.1109/TII.2016.2628439

Sultan, K., Ruhi, U., & Lakhani, R. (2018). *Conceptualizing Blockchains: Characteristics & Applications.* Retrieved from https://arxiv.org/ftp/arxiv/papers/1806/1806.03693.pdf

Sun, D.-W. (2016). *Computer Vision Technology for Food Quality Evaluation. Academic Press.* Elsevier.

Sun, X., Wandel, S., & Stump, E. (2018). Competitiveness of on-demand air taxis regarding door-to-door travel time: A race through Europe. *Transportation Research Part E, Logistics and Transportation Review*, *119*, 1–18. doi:10.1016/j.tre.2018.09.006

Sun, X., Wu, P., & Hoi, S. C. (2018). Face detection using deep learning: An improved faster RCNN approach. *Neurocomputing*, *299*, 42–50. doi:10.1016/j.neucom.2018.03.030

Supriya, T. A., & Kulkarani, V. (2017). Blockchain and Its Applications – A Detailed Survey. *International Journal of Computers and Applications*, *180*(3), 29–35. doi:10.5120/ijca2017915994

Suvarna, K. (2018). *Review of distributed Ledgers: The technological Advances behind cryptocurrency.* Retrieved from https://www.researchgate.net/publication/323628539_Review_of_Distributed_Ledgers_The_technological_Advances_behind_cryptocurrency

Swan, M. (2013). The Quantified Self: Fundamental Disruption in Big Data Science and Biological Discovery. *Big Data*, *1*(2), 85–99. doi:10.1089/big.2012.0002 PubMed

Szabo, N. (1997). *The Ideas of Smart Contracts.* Retrieved from http://www.fon.hum.uva.nl/rob/Courses/InformationInSpeech/CDROM/Literature/LOTwinterschool2006/szabo.best.vwh.net/idea.html

Taeihagh, A., & Si MinLim, H. (2018). Governing autonomous vehicles: emerging responses for safety, liability, privacy, cybersecurity, and industry risks. *Journal Transport Reviews, 39*(1). Retrieved from https://www.tandfonline.com/doi/ref/10.1080/01441647.2018.1494640?scroll=top

Tag, B., Holz, C., Lukowicz, P., Augereau, O., Uema, Y., & Kunze, K. (2018). EyeWear 2018: Second Workshop on EyeWear Computing. *UbiComp '18 Proceedings of the 2018 ACM International Joint Conference and 2018 International Symposium on Pervasive and Ubiquitous Computing and Wearable Computers*, 964-967.

Tahavori, F., Stack, E., Agarwal, V., Burnett, M., Ashburn, A., Hoseinitabatabaei, S. A., & Harwin, W. (2017). Physical activity recognition of elderly people and people with parkinson's (PwP) during standard mobility tests using wearable sensors. 2017 International Smart Cities Conference (ISC2), 1–4. doi:10.1109/ISC2.2017.8090858

Tamaddoni-Nezhad, A., Milani, G. A., Raybould, A., Muggleton, S., & Bohan, D. A. (2013). Construction and Validation of Food Webs Using Logic-Based Machine Learning and Text Mining. In G. Woodward & D. A. Bohan (Eds.), Advances in Ecological Research, Vol 49: Ecological Networks in an Agricultural World (pp. 225–289). Academic Press. doi:10.1016/B978-0-12-420002-9.00004-4

Tan, D. S., Leong, R. N., Laguna, A. F., Ngo, C. A., Lao, A., Amalin, D. M., & Alvindia, D. G. (2018). AuToDiDAC: Automated Tool for Disease Detection and Assessment for Cacao Black Pod Rot. *Crop Protection*, *103*, 98–102. doi:10.1016/j.cropro.2017.09.017

Tavakoli, M., Viegas, C., Sgrigna, L., & de Almeida, A. (2018). SCALA: Scalable Modular Rail based Multi-agent Robotic System for Fine Manipulation over Large Workspaces. *Journal of Intelligent & Robotic Systems*, *89*, 421. doi:10.100710846-017-0560-3

Techopedia. (2019). *Disruptive Technology.* Retrieved from https://www.techopedia.com/definition/14341/disruptive-technology

Tegmark, M. (2018). An Open Letter: Research Priorities for Robust and Beneficial Artificial Intelligence. *Future of Life Institute*. Retrieved from https://futureoflife.org/ai-open-letter/

Tesfaye, K., Sonder, K., Cairns, J., Magorokosho, C., Tarekegn, A., Kassie, G. T., ... Erenstein, O. (2016). Targeting Drought-Tolerant Maize Varieties in Southern Africa: A Geospatial Crop Modeling Approach Using Big Data. *International Food and Agribusiness Management Review, 19*(A, SI), 75–92.

Tewari, S. (2015). *670 Million In Rural Areas Live On Rs 33 Per Day*. Retrieved from https://archive.indiaspend.com/cover-story/670-million-in-rural-areas-live-on-rs-33-per-day-79600

Tham, J. (2018). Persuasive-Pervasive Technology: Rhetorical Strategies in Wearables Advertising. *International Journal of Semiotics and Visual Rhetoric*. doi:10.4018/IJSVR.2018010104

Thenkabail, P. S., Lyon, J. G., & Huete, A. (2012). Advances in Hyperspectral Remote Sensing of Vegetation and Agricultural Croplands. In P. S. Thenkabail, J. G. Lyon, & A. Huete (Eds.), Hyperspectral Remote Sensing of Vegetation (pp. 3–35). Academic Press. Retrieved from https://pubs.er.usgs.gov/publication/70098951

Thorsby, J., Stowers, G. N. L., Wolslegel, K., & Tumbuan, E. (2017). Understanding the content and features of open data portals in American cities. *Government Information Quarterly, 34*(1), 53–61. doi:10.1016/j.giq.2016.07.001

Tien, J. M. (2013). Big Data: Unleashing information. *Journal of Systems Science and Systems Engineering, 22*(2), 127–151. doi:10.1007/s11518-013-5219-4

Toklu, S., & Şimşek, M. (2018). Two-Layer Approach for Mixed High-Rate and Low-Rate Distributed Denial of Service (DDoS) Attack Detection and Filtering. *Arabian Journal for Science and Engineering, 43*(12), 7923–7931. doi:10.100713369-018-3236-9

Torkashvand, A. M., Ahmadi, A., & Nikravesh, N. L. (2017). Prediction of kiwifruit firmness using fruit mineral nutrient concentration by artificial neural network (ANN) and multiple linear regressions (MLR). *Journal of Integrative Agriculture, 16*(7), 1634–1644. doi:10.1016/S2095-3119(16)61546-0

Torres-Sospedra, J., & Nebot, P. (2014). Two-stage procedure based on smoothed ensembles of neural networks applied to weed detection in orange groves. *Biosystems Engineering, 123*, 40–55. doi:10.1016/j.biosystemseng.2014.05.005

Trace Genomics. (2018). *Trace Genomics*. Retrieved December 17, 2018, from https://www.tracegenomics.com/#/

Transparencymarketresearch. (2018). *Blockchain Technology Market to reach US$20 billion by 2024 – TMR*. Retrieved from https://globenewswire.com/news-release/2018/10/26/1627765/0/en/Blockchain-Technology-Market-to-reach-US-20-billion-by-2024-TMR.html

Trautmann, S., Rehm, J., & Wittchen, H. (2016). The economic costs of mental disorders: Do our societies react appropriately to the burden of mental disorders? *Science and Society*. PMID:27491723

Tripathy, A. K., Adinarayana, J., Vijayalakshmi, K., Merchant, S. N., Desai, U. B., Ninomiya, S., ... Kiura, T. (2014). Knowledge discovery and Leaf Spot dynamics of groundnut crop through wireless sensor network and data mining techniques. *Computers and Electronics in Agriculture, 107*, 104–114. doi:10.1016/j.compag.2014.05.009

Tsai, C. F., Hsu, Y. F., Lin, C. Y., & Lin, W. Y. (2009). Intrusion detection by machine learning: A review. *Expert Systems with Applications, 36*(10), 11994–12000. doi:10.1016/j.eswa.2009.05.029

Turing, A. M. (1950). Computing Machinery and Intelligence. *Mind, 49*(236), 433–460. doi:10.1093/mind/LIX.236.433

Uliyar, S. (2017). *A Primer: oracle intelligent bots - powered by artificial intelligence*. Oracle. Retrieved from http://www.oracle.com/us/technologies/mobile/chatbots-primer-3899595.pdf

Ullah, M. R., Bhuiyan, M. A. R., & Das, A. K. (2017). IHEMHA: Interactive healthcare system design with emotion computing and medical history analysis. *2017 6th International Conference on Informatics, Electronics and Vision 2017 7th International Symposium in Computational Medical and Health Technology (ICIEV-ISCMHT)*, 1–8. 10.1109/ICIEV.2017.8338606

Um, J. (2019). *Imaging Sensors. Drones as Cyber-Physical Systems.* Singapore: Springer. doi:10.1007/978-981-13-3741-3_6

UNDESA. (2018). *Transforming our world: the 2030 Agenda for Sustainable Development.* Retrieved from https://sustainabledevelopment.un.org/post2015/transformingourworld

Upwork. (2019). *Overview of artificial intelligence and natural language processing.* Retrieved from https://www.upwork.com/hiring/for-clients/artificial-intelligence-and-natural-language-processing-in-big-data/

Ureta, C., Gonzalez-Salazar, C., Gonzalez, E. J., Alvarez-Buylla, E. R., & Martinez-Meyer, E. (2013). Environmental and social factors account for Mexican maize richness and distribution: A data mining approach. *Agriculture, Ecosystems & Environment, 179*, 25–34. doi:10.1016/j.agee.2013.06.017

USDEO Department of Education Office of Planning, Evaluation and Policy Development. (2009). Implementing data-informed decision making in schools: Teacher access, supports, and use. United States Department of Education. (ERIC Document Reproduction Service No. ED504191)

Van den Hoven van Genderen, R. (2019). Does Future Society Need Legal Personhood for Robots and AI? In Artificial Intelligence in Medical Imaging (pp. 257-290). Springer.

Van Wynsberghe, A., & Robbins, S. (2018). *Critiquing the Reasons for Making Artificial Moral Agents.* Science and Engineering Ethics. Springer Link. doi:10.100711948-018-0030-8

Varade, R. R., Dhotre, M. R., & Pahurkar, A. B. (2013). A Survey on Various Median Filtering Techniques for Removal of Impulse Noise from Digital Images. *IJARCET, 2*(2), 606–609. Retrieved from https://pdfs.semanticscholar.org/030b/45e69b576f99c5c491cf8e58f911e164908c.pdf

Varrasi, S., Lucas, A., Soranzo, A., McNamara, J., & Di Nuovo, A. (2018). IBM Cloud Services Enhance Automatic Cognitive Assessment via Human-Robot Interaction. In G. Carbone, M. Ceccarelli, & D. Pisla (Eds.), *New Trends in Medical and Service Robotics. Mechanisms and Machine Science* (Vol. 65). Cham: Springer. doi:10.1007/978-3-030-00329-6_20

Varsamis, G. (2018). *Disruptive innovation: a weapon that can kneel down giants and a survival tool for difficult times.* Retrieved from https://blog.startuppulse.net/disruptive-innovation-a-weapon-that-can-kneel-down-giants-and-a-survival-tool-for-difficult-times-71af7bffb750

Varshney, K. R., Chen, G. H., Abelson, B., Nowocin, K., Sakhrani, V., Xu, L., & Spatocco, B. L. (2015). *Targeting Villages for Rural Development Using Satellite Image Analysis.* Retrieved from https://www.liebertpub.com/doi/full/10.1089/big.2014.0061

Verscheure, L., Peyrodie, L., Makni, N., Betrouni, N., Maouche, S., & Vermandel, M. (2010). Dijkstra's Algorithm Applied to 3D Skeletonization of the Brain Vascular Tree: Evaluation and Application to Symbolic. *Engineering in Medicine and Biology Society (EMBC), 2010 Annual International Conference of the IEEE*, 3081–3084.

Vijayan, A., Singanamala, H., Nair, B., Medini, C., Nutakki, C., & Diwakar, S. (2013). Classification of Robotic Arm Movement using SVM and Naïve Bayes Classifiers. In *International Conference on Innovative Computing Technology* (pp. 263–268). Academic Press.

Villanova University. (2017). *When Big Data Does Not Work.* Retrieved from https://taxandbusinessonline.villanova.edu/resources-business/infographic-business/when-big-data-doesnt-work.html

Villaronga, E. F., Kieseberg, P., & Li, T. (2018). Humans forget, machines remember: Artificial intelligence and the Right to Be Forgotten. *Computer Law & Security Review*, *34*(2), 304–313. doi:10.1016/j.clsr.2017.08.007

Vincent, J. (2017). *Putin says the nation that leads in AI 'will be the ruler of the world*. Retrieved from https://www.theverge.com/2017/9/4/16251226/russia-ai-putin-rule-the-world

Vincent, L. (1993). Morphological grayscale reconstruction in image analysis: Applications and efficient algorithms. *IEEE Transactions on Image Processing*, *2*(2), 176–201. doi:10.1109/83.217222 PMID:18296207

VineView. (2018). *Aerial Vineyard Mapping - Vigor & Grapevine Disease*. Retrieved December 17, 2018, from VineView website: https://www.vineview.ca/

Virginija Uzdanaviciute, R. B. (2011). Ontology-based Foundations for Data Integration. *The First International Conference on Business Intelligence and Technology*.

Von Solms, R., & Van Niekerk, J. (2013). From information security to cyber security. *Computers & Security*, *38*, 97-102.

Wachter, S., Mittelstadt, B., & Floridi, L. (2017). Transparent, explainable, and accountable AI for robotics. *Science Robotics*, *2*(6). doi:10.1126cirobotics.aan6080

Waga, D., & Rabah, K. (2014). Environmental Conditions' Big Data Management and Cloud Computing Analytics for Sustainable Agriculture. *World Journal of Computer Application and Technology*, *9*.

Wahbeh, A., Sarnikar, S., & El-Gayar, O. (2016). Improving analysts' domain knowledge for the requirements elicitation phase: A socio-technical perspective. Presented at the AMCIS 2016: Surfing the IT Innovation Wave - 22nd Americas Conference on Information Systems.

Wakefield, K. (2018). *Predictive analytics and machine learning*. Retrieved November 14, 2018, from https://www.sas.com/en_gb/insights/articles/analytics/a-guide-to-predictive-analytics-and-machine-learning.html

Walinjkar, A., & Woods, J. (2017). ECG classification and prognostic approach towards personalized healthcare. 2017 International Conference On Social Media, Wearable and Web Analytics (Social Media), 1–8. doi:10.1109/SOCIALMEDIA.2017.8057360

Walsh, B. (2014). *Google's flue project shows the failings of big data*. Retrieved from http://time.com/23782/google-flu-trends-big-data-problems/

Wang, B. (2016). Nvidia Xavier chip 20 trillion operations per second of deep learning performance and uses 20 watts which means 50 chips would be a petaOP at a kilowatt. *NextBigFuture*. Retrieved from https://www.nextbigfuture.com/2016/11/nvidia-xavier-chip-20-trillion.html

Wang, L., Zhou, X., Zhu, X., & Guo, W. (2017). Estimation of leaf nitrogen concentration in wheat using the MK-SVR algorithm and satellite remote sensing data. *Computers and Electronics in Agriculture*, *140*, 321–337. doi:10.1016/j.compag.2017.05.023

Wang, Li'ai, Zhou, X., Zhu, X., Dong, Z., & Guo, W. (2016). Estimation of biomass in wheat using random forest regression algorithm and remote sensing data. *Crop Journal*, *4*(3), 212–219. doi:10.1016/j.cj.2016.01.008

Wang, Y. (2016). Deep reasoning and thinking beyond deep learning by cognitive robots and brain-inspired systems. *2016 IEEE 15th International Conference on Cognitive Informatics Cognitive Computing (ICCI*CC)*, 3–3. 10.1109/ICCI-CC.2016.7862095

Wang, H., Xu, Z., & Pedrycz, W. (2017). An overview on the roles of fuzzy set techniques in big data processing: Trends, challenges and opportunities. *Knowledge-Based Systems*, *118*, 15–30. doi:10.1016/j.knosys.2016.11.008

Wang, L., Niu, Z., Kisi, O., Li, C., & Yu, D. (2017). Pan evaporation modeling using four different heuristic approaches. *Computers and Electronics in Agriculture*, *140*, 203–213. doi:10.1016/j.compag.2017.05.036

Wang, M., Cui, Y., Wang, X., Xiao, S., & Jiang, J. (2018). Machine learning for networking: Workflow, advances and opportunities. *IEEE Network*, *32*(2), 92–99. doi:10.1109/MNET.2017.1700200

Wang, P., Liu, B., & Hong, T. (2016). Electric load forecasting with recency effect: A big data approach. *International Journal of Forecasting*, *32*(3), 585–597. doi:10.1016/j.ijforecast.2015.09.006

Wang, T., Liu, M., Zhu, J.-Y., Tao, A., Kautz, J., & Catanzaro, B. (2018). High-Resolution Image Synthesis and Semantic Manipulation With Conditional GANs. *The IEEE Conference on Computer Vision and Pattern Recognition (CVPR)*, 8798-8807. 10.1109/CVPR.2018.00917

Wang, X., Liu, X., Pedrycz, W., & Zhang, L. (2015). Fuzzy rule based decision trees. *Pattern Recognition*, *48*(1), 50–59. doi:10.1016/j.patcog.2014.08.001 PMID:25395692

Wang, Y., Zang, H., Qiu, C. H., & Xia, S. R. (2018). A Novel Feature Selection Method Based on Extreme Learning Machine and Fractional-Order Darwinian PSO. *Computational Intelligence and Neuroscience*. doi:10.1155/2018/5078268

Wang, Z., Hu, M., & Zhai, G. (2018). Application of Deep Learning Architectures for Accurate and Rapid Detection of Internal Mechanical Damage of Blueberry Using Hyperspectral Transmittance Data. *Sensors (Basel)*, *18*(4), 1126. doi:10.339018041126 PMID:29642454

Weichecki, A. (2017). *Data Science Skill Set*. Retrieved from http://datasc.pl/2017/08/13/data-sciensit-wg-philippa-diesinger/

Weiss, G. M., Lockhart, J. W., Pulickal, T. T., McHugh, P. T., Ronan, I. H., & Timko, J. L. (2016). Actitracker: A Smartphone-Based Activity Recognition System for Improving Health and Well-Being. 2016 IEEE International Conference on Data Science and Advanced Analytics (DSAA), 682–688. doi:10.1109/DSAA.2016.89

Weiss, S. M., & Indurkhya, N. (1997). *Predictive data mining: a practical guide*. Retrieved from https://www.elsevier.com/books/predictive-data-mining/weiss/978-0-08-051465-9

Welch, C. (2018). *Google just gave a stunning demo of Assistant making an actual phone call*. Retrieved from https://www.theverge.com/2018/5/8/17332070/google-assistant-makes-phone-call-demo-duplex-io-2018

Weltzien, C. (2016). *Digital agriculture – or why agriculture 4.0 still offers only modest returns*. Academic Press.

Whitmore, A., Agarwal, A., & Da Xu, L. (2015). The Internet of Things-A survey of topics and trends. *Information Systems Frontiers*, *17*(2), 261–274. doi:10.100710796-014-9489-2

WHO. (2018). Non communicable diseases. Retrieved September 25, 2018, from World Health Organization website: http://www.who.int/news-room/fact-sheets/detail/noncommunicable-diseases

Wiggins, A., & Crowston, K. (2011). From conservation to crowdsourcing: A typology of citizen science. *Proceedings of the Annual Hawaii International Conference on System Sciences*. 10.1109/HICSS.2011.207

WikipediaB. (2018). Retrieved from https://en.wikipedia.org/wiki/Blockchain

Williams, A. (2016). *IBM to open first blockchain innovation centre in Singapore, to create applications and grow new markets in finance and trade*. Retrieved from https://www.straitstimes.com/business/economy/ibm-to-open-first-blockchain-innovation-centre-in-singapore-to-create-applications

Winskell, K., Singleton, R., & Sabben, G. (2018). Enabling Analysis of Big, Thick, Long, and Wide Data: Data Management for the Analysis of a Large Longitudinal and Cross-National Narrative Data Set. *Qualitative Health Research*, *28*(10), 1629–1639. doi:10.1177/1049732318759658 PMID:29557295

Witten, I. H., Frank, E., Hall, M. A., & Christopher, J. P. (2017). *Data mining: Practical machine learning tools and techniques*. Cambridge, MA: Morgan Kaufman-Elsevier.

Witten, I. H., Frank, E., Hall, M. A., & Pal, C. J. (2016). *Data Mining: Practical machine learning tools and techniques*. Morgan Kaufmann.

Wolfert, S., Ge, L., Verdouw, C., & Bogaardt, M.-J. (2017). Big Data in Smart Farming – A review. *Agricultural Systems*, *153*, 69–80. doi:10.1016/j.agsy.2017.01.023

Wolfert, S., Goense, D., & Sorensen, C. A. G. (2014). A Future Internet Collaboration Platform for Safe and Healthy Food from Farm to Fork. *2014 Annual SRII Global Conference*, 266–273. 10.1109/SRII.2014.47

Woodard, J. (2016). Big data and Ag-Analytics An open source, open data platform for agricultural & environmental finance, insurance, and risk. *Agricultural Finance Review*, *76*(1), 15–26. doi:10.1108/AFR-03-2016-0018

Woods, T. (2019). *'Age-Tech': The Next Frontier Market For Technology Disruption*. Intuition Robotics in Forbes.

Woo, J., Shin, S. J., Seo, W., & Meilanitasari, P. (2018). Developing a big data analytics platform for manufacturing systems: Architecture, method, and implementation. *International Journal of Advanced Manufacturing Technology*, *99*(9-12), 2193–2217. doi:10.100700170-018-2416-9

Worldatlas. (2018). *The World's Leading Producers of Tomatoes. Economics*. Retrieved from https://www.worldatlas.com/articles/which-are-the-world-s-leading-tomato-producing-countries.html

Worldometers. (2018). *United Nations. DESA/ Population Division. World Population Clock*. Retrieved from http://www.worldometers.info/world-population/

Wu, J., Yang, S. X., & Tian, F. (2014). A novel intelligent control system for flue-curing barns based on real-time image features. *Biosystems Engineering*, *123*, 77–90. doi:10.1016/j.biosystemseng.2014.05.008

Xiang, Z., Magnini, V. P., & Fesenmaier, D. R. (2015). Information technology and consumer behavior in travel and tourism: Insights from travel planning using the internet. *Journal of Retailing and Consumer Services*, *22*, 244–249. doi:10.1016/j.jretconser.2014.08.005

Xiaomeng, C., Lyu, Z., & Terpenny, J. (2015). Ontology development and optimization for data integration and decision-making in product design and obsolescence management. In Ontology Modeling in Physical Asset Integrity Management (pp. 87–132). Academic Press. doi:10.1007/978-3-319-15326-1_4

Xiao, Y., Mignolet, C., Mari, J.-F., & Benoit, M. (2014). Modeling the spatial distribution of crop sequences at a large regional scale using land-cover survey data: A case from France. *Computers and Electronics in Agriculture*, *102*, 51–63. doi:10.1016/j.compag.2014.01.010

Xuan, S., Man, D., Yang, W., Wang, W., Zhao, J., & Yu, M. (2018). Identification of unknown operating system type of Internet of Things terminal device based on RIPPER. *International Journal of Distributed Sensor Networks*, *14*(10). doi:10.1177/1550147718806707

Yaga, D., Mell, P., Roby, N., & Scarfone, K. (2018). *Blockchain Technology Overview*. Retrieved from https://nvlpubs.nist.gov/nistpubs/ir/2018/NIST.IR.8202.pdf

Yang, L., Qi, J., Song, D., Xiao, J., Han, J., & Xia, Y. (2016). Survey of Robot 3D Path Planning Algorithms. *Journal of Control Science and Engineering*, (5): 22.

Yli-Huumo, J., Ko, D., Choi, S., Park, S., & Smolander, K. (2016). Where is Current Research on Blockchain Technology?- A Systematic Review. *PLoS One*, *11*(10). doi:10.1371/journal.pone.0163477 PMID:27695049

Yong, H., Lee, C., & Wang, D. (2018). *DeepBrain Chain: Artificial Intelligence Computing Platform Driven by Blockchain*. White Paper of DeepBrain Chain, Version 1.1.0.

Yongting, T., & Jun, Z. (2017). Automatic apple recognition based on the fusion of color and 3D feature for robotic fruit picking. *Computers and Electronics in Agriculture, 142*(A), 388–396. doi:10.1016/j.compag.2017.09.019

Yuan, B., & Herbert, J. (2014). A Cloud-Based Mobile Data Analytics Framework: Case Study of Activity Recognition Using Smartphone. 2014 2nd IEEE International Conference on Mobile Cloud Computing, Services, and Engineering, 220–227. 10.1109/MobileCloud.2014.29

Yuce, H., & Avci, K. (2017). Establishment of diagnosing faults and monitoring system with neural networks in air conditioning systems. *The Journal of Cognitive Systems*, *2*(1), 63–69.

Yu, H., & Wilamowski, B. M. (2010). Levenberg–Marquardt Training 12.1. *Industrial Electronics Handbook*, *5*(12), 1–16.

Zafar, M. N., & Mohanta, J. (2018). Methodology for Path Planning and Optimization of Mobile Robots: A review. *International Conference on Robotics and Smart Manufacturing*, *133*, 141–152. 10.1016/j.procs.2018.07.018

Zangeneh, M., Omid, M., & Akram, A. (2010). Assessment of agricultural mechanization status of potato production by means of artificial Neural Network model. *Australian Journal of Crop Science*, *4*(5), 372–377.

Zhang, J., Li, K., & Yao, C. (2018). Event-based Summarization for Scientific Literature in Chinese. 2017 International Conference on Identification, Information and Knowledge in the Internet of Things. *Procedia Computer Science*, 88-92.

Zhang, R., Xue, R., & Liu, L. (2019). Security and Privacy on Blockchain. *ACM Computing Surveys, 1*(1), 1-34. Retrieved from https://arxiv.org/pdf/1903.07602.pdf

Zhang, N., Mi, X., Feng, X., Wang, X., Tian, Y., & Qian, F. (2018). *Understanding and Mitigating the Security Risks of Voice-Controlled Third-Party Skills on Amazon Alexa and Google Home. Computer Science. Cryptography and Security*. Cornell University.

Zhang, Q., Yang, L. T., Castiglione, A., Chen, Z., & Li, P. (2019). Secure weighted possibilistic c-means algorithm on cloud for clustering big data. *Information Sciences*, *479*, 515–525. doi:10.1016/j.ins.2018.02.013

Zhang, Q., Zhou, D., & Zeng, X. (2017). Highly wearable cuff-less blood pressure and heart rate monitoring with single-arm electrocardiogram and photoplethysmogram signals. *Biomedical Engineering Online*, *16*(1), 23. doi:10.1186/s12938-017-0317-z PubMed

Zhang, W., Daim, T., & Zhang, Q. (2018). Understanding the disruptive business model innovation of E-business microcredit: A comparative case study in China. *Technology Analysis and Strategic Management*, *30*(7), 765–777. doi:10.1080/09537325.2017.1376047

Zhang, X., Li, R., Zhang, B., Yang, Y., Guo, J., & Ji, X. (2019). An instance-based learning recommendation algorithm of imbalance handling methods. *Applied Mathematics and Computation*, *351*, 204–218. doi:10.1016/j.amc.2018.12.020

Zhang, X., Qiao, Y., Meng, F., Fan, C., & Zhang, M. (2018). Identification of Maize Leaf Diseases Using Improved Deep Convolutional Neural Networks. *IEEE Access: Practical Innovations, Open Solutions*, *6*, 30370–30377. doi:10.1109/ACCESS.2018.2844405

Zhang, Y. Z., Huang, T., & Bompard, E. F. (2018). Big data analytics in smart grids: A review. *Energy Informatics*, *1*(8), 1–24.

Zhang, Z., Gong, Y., & Wang, Z. (2018). Accessible remote sensing data based reference evapotranspiration estimation modelling. *Agricultural Water Management*, *210*, 59–69. doi:10.1016/j.agwat.2018.07.039

Zhao, W. (2017). *Daimler AG Issues €100 Million Corporate Bond in Blockchain Trial*. Retrieved from https://www.coindesk.com/daimler-ag-issues-e100-million-corporate-bond-blockchain-trial

Zhao, W. (2017). *Europe's Second Largest Port Launches Blockchain Logistics Pilot*. Retrieved from https://www.coindesk.com/europes-second-largest-port-launches-blockchain-logistics-pilot

Zhao, Z., & Zhang, X. (2016). Artificial intelligence applications in power system. Advances in intelligent systems research. In *2nd International Conference on Artificial Intelligence and Industrial Engineering*, (133, 158-161). Atlantis Press.

Zheng, Z., Xie, S., Dai, H. N., & Wang, H. (2017). An Overview of Blockchain Technology: Architecture, Consensus, and Future Trends. *IEEE 6th International Congress on Big Data Congress*.

Zheng, Z., Xie, S., Dai, H. N., & Wang, H. (2018). Blockchain challenges and opportunities: A Survey. *International Journal of Web and Grid Services*, *14*(4), 352–375. doi:10.1504/IJWGS.2018.095647

Zhong, L., Hawkins, T., Holland, K., Gong, P., & Biging, G. (2009). Satellite imagery can support water planning in the Central Valley. *California Agriculture*, *63*(4), 220–224. doi:10.3733/ca.v063n04p220

Zhong, T. Y., Zhang, X. Y., & Huang, X. J. (2009). Simulation of farmer decision on land use conversions using decision tree method in Jiangsu Province, China. *Spanish Journal of Agricultural Research*, *7*(3), 687–698. doi:10.5424jar/2009073-454

Zhou, K., Fu, C., & Yang, S. (2016). Big data driven smart energy management: From big data to big insights. *Renewable & Sustainable Energy Reviews*, *56*, 215–225. doi:10.1016/j.rser.2015.11.050

Zhou, L. (2007). Ontology learning: State of the art and open issues. *Information Technology Management*, *8*(3), 241–252. doi:10.100710799-007-0019-5

Zhou, X., Yang, C., & Yu, W. (2013). Moving Object Detection by Detecting Contiguous Outliers in the Low-Rank Representation. *IEEE Transactions on Pattern Analysis and Machine Intelligence*, *35*, 1–30. Retrieved from https://ieeexplore.ieee.org/abstract/document/6216381 PMID:22689075

Zhu, X., & Choulli, E. (2018). Acquisition and communication system for condition data of transmission line of smart distribution network. *Journal of Intelligent & Fuzzy Systems*, 1-14.

Zhu, N., Liu, X., Liu, Z., Hu, K., Wang, Y., Tan, J., ... Guo, Y. (2018). Deep learning for smart agriculture: Concepts, tools, applications, and opportunities. *International Journal of Agricultural and Biological Engineering*, *11*(4), 21–28. doi:10.25165/j.ijabe.20181104.4475

Zhu, Y. (2011). Automatic detection of anomalies in blood glucose using a machine learning approach. *Journal of Communications and Networks (Seoul)*, *13*(2), 125–131. doi:10.1109/JCN.2011.6157411

Ziegler, P., & Dittrich, K. R. (n.d.). Data Integration — Problems, Approaches, and Perspectives. *Conceptual Modelling in Information Systems Engineering*, 39–58. doi:10.1007/978-3-540-72677-7_3

Zimmer, M., & Kurlanda, K. K. (2017). *Internet research ethics for the social age: New challenges, cases, and contexts*. Peter Lang International Academic Publishers. doi:10.3726/b11077

About the Contributors

Moses Strydom is a retired Professor and recent Chair of the department of of Mechanical and Industrial Engineering at the University of South Africa. An alumni of the University of Perpignan, France (PhD) and New Mexico State University (MSc), Moses is bilingual, French-English, and for the past 30 years, as an academic, has worked in several universities in Africa, France and the USA. His research interestsinclude computational/experimental fluid dynamics, hydrogen fuel cells; big data and robotics, and holographic technology in m-learning. He has published in several research journals and books, and presented conference papers, both nationally and internationally.

Sheryl Buckley is the Director of the School of Computing in the College of Science Engineering & Technology at Unisa. She specializes in Business Information Systems, E-learning, Communities of Practice, Information science and Knowledge Management. Sheryl is a member of a number of professional societies. She has presented papers nationally and internationally.

* * *

Tahir Cetin Akinci received the B.Sc. degree in Electrical Engineering. His master's and Ph.D. degrees, Institute of Pure and Applied Science from Marmara University, Istanbul, Turkey. He is an associate professor in the Department of Electrical Engineering at Istanbul Technical University (ITU), Istanbul, Turkey. His research interests are signal processing, data mining, intelligent systems, ferroresonance phenomenon, artificial neural networks, and renewable energy sources.

Loknath Sai Ambati is a doctoral student pursuing PHD degree in Information Systems with specialization in analytics and decision support at Dakota State University. Loknath is currently working as graduate research assistant at Dakota State University, has extensive experience in machine learning, data analytics and management skills. Prior to joining PHD, Loknath worked as analytics developer at Baylor Scott and White Health and holds a masters degree in analytics from Dakota State University.

Gopala Krishna Behara is Distinguished Member of Technical Staff an Lead Enterprise Architect at Wipro Technologies. Certified Open Group TOGAF, AWS Solution Architect, IBM Cloud Solutions. Associated with Open Group for long time and contributed various blogs, standards around architecture. Serves as an Advisory Architect and Mentor on Enterprise Architecture, Application Portfolio Rationalization and Architecture Assurance initiatives and continues to work as a Subject Matter Expert and Author. He has worked on multiple architecture transformation engagements in the USA, UK, Europe,

Asia Pacific and Middle East Regions that presented a phased roadmap to transformation that maximized the business value, while minimizing costs and risks. Published White Papers, blogs and articles in International Journals in SOA, BPM, Open Source and Next Generation Technologies & e-Governance space. Published books titled "Enterprise Architecture Practitioner Hand Book", "Next Generation Enterprise Reference Architecture For Connected Government" and "Microservices Practitioner Guide". Recipient of EA Hall of Fame - Individual Leadership in EA Practice, Promotion and Professionalization Award, 13th Annual Enterprise Architecture Conference, Washington, DC, USA.

Ouhbi Brahim received his BS degree in Statistics and Computer science from Mohamed V Science Faculty, Morocco. He obtained his Master degree from Paris VI University in Probability and Applications and PhD degree in Systms Control Engineering from Compiegne's Technological University and Habilitation degree in Systms Control Science from Mohammadia Engineering School. Currently. His current research interests include data mining, Decision Support system (DSS), Recommender Systems (RS), and web intelligence (IWeb) & Social Networks (SN). He has published over 80 papers in many leading international journals such as Applied Soft Computing, Communications in Computer and Information Science, Studies in computational Intelligence, Inteligencia Artificial, Journal of Intelligent Information Systems , Renewable and Sustainable Energy Reviews, Inter. Journal of Web Information Systems, Journal of mobile and multimedia, Stat. Inf. Stoch., Inter. Journal on Artificial Intelligence Tools, RAIRO Oper. Res.,Statist. and Probab. Lett., Statistical Planning and Inference, Inter. J. Perf. Eng., Communications of the ACM, and others. He was the chair of many sessions in many leading conferences like IIWAS, NGNS, IWAP, CIST. He is a reviewer in many leading international journals such as Information Sciences, Expert Systems with application, Inter. Journal of Information Technology & Decision Making ; Commun. Stat.-Theor. Methods; Journal Planning Statistics and Inference; Methodology and Computing in Applied Probability; Reliability Engineering and Safety; Stochastic Models. Actually, he is a full Professor at National High School of Arts & Crafts at Moulay Ismail University in Meknès city, Morocco. He has supervised 8 PhD theses on stochastic models for Data mining and Knowledge Discovery and Reliability. He is a supervisor of 6 PhD theses on Intelligent DSS, RS and Community detection in SN and Deep learning in Big Data context.

Omar El-Gayar, Ph.D. is a Professor of Information Systems at Dakota State University. Dr. El-Gayar has an extensive administrative experience at the college and university levels as the Dean for the College of Information Technology, United Arab Emirates University (UAEU) and the Founding Dean of Graduate Studies and Research, Dakota State University. His research interests include: analytics, business intelligence, and decision support with applications in problem domain areas such as healthcare, environmental management, and security planning and management. His inter-disciplinary educational background and training is in information technology, computer science, economics, and operations research. Dr. El-Gayar's industry experience includes working as an analyst, modeler, and programmer. His numerous publications appear in various information technology related fields. He is a member of AIS, ACM, INFORMS, and DSI.

Harry J Foxwell, Ph.D. is an Associate Professor in the Department of Information Sciences and Technology. Dr. Foxwell is a graduate of George Mason University, and taught at Mason for several years as an Adjunct before joining IS&T as a full-time Instructional Faculty member. He previously worked as a Principal System Engineer for the Oracle Corporation and Sun Microsystems, specializing in Cloud Computing infrastructure and Operating Systems. Dr. Foxwell earned his Ph.D. in Information Technology from George Mason University, an M.S. in Applied Statistics from Villanova University, and a B.A. in Mathematics from Franklin and Marshall College. He is also a Vietnam War veteran, having served in the U.S. Army's First Infantry Division as a Platoon Sergeant.

Bouchra Frikh has received her BS from University Sidi Mohamed Ben Abdallah in Computer Science in 1994 and her MS from ENSIAS at Rabat in 1999. She gets her PhD in Information Retrieval science in 2003. In 2012, Pr. B. Frikh gets her Habilitation degree from the University Sidi Mohamed Ben Abdallah-Fez,-Morocco in Web Mining field. Currently, she is a Professor at High school of Technology-Fez. Her research interests include Big Data, Semantic Web, Decision support system, Recommender systems, Knowledge management, informatics for energy and analysis of social networks. In these areas, she has published over 50 combined refereed journals and conference manuscripts. Pr B. Frikh is a Program Committee member in many workshops and colloquiums such as International Colloquium on Information Science and Technology (CIST'2012, CIST' 2014, CIST'2016 and CIST' 2018) at Morocco, Interactive Mobile & Computer Aided Learning Congress (IMCL'2012 in Jordan), (IMCL'2014 in Greece), (IMCL'2015 in Greece), International conference on Big Data and Advanced Wireless technologies 2016 in Bulgaria and International Conference On Intelligent Computing in Data Sciences (ICDS 2018). Pr B.Frikh is reviewer in two journals. She is also a member of a national project dedicated to Decision Support Systems applied to Renewable Energy.

Anupama Hoskoppa Sundaramurthy is currently working as an associate professor. My area of interest is Artificial Intelligence and Machine Learning.

Ferdaous Hdioud has received her BS from the faculty of science and Technology in FEZ (FSTF) in 2010. She obtained her MS from the University of Sidi Mohammed Ben Abdellah (USMBA) and her PhD in Computer Science in 2017. Her research interests cover Recommender Systems in industrial Big Data context. She has published in this area, many manuscripts, as well as, she participated in many conferences and colloquiums such as International Colloquium on Information Science and Technology (Cist'12) at Morocco, International Conference on Information Integration and Web-based Application & Services (Iiwas'13), Next Generation Networks and Services (NGNS'14) and Intelligent Distributed Computing (IDC'15).

Tirumala Khandrika is Senior Architect in the Enterprise Architecture practice of Wipro. Has 17 years of experience in software engineering which includes software architecture, design and development for internet/intranet, client server applications and business application systems for clients in USA, Europe and Asia. Extensive experience in architecting, designing and developing enterprise and web based applications in BFS, manufacturing, aerospace, healthcare domains using Java/JEE, Spring and SOA platforms Experience in customer relationship and successfully handled engagements in gaining the confidence of the customer.

Imadeddine Mountasser received his BS degree in Mathematical and Computer science from Dhar EL-Mehraz Science Faculty, Morocco. He obtained his Master degree from University Mohamed Ben Abdallah University in Decisional Information Systems. Actually, he is a PhD student specialized on Knowledge Management Systems in Big Data Environment.

Tawanda Mushiri is a holder of BSc Mechanical Engineering (UZ), Master of Science in Manufacturing Systems and Operations Management (MSOM) (UZ) a PhD in Automation, Robotics and Artificial Intelligence (U.J) of machinery monitoring systems. He is currently a Senior Lecturer at the University of Zimbabwe teaching Machine Dynamics, Robotics, Solid Mechanics and Finite Element Analysis. He is also the coordinator of Undergraduate projects and Master of Science in Manufacturing Systems and Operations Management (MSOM). Tawanda has supervised more than 100 students' undergraduate projects and 1 Masters Student to completion. He has published 2 Academic Textbooks, 3 Chapters in a book, 14 Journals and 84 Conference Papers plus a Patents in highly accredited publishers. He has done a lot of commercial projects at the University of Zimbabwe. He is a reviewer of 4 journals highly accredited. He has been invited as a keynote speaker in workshops and seminars. Beyond work and at a personal level, Tawanda enjoys spending time with family, travelling and watching soccer.

Jayapandian Natarajan is a PhD. assistant professor at the Christ University, Department of Computer Science and Engineering, Bangalore, India. His research interest includes information security, cloud computing, and grid computing. Dr. Jayapandian holds a Bachelor of Technology degree in Information Technology from IRTT, Anna University and Master degree in Computer Science and Engineering from KEC, Anna University. He is published various research article in reputed international journals. He is an active reviewer of reputed international journals. He has participated in numerous national and international conferences and has made a remarkable contribution to cloud data security field and publishing several articles.

Nevine Nawar is a Lecturer at the Department of Public Health, School of Medicine, Alexandria University, Egypt. Her research interests emphasize preventive strategies that recognize and address risk factors as key to health promotion and disease prevention. Dr. Nawar taught courses in nutrition, child and maternity health, health education, rural health, health care delivery systems, and communicable diseases. She has engaged in funded research including a World Health Organization funded survey to identify microbial causes for diarrhea among preschool children living in rural communities in Egypt. She was also part of a community-based surveys to recognize health problems in urban Alexandrian communities. Dr. Nawar holds a medical degree and a Master's in Public Health from Alexandria University. She obtained her Ph.D. in Public Health also from Alexandria University through a joint-supervision with Harvard University (channel system).

Martinson Ofori, MS, EMBA. is a Computer Science master's student at Dakota State University. Generally, his research interests cover Finance, Agriculture, Developing Economies, and Computer Science. He has a background in Finance and Software Development. Martinson has immense industry experience in Telecommunication and Banking.

Gayathri Rajendran received the MCA degree in computer science from Pondicherry University in 2016. She has qualified the CBSE-UGC NET (National Eligibility Test) and SET (State Eligibility Test) in 2018. Her research interest includes Artificial Intelligence and Machine Learning. She has published research papers in international conference and journals.

Dhanalakshmi Senthilkumar had completed Ph.D Degree in Anna University, Chennai-25. 16 years teaching experience in engineering field.

Marcus Tanque is a principal technologist and senior cyber strategist with proven expertise in operations research, policy-based management, regulatory compliance, enterprise risk management, information security & engineering strategies. Dr. Tanque is a published author/editor, a technology researcher, and practitioner with proven business and technical knowledge. Marcus' research interest focuses on IT engineering, artificial intelligence, policy management, data science, machine learning, systems security, human-computer interaction, management information systems, cognitive science, deep learning, cloud solutions, cyber resilience, cloud database systems, smart intelligent systems, IT operations management, risk management, cryptographic solutions, information management, and cyber strategies. Dr. Tanque received a Ph.D. in Information Technology with a dual specialization in Information Assurance & Security and a master's degree in information systems engineering. Marcus is an active member of several editorial advisory boards; an adjunct assistant professor for the computer networks and cybersecurity programs.

Uma Vijayasundaram received the M.Tech, and PhD degrees in computer science from Pondicherry University in 2007 and 2014 respectively. She was awarded the Pondicherry University gold medal for M.Tech. degree in Distributed Computing Systems. She has more than 10 years of teaching experience at PG level. Her research interest includes Data mining, knowledge representation and reasoning (spatial and temporal knowledge) and sentiment analysis. She has authored and co-authored more than 20 peer-reviewed journal papers, which includes publications in Springer and Inderscience. She has also authored 3 chapters in various books published by IGI Global. She has also authored a book on temporal knowledge representation and reasoning. She received the Best Paper Award in International Conference on Digital Factory in the year 2008.

Index

A

B

C

D

E

Electric Power System 254
Electronic Health Record (EHR) 129, 161
Elliptic Curve Diffie- Hellman-Merkle (ECDHM) 173
ELM 210

F

feature extraction 228-229, 232-233, 245
FFNN 84, 91
FOL 78, 91
food sustainability 179, 189

G

grading 216-223, 225, 227, 231
Gross Domestic Product (GDP) 105, 129
GUI 96, 99

H

hash functions 266
health 2, 17, 19, 25, 30, 32, 94-95, 101, 104-108, 110, 113-120, 129, 157, 161-164, 167, 188, 223, 242, 246, 320, 325-326
Health Intervention 129
HNN 84, 92
Hyderledger Fabric 288

I

IEEE 109, 167, 173, 180
image processing 93, 95, 101, 216, 231-232
image segmentation 224, 228
Intraocular Pressure 114, 129
IoT 45, 117, 152, 164, 167, 188, 190, 210, 300, 302, 309
ITU 167, 173

L

ledger 130-136, 139, 141, 145-146, 150, 156-157, 159-160, 162, 173, 261-262, 264, 271-272, 274-280, 284, 288
LR 73, 85-86, 92
LSTM 87, 92
LTL 78, 92
LTLMoP 87, 92

M

machine learning 9, 17, 20, 25-27, 32-33, 53, 71-73, 84-86, 88, 93-96, 100-101, 104-105, 108, 110, 145, 153-154, 158, 167, 177-180, 184-187, 220, 237, 240, 243-245, 248, 283-284, 291, 293-301, 304-305, 309, 314, 321, 325, 334
machine learning algorithms 71-73, 86, 88, 178, 248, 291, 293, 295, 300, 309
Man-in-the-Middle Attack (MITM) 303, 315
market 2, 10, 27, 29, 105, 116-117, 119-120, 131, 133-134, 158, 173, 177-178, 185, 190, 216, 218-219, 221, 225, 227, 249, 278, 281, 283-284, 291, 320-321, 325
Matlab 216, 228, 231-233, 334
merkle tree 264, 267, 284, 288
MLP 85-86, 92, 96

N

NB 85-86, 92, 228
network attack 291, 306
network security 292-293, 302-303, 309, 315
NIST 167, 173, 325

O

ontology-based integration 46-48, 61-62, 64
Open Data integration 45, 48-50, 62
Open Tourism Data 44, 46-47, 51, 54, 61-64
operationalization 322-324, 328-329
optimization 29, 54, 82, 84, 86, 151-152, 177, 183, 192, 240-241, 244-246, 249-251, 254-255, 294-295, 299-300

P

participants 130, 133-134, 150, 152, 164, 173, 267, 274-275, 277, 288
path planning algorithms 71-73, 75, 80-81, 88
PDDL 78, 92
Permissioned Blockchain 136, 276-277, 288-289
permissionless Blockchain 136, 276-277, 289
Personal Health Record (PHR) 129
planning 44-49, 60-61, 63-64, 71-78, 80-86, 88, 92, 95, 131, 157, 176-177, 184-185, 188, 192, 248, 250-251, 254-255, 281, 326-327, 335
planning languages 72, 78, 88
precision agriculture 175-176, 182, 188
PRM 81, 92
project management 325

Ensure Quality Research is Introduced to the Academic Community

Become an IGI Global Reviewer for Authored Book Projects

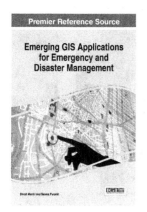

Premier Reference Source

Emerging GIS Applications for Emergency and Disaster Management

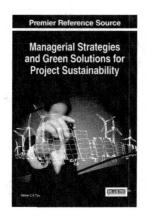

Premier Reference Source

Managerial Strategies and Green Solutions for Project Sustainability

Premier Reference Source

Comparative Approaches to Using R and Python for Statistical Data Analysis

Premier Reference Source

Solutions for High-Touch Communications in a High-Tech World

The overall success of an authored book project is dependent on quality and timely reviews.

In this competitive age of scholarly publishing, constructive and timely feedback significantly expedites the turnaround time of manuscripts from submission to acceptance, allowing the publication and discovery of forward-thinking research at a much more expeditious rate. Several IGI Global authored book projects are currently seeking highly-qualified experts in the field to fill vacancies on their respective editorial review boards:

Applications and Inquiries may be sent to:
development@igi-global.com

Applicants must have a doctorate (or an equivalent degree) as well as publishing and reviewing experience. Reviewers are asked to complete the open-ended evaluation questions with as much detail as possible in a timely, collegial, and constructive manner. All reviewers' tenures run for one-year terms on the editorial review boards and are expected to complete at least three reviews per term. Upon successful completion of this term, reviewers can be considered for an additional term.

If you have a colleague that may be interested in this opportunity, we encourage you to share this information with them.